Wolfram Saenger

Principles of
Nucleic Acid Structure

With 227 Figures

Springer-Verlag
New York Berlin Heidelberg
London Paris Tokyo

Wolfram Saenger
Institut fur Kristallographie
Freie Universität
1000 Berlin 33
Federal Republic of Germany

Series Editor:

Charles R. Cantor
Department of Human Genetics and Development
College of Physicians & Surgeons of Columbia University
New York, NY 10032
USA

Library of Congress Cataloging in Publication Data
Saenger, Wolfram.
    Principles of nucleic structure.
    (Springer advanced texts in chemistry)
    1. Nucleic acids.   I. Title.   II. Series.
QD433.S24   1983   547.7'9   82-19445

Printed on acid-free paper

Typeset by Progressive Typographers, Emigsville, Pennsylvania.
Printed and bound by R.R. Donnelley and Sons, Crawfordsville, Indiana.
Printed in the United States of America.

9 8 7 6 5 4 3 2   (Second corrected printing, 1988)

ISBN 0-387-90762-9 Springer-Verlag New York Berlin Heidelberg (Hardcover Edition)
ISBN 3-540-90762-9 Springer-Verlag Berlin Heidelberg New York (Hardcover Edition)
ISBN 0-387-90761-0 Springer-Verlag New York Heidelberg Berlin (Softcover Edition)
ISBN 3-540-90761-0 Springer-Verlag Berlin Heidelberg New York (Softcover Edition)

Springer Advanced Texts in Chemistry

Charles R. Cantor, Editor

Springer Advanced Texts in Chemistry

Series Editor: Charles R. Cantor

Principles of Protein Structure
G.E. Schulz and *R.H. Schirmer*

Bioorganic Chemistry: A Chemical Approach to Enzyme Action, Second Edition
H. Dugas

Protein Purification: Principles and Practice, Second Edition
R.K. Scopes

Principles of Nucleic Acid Structure
W. Saenger

Biomembranes: Molecular Structure and Function
R.B. Gennis

to
Barbara
Nicole
Jörg

DRAWN FOR TIBS
BY AB TULP

Reprinted with permission from A. M. Campbell (1978), Straightening out the supercoil. *Trends Biochem. Sci.* 104–108.

Series Preface

New textbooks at all levels of chemistry appear with great regularity. Some fields like basic biochemistry, organic reaction mechanisms, and chemical thermodynamics are well represented by many excellent texts, and new or revised editions are published sufficiently often to keep up with progress in research. However, some areas of chemistry, especially many of those taught at the graduate level, suffer from a real lack of up-to-date textbooks. The most serious needs occur in fields that are rapidly changing. Textbooks in these subjects usually have to be written by scientists actually involved in the research which is advancing the field. It is not often easy to persuade such individuals to set time aside to help spread the knowledge they have accumulated. Our goal, in this series, is to pinpoint areas of chemistry where recent progress has outpaced what is covered in any available textbooks, and then seek out and persuade experts in these fields to produce relatively concise but instructive introductions to their fields. These should serve the needs of one semester or one quarter graduate courses in chemistry and biochemistry. In some cases the availability of texts in active research areas should help stimulate the creation of new courses.

New York   CHARLES R. CANTOR

Preface

This monograph is based on a review on polynucleotide structures written for a book series in 1976. The respective volume was, however, never published and the manuscript was kept in the drawer until Springer-Verlag decided to accept it as part of the Advanced Texts in Chemistry. When I looked at my 1976 manuscript again in 1980, I found that many views had changed. Left-handed DNA, above all, had been introduced and changed the notion of the occurrence of exclusively right-handed double-helices. As a matter of fact, the whole story had to be rewritten.

But even then, during the past two years many new discoveries were made and it often appeared to me that I was trying to catch a running train. Research on DNA structure is now booming, caused in part by the availability of synthetic oligonucleotides of defined sequence. They can be crystallized either per se or as complexes with certain metal ions, drugs, and proteins binding to them. And with the proteins again a new field has opened. We can now look at the three-dimensional structures of quite a number of DNA- and RNA-binding proteins. Work is in progress with more complicated systems involving the respective complexes, made possible by recent advances in crystallographic techniques.

I cannot close this short preface without thanking those who have contributed enthusiastically to this book. There are, above all, Meinhard Heidrich and Manfred Steifa who operated the Cambridge Data File to retrieve atomic coordinates. These were used in a combination of programs SCHAKAL (Ref. 1376) and ORTEP (Ref. 1377) to do most of the artistic plots of molecules displayed in this book. Drawings which could not be prepared on a computer came from the hand of Ludwig Kolb who did a magnificent and superb job on the rough drawings I gave him.

If writing a book, one has to have available a good library and people associated with it. The library at the Max-Planck-Institut für Experimentelle Medizin is one of the best I have encountered and Mrs. M. Henschel, B. Küttner and Mr. W. Gellert helped to make available the more exotic journals. Finally, writing and rewriting many versions of the manuscript was done patiently by several secretaries, Chantal Celli, Jenny Glissmeyer and Petra Große contributing predominantly.

Last, but certainly not least, I would like to thank my scientific teacher Professor Friedrich Cramer for his guidance and support during the past 18 years.

Göttingen, August 1983   WOLFRAM SAENGER

Contents

Chapter 4

Chapter 5

Chapter 6

Chapter 7

Modified Nucleosides and Nucleotides; Nucleoside Di- and Triphosphates; Coenzymes and Antibiotics   159

Chapter 8

Metal Ion Binding to Nucleic Acids   201

Chapter 9

Polymorphism of DNA versus Structural Conservatism of RNA: Classification of A-, B-, and Z-Type Double Helices   220

Chapter 13

Synthetic, Homopolymer Nucleic Acids Structures   298

Chapter 14

Hypotheses and Speculations: Side-by-Side Model, Kinky DNA, and "Vertical" Double Helix   321

Chapter 15

tRNA—A Treasury of Stereochemical Information   331

Chapter 16

Intercalation   350

Chapter 17

Water and Nucleic Acids   368

Chapter 18

Protein–Nucleic Acid Interaction   385

Chapter 19

Higher Organization of DNA   432

Acknowledgments

Acknowledgment is gratefully made for permission to reproduce material as follows:

Figures 6-3, 6-15, 6-16, 6-17, 6-18, 11-4, 11-5, 11-6, 12-1, 12-3, 12-4, 12-8, 13-14, 15-2, 15-3, 16-3, 17-6, 17-7, and 18-11 are reprinted by permission from the *Journal of Molecular Biology,* copyright Academic Press Inc. (London) Ltd.

Table 3-2 is reprinted by permission from the *Journal of Theoretical Biology,* copyright Academic Press Inc. (London) Ltd.

Reprinted with permission from the *Journal of the American Chemical Society* are Figure 6-10, copyright 1967 American Chemical Society; Figures 2-8 and 15-10, copyright 1972 American Chemical Society; Figures 3-8 and 7-6, copyright 1977 American Chemical Society; Figures 4-6 and 13-13, copyright 1978 American Chemical Society.

Reprinted by permission from *Nature* are Figure 18-7, copyright 1955 Macmillan Journals Ltd; Figure 10-5, copyright 1971 Macmillan Journals Ltd; Table 6-10, copyright 1973 Macmillan Journals Ltd; Figure 4-24, copyright 1977 Macmillan Journals Ltd; Figure 4-29a and Table 4-4, copyright 1978 Macmillan Journals Ltd; Figure 17-10, copyright 1979 Macmillan Journals Ltd; Figures 11-3, 16-7, and 16-8, copyright 1980 Macmillan Journals Ltd; Figures 18-17 and 18-18, copyright 1981 Macmillan Journals Ltd; Figures 8-5 and 17-12, copyright 1982 Macmillan Journals Ltd.

Reprinted by permission from *Science* is Figure 1-3, copyright 1969 by the American Association for the Advancement of Science.

Reprinted by permission of the authors from *Proceedings of the National Academy of Science, U.S.A.* are Figures 3-5, 6-5, 6-6, 14-2, 14-3, 14-5, 16-6, 18-3, 18-5, 18-12, 19-2, and 19-7.

Figures 4-27, 6-14, 6-20, and 16-5 are reprinted by permission from *Biopolymers,* copyright John Wiley and Sons. Figure 15-1 is reprinted by permission from *Advances in Enzymology,* copyright John Wiley and Sons.

Figures 19-4, 19-6, 19-8 are from "The Nucleosome" by Roger D. Kornberg and Aaron Klug, copyright 1981 by Scientific American, Inc. All rights reserved. Box 19-3 is from "Supercoiled DNA" by William R. Bauer, F.H.C. Crick, and James H. White, copyright 1980 by Scientific American, Inc. All rights reserved.

Figure 19-1 is reprinted by permission from *Cell,* copyright M.I.T. Press.

Chapter 1

Why Study Nucleotide and Nucleic Acid Structure?

Before embarking on a description of nucleotide and nucleic acid structures, let us examine the biological importance of this class of molecules and find out why their structural principles should be known at the atomic level.

Nucleotides have many functions in living organisms. The hereditary material, deoxyribonucleic acid (DNA), is a linear polymer built up of monomeric units, the nucleotides. Even if these units were not constituents of DNA, they would nevertheless be among the most important molecules in biology.

A nucleotide consists of three molecular fragments: sugar, heterocycle, and phosphate. The sugar, ribose or deoxyribose, is in a cyclic, furanoside form and is connected by a β-glycosyl linkage with one of four heterocyclic bases to produce the four normal nucleosides: adenosine, guanosine, cytidine, and thymidine (uridine in ribonucleic acid, RNA). If the 3'- or 5'-hydroxyl group of sugar is phosphorylated, we have a nucleotide. This unit, the nucleotide, is not only the building block of the polynucleotides DNA and RNA but it also exhibits independent functions.

For example, with adenosine derivatives displayed in Figure 1-1, we can show that, depending on chemical modifications, adenosine adapts to several, dramatically different biochemical roles in life. As di- and triphosphates, adenosine acts as an energy pool for many enzymatic processes and for muscle work. The importance of adenosine triphosphate is demonstrated by its turnover rate in humans: about one body weight per day per person. The 3', 5'-cyclic phosphate of adenosine is the "second hormonal messenger," controlling and mediating the activities of peptide hormones. In the form of puromycin, adenosine is a potent inhibitor of protein biosynthesis and as arabino- or 8-azaderivatives, adenosines display antibiotic activities. Adenosine diphosphate, equipped at the terminal phosphate with certain biological molecules, is a constituent of both coenzymes A and NAD$^+$ and is essential for the proper functioning of enzymes which require these cofactors.

To understand the biological function of a nucleotide, we must know its structural features. With the adenosine derivatives mentioned above, we know the *chemical structure* and we know the function. But why and how do they function? Why, for instance, is 8-azaadenosine an antibiotic even though, from a chemical point of view, the isoelectronic substitution of

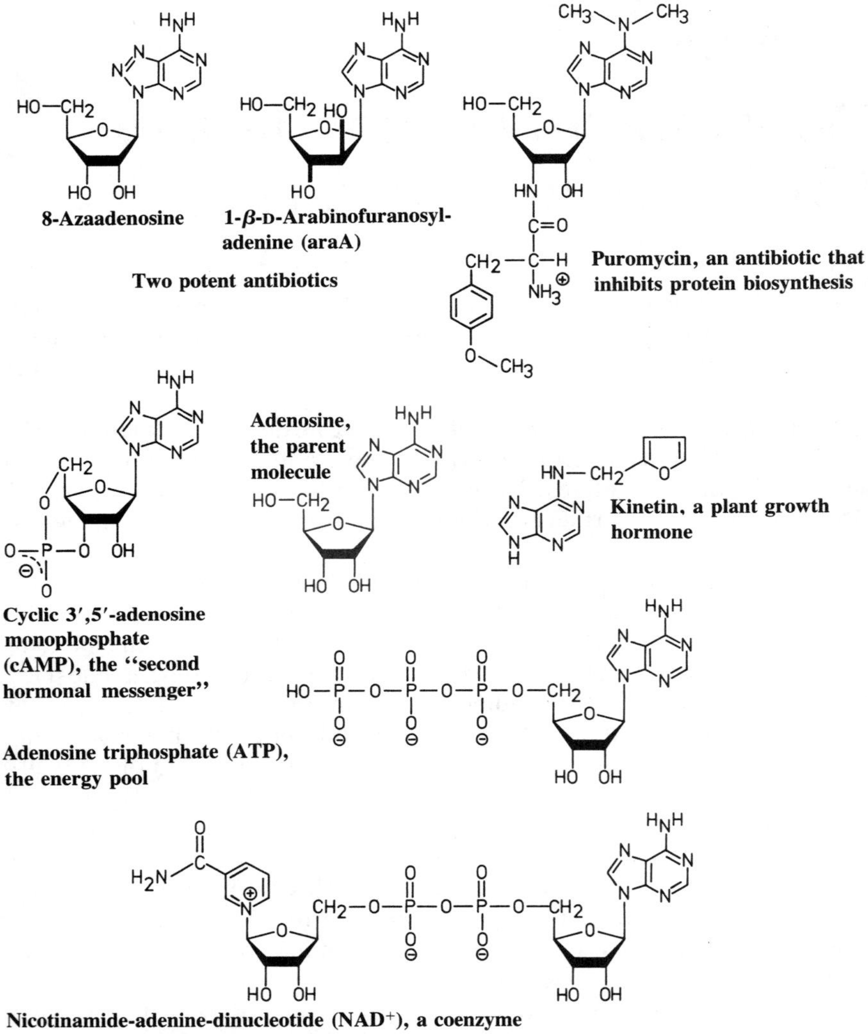

8-Azaadenosine   **1-β-D-Arabinofuranosyl-adenine (araA)**

Two potent antibiotics

Puromycin, an antibiotic that inhibits protein biosynthesis

Adenosine, the parent molecule

Kinetin, a plant growth hormone

Cyclic 3',5'-adenosine monophosphate (cAMP), the "second hormonal messenger"

Adenosine triphosphate (ATP), the energy pool

Nicotinamide-adenine-dinucleotide (NAD⁺), a coenzyme

Figure 1-1. The properties of a biological molecule can change dramatically with only minor chemical modifications. This scheme displays adenosine (color) and some of its derivatives which exhibit very different biochemical behavior.

8-CH by an 8-aza group is expected to have only a modest influence on its chemistry? In this case it is not the electronic but rather the *three-dimensional structure* or conformation of the nucleoside which is changed and which explains 8-azaadenosine's particular properties.

Because nucleosides and nucleotides interact with proteins in all their metabolic or control operations, specific, mutual recognition of the two (or more) reactants is required. This presupposes that the partners involved have well-defined three-dimensional structures which, if we desire to understand functioning at atomic level, we must also know at the atomic level.

Structural features of nucleotides can be studied by several methods. The most powerful of these is X-ray crystallography which, for nucleosides, was pioneered by Furberg who in 1951 published the crystal structure of cytidine (1) (Figure 1-2). In those days, solution of a crystal structure was a major time-consuming task. Thousands of X-ray diffraction intensities

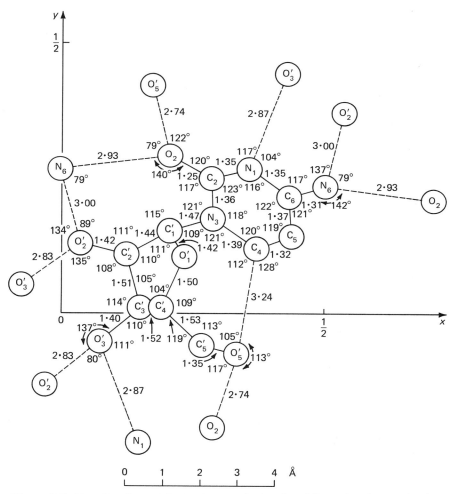

Figure 1-2. Results of crystal structure analysis of cytidine, showing projection along the crystallographic c axis. Bond lengths are given in angstroms; hydrogen bond distances are indicated by broken lines. Note indication of a short contact between C_4 and $O_{5'}$ (present nomenclature: C_6 and $O_{5'}$) which was rediscovered many years later (see section 4.9). From (1).

were measured and electron density maps were computed manually. A year would be needed for a molecule as small as a nucleoside. Modern technology has changed a year into a matter of days, and at present over 600 crystal structure analyses of nucleosides, nucleotides, oligonucleotides, and their individual constituents are available. This large body of data, combined with results from spectroscopic and theoretical studies, enables us to deduce the structural features of nucleotides at an atomic level with confidence.

From nucleotide to DNA double helix. A structural principle that changed the biochemist's world. In 1951, when Furberg published the molecular structure of the first nucleoside, the structure of DNA was still unknown. What was the basis of heredity? How did the genetic material, DNA, reproduce itself from generation to generation? How did protein biosynthesis function and what system was used to relate DNA nucleotide sequence to protein amino acid sequence? These questions could not be answered from chemical data alone, even the data assembled in Todd's

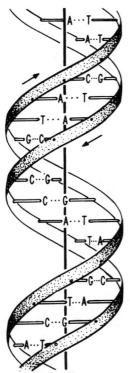

Figure 1-3. Schematic drawing of the DNA double helix. The sugar-phosphate backbones run at the periphery of the helix in antiparallel orientation. Base-pairs (A–T and G–C) drawn symbolically as bars between chains are stacked along the center of the helix. From (7).

Box 1-1. Shortened Version of J. D. Watson's Book, *The Double Helix*.

Hear the song of how the spiral
Complex, twisted, double, chiral
Was discovered. Its construction
Following some shrewd deduction
Was accomplished. From all sides
"Oh tell it as it was" they cried so
Then he took his pen and paper
And composed the helix caper
Writing of the Cambridge popsy
Turned the world of science
  topsy
Turvy like had not before been
Done. The story is no more than
How the structure was
  discovered
Secret of the gene uncovered
Biological prediction
Better than a work of fiction
Stylistically breezy
Everything appeared made easy
And it was. The bases, pairing
Found at last—a small red
  herring
Notwithstanding. How he found
  them
And the hydrogens that bound
  them
Was a stroke more accidental
Than a work experimental
Chemistry was not his calling
He had read *one* book by Pauling
Models of the bases he'd made
With them on his desk top he'd
  played
Shuffling them and putting like
  with
Like he made no lucky strike
  with
Guanine, thymine, adenine. A

Letter came from Pasadena
Horrors: then the day was won
  for
Pauling's model was quite done
  for
Extra atoms he had set in
Where no atoms should be let in
With some phrases less than
  modest
Pauling's model's called the
  oddest
Back now to the basic pairing
How's the structure building
  faring?
Faster, faster goes the race, then
Everything falls into place when
Tautomers which nature chooses
Are the ones our author uses
Two chains (but here we can't be
  sure)
Plus this extra structure feature:
Pauling's outside bases inside
Pauling's inside phosphates
  outside
Now they all were quite ecstatic
Soon became they quite dogmatic
Having nature's secret later
Checked against the x-ray data
This was difficult, for Rosie
Found our scientist rather nosey
Bragg compares the book with
  Pepys. Is
This the verdict of Maurice? His
Thoughts alas we cannot gainsay
Trumpet blowing's not his forte
Tales like this do have a moral
Whether printed whether oral
Find the right man to advise you
Then you'll get a Nobel prize too.

J. Field

From: Journal of Irreproducible Results, **17,** p. 53 (1968).

laboratory which established that nucleotides are linked via 3′,5′-phosphodiester bonds to produce a linear polymer, DNA (2). Crucial additional information came from the work of Chargaff and co-workers (3) showing that, in DNAs of varying base composition, the ratios A/T and G/C are always one. Astbury's first X-ray photographs of fibrous DNA displayed a very strong meridional reflection at 3.4 Å distance (4), hinting at bases stacked upon one another, and Gulland (5), on the basis of electrotrimetric studies, concluded that bases were linked by hydrogen bonding. New X-ray diffraction data, obtained in Wilkins' laboratory, in addition suggested that DNA was a helical molecule and is able to adopt several different conformations depending on the relative humidity and counterion present (6).

All these pieces of the puzzle were put together by Watson and Crick

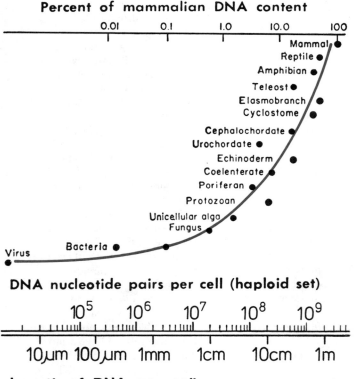

Figure 1-4. Minimum haploid DNA content in species at various levels of organization. DNA length was obtained from the number of nucleotide pairs per cell on the assumption that one nucleotide (base) pair corresponds to 3.4 Å. The ordinate is not a numerical scale and the exact shape of the curve has no significance—it is only meant to show that the amount of DNA increases with the complexity of a species.

(7, 8), who proposed specific, hydrogen-bonded base-pairs of adenine with thymine (A-T) and guanine with cytosine (G-C) to explain Chargaff's and Gulland's data. These base-pairs are stacked like rolled coins at 3.4 Å distance as shown by Astbury's and Wilkins' X-ray data, and right-handed rotation about 36° between adjacent base-pairs produces a double helix with 10 base-pairs per turn (Figure 1-3). Adapting dimensions and shapes of individual nucleotides from Furberg's structure of cytidine, a model of the helix was constructed with bases located along the helix axis and sugar-phosphate backbones winding in antiparallel orientation along the periphery.

This discovery, documented in the fabulous book of Watson (9; Box 1-1) and honored in 1962 with the Nobel prize, answered many questions,

Figure 1-5. Electron micrograph of a histone-depleted metaphase chromosome from HeLa cells. It shows the central protein scaffold as an X-shaped contour and part of the surrounding DNA. The bar represents 1 μm or 10,000 Å. From (11).

posed new ones, and led to an explosion of new developments in many areas of biochemistry and genetics. The basis of all this, to emphasize it clearly, was the insight into the structure of the genetic material, DNA, at an atomic level.

From double helix to chromosomes—a few impressive numbers. Knowing the structure of DNA is not the end of the story. In organisms DNA is packaged and organized in higher-ordered forms. Depending on the level of evolution of a living being, the length of the total DNA it contains varies from micrometers to several centimeters (Figure 1-4). The DNA is assembled in simple virus capsids, in prokaryotic cells, or in eukaryotic cell nuclei. In human somatic cells, there are 46 chromosomes, each consisting of a single DNA duplex molecule about 4 cm long. Placing all the chromosomes end to end would yield a DNA string nearly 2 m long for each somatic cell. This 2-m-long DNA is packed with the help of basic polypeptides called histones into a nucleus 0.5 μm in diameter. By way of analogy this corresponds to a string 0.1 mm thick and 100 km long, squeezed into a sphere 25 cm in diameter, with histones ordering and holding the tangle together. If the histones are removed and the chromosome is examined under the electron microscope, the mass of coils shown in Figure 1-5 becomes visible, all assembled around a central protein scaffold.

What is the organization of DNA in chromosomes and how does DNA expression take place? Although we are far from a detailed description of these processes, we can say in summary that the physical and chemical properties of nucleic acids and their constituents are closely associated with their three dimensional structure. Therefore the question posed at the onset can be answered. We have to study principles of nucleotide and nucleic acid structure in order to understand the function of these biological molecules at the atomic level.

Chapter 2

Defining Terms for the Nucleic Acids

In structural chemistry, defining the terms used to describe the geometry of molecules is fundamental in understanding basic principles. This chapter explains the conventions and nomenclature which have been adopted for nucleotides and nucleic acids. Although in nucleic acid research, atom numbering schemes and abbreviations are those recommended by the Commissions of the International Union of Pure and Applied Chemistry (IUPAC) and of the International Union of Biochemistry (IUB) (12–14), definitions of interatomic bond distances, bond angles, and torsion angles were, and still are, complex because of competing and conflicting terminology (15–18). Recently, an IUPAC-IUB subcomission worked out standard definitions (19) which will be explained and used throughout the following text.

2.1 Bases, Nucleosides, Nucleotides, and Nucleic Acids—Nomenclature and Symbols

The basic, repeating motif in deoxyribonucleic acid (DNA) and ribonucleic acid (RNA) is the nucleotide (Figure 2-1). It is composed of a cyclic, furanoside-type sugar (β-D-ribose in RNA or β-D-2'-deoxyribose in DNA) which is phosphorylated in the 5' position and substituted at $C_{1'}$ by one of four different heterocycles attached by a β-glycosyl $C_{1'}$–N linkage. The heterocycles are the purine bases adenine and guanine and the pyrimidine bases cytosine and uracil. The latter is only found in RNA and replaced by the functionally equivalent thymine (5-methyluracil) in DNA (Figure 2-2, Table 2-1).

The free bases adenine (Ade), guanine (Gua), cytosine (Cyt), uracil (Ura), and thymine (Thy) bear a hydrogen atom in positions 9 (purine) and 1 (pyrimidine) which in the nucleosides is replaced by the sugar moiety (Figure 2-2, Table 2-1). Besides the basic nucleosides adenosine (Ado or A), guanosine (Guo or G), cytidine (Cyd or C), uridine (Urd or U), and thymidine (dThd or dT), there exist numerous naturally occurring and chemically synthesized, modified nucleosides. Many of these have antibiotic activity, among them the important class of arabinosides, nucleosides with β-D-arabinose instead of β-D-ribose (Figure 2-2) (20,21).

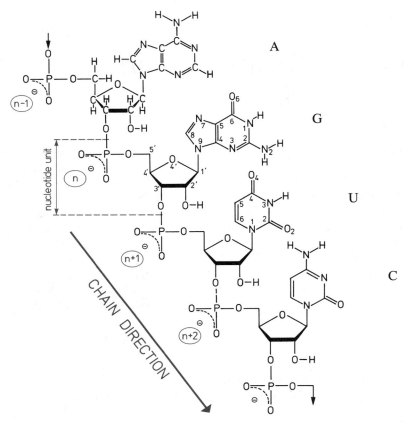

Figure 2-1. Fragment of ribonucleic acid (RNA) with sequence adenosine (A), guanosine (G), uridine (U), cytidine (C) linked by 3′,5′-phosphodiester bonds. Chain direction is from 5′- to 3′-end as shown by arrow. Atom numbering scheme is indicated in one framed nucleotide unit, 5′-GMP. All hydrogen atoms drawn in A and only functional hydrogens in other nucleotides. In short notation, this fragment would be pApGpUpCp or pAGUCp. In deoxyribonucleic acid (DNA), the hydroxyl attached to $C_{2'}$ is replaced by hydrogen and uracil, by thymine.

A nucleotide is a nucleoside phosphorylated at one of the free sugar hydroxyls. Several isomers with phosphate at the 2′, 3′, and 5′ positions are known, e.g., adenylic acid, adenylate or adenosine 5′-monophosphate (5′-AMP), uridine 3′-monophosphate (3′-UMP). The phosphate group can form a cyclic diester with 2′, 3′-hydroxyls in ribonucleotides and with 3′, 5′-hydroxyls in ribo- and deoxyribonucleotides, leading to the biologically important "second messenger" adenosine 3′,5′-cyclic phosphate (3′,5′-AMP, cAMP, A > p). In the "energy-rich" di- and triphosphates, the oligophosphate residue is attached to the 5′ position of the sugar to form adenosine triphosphate (ATP, pppA, Ado-5′PPP), which hydrolyzes to

Table 2-1. Abbreviations and Symbols for Bases, Nucleosides, and Nucleotides[a]

Base		Nucleoside		Nucleotide		Other examples
Name	Symbol	Name	Symbol	Name	Symbol	
			1. Ribonucleosides and -nucleotides			
Uracil	Ura	Uridine	Urd or U	Uridylic acid	5′-UMP or pU	Uridine 2′-monophosphate (2′-UMP)
						Uridine 3′-monophosphate (3′-UMP, Up)
Cytosine	Cyt	Cytidine	Cyd or C	Cytidylic acid	5′-CMP or pC	Cytidine diphosphate (CDP, ppC)
						Cytidylyl-(3′,5′)-uridine (CpU)
Adenine	Ade	Adenosine	Ado or A	Adenylic acid	5′-AMP or pA	Adenosine triphosphate (ATP, pppA)
Guanine	Gua	Guanosine	Guo or G	Guanylic acid	5′-GMP or pG	Guanosine 2′,3′-cyclic phosphate
						(2′,3′-GMP, G>p, cGMP)
			2. 2′-Deoxyribonucleosides and -nucleotides[b]			
Thymine	Thy	Deoxythymidine[c]	dThd or dT	Deoxythymidylic acid	5′-dTMP or pdT	Polyadenylic acid (poly A); alternate
Cytosine	Cyt	Deoxycytidine	dCyd or dC	Deoxycytidylic acid	5′-dCMP or pdC	copolymer of dA and dT, Poly(deoxy-adenylate–deoxythymidylate), poly
Adenine	Ade	Deoxyadenosine	dAdo or dA	Deoxyadenylic acid	5′-dAMP or pdA	[d(A-T)], or poly(dA–dT) or (dA–dT)
Guanine	Gua	Deoxyguanosine	dGuo or dG	Deoxyguanylic acid	5′-dGMP or pdG	or d(A-T)$_n$; the same but randomly

distributed dA,dT: replace hyphen by comma, poly d(A,T) etc. A complex between poly(A) and poly(U) is designated poly(A)·poly(U).
Alanine-specific transfer RNA from *E. coli* (tRNA^{Ala}(*E. coli*)).

[a] Adapted from (38).

[b] The symbols for 2′-deoxyribonucleosides and -tides are as for ribonucleosides and -tides with the prefix d.

[c] Since thymidine occurs as a ribonucleoside in tRNA, use of the prefixes d for deoxyribose and r for ribose is recommended (12).

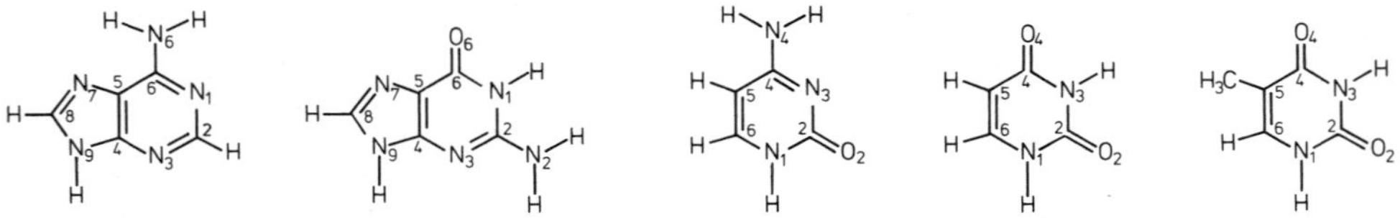

Adenine Ade | Guanine Gua | Cytosine Cyd | Uracil Ura | Thymine Thy

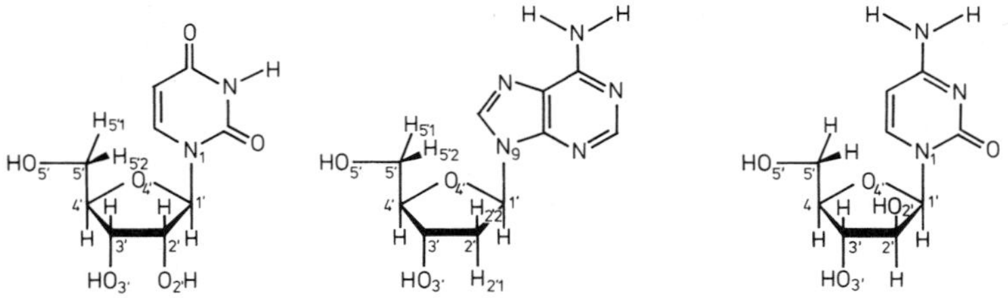

Uridine Urd, U | Deoxyadenosine dAdo, dA | Arabinocytidine araCyd, araC

Figure 2-2. Chemical structure of some bases, nucleosides, nucleotides, and the coenzyme NAD^+.

adenosine diphosphate (ADP, ppA) and to the monophosphate, adenylic acid (5′-AMP). Other nucleotides likewise form oligophosphates which play important roles in polynucleotide biosynthesis. A special case of a diphosphate is the coenzyme nicotinamide adenine dinucleotide (NAD^+) which contains nicotinamide riboside (Nir) and Ado separated by a pyrophosphate group, Ado-5′PP5′-Nir (12) (Figure 2-2).

In oligo- and polynucleotides, the individual nucleotides are ordinarily linked via 3′,5′-phosphodiester bonds. A trinucleoside diphosphate with the sequence G–C–U is called guanylyl-3′,5′-cytidylyl-3′,5′-uridine and abbreviated GpCpU or GCU, with G the 5′ end and U the 3′ end of the oligomer. Oligonucleotides with 2′,5′-phosphodiester bonds have not been observed in nature but are available synthetically. They are possible only in the ribo-series and have recently roused interest because of their interferon-inducing acitivity.

Synthetic, *homopolymer nucleic acids* are called, for example, poly(uridylate) or poly(deoxyadenylate) or in short poly(U) or poly(dA). A *heteropolymer* with alternating sequence, poly(dA-dT), is called poly(deoxyadenylate-thymidylate). If the nucleotides dA and dT are randomly distributed over the chain, a comma replaces the hyphen: poly(dA,dT). Poly(U) and poly(A) can form 1 : 1 or 2 : 1 complexes indicated by a dot between the names, poly(A) · 2poly(U). In designating different amino acid-specific transfer RNAs, for example, $tRNA^{Phe}$ (yeast),

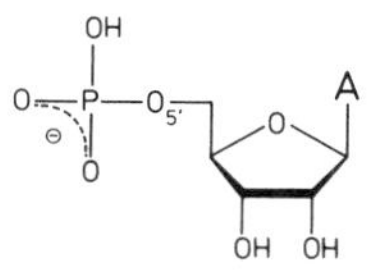

Adenylic acid, 5'-AMP, pA
Adenosine 5'-phosphate

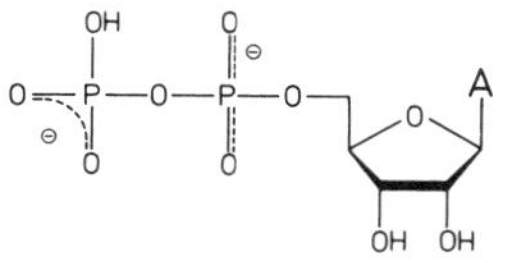

Adenosine diphosphate, ADP, ppA

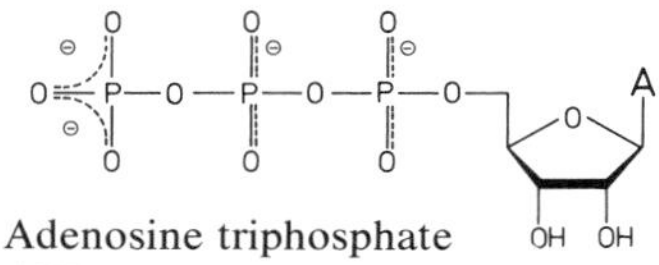

Adenosine triphosphate
ATP, pppA

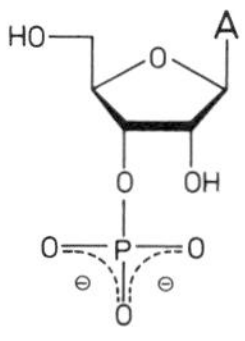

Adenosine 3'-phosphate
3'-AMP, Ap

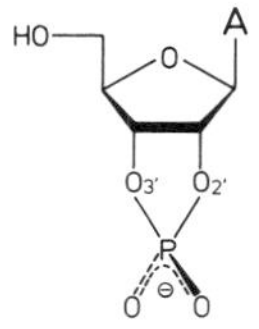

Adenosine 3',5'-cyclic phosphate,
cyclic phosphate, 2'-3'-AMP
A > p

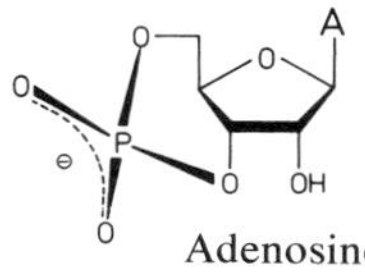

Adenosine-3',5'-cyclic phosphate,
3',5'-AMP, "cyclic AMP"

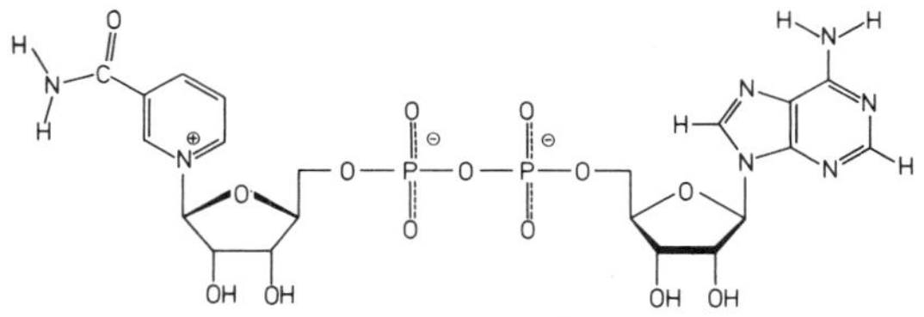

Nicotinamide adenine dinucleotide, NAD$^+$

Figure 2-2. (*Continued*)

the superscript indicates specificity for phenylalanine and the source (yeast) from which it was isolated is shown (Table 2-1).

2.2 Atomic Numbering Scheme

Figures 2-1 to 2-3 indicate the atomic numbering scheme adopted for nucleotides. Sugar atoms are distinguished from base atoms by a prime. The counting direction of atoms of a polynucleotide backbone is described by the sequence $P \rightarrow O_{5'} \rightarrow C_{5'} \rightarrow C_{4'} \rightarrow C_{3'} \rightarrow O_{3'} \rightarrow P$. In a sugar ring, the num-

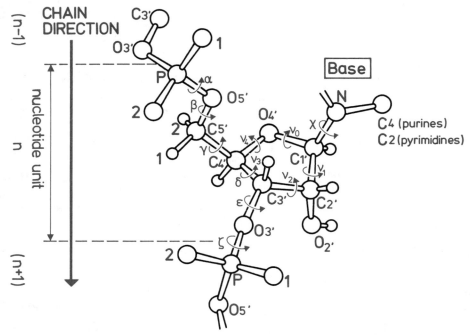

Figure 2-3. Atomic numbering scheme and definition of torsion angles for a polyribonucleotide chain. Counting of nucleotides is from top to bottom, i.e., in the direction $O_{5'} \to O_{3'}$. Hydrogens at $C_{5'}$ and oxygens at P are differentiated by 1 and 2 according to the rule given in the text. In deoxyribose, the hydrogen replacing $O_{2'}$ is labeled 1, the other one, 2. For a full description of torsion angles, see Table 2-2.

bering sequence is $C_{1'} \to C_{2'} \to C_{3'} \to C_{4'} \to O_{4'} \to C_{1'}$ [note that the previously used designation $O_{1'}$ is now replaced by $O_{4'}$ (19)].

The two hydrogen atoms at $C_{5'}$ and at deoxyribose $C_{2'}$, and the two free oxygen atoms at phosphorus, are labeled 1 and 2. Looking along $O_{5'} \to C_{5'}$ in the direction of the main chain, a clockwise counting gives $C_{4'}$, $H_{5'1}$, $H_{5'2}$ and in the same manner, looking along $O_{3'} \to P \to O_{5'}$ a clockwise counting defines O_1 and O_2 at the phosphorus (Figure 2-3).

2.3 Torsion Angles and Their Ranges

The three-dimensional structure of a molecule is characterized by bond lengths, bond angles, and rotations of groups of atoms about bonds. These rotations about a central bond B–C are described by torsion angles involving four atoms in sequence, A–B–C–D. The *torsion angle*, θ in Figure 2-4, is defined as the angle between projected bonds A–B and C–D when looking along the central bond either in direction B→C or in the op-

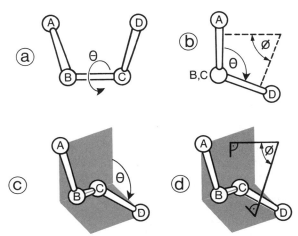

Figure 2-4. Definition of *torsion* and *dihedral angles*. (a) *Torsion angle* θ (A–B–C –D) describing orientations of bonds A–B and C–D with respect to the central bond B–C. (b) View along B→C. θ is the torsion angle between the projected bonds A–B and C–D; the complement φ is called the *dihedral angle*. If A–B and C–D are *cis*-planar (coinciding in projection), angles θ and φ are 0°; they are counted positive if the far bond C–D rotates clockwise with respect to the near bond A–B. (c) θ is defined as the angle between planes A–B–C and B–C–D. (d) The *dihedral angle* φ represents the angle between normals to these planes.

posite sense C→B. It is defined as 0° if A–B and C–D are eclipsed (*cis* and coplanar) and the sign is positive if the far bond is rotated clockwise with respect to the near bond. The torsion angle is usually reported either in the range 0° to 360° or − 180° to + 180° (19,22).

Rather than describing a torsion angle in terms of an angle between projected bonds, it can also be formulated as angle between the two planes containing atoms A, B, C and B, C, D. Another definition uses the angle between the *normals* to these planes. This *dihedral* angle φ (in Figure 2-4) is in fact the complement of the torsion angle. *In the literature, torsion and dihedral angles are frequently used synonymously although, in a strict sense, they should not be confused.*

In molecules with rotational freedom about single bonds, usually not all feasible torsion angles are assumed but rather certain sterically allowed conformations are preferred. Therefore, it is often convenient to describe a structure with torsion angle ranges. The ranges commonly used in organic chemistry are those proposed by Klyne and Prelog, *viz. syn* (∼0°), *anti* (∼180°), ±*synclinal* (∼±60°) and ±*anticlinal* (∼±120°) (23,24). In spectroscopic and crystallographic publications, the notation *cis* (∼0°), *trans* (∼180°), ±*gauche* (∼±60°) is most frequently employed. These two conventions are illustrated in Figure 2-5. As the IUPAC-IUB subcommission on nucleotide nomenclature has recommended the Klyne-Prelog torsion angle ranges, these will be used throughout this text.

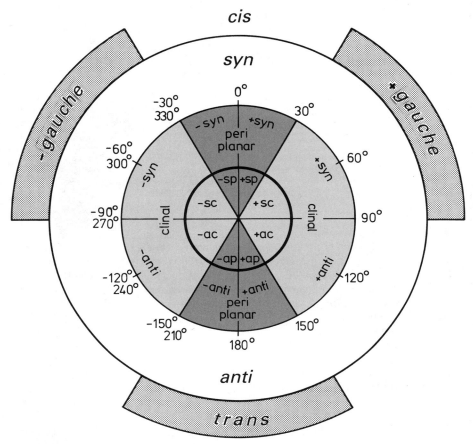

Figure 2-5. Pictorial correlation of torsion angle ranges used by spectroscopists (*cis, trans, +gauche, −gauche*) with ranges defined by Klyne and Prelog (*syn* or *synperiplanar, anti* or *antiperiplanar, +synclinal, −synclinal, +anticlinal, −anticlinal*). The latter have been recommended by the IUPAC–IUB commission on nucleotide nomenclature (19) and are used in this text. The terms *syn, anti* have a special meaning in nucleotide stereochemistry (see Figure 2-10).

2.4 Definitions of Torsion Angles in Nucleotides

The conformation of the sugar-phosphate backbone following the sequential numbering of atoms $P{\rightarrow}O_{5'}{\rightarrow}C_{5'}{\rightarrow}C_{4'}$ etc. is defined by torsion angles α, β, γ, δ, ϵ, ζ in alphabetical order, with α the $P{\rightarrow}O_{5'}$ rotation (see Table 2-2 and Figure 2-3). The endocyclic torsion angles of the sugar are denoted ν_0 to ν_4; the orientation of the base relative to the sugar is given

Table 2-2. Definition of Torsion Angles in Nucleotides [From (16).]^a

Torsion angle	Atoms involved
α	$_{(n-1)}O_{3'}-P-O_{5'}-C_{5'}$
β	$P-O_{5'}-C_{5'}-C_{4'}$
γ	$O_{5'}-C_{5'}-C_{4'}-C_{3'}$
δ	$C_{5'}-C_{4'}-C_{3'}-O_{3'}$
ϵ	$C_{4'}-C_{3'}-O_{3'}-P$
ζ	$C_{3'}-O_{3'}-P-O_{5'(n+1)}$
χ	$O_{4'}-C_{1'}-N_1-C_2$ (pyrimidines)
	$O_{4'}-C_{1'}-N_9-C_4$ (purines)
ν_0	$C_{4'}-O_{4'}-C_{1'}-C_{2'}$
ν_1	$O_{4'}-C_{1'}-C_{2'}-C_{3'}$
ν_2	$C_{1'}-C_{2'}-C_{3'}-C_{4'}$
ν_3	$C_{2'}-C_{3'}-C_{4'}-O_{4'}$
ν_4	$C_{3'}-C_{4'}-O_{4'}-C_{1'}$

^{a} Atoms designated $(n-1)$ and $(n+1)$ belong to adjacent units.

by χ. Note that ν_3 and δ describe orientations about the same bond $C_{3'}-C_{4'}$, one endocyclic, the other exocyclic. They are therefore correlated, as outlined in greater detail in Chapter 4.

2.5 Sugar Puckering Modes: The Pseudorotation Cycle

The five-membered furanose ring is generally nonplanar. It can be puckered in an envelope (E) form with four atoms in a plane and the fifth atom is out by 0.5 Å; or in a twist (T) form with two adjacent atoms displaced on opposite sides of a plane through the other three atoms (25) (Figure 2-6). Atoms displaced from these three- or four-atom planes and on the same side as $C_{5'}$ are called *endo;* those on the opposite side are called *exo* (26) (Figure 2-7). Sugar puckering modes are defined accordingly. It should be noted that because the transition between E and T forms is facile, usually the atoms defining the four-atom planes are not perfectly coplanar, and displacements from a three-atom plane are symmetrical only in rare cases. Therefore, the largest deviation from planarity is called major puckering; the lesser deviation is called minor.

In an abbreviated notation, superscripts for *endo* atoms and subscripts for *exo* atoms precede or follow the letter E or T, depending on major or minor puckering. An unsymmetrical $C_{3'}-endo-C_{2'}-exo$ twist

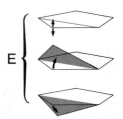

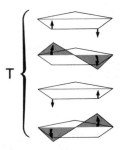

Figure 2-6. Puckering of five-membered ring into envelope (E) and twist (T) forms. In E, four of the five atoms are coplanar and one deviates from this plane; in T, three atoms are coplanar and the other two lie on opposite sides of this plane.

with major $C_{3'}$ and minor $C_{2'}$ puckering is represented as $^3T_2$. The same twist in symmetrical form is designated as ^3_2T. A $C_{2'}-endo$ envelope is given as ^2E, a $C_{3'}-exo$ envelope as $_3E$ (Figure 2-7).

The pseudorotation cycle. The preceding methods for describing sugar

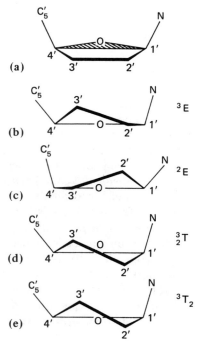

Figure 2-7. Definition of sugar puckering modes. (a) Starting position with flat five-membered sugar, a situation never observed. Plane $C_{1'}-O_{4'}-C_{4'}$ is shown hatched. (b-e) View with this plane perpendicular to the paper. (b) Envelope $C_{3'}-endo$, ^3E. (c) Envelope $C_{2'}-endo$. ^2E. (d) Symmetrical twist or half-chair $C_{2'}-exo-C_{3'}-endo$. ^3_2T. (e) Unsymmetrical twist with major $C_{3'}-endo$ and minor $C_{2'}-exo$ pucker, $^3T_2$.

puckering are only approximate and, if intermediate twist modes are considered, they are inadequate. Therefore, the puckering of five-membered rings has been treated analytically in a form elegantly represented by the concept of pseudorotation, first introduced for cyclopentane (27,28) and later applied to substituted furanose (29).

In cyclopentane, conformational changes do not proceed via a planar intermediate but the maximum pucker rotates virtually without potential energy barriers, giving rise to an "infinite" number of conformations. These can be described in terms of the maximum torsion angle (degree of pucker), ν_{max}, and the pseudorotation phase angle, P (30,31) (Figure 2-8). If the five-membered ring is unsymmetrically substituted as in nucleotides, potential energy thresholds are created which limit the pseudorotation and lead to preferred puckering modes (Chapter 4).

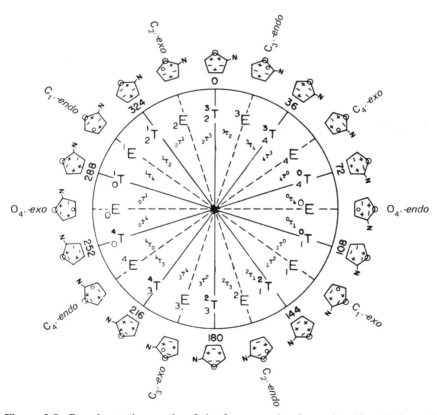

Figure 2-8. Pseudorotation cycle of the furanose ring in nucleosides. Values of phase angles given in multiples of 36°. Envelope E and twist T forms alternate every 18°. After rotation by 180° the mirror image of the starting position is found. On the periphery of the cycle, riboses with signs of endocyclic torsion angles are indicated. (+) Positive, (−) Negative, (0) angle at 0°. From (31).

In nucleotides, the pseudorotation phase angle P is calculated from the endocyclic sugar torsion angles according to (31):[*]

$$\tan P = \frac{(\nu_4 + \nu_1) - (\nu_3 + \nu_0)}{2 \cdot \nu_2 \cdot (\sin 36° + \sin 72°)}. \tag{2-1}$$

The phase angle $P = 0°$ is defined such that torsion angle ν_2 is maximally positive, corresponding to a symmetrical $C_{2'}-exo-C_{3'}-endo$ twist form, ${}^3_2T$ (Figure 2-7d). The mirror image, $C_{2'}-endo-C_{3'}-exo$, ${}^2_3T$, is represented by $P = 180°$ (Figure 2-8). The maximum torsion angle ν_{max} which describes the maximum out-of-plane pucker, is given by:

$$\nu_{max} = \nu_2 / \cos P. \tag{2-2}$$

Figure 2-9 displays the theoretical changes of the five torsion angles during one full pseudorotation cycle (32). At every phase angle P, the sum of the positive torsion angles equals that of the negative torsion angles; i.e., the sum of the five angles is zero:

$$\nu_0 + \nu_1 + \nu_2 + \nu_3 + \nu_4 = 0. \tag{2-4}$$

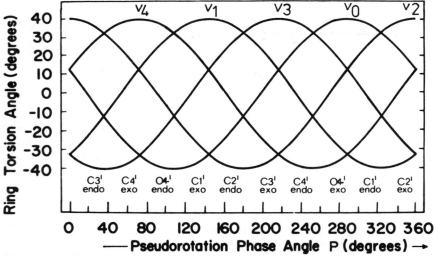

Figure 2.9. Changes of the five endocyclic furanose torsion angles during one full pseudorotation cycle. After (32).

---

[*]in the original publication (31), the torsion angles angles are defined differently and denoted θ_n. The transformation from the torsion angles used here to those in ref. (31) is: $\nu_0 = \theta_3$, $\nu_1 = \theta_4$, $\nu_2 = \theta_0$, $\nu_3 = \theta_1$, $\nu_4 = \theta_2$.

In the pseudorotation cycle (Figure 2-8), note that envelope (E) and twist (T) conformations alternate every 18°. T is observed at even multiples of 18° and E is found at odd multiples. **In nucleotide structure analysis two ranges of pseudorotation phase angles are preferred: $C_{3'}-endo$ at $0° \leq P \leq 36°$ (in the "north" of the cycle, or N) and $C_{2'}-endo$ at $144° \leq P \leq 190°$ ("south," or S).**

2.6 *syn / anti* Orientation About the Glycosyl Bond

Relative to the sugar moiety, the base can adopt two main orientations about the glycosl $C_{1'}-N$ link, called *syn* and *anti* (33,34, Figure 2-10). They are defined by torsion angle χ described in Figure 2-3 and in Table 2.2:

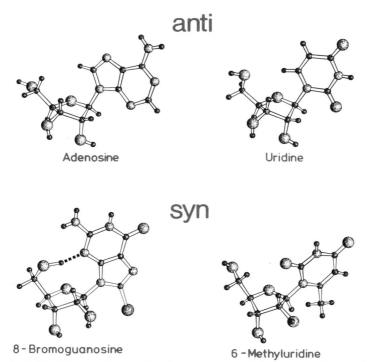

anti

Adenosine   Uridine

syn

8 - Bromoguanosine
6 - Methyluridine

Figure 2-10.(a) Diagram illustrating how the overall geometry of a nucleoside changes if bases are in *syn* or in *anti* orientation. Shown are adenosine (39) and uridine (40) in *anti* conformation whereas 8-bromoguanosine (41) and 6-methyluridine (42) are *syn* due to their bulky substituents *ortho* to the glycosyl link. In 8-bromoguanosine, an intramolecular $O_{5'}H\cdots N_3$ hydrogen bond indicated by the broken line stabilizes the *syn* conformation. Note sugar puckerings, $C_{3'}-endo$ preferred for *anti* but $C_{2'}-endo$ for *syn* nucleosides. Spheres of increasing size represent H, C, N, O atoms; atomic positions obtained from crystal structure analyses.

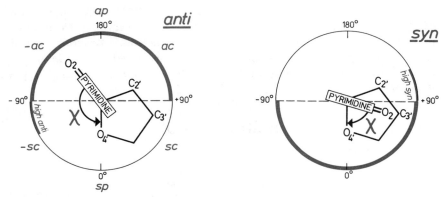

Figure 2-10. (b) Definition of *anti* and *syn* conformational ranges according to ref. (19), shown for pyrimidine nucleoside, χ is defined as torsion angle $O_{4'}-C_{1'}-N_1-C_2$ (Table 2-2). The pyrimidine base is toward the viewer; the base is rotated relative to the sugar. The *high-anti* (*-sc*) range with $\chi \sim -90°$ is actually part of *syn* and the *high-syn* (*+ac*) range is part of *anti*.

$$\chi \begin{cases} O_{4'}-C_{1'}-N_9-C_4 \text{ (purine nucleosides)} \\ O_{4'}-C_{1'}-N_1-C_2 \text{ (pyrimidine nucleosides)} \end{cases}$$

In *anti*, the bulk of the hetercycles, i.e., the six-membered pyrimidine ring in purines and O_2 in pyrimidines, is pointing away from the sugar, and in *syn* it is over or toward the sugar (Figure 2-10). *In Klyne–Prelog notation, anti is ap and syn is sp.*

In the literature, other definitions for χ have been used: their relationship to the present χ is given in Table 2-3 (35).

A variant of *anti* has been observed in some crystal structures of naturally occurring and chemically modified nucleosides. In this particular conformation, the bond $C_{1'}-C_{2'}$ is nearly eclipsed with N_1-C_6 in pyrimi-

Table 2-3. Conversion of Different Definitions for Torsion Angle about the Glycosyl $C_{1'}-N$ Linkage [$\chi_{present} = \chi_{other} + $ Difference]. Differences according to (35).

	Present χ	Other definition	Difference between present and other
Purine		$O_{4'}-C_{1'}-N_9-C_8$	$+180°$
Pyrimidine		$O_{4'}-C_{1'}-N_1-C_6$	
Purine	$O_{4'}-C_{1'}-N_9-C_4$	$C_{2'}-C_{1'}-N_9-C_8$	$-62.5°$
Pyrimidine	$O_{4'}-C_{1'}-N_1-C_2$	$C_{2'}-C_{1'}-N_1-C_6$	
Purine		$C_{2'}-C_{1'}-N_9-C_4$	$+116.5°$
Pyrimidine		$C_{2'}-C_{1'}-N_1-C_2$	

dine or N_9-C_8 in purine nucleosides. This extreme or *high anti* (36) position of the base with $\chi \sim$ -90° corresponds, in Klyne-Prelog notation, to $-sc$. In our definition of χ, the term *high anti* actually denotes a torsion angle lower than *anti*. To avoid confusion we will add a $-sc$ in parentheses, *high anti* $(-sc)$.

2.7 Orientation About the $C_{4'}-C_{5'}$ Bond

Rotation about the exocyclic $C_{4'}-C_{5'}$ bond allows $O_{5'}$ to assume different positions relative to the furanose. Three main conformations with all substituents in staggered positions are possible (Figure 2-11). It is common usage to describe these three conformations with the two torsion angles ϕ_{oo} $(O_{5'}-C_{5'}-C_{4'}-O_{4'})$ and ϕ_{oc} $(O_{5'}-C_{5'}-C_{4'}-C_{3'})$ (37) or to use the angular ranges $(+)$ or $(-)$ *gauche* or *trans*, leading to the three conformations

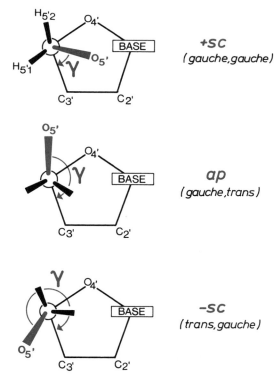

Figure 2-11. Definition of torsion angle ranges about the $C_{4'}-C_{5'}$ bond, looking in the direction $C_{5'} \rightarrow C_{4'}$.

Table 2-4. Different Definitions for Torsion Angle
γ about Exocyclic $C_{4'}-C_{5'}$ Bond

Present γ	Other definitions for orientation about the $C_{4'}-C_{5'}$ bond
$+sc$	gauche,gauche; (+)gauche; gg; (+)g
$-sc$	trans,gauche; trans; tg; t
ap	gauche,trans; (−)gauche; gt; (−)g

given in Table 2-4. According to the IUPAC-IUB recommendation (19), it is only necessary to state torsion angle $\gamma \ (= \phi_{oc})$ in order to define the orientation about the $C_{4'}-C_{5'}$ bond (Table 2-4).

Terms introduced thus far describe the principal conformation features of nucleotides in monomers and also in polymers. However, polymers involve additional parameters if helical symmetry and basepairing are considered.

2.8 Helical Parameters: Hydrogen Bonding Between Bases

Polynucleotides in helical arrangement display order which can be expressed in terms of helical parameters. The pitch of a helix is the distance traveled along the helix axis for one complete helix turn (360° rotation). The pitch height P relates the number n of nucleotides in one turn and the unit height, h, defining the translation per residue along the helix axis, by $P = n \cdot h$. For integral helices n is an integer, but, in general, it can be any number. The unit twist is $t = 360°/n$ and is the rotation between one nucleotide and its nearest neighbor (Figure 2-12).

In a double helix, the two polynucleotide chains are connected by complementary base-pairing (7,33) (Figure 2-13). The A:U and G:C base-pairs are similar in shape, or isomorphous, with practically identical $C_{1'} \ldots C_{1'}$ distances. A pseudodyad axis relates glycosyl $C_{1'}-N$ linkages and, attached to them, sugar-phosphate backbones which, by this dyad operation, are in antiparallel orientation. In general, therefore, DNA and RNA double helices display pseudodyad symmetry.

Base-pairs are usually not centered on the helix axis but rather are displaced by distance D (Figure 2-14). Moreover, they are not exactly perpendicular to the helix axis but are inclined with tilt angle θ_τ and roll angle θ_R (Figure 2-14). Bases in a base-pair are not necessarily coplanar; they can have a propeller twist θ^P defined by a rotation about the roll axis or by a dihedral angle between the normals to the base planes.

Double-helical RNAs and DNAs exhibit characteristic grooves. Because

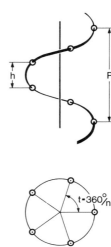

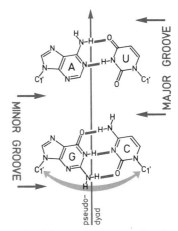

Figure 2-12. Definition of helical parameters pitch P, axial rise per residue h, and unit twist t, shown for a right-handed helix with $n = 5$ residues per turn.

the two glycosyl bonds branch off from one side of the base-pairs and because base-pairs are displaced by D from the helix axis, the outer envelope of the double helix is not cylindrically smooth but can display two grooves of different width and depth. The minor groove is at the O_2 (pyrimidine) or N_3 (purine) side of the base-pair and the major groove is on the opposite side (Figure 2-13). In all cases where a twofold or other sym-

Figure 2-13. Schematic description for A:U(T) and G:C base-pairs occurring in RNA (DNA). Hydrogen bonds $N-H\cdots N$ and $N-H\cdots O$ indicated by colored solid lines. Pseudodyad in thin colored line relates glycosyl $C_{1'}-N$ bond of one base to that of its hydrogen-bonded partner (double arrow in light color). Minor and major groove sides of base-pairs are defined.

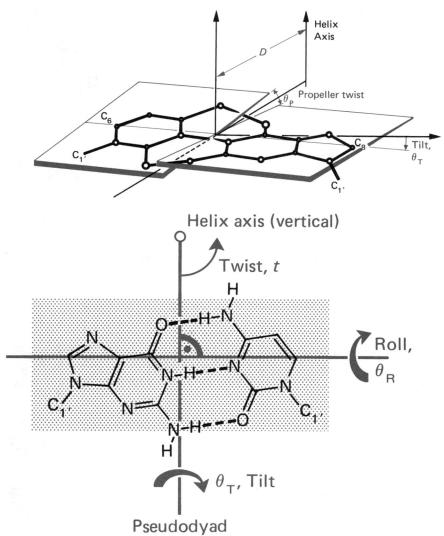

Figure 2-14. (Top) The two bases in a base-pair are not exactly coplanar but display propeller twist θ_P, defined as dihedral angle between individual base planes. Looking down the long axis of a base-pair, θ_P is defined positive if the near base is rotated clockwise with respect to the far one. (Bottom) The orientation of a base-pair with respect to the helix axis is given by rotational twist t, and by tilt θ_T and roll θ_R of a mean base-pair plane (stippled). Reference line for t and θ_T is the pseudodyad (see Figure 2-13), and for θ_R, a vertical to the pseudodyad passing approximately through C_6 (pyrimidine) and C_8 (purine) is taken. Signs of angles θ_T and θ_R is positive for clockwise rotation as indicated by curved, thick arrows. Reference line for 0° is paper plane (or vertical to helix axis). Positive rotation θ_T dips base to the right (cytosine) below paper plane, and for θ_R, atoms in major grove side (top of stippled area) are dipped below paper plane.

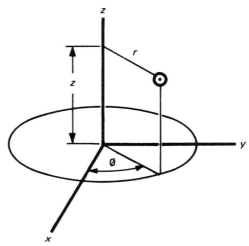

Figure 2-15. Definition of cylindrical coordinates r, ϕ, z.

metry axis coincides with the helix axis as in $[\text{poly}(AH^+)]_2$ or in $[\text{poly}(G)]_4$, there is only one type of groove (see Section 13.3 and 13.6).

Atomic coordinates of a helical molecule are given in cylindrical notation r, ϕ, z, **with r the radial distance from the helix axis, ϕ the rotation angle from a given origin, and z the axial rise,** as described in Figure 2-15. In double helices with twofold axes perpendicular to the helix axis and relating the two chains, origin points $\phi = 0°$ and $z = 0$ are generally defined by a pseudo-twofold axis.

Hydrogen bonding N-H$\cdots$N or N-H$\cdots$O in a base-pair (Figure 2-13) **is indicated by denoting first the donor and then the acceptor atom.** In a G:C base-pair, the hydrogen bonding is given $GN_1:CN_3$, $GN_2:CO_2$ and $CN_4:GO_6$.

Summary

In this chapter, nomenclature used throughout the text is defined. This nomenclature follows recent recommendations of a subcommittee set up by IUPAC-IUB and differs in some parts from nomenclatures used by different individuals.

Designation of torsion angle ranges is according to Klyne and Prelog, with the commonly used $+gauche$, $-gauche$, and $trans$ now replaced by $+syn\ clinal\ (+sc)$, $-syn\ clinal\ (-sc)$, and $anti\ periplanar\ (ap)$. Along the polynucleotide backbone, torsion angles are called Greek alphabet α, β, γ, δ, ϵ, ζ and counted from P-$O_{5'}$ down the chain to $O_{3'}$-P. Within the sugar rings, torsion angles are designated ν_0 to ν_4, and χ indicates torsion about glycosyl link, $O_{4'}-C_{1'}-N_1-C_2$ in pyrimidine and $O_{4'}-C_{1'}-N_9-C_4$ in purine systems.

The pseudorotation description of sugar puckering is explained and compared with $C_{2'}-endo$ or $C_{3'}-endo$ nomenclature. Also, geometry of base-pairing, hydrogen bonding, and nucleic acid double helices are defined.

Chapter 3

Methods: X-Ray Crystallography, Potential Energy Calculations, and Spectroscopy

Nucleoside, nucleotide and nucleic acid structures have been elucidated mainly by X-ray crystallographic techniques. These methods provide information at different levels of detail depending on the molecular weight and organization of the material into crystalline or quasi-crystalline matrices. Readers not familiar with X-ray diffraction methods should be aware of the basic principles; they will then understand why the results of X-ray crystal structure analyses of small molecules up to the oligonucleotide level are unambiguous and why those of crystalline macromolecules like tRNA can, in some close details, be a matter of interpretation. Fiber diffraction analysis of quasi-crystalline polymers like DNA and RNA gives only approximate, overall structural information and cannot yield a satisfying structural model without additional data obtained by other methods.

Potential energy calculations allow us to derive a general picture of the flexibility of a molecule. They are frequently used in structure refinement of macromolecules and are a prerequisite for modern fiber diffraction analysis.

Spectroscopic methods, with emphasis on NMR, are briefly discussed because they complement X-ray crystallographic results which are inherently limited to the solid state and shed light on the dynamic structure of a molecule in solution. It is the interplay of crystallographic and spectroscopic data which, combined with theoretical considerations, leads to a comprehensive understanding of molecular structure and properties.

3.1 Crystal Structure Analysis of Small Molecules

The term "small molecules" is used here to distinguish molecules with molecular weights less than 2000 daltons from globular or fibrous macromolecules. The reason for this distinction is that X-ray diffraction methods and results differ considerably. Resolution is a function of crystal quality and in small-molecule crystallography is usually better than 1 Å, leading to clear separation of electron densities of individual atoms which are located within 0.002 Å uncertainty. For crystals of globular macromolecules, resolution is typically in the range 2-3 Å. This means that electron density is not well defined for each atom and rather envelops chains of atoms whose locations can only be traced to about 0.25 Å accuracy by means of model fitting. For fibrous macromolecules such as DNA, fiber diffraction yields only unit cell constants and helix parameters; full interpretation of the data is achieved with the help of model building studies and inclusion of other chemical and physical (spectroscopic) information.

In a crystal, molecules (and sometimes solvent molecules) are regularly arranged in a three-dimensional matrix. The basic motif is the asymmetric unit which, by the operation of space group symmetry elements, is repeated to build up the unit cell. Translation of the unit cell in all three dimensions creates the crystalline array (Figure 3-1) (43).

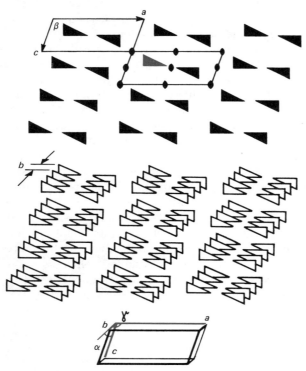

Figure 3-1. Arrangement of molecules in a crystalline array. (Top) Two-dimensional representation of one asymmetric unit (in color), which through operation of the twofold axis perpendicular to the paper (symbol ●) is transformed into another (black) unit, thus filling the unit cell of dimensions a, c, β. Extending this array in the third dimension leads to the crystal lattice (middle). The crystal unit cell (bottom) is described by cell axes a, b, c and by angles $\alpha(<b,c)$, $\beta(<a,c)$, $\gamma(<a,b)$.

The ultimate goal of an X-ray crystal structure analysis is to locate all the atoms in an asymmetric unit. In order to derive unit cell parameters and space group, a crystal of dimensions 0.1 to 0.5 mm is mounted in a Weissenberg or precession camera. A finely collimated, monochromatic X-ray beam illuminates the crystal, and by simultaneous, synchronized movement of crystal and photographic film, a great number of lattice planes within the crystal are brought into reflecting position according to Bragg's law [Eq. (3-1), Box 3-1].

Figure 3-2 illustrates a typical diffraction pattern. Its symmetry and systematic extinctions (see legend to Figure 3-2) furnish information about space group symmetry, and the distances between reflections give

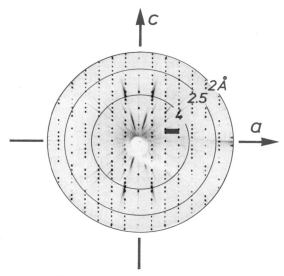

Figure 3-2. X-ray diffraction pattern of the trinucleoside diphosphate adenylyl-(3',5')-adenylyl-(3',5')-adenosine (ApApA). Precession photograph shows enlarged reciprocal unit cell (hatched). Glancing angle θ is derived from distance on film and known crystal-to-film separation; spacings are obtained via Bragg's law [Eq. (3-1)], furnishing the real unit cell dimensions $a = 14.155$ Å, $c = 44.00$ Å. On axis a, every second reflection, and on axis c, only every fourth reflection, are present; the others are systematically absent and suggest two- and fourfold screw axes. These and other photographic details indicate that the space group is tetragonal, $P4_12_12$ (or the enantiomorphic $P4_32_12$). Because in this space group all angles defining the unit cell are 90°, the pattern is mirror-symmetrical about a and c. Modulation of intensity in diffraction spots contains information on atomic distribution within the unit cell. This photograph presents only one section of the reciprocal crystal lattice. In reality a sphere of reflections is present; in this case 3260 data to a resolution of 0.95 Å (160) were measured. Resolution ranges (Å) are indicated by circles. Photograph taken by Dr. P. C. Manor with CuKα radiation, 60 mm crystal-to-film distance.

Box 3-1. A Few Formulas Describing Bragg's Law, Electron Densities, Structure Factors, and R-Factor.

Bragg's Law In the diffraction pattern of a crystal as shown in Figure 3-2, many reflections are seen. They are individually numbered with Miller indices *hkl*, where *h*, *k*, and *l* are three integers. For each reflection this index triple also defines the set of parallel crystal lattice planes which are, for that particular reflection, in reflecting position. The corresponding glancing angle θ_{hkl} depends on the vertical distance d_{hkl} between these planes, according to Bragg's law:

$$2d_{hkl} \cdot \sin \theta_{hkl} = n \cdot \lambda$$

or
$$d_{hkl} = (n \cdot \lambda)/(2 \cdot \sin \theta_{hkl}). \tag{3-1}$$

In this equation, *n* gives the order of reflection *hkl*. The wavelength λ is 1.542 Å for the commonly used CuK_α radiation. Note that the smaller d_{hkl} (i.e., the higher the resolution attained), the wider θ_{hkl}. Because in crystals of macromolecules like transfer RNA, reflections fade away toward the edge of a diffraction pattern (Figure 3-5), resolution (d_{hkl}) is, in general, limited to the 2–3 Å range.

Electron Density An X-ray reflection *hkl* is, in fact, a wave with wavelength λ, amplitude $|F_{hkl}|$ (called *structure amplitude*), and a phase angle α_{hkl}. The latter cannot be measured and is derived by "direct" or "Patterson" (heavy atom) methods. The electron density ρ_{XYZ} at a certain point XYZ in the unit cell can then be calculated:

$$\rho_{XYZ} = \frac{1}{V} \sum_h \sum_k \sum_l |F_{hkl}| \cdot e^{i\alpha_{hkl}} \cdot e^{-2\pi i(hX+kY+lZ)}. \tag{3-2}$$

V is the volume of the unit cell and the triple summation is over all measured data, in general a few thousand. Because the unit cell is divided into several thousand grid points XYZ for calculation of the electron density distribution, extensive calculations on fast computers are necessary.

Structure Factor The electron density obtained allows us to derive atomic positions. These, associated with the proper scattering factor f_j (a number listed in tables and giving the X-ray scattering power of atom *j*), are used to calculate the structure factor F_{hkl}:

$$F_{hkl,\text{calc}} = \sum_{j=1}^{m} f_j \cdot e^{2\pi i(hx_j+ky_j+lz_j)} \cdot e^{-\text{Temp}_{j,hkl}}. \tag{3-3}$$

Summation is over all *m* atoms in the unit cell and $\text{Temp}_{j,hkl}$ is a temperature factor accounting for vibrations of atoms in the crystal. The vibrations can be treated either isotropically, with a single term, or anisotropically with a 3×3 tensor. In the first case, the atoms are represented as spheres, and in the latter, as ellipsoids, with volumes directly associated with vibrational amplitudes.

The structure factor F_{hkl} describes the properties of the reflection hkl and can be rewritten in the form

$$F_{hkl,\text{calc}} = |F_{hkl,\text{calc}}| \cdot e^{i\alpha_{hkl}}, \tag{3-4}$$

where $|F_{hkl,\text{calc}}|$ is a quantity that can be compared with $|F_{hkl,\text{obs}}|$ derived directly from the measured (observed) intensity data I_{hkl}.

Reliability Index R Structure amplitudes $|F_{hkl,\text{calc}}|$ calculated from the obtained molecular model using Equation 3-3 can be fitted to the observed quantities $|F_{hkl,\text{obs}}|$ by least-squares adjustment of atomic coordinates and temperature factors to yield an improved model. In publications, the goodness-of-fit is usually quoted in the form of the reliability index, R,

$$R = \sum_{hkl} ||F_{\text{obs}}| - |F_{\text{calc}}|| \bigg/ \sum_{hkl} |F_{\text{obs}}|, \tag{3-5}$$

where summation is over all data.

Structure factor $F(R, \zeta)$ for helical molecules. In fiber diffraction analysis, atomic coordinates are expressed in radial coordinates r, ϕ, z (Figure 2-15) rather than in Cartesian system and therefore, Bessel function terms are introduced (95 to 100). For the cylindrically averaged structure factor, we have:

$$F(R, \zeta) = \sum_{n} \sum_{j} f_j \cdot J_n(2\pi R r_j) \cdot e^{2\pi i(lz_j - n\phi_j)} \tag{3-6}$$

where R, ζ are radial and axial coordinates in reciprocal space. For each repeating subunit, r_j, Φ_j, z_j denote coordinates in real space of atom j with scattering factor f_j corrected for water of hydration (101). J_n is the n-th order Bessel function, the value of n being related to layer line number l, helix repeat c, pitch height P and axial rise per residue, h, by the selection rule

$$\zeta = l/c = n/P + m/h \tag{3-7}$$

with m any positive or negative integer.

---

unit cell dimensions. Each reflection is designated by Miller indices hkl which describe a set of lattice planes in reflecting position for this reflection to occur. In Equation (3-1), the distance between these planes, d_{hkl}, is reciprocally proportional to the sine of glancing angle θ_{hkl}; thus, large separations between reflections correspond to small unit cell constants and vice versa, and the diffraction pattern represents an enlarged picture of the so-called reciprocal unit cell (44–47).

The unit cell content, i.e., the positions of atoms within the asymmetric unit, is encoded in the modulation of reflection intensities. All the beams hkl diffracted from a crystal are characterized by an amplitude $|F_{hkl}|$ proportional to the square root of the intensity I_{hkl} and by a phase angle α_{hkl} (Figure 3-3). While the intensity can be measured either by densitometric scanning of the photographic film or by means of a scintillation counter mounted on an automated four-circle diffractometer (48), the phase information is lost and distresses the crystallographer as the well-known *"phase problem."*

The phase angles can, however, be estimated either with "direct methods" which utilize the phase information contained in the distribution of intensities or with the help of "heavy atoms" applying Patterson techniques (44–47). If, then, amplitudes $|F_{hkl}|$ and corresponding phase angles α_{hkl} are known, the electron density distribution (the Fourier transform) for the unit cell can be calculated.

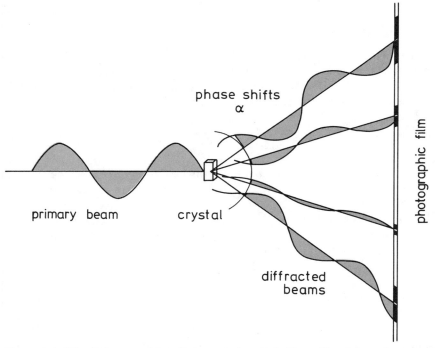

Figure 3-3. The "phase problem." A monochromatic X-ray illuminates the crystal and is diffracted into a multitude of rays (hkl) with different phase angles α_{hkl} and intensities I_{hkl}. The latter can be determined from blackening of the film but the phase angles cannot be measured. Both types of information are necessary to calculate the electron density using Equation (3-2).

Equation (3-2) is the Fourier transform equation which relates the electron density ρ_{XYZ} at location X, Y, Z to measured structure amplitudes $|F_{hkl}|$ and corresponding phase angles α_{hkl} (Box 3-1). In practice, the unit cell is subdivided (rastered) into grid points and at each of these points X, Y, Z a computer is used to calculate the electron density and draw a contour map of it (Figure 3-4).

The electron density reveals atomic positions of the whole structure or, since initial phase angles are often only approximate, part of it. In this latter case, the known part of the structure can be used to obtain improved estimates of the phase angles α_{hkl} and to iterate to a full structure solution. The coordinates x_j, y_j, z_j of the atoms finally located are refined by further computation using, generally, full matrix least-squares methods to minimize the squares of differences between measured and calculated structure amplitudes, $|F_{obs}|$ and $|F_{calc}|$. The latter are derived from atomic coordinates of the molecular model according to Equations (3-3, 3-4) (see Box 3-1).

At the end of a refinement procedure a "difference electron density" is computed with coefficient $|F_{obs} - F_{calc}|_{hkl}$ instead of $|F_{hkl}|$ in Equation (3-2). This is done to locate any atoms omitted from the molecular model, i.e., in the calculation of F_{calc}. Typically, these atoms are hydrogen or atoms of another element disordered over several, fractionally occupied sites (Figure 3-4b). Because the hydrogen atoms with only one electron diffract X-rays rather weakly compared to second and higher row elements of the periodic table, the standard deviations of their atomic coordinates are usually a factor 10 worse than for other atoms. Very accurate hydrogen positions can be obtained using neutron diffraction data. Neutrons locate atomic nuclei rather than electrons and, moreover, interact strongly with hydrogen atoms.

Equation (3-5), Box 3-1, is an assessment of the goodness-of-fit between structure amplitudes $|F_{hkl}|$ calculated on the basis of the molecular model [Equations (3-3, 3-4) in Box 3-1] and the observed (measured) data. R factors are in the range 0.03 to 0.07 for well-refined crystal structures. Typically, about 100 reflection intensities are measured for each nonhydrogen atom of the asymmetric unit, yielding a measurements/parameter ratio of about 10:1 because there are 10 parameters per atom. These include the three positional coordinates, x, y, z, one occupation factor indicating the fraction of an atom present in case of disorder (1.0 for full occupancy) and in addition six temperature factors. The latter are introduced to describe vibrations of atoms in the crystal lattice which are, in general, of ellipsoidal form and characterized by the lengths of the three main axes of the ellipsoids and by three cosine terms giving their orientation in the crystal unit cell. To reasonable approximation, the ellipsoids can be described by spheres of comparable volume. Then only one temperature factor is required and the measurements/parameter ratio increases to about 20:1; see Equation (3-3). The number of measurable reflection intensities depends

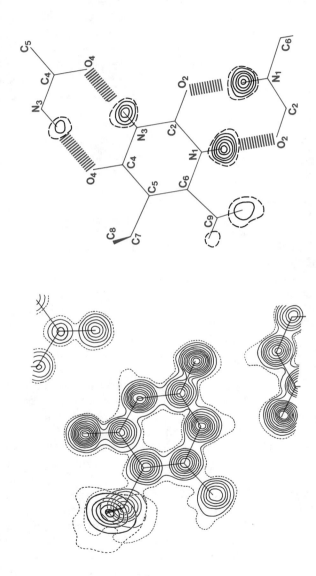

Figure 3.4. (Left) Electron density ρ_{XYZ} calculated in the plane of the heterocycle of 5-ethyl-6-methyluracil for data extending to about 0.8 Å resolution (161). Contours are at intervals of $1\ e\ \text{Å}^{-3}$ beginning with $1\ e\ \text{Å}^{-3}$ (dashed). Second row atoms are clearly seen; hydrogens appear as bulges at N-atom positions. A line structure of the molecule is superimposed in color; see Figure 3-4 (right) for atom designation. **(Right)** Difference electron density corresponding to Figure 3-4 (left), calculated with coefficients $|F_{obs} - F_{calc}|$, where $|F_{obs}|$ are measured structure amplitudes and $|F_{calc}|$ those measured from second row atom contributions only (161). Hydrogens show up as strong peaks, contours as in Figure 3-4 (left); the line structure of molecule and atom numbering is superimposed. Hydrogen bonds are indicated by hatched bars; hydrogen atoms at ethyl and methyl groups are above the molecular plane and therefore not drawn.

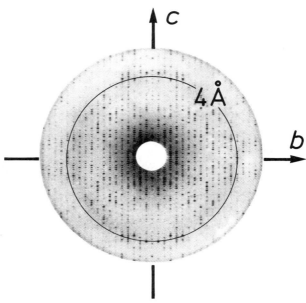

Figure 3-5. X-ray diffraction pattern of phenylalanine-specific tRNA from yeast, tRNA$_{yeast}^{Phe}$. Resolution ranges (Å) indicated as in Figure 3-2. From spacings between reflections, cell dimensions $a = 33.2$ Å, $b = 56.1$ Å, $c = 161$ Å are derived; systematic absences of every second reflection on c and a axes and symmetry of the diffraction pattern indicate orthorhombic space group $P2_122_1$. Note that with increasing resolution, intensities blur out and fade away, limiting resolution to about 2.3 Å. From (162).

on the resolution of the diffraction pattern, i.e., that 2 θ range beyond which diffraction fades away (Figure 3-5). The resolution depends on the quality of the crystals and thermal motion of atoms within the lattice and for small molecules is generally <0.8 Å; i.e., atoms bonded with a typical distance of about 1.54 Å are clearly resolved (Figure 3-4).

Results of a successful, well-refined crystal structure analysis of a small molecule yield information on interatomic distances and angles with standard deviations around 0.005 to 0.01 Å and 0.3° to 1°. The conformation of a molecule is determined by torsion angles (defined in Chapter 2), accurate to 0.8° to 2°. Based on the atomic arrangement in the asymmetric unit, application of space group symmetries and unit cell translations generate the whole crystal structure and reveal the molecular packing (Figure 3-6). This overall picture gives insight into molecular self-organization and intermolecular interactions such as hydrogen bonding, stacking of heterocycles, and dipolar, van der Waals and electrostatic forces. Presently, X-ray crystallography can yield precise structures for molecules of weight up to 2,000 daltons, corresponding to hexanucleotides if an average molecular weight of 350 daltons per nucleotide is assumed.

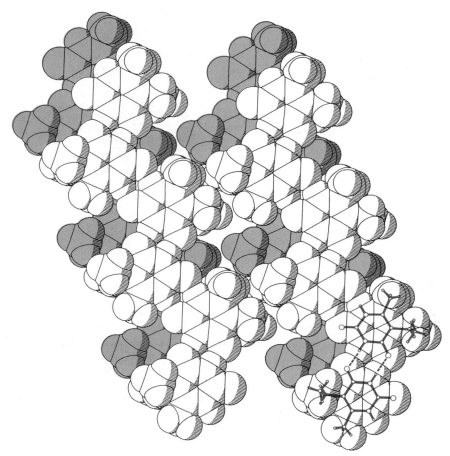

Figure 3-6. Packing of molecules of 5-ethyl-6-methyluracil (161) within the crystal lattice, showing van der Waals space filling of atoms. Spheres of hydrogen atoms are flattened if involved in N-H···O hydrogen bonds. The diagram was drawn with programs and data from the Cambridge Data File (163).

3.2 Potential Energy Calculations

Once the crystal structure of a molecule is known, the information can be utilized to compute potential energies. These are of main interest if the molecule has internal freedom of rotation about interatomic bonds. In mono- and polynucleotides, such rotations occur about glycosyl, phosphodiester, internal and external ($C_{4'}-C_{5'}$) sugar ring bonds.

Table 3-1. Selection of van der Waals Radii Defined as Distances of Closest Approach (157) and Allinger's Minimum Potential Contacts [From (158)]

Type of atom	Radius (Å)	
	Closest approach	Minimum potential contacts
H	1.20	1.50
O (single or double bonded)	1.52 1.40[a]	1.65
C (alipathic)	1.70	1.75 (sp^3)
C (aromatic)	1.70	1.85 (sp^2, sp)
N (aliphatic or aromatic)	1.55 1.50[a]	1.70
P	1.80 1.90[a]	2.05
=S	1.75 ⎫ 1.85[a]	2.00
—S—	1.80 ⎭	
F	1.35[a]	1.60
Cl	1.80[a]	1.95
Br	1.95[a]	2.10
I	2.15[a]	2.25

[a] Taken from Ref. (159). These values and H,C of "closest approach" are most commonly used.

In potential energy calculations, either one or more rotational parameters are varied at intervals and the interatomic distances r_{ij} between atom pairs i, j are computed. Initially, calculations were based on the assumption that there was no attraction if $r_{ij} > r_0$ but infinite repulsion if $r_{ij} \leq r_0$, with r_0 the sum of van der Waals radii of atoms i and j (Table 3-1, Figure 3-7) (50–53). This "hard-sphere potential" can be used for preliminary assessment of rotational freedom of a molecule (18,34,54) but present-day studies involve more realistic potential energy functions (55–64).

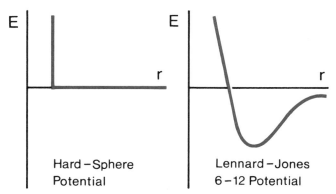

Figure 3-7. Comparison between hard-sphere potential (left) and Lennard-Jones "6-12" nonbonding potential (right).

The total energy V_{tot} between nonbonded atoms of a molecule is composed of individual contributions from dispersion (London attraction V_a), valence repulsion (V_r), electrostatic Coulombic interaction (V_{es}), and intrinsic torsional potentials (V_{tor}). V_a and V_r are usually combined in the Buckingham "6-exp" or Lennard-Jones "6-12" nonbonding potential,

$$V_{nb} = -A_{ij}/r_{ij}^6 + B_{ij}/r_{ij}^{12}, \qquad (3\text{-}8)$$

where parameter A_{ij} depends on the types of atoms i, j involved and may be evaluated from atomic polarizabilities and the effective number of valence electrons using the Slater-Kirkwood equation (60,65). B_{ij} can be evaluated from A_{ij} taking into account r_{ij}, the sum of van der Waals radii of the two interacting atoms i and j. Selected values are listed in Table 3-2 and in a modified form in (63).

The electrostatic potential V_{es} in (kcal/mole) between two charged atoms at r_{ij} (Å) distance is approximated by simple charge-charge interactions and described as

$$V_{es} = 322 \cdot e_i \cdot e_j/\epsilon' \cdot r_{ij}. \qquad (3\text{-}9)$$

The (partial) electronic charges e_i and e_j are given in electron charge units e_o and due to σ and π contributions which can be calculated with quantum chemical techniques using, for example, self-consistent force

Table 3-2. Parameters A_{ij}, B_{ij} for Various Atom Pairs i, j Used in Lennard-Jones "6–12" Potential Function[a]

Atom pair	r_{min} (Å)	A_{ij} (kcal·Å^6·mole^{-1})	B_{ij} (kcal·Å^{12}·mole^{-1})
H–H	2.40	46.7	4.46×10^3
H–C	2.90	128	3.80×10^4
H–N	2.75	125	2.70×10^4
H–O	2.72	124	2.51×10^4
H–P	3.10	346	1.54×10^5
C–C	3.40	370	2.86×10^5
C–N	3.25	366	2.16×10^5
C–O	3.22	367	2.05×10^5
C–P	3.60	1000	1.09×10^6
N–N	3.10	363	1.61×10^5
N–O	3.07	365	1.53×10^5
N–P	3.45	990	8.35×10^5
O–O	3.04	367	1.45×10^5
O–P	3.42	995	7.96×10^5
P–P	3.80	2710	4.08×10^6

[a] r_{min} denotes the minimum van der Waals contact distance between atoms i and j. More data involving halogen atoms are given in (65), the source of this list.

Table 3-3. Approximate Rotation Barriers $V_0{}^a$

Rotation about bond	V_0 (kcal/mol)
C–C	3.11
C–NH$_2$	1.75
C–OH	0.95
C–NO$_2$	0.003 (usually taken for glycosyl bond rotation)

^{a} For more details see Ref. (73).

field methods in the CNDO/2 (Complete Neglect of Differential Overlap) approximation (66). Other schemes to evaluate π and σ charges individually are those developed by Hückel (67) and by Del Re (68,69). Charge distributions in nucleosides and nucleotides are given in References (70–72) and summarized in Figure 5-1. For the dimensionless "dielectric constant," ϵ', which accounts for charge shielding by solvent, an average value of 3.5 to 4.0 is usually assumed.

The torsional potentials

$$V_{tor} = (V_0/2) \cdot (1 + \cos n \cdot \theta) \tag{3-10}$$

are due to exchange interactions of electrons in bonds attached to the bond about which rotation θ occurs, and, in addition, to van der Waals forces. The integer n gives the periodicity of the rotational barrier; values of V_0 are presented in Table 3-3 (73).

The combined potential energies,

$$V_{tot} = V_{nb} + V_{es} + V_{tor}, \tag{3-11}$$

often termed "classical potential energies" or "semiempirical potentials," are evaluated and presented diagrammatically as function of torsion angles (see Figure 4-27).

Frequently, deformations of bond lengths and particularly of bond angles occur which are not considered in the above methods. Treatment of such deviations from "ideal" geometry is possible with the consistent force field method (CFF) (64,74), which produces a more realistic picture of the molecular flexibility. At yet a higher level of sophistication, quantum chemical calculations, mostly of the form described by the extended Hückel method (75,76), by CNDO/2 (66) or PCILO (Perturbative Configuration Interactions using Localized Orbitals) (77,78), as well as "ab initio" methods (79), are applied. These are, however, restricted to small molecules like mononucleotides and yield, besides torsional potentials, information on charge distribution, electrostatic and ionization potentials, and dipole moments. For larger systems, these methods cannot be used and are replaced by the semiempirical potentials described by V_{nb}, V_{es}, V_{tor}, or by smaller molecular units serving as model compounds.

3.3 Crystallography of Macromolecules

tRNAs with a molecular weight around 26,000 daltons can be crystallized from aqueous medium by the addition of alcohols or salts. Like protein crystals, tRNA crystals contain between 30 and 80% solvent and can almost be considered a concentrated solution. Owing to the weak intermolecular interactions stabilizing these crystals, they are very sensitive to changes of environment and must be kept under stable conditions (47,80,81).

Methods for space group determination and X-ray data collection are the same as for small molecules, but the phase problem must be solved using a different approach. This is done by the technique of isomorphous substitution: "heavy" atoms with more than about 50 electrons are introduced into the crystal in such a way that the lattice of tRNA molecules is not disturbed. Heavy atom substitution is achieved either by covalently linking a reagent with tRNA or by soaking crystals in solutions containing such heavy atoms as Pb^{2+}, Sm^{3+}, $AuCl_4^-$, et cetera. The heavy atom compounds diffuse into the solvent-filled interstices between the macromolecules and can bind to certain, definite positions on the tRNA molecules. A successful derivatization is evident from small changes in the intensity distribution of the X-ray diffraction pattern of a crystal; the positions of the intensities, however, remain identical in case of a truly isomorphous substitution. At least two different heavy atom derivatives (or one derivative and the measurement of special "anomalous dispersion" data) are required. Based on the location of the heavy atoms in the unit cell, phase angles can be approximated (82), and are used to calculate an electron density map [Equation (3-2), Box 3-1] which reveals the structure of the macromolecule.

In tRNA (and protein) crystals, intermolecular interactions are rather weak: Especially on the surface of the molecule, atomic groups show extensive thermal movement and sometimes even positional disorder which considerably reduce the resolution of the X-ray pattern (Figure 3-5). The best tRNA crystals diffract to 2.5 Å resolution, corresponding to a total of 8000 data for 1652 nonhydrogen atoms. The data/parameter ratio is about 1 if only atomic coordinates and isotropic temperature factors are considered. Compared with small-molecule crystallography where data/parameter ratios of 10 to 20 are common, this ratio is too low to allow a reasonable refinement of the structural model. This situation can be considerably improved if intact groups such as nucleobase, sugar, and phosphate are used in the refinement process rather than individual atoms and if torsion angles only are varied subject to calculated potential energy, while bond angles and distances are maintained at canonical values known from small-molecule crystallography.

Due to the limited resolution of the X-ray data, atoms cannot be resolved in an electron density map of tRNA. Bases and riboses appear as more or less well defined disks and phosphate groups as spheres of higher density. The interpretation of such electron density maps is achieved either by fitting of a molecular model using a computer-controlled graphical display or with a Richard's comparator (83). The Richard's comparator is a box in which a wire model of the molecule is built manually so as to coincide with the electron density; model and density are optically superimposed by means of a half-silvered mirror positioned between wire model and density map (Figure 3-8). As purine and pyrimidine bases have similar shapes, nucleic acid sequence assignment is more difficult than sequence assignment for proteins where the variable sizes of amino acid side chains allow at least partial sequencing on the basis of X-ray data.

Since the electron density, especially in nonhelical regions, is not always well defined, the interpretation is sometimes a matter of personal intuition. Therefore, for the same molecule under the same crystallographic conditions, different research groups might draw different conclusions concerning fine structural details. Another problem in large-molecule crystallography is that there are several refinement methods to improve the fit of the model to the electron density, and these may not all

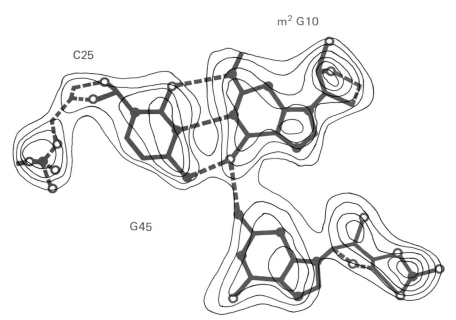

Figure 3-8. Section of electron density of tRNA$_{yeast}^{Phe}$ calculated with 2.5 Å resolution data, the adjusted wire model of the molecule (in color) being shown superimposed. Individual atoms are not resolved in the electron density and interpretation is more ambiguous than in small-molecule crystallography (Figure 3-4). From (164).

converge to the same result, especially for the less well defined regions (84). Besides the macromolecule, there is solvent in the unit cell which due to its fluid nature (except for a few tightly bonded solvent molecules at the surface of the macromolecule) cannot be accounted for properly. Therefore, R factors for well-refined tRNA crystal structures cannot compare with small-molecule analyses and are usually between 0.5 and 0.30 (85–87).

3.4 Fiber Structure Determination

Preparations of polymeric nucleic acids such as double-helical DNA and RNA generally represent a heterogeneous population of molecules. Unlike globular, well-defined tRNA, they cannot be obtained in the form of single crystals suitable for X-ray analysis. While several morphological forms of microcrystalline DNA have been reported (88–91) their X-ray diffraction patterns allow no detailed structural analysis (91). Therefore, the only reasonable means to investigate polymeric nucleic acids is to prepare fibers. These can be drawn from highly concentrated, viscous gels produced by precipitation of the nucleic acid from aqueous buffer through addition of alcohol (92). The rod-like helical nucleic acid molecules form parallel, quasi-crystalline arrays disordered randomly about the fiber axis. Due to this rotational disorder, the diffraction pattern obtained with a monochromatic X-ray beam perpendicular to the fiber axis (Figure 3-9) corresponds to a rotation photograph in single-crystal work and can be interpreted accordingly (93).

Owing to the inherent disorder of molecules in the fiber, X-ray reflections are more or less drawn into arcs and, in general, the resolution does not extend beyond 3 Å. Reflections tend to overlap and the data available from a fiber diffraction pattern are rather limited: Typically between 10 and 100 individual reflections can be used to determine a conventional unit cell. Because of the limited data and the disorder in the crystalline array, experimental methods are insufficient to calculate an electron density distribution. However, the diffraction pattern does contain enough information to indicate the gross structural features of a helical molecule.

Reflection intensities are arranged on layer lines spaced at $1/c$ (Å^{-1}), with c the crystallographic repeat distance along the fiber axis. This repeat corresponds to the helix turn or pitch height, P, for helices with integral numbers of nucleotides per turn as found for all naturally occurring and synthetic DNAs and RNAs (38) except DNA in the C form (94). The spacing of the first meridional reflection in the region 2.6–3.8 Å suggests the vertical separation between nucleotides, h, and the layer line number, l, of this reflection gives the number of nucleotides, n, per turn (95).

The helical diffraction pattern often has the appearance of a cross. This result can be deduced from the structure factor equation for fiber diffraction [Equation (3-6) in Box 3-1] which is analogous to Equation (3-3) in Box 3-1. However, it contains Bessel functions J_n instead of the exponential term as the atomic coordinates are expressed in polar notation (Figure 2-15) (95–101). If n increases, the maximum of the Bessel function contribution moves away from the origin, thus producing the cross-like pattern for different layer lines.

The information on helical parameters and unit cell constants derived from the X-ray diffraction pattern in combination with knowledge of the stereochemistry of the monomeric units and their association via base stacking and hydrogen bonding is sufficient to build a model from which (polar) atomic coordinates can be measured. These are used to compute the diffraction intensity (the helical Fourier transform) for each layer line using Equations (3-6) and (3-7) in Box 3-1. Comparison with the observed X-ray diffraction pattern may indicate an incorrect model. If so, the model can be modified by changing the backbone torsion angles and sugar puckering mode and by minimizing the potential energy. Using linked atoms least-squares methods, final minimization of differences between observed and calculated structure amplitudes is carried out (100,102). Typically the R factors are in the range 0.25 to 0.4.

It should be stressed that, since no experimental phase information is available, a wrong model may not be rejected by the calculations. Three exemplary cases are the structure determinations of poly(C), poly(I), and poly(G). For poly(C), a double-helical model was proposed (103) which later, with additional information and improved X-ray data, was corrected in terms of a single-stranded helix (104) (see Section 13.1). Both poly(I) and poly(G) were originally assumed to form triple-stranded helices (105), yet spectroscopic data combined with new crystallographic results led to the conclusion that these polymers exist as quadruple helices (106) (see section 13.6). The literature questioning (33,107–115) and supporting (116 –122) the double-helical DNA model of Watson and Crick (7,8) is most refreshing to read. In all alternative proposals, guanine-cytosine and adenine-thymine (uracil) base-pairing schemes have been retained. However, two pairing geometries, Watson-Crick and Hoogsteen, have been suggested, and DNA models with side-by-side arrangement instead of intertwining have been envisaged (see Chapter 14).

3.5 Spectroscopic Methods

Structural studies on nucleic acids in solution have been carried out employing UV and IR absorption, circular dichroism (CD), laser Raman, fluorescence, light scattering, and NMR techniques. For reviews, see (123–129). Among these spectroscopic methods, NMR yields the most detailed information and it has been applied to mono-, oligo-, and polynucleotide sys-

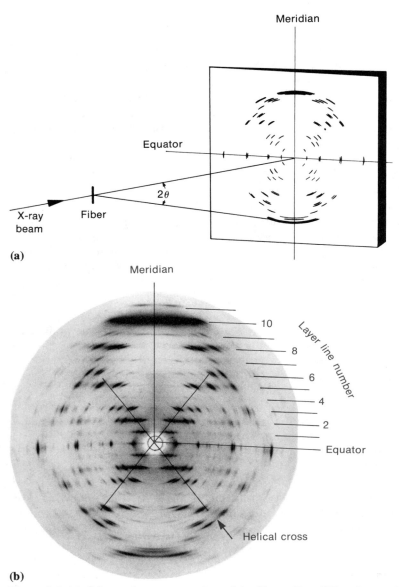

(a)

(b)

Figure 3-9.(a) Schematic representation of the X-ray fiber diffraction method [flat film technique (165)]. The monochromatic X-ray is collimated to 0.1 to 0.2 mm in diameter; fiber axis and film are perpendicular to the X-ray beam. Glancing angle θ and spacings are obtained via Bragg's law, Equation (3-1). **(b)** X-ray diffraction pattern of calf thymus DNA in the B form, courtesy Drs. S. Arnott and R. Chandrasekaran. Typical crosslike intensity distribution suggests helical structure. Separation of layer lines gives pitch height P; strong reflection on meridian (tenth layer line at 3.4 Å distance) indicates that one helix turn contains 10 nucleotides with bases stacked 3.4 Å apart nearly vertical to the helix axis. If the fiber had been exactly vertical to the X-ray beam during exposure, the diagram would be mirror-symmetrical about the equator. It is not because the fiber was tilted slightly to bring the base-pairs into reflecting position in order to produce the strong meridional reflection on the tenth layer line.

A-DNA   B-DNA

C-DNA   D-DNA

Z-DNA

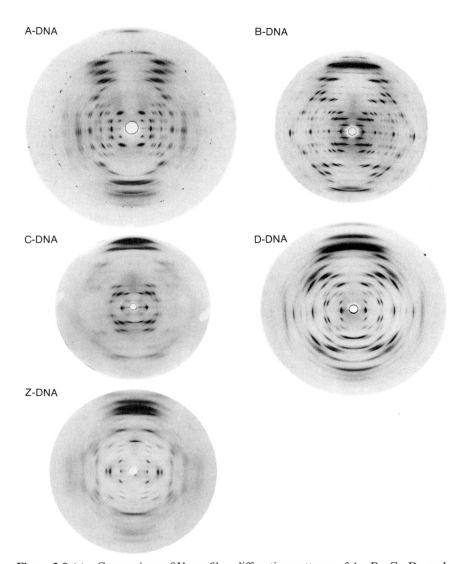

Figure 3-9.(c)   Comparison of X-ray fiber diffraction patterns of A-, B-, C-, D-, and Z-DNA, courtesy of Drs. S. Arnott and R. Chandrasekaran. The ring of reflections in the A-DNA pattern is due to calcite dusted on the DNA fiber for calibration purposes, and corresponds to 3.03 Å. All patterns are on the same scale and fibers were tilted from vertical to the X-ray beam in order to produce the meridional reflection indicating the number of base-pairs per pitch: 11, 10, 9, 8 and 12 for A- to Z-DNA. In Z-DNA, a meridional reflection appears on the sixth layer line in the center of the "helical cross" because the repeat unit is a dinucleotide (pCpG) as outlined in Chapter 12, and not a mononucleotide as in all other DNA forms.

tems using ^{1}H, ^{13}C, ^{15}N, and ^{31}P techniques (129–132). The three NMR parameters chemical shift (ϑ), spin-spin coupling (J), and spin-lattice relaxation time (T_1) are interpreted in terms of conformational characteristics and intermolecular interactions. Hydrogen bonding and base stacking are deduced from chemical shift data (ϑ) which are sensitive to chemical environment of stacked bases if isoshielding curves are applied. These are calculated by quantum chemical methods and give a picture of the magnetic current surrounding an aromatic residue (133–135).

Details on molecular conformation can be obtained from spin-spin coupling constants J. These are related to the torsion angle θ defined by X–C–C–Y or, if phosphorus is involved, by X–C–O–P with X and Y suitable atoms like ^{1}H, ^{13}C, ^{31}P in vicinal position, i.e., attached to two adjacent carbon atoms. Using Karplus' equation (136) in the form

$$^3J_{\text{vic}} = A \cos^2\theta - B \cos \theta + C \text{ [Hz]}, \qquad (3\text{-}12)$$

where the number 3 in $^3J_{\text{vic}}$ denotes the three bonds involved in the group X–C–C–Y and A, B, C are constants chosen according to the substituents at the central atoms of the torsion angle, θ can be calculated. For $^3J_{\text{HH}}$ coupling in vicinal groups H–C–C–H occurring in nucleoside sugar moieties, average constants

$$A = 10.2, B = 0.8, C = 0.0 \text{ [Hz]}$$

have been proposed based on a survey of existing NMR data on these molecules (130).

The pseudorotation concept (see Section 2.5) can be utilized to interpret coupling constants in terms of the two states $C_{2'}$-*endo* (S) and $C_{3'}$-*endo* (N), yielding the populations of N and S conformers (130–132,137). As a rule of thumb, the percentage S and percentage N conformers correspond to $10 \times {}^3J_{1'2'}$ and to $10 \times {}^3J_{3'4'}$ respectively, and the N⇌S equilibrium constant K_{eq} is to a good approximation given by (137)

$$K_{\text{eq}} \sim {}^3J_{1'2'}/{}^3J_{3'4'}. \qquad (3\text{-}13)$$

For more detailed and accurate studies, all three individual coupling constants $^3J_{1'2'}$, $^3J_{2'3'}$, and $^3J_{3'4'}$ are employed and charts relating coupling constants and conformational characteristics are used (130,138,139).*

Conformers about the $C_{4'}$–$C_{5'}$ **bond are determined using** $^3J_{4'5'}$ (130–132). In $-sc$ and ap conformations the same spatial relationship exists between protons at $C_{4'}$ and $C_{5'}$ (Figure 2-11); They cannot be distinguished by NMR methods and are reported as a $-sc$, ap blend. The $+sc$ conformation, however, is well defined and therefore populations of the two basic orientations of the $O_{5'}$ atom, "above" or "away" from the sugar

---

* Strictly speaking, the Karplus equation is not unique for ^{1}H^{1}H vicinal coupling if torsion angles H–C–C–H are close to *syn-periplanar* ($\theta < 90°$), and criticism concerning its application has been raised (140,141).

five-membered ring, can be derived. A similar situation is encountered for rotation about the $C_{5'}-O_{5'}$ bond in 5'-nucleotides if coupling between ^1H and ^{31}P is measured. In 3'-phosphates, the determination of the conformation about $C_{3'}-O_{3'}$ is more difficult because there is only one proton attached to $C_{3'}$ to be utilized (130–132).

A way out of the dilemma is offered by vicinal $^3J(^{31}P, ^{13}C)$ coupling in case of ^{31}P and ^{13}C for atomic groups $P-O_{3'}-C_{3'}-C_{2'}$ or $-C_{4'}$ and for $P-O_{5'}-C_{5'}-C_{4'}$ yielding information regarding both C–O ester bonds. Additionally, two-bond coupling $^2J(^{31}P, ^{13}C)$ for geminal bonds, i.e., attached to the same central atom as in groups $P-O_{3'}-C_{3'}$ and $P-O_{5'}-C_{5'}$, can also be employed (130–132,142–144). NMR measurements thus yield sugar puckering modes as well as backbone torsion angles about C–C and C–O bonds. The orientation about the P–O bonds, however, is not directly accessible because there are no nuclei to produce suitable coupling constants (145,146). In this case, other methods such as trends in chemical shifts with increasing temperatures must be employed (147) and can yield at least qualitative assignments.

The orientation of the base relative to the sugar moiety has been analyzed several ways: (i) Spin relaxation times T_1 (148), (ii) nuclear Overhauser effect (NOE) methods which take advantage of double resonance between closely located protons (149,150), (iii) measurements of pH effects, (iv) perturbation of sugar proton chemical shifts by the base, and (v) pseudo-contact shifts (151–153,138) have been used. Results of these studies do not agree satisfactorily, the first two methods estimating a higher population of the *syn* conformer than the last three. More direct information is obtained by replacing uridine C_2 or C_6 base atoms with ^{13}C and determining the $^3J^{13}C^1H$ coupling constant (154).

If all the details of an NMR analysis on a nucleotide are combined, a picture of the structural characteristics of the molecule emerges. The NMR method has been refined and has become increasingly sensitive in the past ten years so that oligo- and polymeric systems can be successfully treated (130–132,145–147,155,156).

Ultraviolet, circular dichroism, and infrared spectra of nucleotides are used as tools to monitor structural changes. Ultraviolet (UV), circular dichroism (CD), and infrared (IR) spectra of nucleotides and of their polymers are valuable indicators displaying changes in chemical composition or in three-dimensional structure. Ultraviolet spectra, for instance, are sensitive to base-stacking effects in polynucleotides and the observed reduction in UV absorption (hyperchromicity) has been used profitably to investigate helix⇌coil transitions (Chapter 6). Furthermore, UV spectra reflect the electronic state of a heterocycle and thus show if chemical modifications or protonations/deprotonations have taken place.

Optical rotatory dispersion (ORD) or, better, circular dichroism (CD) spectra are related to UV spectra. This is because in the UV absorption band region the chromophore (base), if substituted by a chiral moiety like a furanose, displays a Cotton effect. CD measurements can therefore be applied to determine asymmetric surroundings of a chromophore and are especially useful for monitoring changes of local environment, for instance, if helix⇌coil transitions occur.

Finally, IR spectra indicate vibrational states of molecules. In the double-bond stretching region, 1500 to 1750 cm^{-1}, they are sensitive to base substituents and therefore can be applied to study keto⇌enol tautomerism (Section 5.4) and base-base associations through hydrogen bonding. A disadvantage of IR methods is that within the spectral region of interest, water absorbs and therefore all measurements have to be carried out in D_2O. Furthermore, molar absorptivities are about an order of magnitude lower than those with UV absorption. If, however, laser Raman spectroscopy is applied, this method is considerably extended and new information on molecular vibrations is obtained. Recently, remarkable progress has been made in these techniques and the reader is referred to (127,128) for further study.

Summary

In nucleic acid structure research, three different X-ray crystallographic techniques are applied. Smaller mono- or oligonucleotides are studied by conventional single crystal X-ray analysis and for well-defined globular macromolecules such as tRNA, protein crystallographic methods are used. In both kinds of analysis, information on structure amplitudes and phase angles of diffracted beams can be derived from experimental data and an unambiguous solution is obtained. From polymer DNA and RNA, only quasi-crystalline fibers can be drawn, which due to the inherent orientational disorder give relatively poor diffraction data. Based on stereochemical knowledge of monomer and oligomer units, models are developed and their calculated Fourier transforms compared with the diffraction pattern, but pitfalls cannot be excluded. Potential energy calculations indicate conformational flexibility of a molecule and are useful tools in refinement of macromolecular and polymer structures where X-ray data are limited.

Among all spectroscopic methods, NMR is the most powerful and yields detailed information on conformation of mono-, oligo-, and polymeric nucleotides in solution. These data, combined with results from crystallographic analyses and theoretical calculations using the classical potential energy approach, yield a comprehensive picture of the structural features of a molecule.

Chapter 4

Structures and Conformational Properties of Bases, Furanose Sugars, and Phosphate Groups

Because the molecular geometry and conformational properties of bases, nucleosides, and nucleotides are intrinsically related, they are discussed together in this chapter. Foundations will be laid for understanding the main structural features of the building blocks of nucleic acids which are a prerequisite for the interpretation of oligo- and polynucleotide structure. Several review articles on nucleoside and nucleotide structure and conformation have been published, describing results obtained from crystallographic studies (35,166–169) and their detailed analyses (170–172), from spectroscopic data (130,135,144,173–178) or from classical potential energy and quantum chemical calculations (70,179–182). The general physicochemical and biochemical properties of nucleosides, nucleotides, and nucleic acids are treated in several monographs and book series (123–125,183-190).

4.1 Geometry of Bases

Pyrimidine and purine bases have been investigated as individual units and as moieties substituted at the glycosyl nitrogen, N_1 in pyrimidines and N_9 in purines. In both substituted and unsubstituted bases of each kind, the main structural features are similar except for details concerning the environment of the glycosyl nitrogens. Geometric data for bases are displayed in Figure 4-1 (191,192), all substituted at the glycosyl position because only these are relevant to nucleoside and nucleotide structures. It must be emphasized that in all the nucleoside and nucleotide crystal structures thus far published, *only* the tautomeric forms of bases shown in Figures 2-1 and 2-2 have been observed. The substituents —NH_2 and =O are in the amino and keto forms, never in the imino or enol configuration. Calculations and spectroscopic data demonstrated that very small amounts of other tautomeric forms can exist in solution (see Chapter 5).

In general, the pyrimidine and purine heterocycles of the bases are

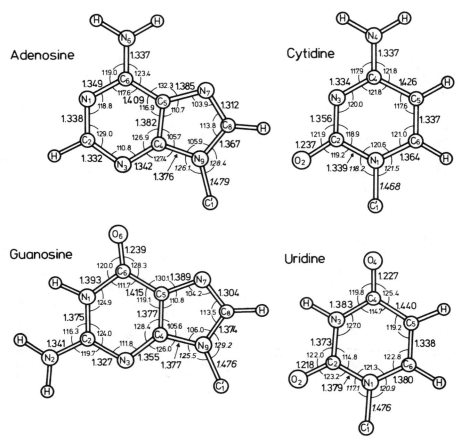

Figure 4-1. Average bond angles (°) and distances (Å) in 1-substituted uracil and cytosine and in 9-substituted adenine and guanine. Data are from (191); standard deviations are defined as $\sigma = [\Sigma(\bar{X}_i - X)^2/(N - 1)]^{1/2}$, where X_i and $\bar{X}$ individual and mean values, and N is the number of observations. For angles, σ's are in the range 0.3° to 1°, and for distances, 0.005 Å to 0.016 Å. N is 32 for uracil, 14 for cytosine, 21 for adenine, and 7 for guanine. Data in italic involve glycosyl C–N linkages taken from (192). They are less accurate because they depend on orientation of base and on furanose pucker.

planar.* Slight deviations (<0.1 Å) of ring atoms and significant deviations (<0.3 Å) of ring substituents from calculated least-squares planes defined by ring atoms are not systematic and reflect influences of crystal packing forces. In solution, the time-average geometry of bases and attached substituent atoms will be planar.

---

* In some crystal structures of purine-nucleosides, a dihedral angle of 0.5° to 1° has been observed between normals to pyrimidine and imidazole planes; i.e., purines can adopt a butterfly-like structure slightly folded about C_4–C_5 (193–195).

   **Amino groups in purine bases and in cytosine are integrated into the res-
onance systems,** as suggested by exocyclic $C-NH_2$ bond distances of
about 1.34(1) Å (Figure 4-1). These distances are considerably shorter
than aliphatic $C-N$ single bond of 1.472(5)Å lengths (196,197) (Table
4-1), and they show partial double-bond character (198) such as that found
in conjugated heterocyclic systems and in single bonds of $\searrow N-C=O$
groups (Table 4-1). Double-bond orders are between 0.41 and 0.47 (70,
109), i.e., about halfway between single and double bonds.

   The partial double bonds force the amidine-like amino groups to be co-
planar with the attached heterocycles and rotation about the exocyclic $C-NH_2$ bonds is hindered. The height of the rotation barrier calculated by
quantum chemical methods (CNDO/2) is 15 to 25 kcal/mole, depending on
the N-substituents, hydrogen or methyl groups, in cytosine (200–202), adenine (203), and guanine (203) derivatives. NMR-spectroscopic data seem
to indicate that the barrier to rotation of the amino group in purine bases is
lower that that in cytosine (204) for which a value of 15 to 18 kcal/mole
was derived (205), in close agreement with the theoretical results. Because of the high barrier, the rate of rotation about the exocyclic C-N
groups in N_4-methylcytosine is very slow, around 90 to 270 sec^{-1}, depending on the temperature in the range between 20° and 30°C, and on the
solvent (204–206).

   It is noticeable that even in a sterically hindered, overcrowded cytosine derivative, $1,5,N_4,N_4$-tetramethylcytosine (Figure 4-2), all atoms
are essentially coplanar. The repulsion between neighboring methyl substituents at C_5 and N_4 is *not* relieved by rotation of the dimethylamino
group about the C_4-N_4 bond; instead, widening of the bond angles is observed (207,208).

   **Compared with the $C-NH_2$ bonds, the exocyclic $C=O$ bonds of guanine
and of the pyrimidine bases display even more double-bond character** and
exhibit double-bond orders in the range 0.84 to 0.89 (70,199). The $C-O$
bond lengths of 1.22(1) to 1.24(1) Å (Figure 4-1) are nearly the same as a
standard $C=O$ double bond, 1.215(5) Å (Table 4-1) and thus reflect the
presence of the lactam tautomeric form.

   **Endocyclic bond distances of purine and pyrimidine bases are consistent
with the number of resonance structures that can be drawn** (192) and agree
with distances and bond orders calculated by quantum chemical methods
(199). Thus C_2-N_3 in purines is shorter than the corresponding N_1-C_2 in
pyrimidines and the pyrimidine C_5-C_6 distance is shorter than the C_4-C_5
counterpart in purines. In the purine series, C_8-N_9 has less double-bond
character than N_7-C_8 and the bond distances differ accordingly.

   The endocyclic valence angles at N atoms show an interesting trend (209,
210). In pyrimidine rings, the internal angle at unsubstituted N is about 6°
to 8° smaller than at substituted N [compare the angle $C_6-N_1-C_2$ in adenine against guanine and the angle $C_2-N_3-C_4$ in cytosine against uracil

Table 4-1. Some Selected Bond Distances Relevant in
Nucleotide Geometry [From (196,197).]

	Distance (Å)
C—C single bond, paraffinic	1.537 ± 0.005
in C—C=C	1.510 ± 0.005
in C—C=O	1.506 ± 0.005
C—C aromatic	1.394 ± 0.005
C=C double bond, paraffinic	1.335 ± 0.005
in C=C—C=O	1.36 ± 0.01
in C=C—C=C	1.336 ± 0.005
C—N single bond, paraffinic	1.472 ± 0.005
in C—N=	1.475 ± 0.01
in ＞N—C=O	1.333 ± 0.005
in conjugated heterocyclic	1.339 ± 0.005
C—O single bond, paraffinic	1.426 ± 0.005
aromatic (salicylic acid)	1.36 ± 0.01
carboxylic acids and esters C—O	1.358 ± 0.005
and C=O	1.233 ± 0.005
C=O double bond in aldehydes and ketones	1.215 ± 0.005
in O=C—N＜	1.235 ± 0.005
C—S single bond, paraffinic	1.817 ± 0.005
＞C=S double bond	1.71 ± 0.01
C—F paraffinic	1.379 ± 0.005
aromatic	1.328 ± 0.005
C—Cl paraffinic	1.767 ± 0.005
aromatic	1.70 ± 0.01
C—Br paraffinic	1.938 ± 0.005
aromatic	1.85 ± 0.01
C—I paraffinic	2.139 ± 0.005
aromatic	2.05 ± 0.01
C—H paraffinic	1.096 ± 0.005
aromatic	1.084 ± 0.005
P—O single bond, calculated	1.71
P—O—R ester	1.60 ± 0.01
P—O—H in KH_2PO_4	1.58 ± 0.02
P—$O^{\delta-}$ in KH_2PO_4	1.51 ± 0.02
P=O double bond, calculated	1.40
P—S in $PSCl_3$, etc.	1.86 ± 0.02
P—C	1.841 ± 0.005

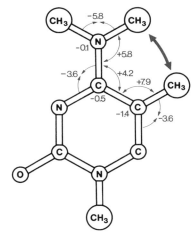

Figure 4-2. Angular difference (°) between averaged 1-substituted cytosine from Figure 4-1 (black) and overcrowded $1,5,N_4,N_4$-tetramethylcytosine (color). In order to relieve steric strain between methyl groups at N_4 and C_5 (double arrow), the bond angles involved are widened. Both molecules are essentially planar. After (207).

and thymine (Figure 4-1)]. This angular difference is 2° to 3° less for angles $C_5-N_7-C_8$ in imidazole rings (192). The widening of the valence angles at N upon substitution is compensated by a corresponding decrease in adjacent $N-C-C$ angles in order to maintain planarity. The same scheme is followed if bases are substituted at N by protonation or if NH groups are deprotonated. For geometrical data of protonated bases, deprotonated uracil (uracil anion), and double protonated adenine, see the compilation (191).

4.2 Preferred Sugar Puckering Modes

The five-membered furanose ring systems in ribo- and deoxyribonucleosides are never planar. Instead they are puckered in either envelope or twist (half-chair) forms (Figures 2-6, 2-7). The nomenclature for the different puckering modes and the concept of pseudorotation are explained in Chapter 2. A detailed analysis of ribose puckering in a large number of nucleosides has shown that the puckering amplitude τ_m displays unimodal distribution, $\langle \tau_m \rangle = 38.6 \pm 3°$. This suggests that τ_m is an intrinsic property of the C- and O-ring structure and is not dependent on type of sugar, base, pseudorotation phase angle P, or torsion angles γ and χ (170).

   The two principal puckering modes are $C_{3'}$-*endo*(N) and $C_{2'}$-*endo* (S). Because the furanose in nucleosides is unsymmetrically substituted, the pseudorotation phase angles calculated for the crystallographically deter-

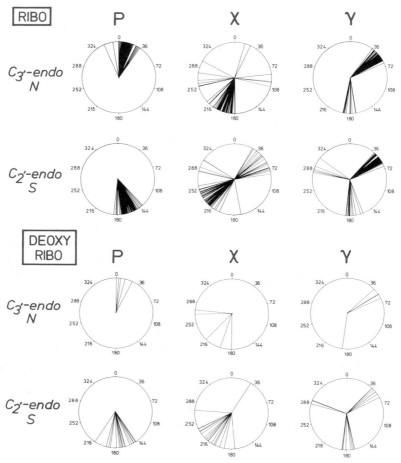

Figure 4-3. Representation of pseudorotation phase angles P and of torsion angles χ about the glycosyl bond and γ about the $C_{4'}-C_{5'}$ bond. Values (°) obtained from 178 crystal structure analyses of ribo- and deoxyribonucleosides and their derivatives are indicated by individual radial lines. $C_{2'}$-*endo* (S) and $C_{3'}$-*endo* (N) sugar puckerings are equally populated in the ribo series, whereas for deoxyribose, $C_{2'}$-*endo* is preferred. In general, $C_{2'}$-*endo* allows more conformational freedom for torsion angles χ and γ as indicated by population of $\chi(syn)$ and $\gamma(-sc)$ ranges, which are either largely suppressed (χ) or not existent (γ) in $C_{3'}$-*endo* (N) pucker. Five data points around $P \sim 90°$ ($O_{4'}$-*endo*) have been omitted: for the derivative displayed in Figure 4-10, $P = 91.3°$, $\tau_m = 45.8°$; for deoxyguanosine 5'-phosphate, $P = 83.3°$, $\tau_m = 40.8°$ (300); for dihydrothymidine, $P = 84.3°$, $\tau_m = 31.6°$ (397); for 5-hydroxy-5, 6-dihydrothymidine, $P = 95.7°$, $\tau_m = 37.6°$ (398); and for 1-β-D-arabinofuranosylthymine, $P = 105.5°$, $\tau_m = 41.4°$ (399). On the other side of the pseudorotation cycle ($P \sim 270°$, $O_{4'}$-*exo*), no data points are found. After (170).

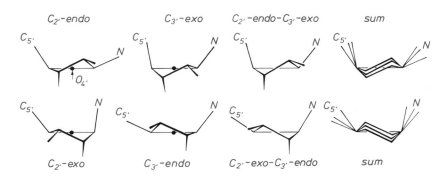

Figure 4-4. Schematic representation of the most frequently observed puckering modes, corresponding to the plot for P in Figure 4-3. Horizontal transitions are continuous and at the same energy levels whereas vertical transitions are separated by a (shallow) energy barrier and describe an N$\rightleftharpoons$S interchange. Note that directions of exocyclic $C_{1'}$–N and $C_{4'}$–$C_{5'}$ bonds are intrinsically related to sugar conformation. From (169).

mined nucleoside and nucleotide structures are not evenly distributed over the pseudorotation cycle (Figure 4-3). Instead, they cluster in two domains centered at $C_{3'}$-endo, $P = -1°$ to $34°$, and at $C_{2'}$-endo, $P = 137°$ to $194°$ (170). According to their geographical position on the cycle, these preferred sugar puckerings are called north (N) and south (S) by spectroscopists, with the understanding that both consist of a family of closely related puckering modes summarized in Figure 4-4. Conformational changes along horizontal lines in Figure 4-4 are equivalent to rotation within a limited range on the pseudorotation cycle. The lack of substantial potential energy difference between envelope and twist forms makes these conversions "continuous." Vertical transitions in Figure 4-4, however, invert the puckering, corresponding to an N$\rightleftharpoons$S interchange on the pseudorotation cycle and to different axial$\rightleftharpoons$equatorial orientations of the sugar substituents (Table 4-2).

   The distribution of crystallographically determined data (Figure 4-3) suggests that the two possible routes from N to S are energetically nonequivalent. The passage via $O_{4'}$-endo is "short" and must represent a relatively low threshold because it is sparsely, yet significantly, populated (legend Figure 4-3). In contrast, the "long" $O_{4'}$-exo route is energetically unfavorable and impassable (170,171). This can be understood on purely geometrical grounds: in the $O_{4'}$-exo puckering mode the base and $C_{5'}$ exocyclic substituents are both axial and are in steric conflict while in

Table 4-2. Orientations of Substituents with Different Sugar Puckering Modes[a]

Ribose pucker	Atom							
	N	$H_{1'}$	$H_{2'}$	$O_{2'}$	$H_{3'}$	$O_{3'}$	$H_{4'}$	$C_{5'}$
$C_{2'}$-endo	e	a	a	e	e	a	b	b
$C_{3'}$-endo	b	b	e	a	a	e	a	e
$O_{4'}$-endo	e	a	b	b	b	b	a	e
$O_{4'}$-exo	a	e	b	b	b	b	e	a

[a] e and a designate quasiequatorial and -axial because equatorial and axial orientations known from six-membered cyclohexane rings are only approximated. b means bisectional, i.e. halfway between a and e.

the $O_{4'}$-endo puckering mode they are equatorially oriented and are farther apart (Figure 4-5, Table 4-2).

The conformational energy of $C_{2'}$-endo and $C_{3'}$-endo furanose puckering modes and the energy barrier separating these two conformational states have been the subject of quantum chemical, PCILO (211), extended Hückel, CNDO (212), and classical potential energy calculations (213–215). These show that for *unsubstituted* ribose and deoxyribose, $C_{2'}$- and $C_{3'}$-endo puckering modes are almost equivalent energetically. The two conformations are separated by nearly symmetrical potential energy barriers of 2 to 4 kcal/mole, which correspond to the unfavorable $O_{4'}$-exo and $O_{4'}$-endo sugar puckerings (212,214). However, if a base is attached to the sugar in a $C_{1'}$-β-N-glycosyl linkage, the energy distribution along the pseudorotation cycle becomes asymmetrical with $C_{2'}$-endo and $C_{3'}$-endo conformations at energy minima and $O_{4'}$-endo at a modest energy barrier of 2.5 to 5 kcal/mole. $O_{4'}$-exo represents a barrier so high that it is practically unsurmountable. All these calculations indicate that the route of lowest energy between conformations follows the pseudorotational path

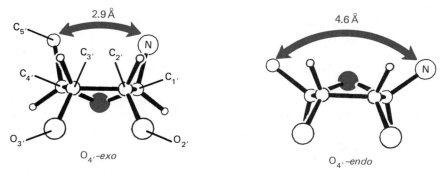

Figure 4-5. Equatorial and axial positioning of substituents N (base) and $C_{5'}$ in sugar puckering modes $O_{4'}$-exo (left) and $O_{4'}$-endo (right). In $O_{4'}$-exo, $C_{5'}$ and N interfere sterically and therefore the $C_{2'}$-endo $\rightleftharpoons C_{3'}$-endo interchange occurs via $O_{4'}$-endo as intermediate.

and that a planar ribose or deoxyribose as transition intermediate is improbable.

In the above computations bond angles and distances were kept at standard values and only torsion angles were varied. In a more recent study (32) using consistent force field methods (74), bond stretching and angle bending were permitted, yielding a picture closer to experimental reality. It was demonstrated that rotation δ about the $C_{3'}$-$C_{4'}$ bond (Figure 2-3) [which is directly correlated with sugar pucker (Figure 4-6)] is even less hindered than expected for rotation about a normal C—C single bond, and that only a modest energy barrier exists for $O_{4'}$-*endo* pucker between the two lowest energy states $C_{2'}$-*endo* and $C_{3'}$-*endo* (Figure 4-6). Energy variation along the full pseudorotation cycle shows, that in a relaxed system with only the endocyclic torsion angle ν_3 about bond $C_{3'}$-$C_{4'}$ constrained to the path of pseudorotation, the energy for $O_{4'}$-*endo* pucker is about 0.6 kcal/mole above the energy computed for $C_{2'}$-*endo* and $C_{3'}$-*endo* states while the barrier for the $O_{4'}$-*exo* route is about 2.5 kcal/mole (Figure 4-7). Calculated energies for ribose and deoxyribose compare to within 0.1 kcal/mole.

Although these results have been severely criticized (170,171) because the acceptable geometries were violated in some cases (angle at $O_{4'}$ increased by 5–6°), a recent careful study revealed that, by and large, the energy profile depicted in Figure 4-6 is reasonable although the energy barriers calculated at $O_{4'}$-*endo* and $O_{4'}$-*exo* were higher, 1.8 and 5.8 kcal/mole for deoxyribose whereas for ribose, 3.8 and 7.5 kcal/mole were found (216).

Why do furanose rings pucker? All these computations give qualitatively consistent results and show that furanose rings should be puckered. What is the basis for the observed puckering preferences? Obviously, the planar furanose is energetically unfavorable because, in this arrangement, all torsion angles are 0° and the substituents attached to carbon atoms are fully eclipsed. The system reduces its energy by puckering, and because the barriers to rotation about C–O bonds are lower than those about C–C bonds (Table 3-3), the furanose-ring adopts conformations with C–O torsion angles ν_0 and ν_4 nearly eclipsed (around 0°) and C–C torsion angles ν_1, ν_2, and ν_3 maximally staggered, resulting in $C_{2'}$-*endo* or -*exo* puckering for $\nu_4 = 0°$ and in $C_{3'}$-*endo* or -*exo* puckering for $\nu_0 = 0°$. In other words (217), puckering of $C_{1'}$ out of the plane of the other four atoms causes rotation about bonds $C_{4'}$-$O_{4'}$ and $C_{2'}$-$C_{3'}$. This relieves contacts between substituents at $C_{2'}$ and $C_{3'}$ atoms although substituents at $C_{3'}$ and $C_{4'}$ remain eclipsed (Figure 4-8). The energetic situation is improved if $C_{2'}$ puckers: rotations then occur about bonds $O_{4'}$–$C_{1'}$ and $C_{3'}$–$C_{4'}$, leaving the torsion angle about $C_{4'}$–$O_{4'}$ (ν_4) at 0° and no substituents in the eclipsed configuration. Displacements of $C_{4'}$ and $C_{3'}$ have an effect similar to that described for $C_{1'}$ and $C_{2'}$. Additionally, C–O–C angles are less flexible than angles with carbon as a vertex (32) and because endocyclic angles at puckered atoms are smaller than angles at in-plane atoms

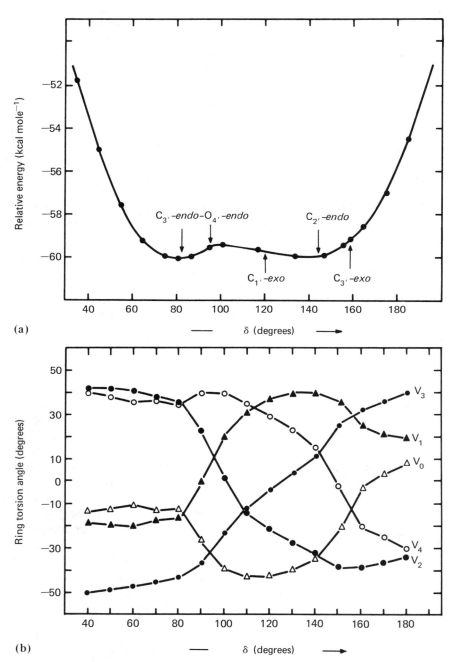

Figure 4-6. (Top) Total energies of furanoses of nucleosides, calculated for $35° \le \delta \le 185°$ using force field methods. Note that δ is directly related to ν_3 and thus determines endocyclic sugar torsion angles (bottom) which, in turn, describe ring puckering. From (32).

(see Section 4.3), this angle-puckering relation again counteracts $O_{4'}$ puckering. In summary, therefore, $C_{2'}$-*endo* and $C_{3'}$-*endo* puckering modes are preferred because nonbonding interactions between furanose ring substituents are at a minimum.

Spectroscopic data indicate rapid $C_{3'}$-*endo*(N)$\rightleftharpoons C_{2'}$-*endo*(S) interconversion. The results of crystal structure analyses and energy calculations are corroborated by spectroscopic data. NMR investigations in solution demonstrate that $C_{2'}$- and $C_{3'}$-*endo* types of conformations are in rapid equilibrium. This can be evaluated from the ratio of vicinal ^1H coupling constants $^3J_{1'2'}$ and $^3J_{3'4'}$ using Equation (3-13). In general terms pyrimidine ribonucleosides and -tides favor $C_{3'}$-*endo*(N) puckering while purine derivatives occur preferentially in the $C_{2'}$-*endo*(S) mode with the understanding that "preferential" means populations in the range 60 to 80%. In the deoxyribose series, a trend to larger proportions of $C_{2'}$-*endo* puckered sugar is observed (31,126,130,138,173–176,218–222). The barrier to interconversion between these two sugar puckering states is so low that it can be measured only under special circumstances in liquid deuterated ammonia at temperatures of $-60°C$ to $+40°C$. For purine ribonucleosides, the average activation energy is 4.7 ± 0.5 kcal/mole and for pyrimidine ribonucleosides it is higher but has thus far escaped experimental measurement (223).

4.3 Factors Affecting Furanose Puckering Modes

NMR spectroscopy shows an influence of the $C_{2'}$ and $C_{3'}$ substituents on the $C_{3'}$-*endo*(N)$\rightleftharpoons C_{2'}$-*endo*(S) equilibrium; the more electronegative substituent prefers an axial orientation (224–228,228a). Systematic surveys of 2'-substituted adenosine (229) and uridine (230) derivatives indicated that the amount of the $C_{3'}$-*endo*(N) conformer increases linearly with the electronegativity of the 2'-substituent (Figure 4-9). This correlation predicts that deoxyribonucleosides and -tides would prefer the $C_{2'}$-*endo* puckering mode because only the slightly electronegative hydrogen is attached at $C_{2'}$. This is indeed observed and suggests that the dipole moment of the furanose ring actually determines its conformational characteristics and that a more electronegative substituent favors an axial orientation.*

These NMR studies were carried out with the nucleosides dissolved in deuterated dimethylsulfoxide or water. However, when adenosines with 2'- or 3'-hydroxyl replaced by amino groups were investigated in deu-

---

* As outlined in (216,228a), it is probable that the *gauche effect* described in Paragraph 4.12 plays an important role in determining the most preferred sugar conformation. This effect directs torsion angles of groups X–C–C–Y with X,Y electronegative substituents like OH, F into + or −*gauche* (+ or −*synclinal*) and tends to avoid *antiperiplanar*. In a ribose, many such groupings are found, viz. $O_{4'}C_{1'}C_{2'}O_{2'}$; $O_{4'}C_{4'}C_{3'}O_{3'}$; (base)-$C_{1'}C_{2'}O_{2'}$.

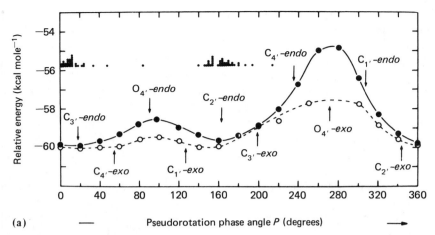

(a)   ——   Pseudorotation phase angle P (degrees)   ⟶

Figure 4-7.(a) Variation of total energy with pseudorotation phase angle P. Solid line calculated for all five endocyclic torsion angles ν_n constricted to the pseudorotation path, dashed line for only ν_3 constrained and the other torsion angles allowed to relax and to move off the "ideal lines" described in Figure 2-9. Dot histograms give phase angles P from nucleoside crystal structures. From (32); for newer, better energy values which, however, do not alter the general picture given here, see (216).

terated ammonia (ND_3) as solvent, the reverse effect was observed (231). These studies indicate either that solvent molecules contribute vitally in the stabilization of the sugar conformation, or that 2′- or 3′-amino substituted nucleosides are not good reference molecules. The latter explanation may be true as the crystal structure of β-D-2′-amino-2′deoxyadenosine (232) displays intramolecular $N_{2'}$-H$\cdots O_{3'}$ hydrogen bonding not observed in other derivatives [except for one modified nucleoside with unusual $O_{4'}$-*endo* puckering mode (Figure 4-10), (233)].

The sugar puckering modes are also influenced by modifications of the base (173). This is especially true if purine C_8 or pyrimidine C_6 is substituted by bulky groups which shift the *syn* ⇌ *anti* equilibrium about the glycosyl link toward *syn* (see Section 4.5). In general, the *syn* conformation correlates with $C_{2'}$-*endo*(S) sugar pucker, especially in the purine series. Also, cyclizations between $O_{2'}$ and $O_{3'}$ or between $O_{3'}$ and $O_{5'}$ or intramolecular bond formation between sugar and base (as in the case of $O_{2',2}$-cyclouridine) limit and determine the puckering preferences of furanoses (126,170–172); see Chapter 7. And if the base is removed and replaced by H or by $-O-CH_3$, NMR studies (233a) show that the furanose ring strongly prefers the $C_{3'}$-*endo* form.

An interesting conformational feature is displayed by polymer ribo- and deoxyribonucleotides (Chapter 11). In oligo- and polynucleotides, individual nucleotides are not independent of each other but they interact by

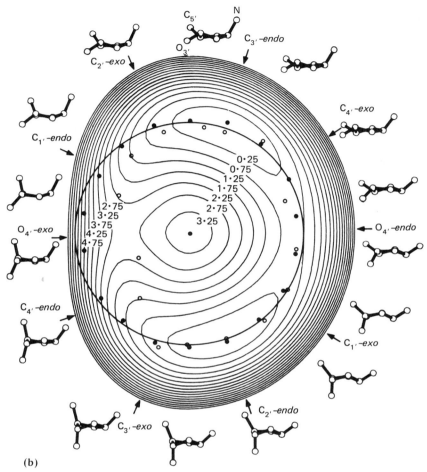

(b)

Figure 4-7.(b) Another, more complete description of Figure 4-7a, giving the energy map of furanose conformations as a function of pseudorotation phase angle P. The latter is counted in a clockwise sense with $P = 0°$ in vertical position. The puckering amplitude q corresponds to ν_{max} defined in Equation (2-3) and increases radially from the central dot representing a planar furanose. The ring indicates the path of true pseudorotation with $q = 0.4$ Å; filled circles describe energies calculated when all five torsion angles ν_n are constrained to the path of pseudorotation; empty circles correspond to only ν_3 constrained and other torsion angles relaxed. Note large differences at $P \sim 270°$ ($O_{4'}$-*exo* puckering). Drawn molecules show relaxed geometries and belong to nearest empty circles in the energy plot. The nomenclature and orientation are given on the molecule $p = 0°$ (top of circle). Numbers in the diagram indicate energies (kcal/mole) above global minimum. From (32); newer energy values in (216) do not alter the overall picture given here.

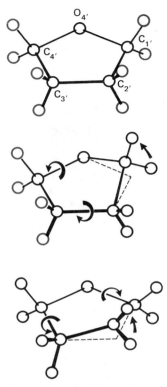

Figure 4-8. Avoiding eclipsed conformations in a furanose ring through puckering. (Top) Planar furanose, all substituents in color fully eclipsed; (middle) $C_{1'}$-*endo* puckering, color substituents at $C_{3'}$, $C_{4'}$ still eclipsed; (bottom) $C_{2'}$-*endo* puckering with no substituents eclipsed. The bottom situation also holds for $C_{2'}$-*exo*, for $C_{3'}$-*endo,* and for $C_{3'}$-*exo* puckering modes. After (217).

base stacking and preferentially are organized into (right-handed) helical form. DNA as a representative of deoxyribonucleotides behaves very differently with respect to RNA, the ribonucleotide analogue. These macromolecular properties can probably in part be traced back to the properties of the monomeric units. DNA adopts two main conformations, $C_{3'}$-*endo* and $C_{3'}$-*exo* (a variant of $C_{2'}$-*endo*), and transition from one to the other can be induced by changing the water activity of the medium. This does not apply to RNA, which under all conditions maintains a $C_{3'}$-*endo* form. There have been a number of proposals trying to explain these different properties in terms of influence of the $O_{2'}H$ hydroxyl present in RNA but not in DNA. A direct hydrogen bond between $O_{2'}H$ as the donor and $O_{4'}$ of the adjacent nucleotide as the acceptor has been advocated as stabilizing factor in RNAs on the basis of model building studies (234,235), IR-spectroscopic data (236), NMR data (237), and on the basis of the crystal structure of tRNA which seems to support at least weak $O_{2'}-H\cdots O_{4'}$ hy-

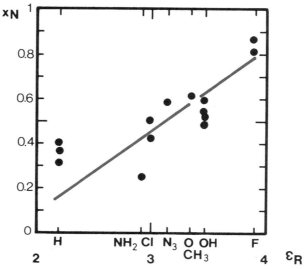

Figure 4-9. Correlation of mole fraction X_N of $C_{3'}$-*endo*(N) puckered pyrimidine nucleosides substituted at $C_{2'}$ by different groups with electronegativities ϵ_R in the range 2.1(H) to 4.0(F). After (230).

drogen bonding in some instances (87) (Figure 4-11a). However, $O_{2'}$ methylation (238–240) or replacement of $O_{2'}$ by chlorine (241), fluorine (242–244), or the azido group (245) does not markedly reduce the conformational stability of synthetic RNAs and therefore the $O_{2'}-H\cdots O_{4'}$ hydrogen bond cannot be a crucial, structure-stabilizing factor. More recent NMR data suggest a water-mediated hydrogen-bonded bridge between $O_{2'}H$ and $3'$-phosphate which could well account for at least part of the conformation rigidity of RNAs (246,247) (Figure 4-11b). As $2'$-chloro- or $2'$-fluoro-substituents in RNAs cannot form such hydrogen bonds and can at best be regarded as weak hydrogen bond acceptors, the stabilizing effect in these instances must be explained by the strong electronegativity of these substituents which favor $C_{3'}$-*endo* puckering (229,230) (Figure 4-9). Or could it be that simply steric hindrance of the $O_{2'}H$ hydroxyl group prohibits other puckering modes as concluded from NMR data (248)? In summary, the forces determining the $C_{3'}$-*endo* stabilization of RNA secondary structure remain unclear, and probably several factors contribute.

4.4 Bond Distances and Angles in Furanoses

Electron diffraction (249,250), microwave (251), and theoretical methods (252) have been employed to investigate pseudorotational characteristics and bonding geometry of tetrahydrofuran, the unsubstituted prototype of

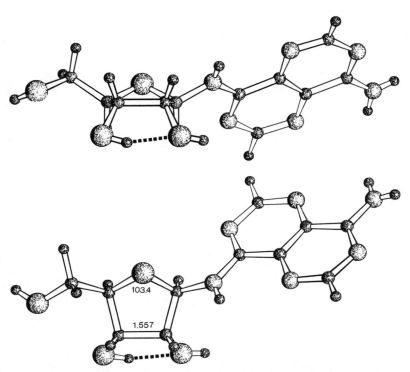

Figure 4-10. Molecular structure of 6-amino-10-(β-D-ribofuranosylamino)pyrimido [5,4-*d*] pyrimidine, a nucleoside derivative in the usual $O_{4'}$-*endo* puckering mode (two views). Owing to eclipsed bonds $C_{2'}-O_{2'}$ and $C_{3'}-O_{3'}$, an intramolecular $O_{3'}H\cdots O_{2'}$ hydrogen bond (dashed line) is formed which is never observed in sugar puckering modes where $C_{2'}-O_{2'}$ and $C_{3'}-O_{3'}$ bonds are staggered. In addition, the $C_{2'}-C_{3'}$ bond length, 1.557 Å, is longer than usually observed, and the $C_{1'}-O_{4'}-C_{4'}$ angle, 103.4°, is smaller; compare data in Figure 4-13. After (233).

the five-membered sugar ring of nucleosides. The tetrahydrofuran bond distances and angles are shown in Figure 4-12. The distances agree well with standard C–C and C–O values (Table 4-1), but the angles are significantly smaller than sp^3-tetrahedral, 109.5° and also less than the 108° required of a planar five-membered ring with sum of angles 5 × 108° = 540°. The sum of angles in tetrahydrofuran, 524.9°, is 15.1° less than the 540° in a flat five-membered ring and caused by out-of-plane puckering of atoms, as can be shown analytically (253,254). The angles are not evenly reduced, but rather follow the trend C–O–C > C–C–O > C–C–C because angle bending forces also decrease in this order (255).

In nucleoside furanoses, C–C bond distances decrease only by approximately 0.02 Å as torsion angles about these bonds vary from 0° to 40° and nonbonded van der Waals interactions are relieved. For C–O bond dis-

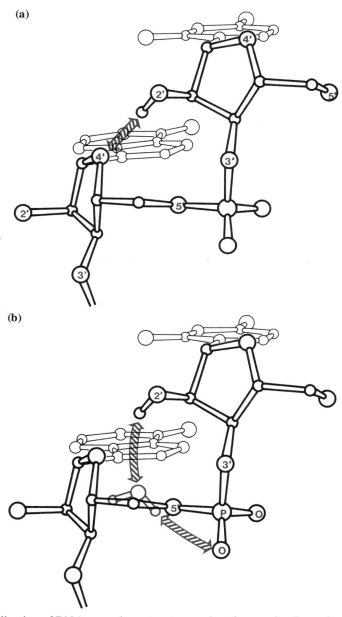

Figure 4-11. Stabilization of RNA secondary structure and preference for $C_{3'}$-*endo* puckering in RNA by (a) intrastrand $O_{2'}H\cdots O_{4'}$ hydrogen bond and (b) water-mediated $O_{2'}\cdots$phosphate interaction. See text for discussion and Figure 4-9 for another explanation.

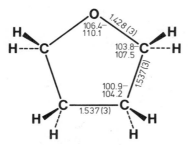

Figure 4-12. Geometry of the unsubstituted furanose five-membered ring, tetrahydrofuran, as obtained from electron diffraction data. Standard deviations of C–C and C–O bonds are in parentheses; for valence angles only ranges can be given. Averaged data from (249,250).

tances, no such dependence is observed (172). Variations of bond angles are more pronounced, between 3° and 5°, if the pseudorotation cycle is followed, as described in four different detailed studies (170–172,216). Here we will focus on the main regions of the pseudorotation cycle, $C_{2'}$-*endo*(S) and $C_{3'}$-*endo*(N), geometrical data for which are given in Figure 4-13.

The distribution of endocyclic angles in furanoses is similar to that in tetrahydrofuran, with the sum of angles a comparable 523°, suggesting that ring closure and puckering effects are more influential than carbon atom substitution. All the endocyclic C–C bond distances in nucleoside furanoses are significantly shorter than ideal C–C single bonds of 1.537 Å (Table 4-1). This is probably due to substitution of the carbon atoms by the base and by $O_{4'}$ at $C_{1'}$, and the oxygen(s) at $C_{3'}$ (and $C_{2'}$). In α-D-glucose a similar shortening of C–C bonds is observed (256). In all nucleoside crystal structures, $C_{1'}$-$O_{4'}$ is about 0.03 Å shorter than $C_{4'}$-$O_{4'}$ because $O_{4'}$ is conjugated with the base attached at $C_{1'}$.

The exocyclic angles at $C_{1'}$ and $C_{4'}$ are nearly tetrahedral (~109°) but those involving $C_{2'}$ and $C_{3'}$ increase to between 112.6 and 116.8°. A trend is observed with exocyclic angles having $C_{2'}$ and $C_{3'}$ as vertices. In $C_{2'}$-*endo* puckering mode, C–C–$O_{2'}$ angles increase and the $C_{2'}$-$O_{2'}$ distance decreases. A corresponding angle-distance relation is observed for $C_{3'}$-*endo* puckering, and suggests that the out-of-plane atom has more sp^2-character than the in-plane atoms (257).

In this context, the structure of 6-amino-10-(β-D-ribofuranosylamino)pyrimido-[5,4d] pyrimidine is of interest (233) (Figure 4-10). This is the only (base-modified) ribonucleoside thus far found in the rare $O_{4'}$-*endo* puckering mode. The heterocycle is bonded to atom $C_{1'}$ via a secondary amino group. The angle at the puckered out-of-plane $O_{4'}$ atom, 103.4(4)°, is reduced by 6° relative to $C_{2'}$- or $C_{3'}$-*endo* puckered riboses, an unfavorable situation if the tendency for C–O–C angles to adopt tetrahedral values (109.5°) is considered. Furthermore, since $C_{2'}$-$O_{2'}$ and $C_{3'}$-$O_{3'}$ are *syn-periplanar* with torsion angle $O_{2'}$-$C_{2'}$-$C_{3'}$-$O_{3'}$ eclipsed (3.2°), the $C_{2'}-C_{3'}$ bond, 1.557(4) Å, is lengthened beyond the normal value of a $C-C$ single bond (Table 4-1). The $O_{4'}$-*endo* puckering as transition intermediate between $C_{2'}$- and $C_{3'}$-*endo,* therefore, corresponds to a strained conformational state as indicated by theoretical methods (Section 4.2).

4.5 *syn-anti* Conformation

The rotational position of the base relative to the sugar is sterically restricted and two conformational states are preferred, *syn* **and** *anti* (33,34) (Figure 2-10). As outlined in Chapter 2, in the *anti* conformation there is no particular steric hindrance between sugar and base but in *syn,* the bulky part of the base is located over the sugar, giving rise to close interatomic contacts. These can be relieved if the sugar adopts the $C_{2'}$-*endo* pucker: in this form, base and $C_{5'}$ atom are in equatorial orientation and moved apart (Figure 4-4). Crystal structure analyses of *syn*-nucleosides show a preference for $C_{2'}$-*endo* puckered sugars (42,195). Scrutiny of all nucleoside crystal data suggests that purine nucleosides with $C_{2'}$-*endo* pucker adopt both *syn* and *anti* forms in nearly equal distribution but $C_{3'}$-*endo* puckering shifts the orientation about the glycosyl bond to *anti* (170). For pyrimidine ribonucleosides, only in rare cases the *syn* form is found and it occurs with both $C_{2'}$- and $C_{3'}$-*endo* sugars. The *anti* conformation greatly dominates and is associated with $C_{3'}$-*endo* for riboses while pyrimidine deoxyribonucleosides prefer $C_{2'}$-*endo* over $C_{3'}$-*endo* puckering (35,167–169).

In one case, 4-thiouridine, crystallization from aqueous solution led to a *syn* conformation (258) but if butyric acid was used as solvent, 4-thiouridine crystallized in the standard *anti* form (259) (Figure 4-14). The main difference between these two crystal structures is the inclusion of water of hydration in the crystal lattice of the *syn* form which stabilizes the molecular packing through hydrogen bonding. If dissolved in aqueous medium, 4-thiouridine occurs preferentially in the *anti* conformation (260–262). This example shows that care is necessary in drawing general conclusions concerning the conformation of a flexible molecule when only one crystal structure is available.

For pyrimidine nucleosides, the *anti* **orientation of the base about the glycosyl bond is finely tuned by the sugar pucker** (263). The torsion angle about the $C_{1'}-N$ bond, χ, is correlated with the $C_{2'}$-*endo* or $C_{3'}$-*endo* pucker according to

$$-180° \leq \chi \leq -138° \text{ for } C_{3'}\text{-}endo,$$

$$-144° \leq \chi \leq -115° \text{ for } C_{2'}\text{-}endo.$$

This angular adjustment with sugar conformation is probably due to steric hindrance. Thus, in $C_{3'}$-*endo* puckering, the hydrogen at $C_{3'}$ is oriented axially (Table 4-2) and interacts with C_6-H of the base, leading to χ

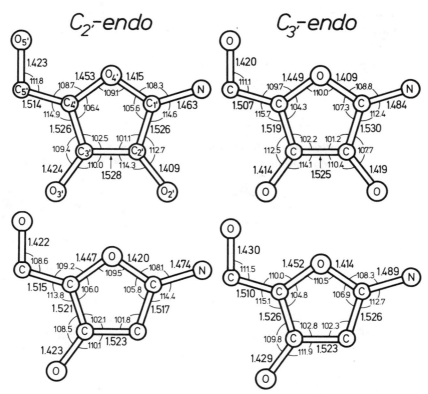

Figure 4-13. (a) Geometrical data for ribose and deoxyribose in nucleosides with $C_{2'}$-*endo* and $C_{3'}$-*endo* puckering. For other averaged data, see (392). Data are averages obtained from well-refined crystal structures ($R < 0.08$); standard deviations σ calculated according to formula given in legend to Figure 4-1. For $C_{2'}$-*endo* and $C_{3'}$-*endo* riboses, $N = 25$, $\sigma_{\text{dist.}} = 0.006$ to 0.014 Å, $\sigma_{\text{angles}} = 0.6$ to $2.3°$; for $C_{2'}$-*endo* deoxyribose, $N = 7$, $\sigma_{\text{dist.}} = 0.009$ to 0.018 Å, $\sigma_{\text{angles}} = 0.5$ to 3.2 Å, for $C_{3'}$-*endo* deoxyribose, $N = 8$, $\sigma_{\text{dist.}} = 0.007$ to 0.024 Å, $\sigma_{\text{angles}} = 0.5$ to $2.1°$.

around 180°. In $C_{2'}$-*endo*, $H_{3'}$ is in an equatorial position and the base can rotate into the $-ac$ range until its O_2 keto oxygen contacts $H_{1'}$ (264). In the purine series, these interactions between base and sugar are less severe because the imidazole ring geometry places C_8-H and N_3 atoms away from the sugar atoms and therefore the $-sc$ range becomes accessible. These differences in torsion angle ranges are systematically observed in the dodecanucleotide d(CGCGAATTCGCG) and have suggested the principle of anticorrelation to be discussed in Section 11.2.

Another structural relation concerns the dependence of the glycosyl $C_{1'}$–N bond length on the χ torsion angle. For a number of pyrimidine nucleosides (265,266), the $C_{1'}$–N bond length is a maximum (1.52 Å) if χ is near 180° and decreases almost linearly to 1.48 Å as χ approaches $-140°$. This

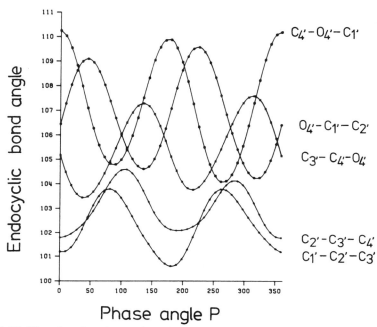

Phase angle P

Figure 4-13.(b) Plot showing that endocyclic bond angles in furanoses vary with phase angle of pseudorotation, *P*. For that reason, data in Figure 4-13a should be considered as idealized. From (170).

finding shows that, at least for pyrimidine nucleosides, hydrogen atoms at $C_{3'}$ and $C_{2'}$ are not the only barrier to rotation (34). Atom $O_{4'}$ also plays a role; the lengthening of $C_{1'}-N$ reflects steric interactions between $O_{4'}$ and the hydrogen atom attached to C_6.

For purine nucleosides, such dependence is not as obvious as in the pyrimidine series, because the steric interference is reduced. This leads to a more constant glycosyl bond length of about 1.46 Å, significantly shorter than the average 1.49 Å for pyrimidine bases (265).

syn⇌*anti* **equilibrium in solution.** Results of crystallographic studies on the *syn/anti* conformation in nucleosides are supported and supplemented by a vast body of NMR- and CD-spectroscopic data which suggest that, in solution, nucleosides undergo rapid *syn*⇌*anti* interchange (126,141,144,145,174,175,178,219,261,262,267–275). Ultrasonic relaxation studies showed that purine nucleosides exhibit a relaxation process in the nanosecond time range which is probably associated with rotation of the base about the glycosyl bond (276,277). Temperature dependence of the relaxation yields thermodynamic parameters for the *syn/anti* interconversion. For pyrimidine nucleosides, no absorption can be measured in the frequency range 10 to 250 MHz because they are conformationally more restricted than the purine analogs (276).

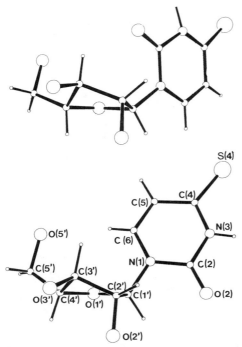

Figure 4-14. Molecular structures of 4-thiouridine, one of the rare nucleosides occurring in tRNA. 4-Thiouridine crystallized from water in the *syn* conformation (top), and from butyric acid in the usually preferred *anti* form (bottom). In *anti*, the sugar pucker is an ideal $C_{3'}$-*endo* envelope, ^3E, but twisted $C_{3'}$-*endo*, $C_{4'}$-*exo*, $^3T_4$, in *syn*. In *syn*, the $O_{5'}\cdots O_2$ distance is too long (>3.5 Å) to form an intramolecular hydrogen bond. Hydroxyl H's not drawn. From (259).

NMR data also suggest that the type of nucleoside, sugar pucker, and orientation about the glycosyl link are correlated. In purine nucleosides, both *syn* and *anti* conformers are about equally abundant but for the pyrimidine analogs, *anti* dominates over *syn*. In general, $C_{3'}$-*endo*(N) sugar puckering is associated with the base in the *anti* position for ribonucleosides but deoxyribonucleosides prefer $C_{2'}$-*endo* (S) puckering. Finally, in purine nucleosides, the *syn* conformation is accompanied by $C_{2'}$-*endo* puckering but in pyrimidines locked in the *syn* form by methylation at C_6, a trend toward $C_{3'}$-*endo* is observed (278–280) for both sugar types. The picture changes toward a drastic increase in the population of $C_{3'}$-*endo*, χ *anti*, γ +*sc* if nucleosides are phosphorylated at $O_{5'}$ because then interactions between this atom and base are enhanced, as described in Section 4.9. However, it is obvious that phosphorylation of $O_{2'}$ and $O_{3'}$ hydroxyls should have no significant influence on nucleotide conformation since distances between these atoms and base are too long to exert an appreciable effect (281,282).

The energy profiles of the *syn/anti* conformers have been investigated using classical potential energy and quantum chemical calculations (211–215,283–286). In most of these calculations, the base is rotated about the glycosyl $C_{1'}$–N bond and the sugar is held at one of the preferred puckering modes displayed in Figure 4-4. However, we have seen in Section 4.2. that the furanose ring is rather flexible; crystallographic and spectroscopic data both indicate that the rotation of the base is intrinsically associated with sugar pucker. Therefore, in two more sophisticated studies the pseudorotation concept was applied, elaborating the dependence of torsion angle χ on sugar conformation (211,212).

In one of these investigations using quantum chemical (PCILO) methods, the energy contour maps displayed in Figure 4-15 were obtained (211). The maps agree closely with the experimental data and suggest that for purine nucleosides, the *syn-* conformation correlates with $C_{2'}$-*endo* pucker while $C_{3'}$-*endo* would correspond to 1–2 kcal higher energy; for *anti* purine nucleosides, both sugar puckers are equally probable. In pyrimidine nucleosides, both sugar puckering types are of about the same energy for *syn-* conformation, and this is what is found experimentally for 6-methyluridine [$C_{2'}$-*endo* (42)] and 4-thiouridine [$C_{3'}$-*endo*, $C_{4'}$-*exo* (258)]. From crystallographic data on pyrimidine nucleosides in the *anti* conformation, $C_{2'}$-*endo* and $C_{3'}$-*endo* sugar types are about equally distributed but in the deoxyribo series, data cluster at $C_{2'}$-*endo* even though the calculated energy minima are the same.

It is of interest to note that the $C_{2'}$-*endo* $\rightleftharpoons C_{3'}$-*endo* interconversion proceeds via $O_{4'}$-*endo* (Figure 4-15) with a barrier of about 6 kcal/mole; this is appreciably higher than the value computed with force field methods and relaxed sugar geometry (Figure 4-7). Also, the maps for pyrimidine nucleosides indicate that in the *anti* region, χ is close to 180° for $C_{3'}$-*endo* pucker but near $-120°(-ac)$ for $C_{2'}$-*endo*, as discussed above.

4.6 The *high anti* (-*sc*) Conformation

The diagrams in Figure 4-15 suggest that the allowed *anti-* χ angle for purine nucleosides extends from 180° to about $-60°$, i.e., from the *anti* (*ap*) to the *high anti*(-*sc*) range. For pyrimidine nucleosides, χ is more closely restricted to the classical *anti* domain. The reason for this difference is steric hindrance at torsion angles near (or even below) $-120°$ between hydrogen atoms pyrimidine C_6–H or purine C_8–H and the sugar hydrogens. If, however, the pyrimidine C_6–H or the purine C_8–H group is replaced by nitrogen as in the chemically synthesized azanucleosides or in formycin, then this steric interaction is avoided and the nucleosides can

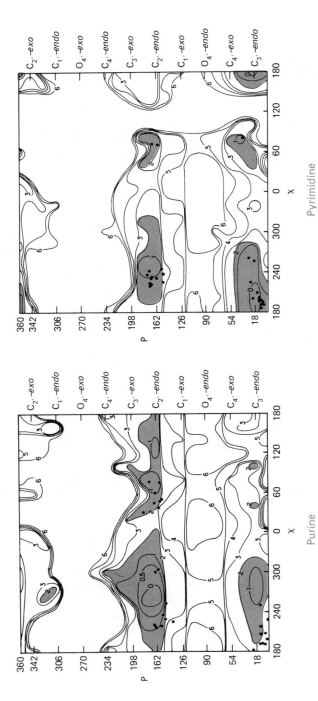

Figure 4-15. Conformational energy maps calculated with PCILO method. Glycosyl torsion angle χ in ribo- and deoxyribonucleosides is correlated with sugar puckering expressed as phase angle of pseudorotation, P. Numbers at isoenergy contour lines give kilocalories per mole above global minimum taken as zero energy; conformationally allowed regions below 1 kcal/mole are indicated in color. Dots mark results from crystal structure analyses. From (211).

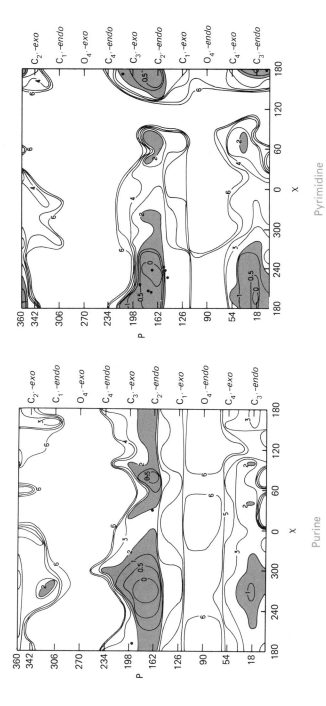

Figure 4-15. (*Continued*)

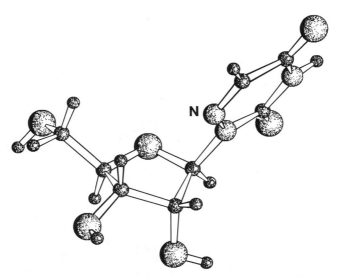

Figure 4-16. Structure of 6-azauridine in the *high anti*($-sc$) conformation. Note near-eclipsed bonds $C_{1'}-C_{2'}$ and N_1-N_6. In normally occurring pyrimidine nucleosides, N_6 is replaced by C_6-H and interferes sterically with $H_{2'}$. The orientation γ about the $C_{4'}-C_{5'}$ bond is *ap* due to electrostatic repulsion $N_6\cdots O_{5'}$. After (287). (Compare with Figure 2-10a.)

occur in the *high anti* form (Figure 4-16). This particular conformation appears to be stabilized through electrostatic repulsion between pyrimidine N_6 or purine N_8 and the $O_{5'}$-hydroxyl group (287–289).

4.7 Factors Affecting the *syn/anti* Conformation: The Exceptional Guanosine

The ratio of *syn/anti* conformers of a nucleoside or nucleotide can be controlled by chemical modifications of sugar and base. Of particular interest are nucleosides substituted at pyrimidine C_6 or at purine C_8 because such structural changes, if properly chosen, influence the electronic characteristics of the nucleosides only slightly. On the other hand, the substituents interact directly with the ribose and shift the *syn/anti* equilibrium to *syn*.

Crystallographic and spectroscopic analyses have shown that 6-methyluridine (42,261,275,278–280), 8-methyladenosine 3'-phosphate (290), and 8-halogenated purine nucleosides like 8-bromoadenosine or -guanosine (41,271,291–294) occur mostly in the *syn*- conformation both in the solid state and in aqueous solution (Figure 2-10a). However, if 8-bromoadenosine is attached to the NAD$^+$ binding site in horse liver alcohol dehydrogenase, it adopts the *anti* conformation just as NAD^+ does (295). Thus, halogenation and, as could be shown crystallographically, introduction of an *n*-butylamino group in the purine C_8 position (296) (Figure 4-17) are not sufficient to prohibit rotation about the glycosyl bond. Bulkier substituents such as *tert*butyl have been devised in order to achieve complete blockage in the *syn* form (Figure 4-18) (297,298).

Another factor stabilizing the *syn* conformer in nucleosides is the free $O_{5'}H$ hydroxyl which can form a hydrogen bond $O_{5'}H \cdots N_3$ (purine) or $O_{5'}H \cdots O_2$(pyrimidine) (Figure 2-10a). Hydrogen bonds of this type have been detected in many crystal structures of *syn* conformers. However, there are also exceptions (42,258), indicating that the favorable proximity of purine N_3 and pyrimidine O_2 to $O_{5'}$ does not inevitably lead to formation of a hydrogen bond. Another type of hydrogen bonding $C-H \cdots O_{5'}$ between base and $O_{5'}$ stabilizes the *anti* orientation of the base, especially in 5'-nucleotides (Section 4.9).

Guanosine nucleosides and nucleotides prefer *syn* orientation. In addition to these stereochemical and energetic factors, the orientation about

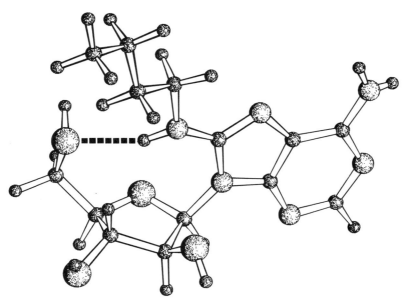

Figure 4-17. 9-β-D-Arabinofuranosyl-8-*n*-butylaminoadenine, a nucleoside crystallizing in the *anti* form although carrying a bulky substituent at C_8. This particular molecule and 8-amino-substituted adenosines in general are stabilized in *anti* by an intramolecular $N_8H \cdots O_{5'}$ hydrogen bond as can be shown by NMR techniques (393) and calculated by quantum chemical methods (394). Otherwise, 8-substituted purine nucleosides occur preferentially in *syn;* see Figure 2-10a. After (296).

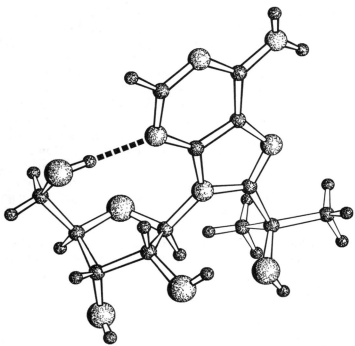

Figure 4-18. The C_8 substituent in 8-hydroxyisopropyladenosine is so bulky that this nucleoside is definitely locked in *syn*. Intramolecular $O_{5'}-H\cdots N_3$ hydrogen bond indicated by dashed line. After (298).

the glycosyl bond is also influenced by the nature of the base. Thus, 5′-guanylic acid appears to be different from the other nucleotides (299). As a deoxy analogue, it crystallizes in the *syn* form not only as a monomer (300,301) but also as a constituent of alternating tetra- and hexanucleotides, d(CpGpCpG) (302) and d(CpGpCpGpCpG) (303), and in the corresponding alternating polymer (304), which all occur as left-handed double helices (Chapter 12). In solution, guanylic acid also prefers the *syn* rather than the *anti* form as can be demonstrated by circular dichroism (for GpU, 305) and NMR (306). Classical potential energy (307,308), extended Hückel (309), and other quantum chemical (310) calculations led to the same conclusions. They showed that van der Waals and electrostatic attractions between the amino group in the 2 position and the 5′-phosphate are associated with delocalization of a $-NH_2$ lone-pair electron into an antibonding $\pi^*_{C_2-N_3}$ orbital and are responsible for the preferred *syn* conformation. A similar effect, although to a lesser extent, should stabilize inosine -5′-phosphate in *syn* relative to *anti* (310) but in three independent crystal studies, only the latter was found (311,312).

4.8 The Orientation About the $C_{4'}$–$C_{5'}$ Bond

Rotation about the exocyclic $C_{4'}$–$C_{5'}$ bond plays a crucial role in positioning the 5′-phosphate group relative to the sugar and base in nucleotides and in nucleic acids. It has been scrutinized in detail by crystallographic (35,166–169), spectroscopic (130,144,173–175,218–222,313), and theoretical methods (212–215,283–286,314–316). In contrast to the orientation χ about the glycosyl bond with two main populations, the rotation γ about the $C_{4'}$-$C_{5'}$ bond follows the classical threefold staggered pattern of ethane (Figure 2-11). There is, however, a similarity between the χ and γ rotations: *syn* and $+sc$ position the base and $O_{5'}$ "over" the ribose whereas *anti* and $-sc$ or *ap* direct base and $O_{5'}$ "away."

The three ranges $+sc$, $-sc$, and *ap* are not uniformly populated because their distribution is dependent on sugar pucker and on base. In crystalline purine nucleosides, $+sc$ and *ap* ranges for torsion angle γ show a similar frequency both in $C_{2'}$-*endo* and $C_{3'}$-*endo* puckers, and $-sc$ is rarely observed only with $C_{2'}$-*endo*. In pyrimidine nucleosides, the preferred conformation is $+sc$ regardless of sugar pucker, and the few cases exhibiting $-sc$ and *ap* are found with $C_{2'}$-*endo* (Figure 4-3).

NMR studies on a series of substituted nucleosides are consistent with these results and suggest a relationship between sugar pucker and rotation about the $C_{4'}$–$C_{5'}$ bond. As NMR methods cannot distinguish between $-sc$ and *ap* conformers (see Chapter 3), only the total population of both is obtained. The data indicate that this mixture is favored in $C_{2'}$-*endo* puckering modes but that in $C_{3'}$-*endo*, the $+sc$ orientation about the $C_{4'}$-$C_{5'}$ bond dominates (173,313). If the base is replaced by —O–CH_3, γ populates $+sc$ and *ap* states but not $-sc$ (233a). Because the $O_{4'}$–$C_{4'}$–$C_{5'}$–$O_{5'}$ torsion angles are then in $+sc$ and $-sc$ ranges, this gives a rather clear indication of the *gauche* effect described in Section 4.12. In this modified nucleoside, the $+sc$ orientation is not preferred as in nucleosides, suggesting that in the latter, base$\cdots O_{5'}$ interactions pull torsion angle γ into the $+sc$ range. Such interactions have been observed, measured, and calculated (see Section 4.9).

In contrast with experimental data, hard-sphere and classical potential calculations give nearly uniform occupation of the three conformers (283). Predictions obtained by quantum chemical (PCILO) methods are more realistic and reveal preference for the $+sc$ form by 1 to 2 kcal/mole, Figure 4-19 (315). The correlation among sugar pucker, base type, and dominating γ is also evident if energy levels of 0.5 to 1 kcal/mole are considered. The energy maps show that pyrimidine nucleosides favor $+sc$ with $C_{3'}$-*endo* rather than $C_{2'}$-*endo*, and that purine nucleosides favor $+sc$ to a less extent and regardless of sugar conformations.

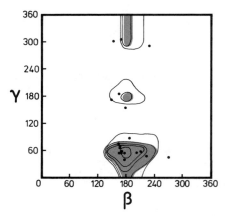

Figure 4-19. Conformational energy map for 5'-ribonucleotides calculated with PCILO method. Sterically allowed regions below 2 kcal/mole are indicated in color. γ is rotated as function of β and torsion angles α, and ϵ, ζ are fixed at 300°, 240°, and 300°. The most preferred range for γ is $+sc$ while ap and $-sc$ correspond to about 2 kcal/mole higher energy. Note that β is restricted to ap; see Section 4-12. Dots give crystallographically observed data points. From (315).

4.9 Factors Influencing the Orientation about the $C_{4'}-C_{5'}$ Bond

The distribution of conformers about the $C_{4'}-C_{5'}$ bond suggests that the type of base plays a role in stabilizing or destabilizing the $\gamma +sc$ conformer. Crystallographic studies reveal that pyrimidine C_6 and purine C_8 are frequently at distances of only 3.1 Å to 3.3 Å from $O_{5'}$, a distance significantly less than the sum of the van der Waals radii of O and H atoms and C–H bond length which amounts to 3.68 Å for the C···O distance (317–319) (Figure 4-20).

In nucleosides with *anti*-oriented bases the $+sc$ conformer is in part stabilized by (base)C–H···$O_{5'}$ "hydrogen bonding" and therefore $+sc$ is favored over the other two staggered forms (320–322). Although H attached to C is not normally considered a hydrogen bond donor, this finding is supported by deuterium-exchange NMR experiments (268,323,324) on purine mono- and oligonucleotides which show that the C_8 proton is partly acidic and interacts with the $O_{5'}$-phosphoester oxygen. Pyrimidine protons at C_5 and C_6 also appear to be considerably acidic (325).

A quantum chemical (CNDO/2) account of the energies involved in C–H···$O_{5'}$ interactions led to the conclusion that between 1.84 and 2.27 kcal/mole stabilize the $+sc$ and *anti* orientations about $C_{4'}-C_{5'}$ and glycosyl bonds (326) of uridine, thymidine, and 5-fluorouridine. In adenosine, a stabilization of only 1.11 kcal/mole was calculated and this agrees with the observation that the $+sc$ conformer is less favored for purine nucleosides than for pyrimidine analogs. Analysis of the forces involved in C–H···O "hydrogen bonding" indicated that it is not only the hydrogen atom that attracts $O_{5'}$. Rather the electron withdrawing groups attached to purine C_8 or pyrimidine C_6 cause electron deficiency at this carbon atom which then additionally attracts $O_{5'}$.

An even greater contribution of C–H···$O_{5'}$ interatomic forces was estimated for 5'-nucleotides because in these, the phosphate attached to $O_{5'}$ renders this atom more electronegative than the $O_{5'}$–H in nucleosides. This conclusion leads directly to the "rigid nucleotide" proposal discussed in Section 4.10.

4.10 The "Rigid Nucleotide"

On the basis of systematic surveys of 5'-nucleotide crystal structures, Sundaralingam observed that a nucleotide is generally less flexible than a nu-

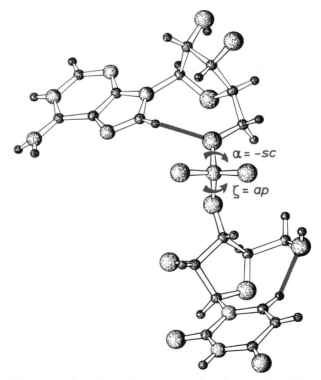

Figure 4-20. Mutual stabilization of conformations about the glycosyl bond (χ, *anti*) and $C_{4'}-C_{5'}$ bond (γ, $+sc$) via (base) C–H···$O_{5'}$ hydrogen bonds (color lines) in the crystallographically determined structure of UpA. After (317,318).

cleoside. This holds in particular for rotations about $C_{4'}-C_{5'}$ and glycosyl bonds. For nucleotides, torsion angles γ and χ adopt values corresponding to $+sc$ and *anti* ranges with both $C_{2'}-endo$ and $C_{3'}-endo$ (S and N) sugar moieties (35,58). The "rigid nucleotide" concept is generally valid for 5'-ribonucleotides, both in isolated form and in oligo- or polynucleotide chains. Some exceptions are known, such as adenosine 5'-phosphate which in the complex with platinum-terpyridyl crystallizes with γ in *ap* (327), and ApA in complex with proflavine, showing *syn* orientation of base and also γ in the *ap* form (328). In the deoxyribo series, deviations have been observed for the monomers 5'-deoxyuridylic and -guanylic acid which both crystallize with γ in the *ap* range (301,329,330), for the tetra- and hexanucleotides d(CpGpCpG) and d(CpGpCpGpCpG) forming left-handed double-helical structures with 5'-guanylic acid in *syn* conformation and γ in the *ap* range (302,303) and also for polymeric [poly d(G–C)] which adopts a structure similar to the oligomer (304).

Spectroscopic data are approximately in agreement with the "rigid nucleotide" concept with the understanding that, in solution, the "rigid" conformation means the most preferred one and one which is nevertheless in equilibrium with other conformations having γ in $-sc$ and ap and χ in *syn* (31,130,218,221,268,269,272,292,330).

Comparative NMR-spectroscopic data for adenosine, its 3'- and 5'-mono- and diphosphorylated derivatives, and oligomeric ApApA (331); quantum chemical considerations (310,332); and the crystal structure of tRNA (333) cast doubts on the validity of the "rigid" nucleotide concept. It is clear that the term "rigid" should not be taken too literally because there are exceptions from the rule both on the mono- and oligonucleotide level, especially in the deoxyribo series. As we shall later see (Chapter 18), the concept breaks down completely if nucleotides associate with binding sites in proteins. Results of tRNA crystal studies also suggest that the term "rigid" should be handled with care. It is of value, however, for predictions on isolated, uncomplexed mono-, oligo-, and polynucleotides. The hypothesis holds very well for right-handed, helical aggregates (334), and it finds support in recent studies on correlated motions between torsion angles (Section 4.13).

4.11 The Phosphate Mono- and Diester Groups and the Pyrophosphate Link: Bonding Characteristics and Geometry

A discussion of the electronic structure of the phosphate group must include consideration of *d*-electrons. In contrast to nitrogen and other first row elements, phosphorus can use vacant *d*- orbitals to form hybridized orbitals for bonding with ligands. In the phosphate group, the four oxygen atoms are linked to the phosphorus by sp^3 σ bonds and consequently the oxygens are in a tetrahedral arrangement. Additionally, the five d-orbitals of phosphorus are employed (Figure 4-21), with $3d_{x^2-y^2}$ and $3d_{z^2}$ orbitals contributing more to bonding than the less extended $3d_{xy}$, $3d_{yz}$, and $3d_{xz}$ orbitals (335–337).

Depending on this extra π-bonding, the ideal P–O single bond distance of 1.71 Å calculated according to the Shomaker-Stevenson rule (338) diminishes linearly with increasing bond order, leading to the extreme value of 1.40 Å for a "pure" double bond (339) (Figure 4-22). In a phosphate group, the π-bond order of each P–O linkage is theoretically ½ because both $3d_{x^2-y^2}$ and $3d_{z^2}$ orbitals are π-bonded to four oxygens and contribute

$$\tfrac{1}{4}\,(d_{x^2-y^2}) + \tfrac{1}{4}\,(d_{z^2}) = \tfrac{1}{2}\,(d).$$

The ideal P–O distance in an unperturbed phosphate group is therefore 1.54 Å (Figure 4-22).

In phosphate groups of inorganic salts, polyphosphates, or organic mono-, di-, and triesters, the sum of the P–O bond distances for any phosphate group is a constant (335). From 45 crystal structures on a variety of such phosphates, the average sum over the four P–O distances is (337)

$$d_{4\times(P-O)} = 6.184 \text{ Å},$$

yielding an average P–O distance of

$$d_{P-O} = 1.546 \text{ Å},$$

close to the ideal P-0 bond distance with π bond order $= 1/2$. These rules show that within the phosphate group, P–O bonds are intimately related and it is therefore possible to put forward formulas describing bond strength in terms of bond distances (340–342). Interestingly, the total bond energy for phosphate groups is

$$E_{PO_4} = 395.8 \text{ kcal/mole},$$

regardless of the kind of compound containing the phosphate group measured (337).

Individual P–O bond lengths in phosphoesters differ considerably. In phosphates, the P–O ester bond in mono- and diesters is between 1.59 and 1.62 Å (Figure 4-23) and significantly shorter than the ideal single bond of 1.71 Å, indicating about 35% π-bond contribution. As a consequence, the valence angle at oxygen, P–O–R, has some sp^2 character and is widened to 118° to 120°. In single protonated phosphomonoesters, the P–OH bond is close to 1.57 Å. This is less than the P–O ester bond distance but more than the distance for the two P–O bonds which carry the negative charge. These two usually differ (from 1.46 to 1.56 Å) due to environmental influences in the crystalline state and they are distinctly longer than the P=O double bond of 1.40 Å. If the phosphomonoester exists as a dication, the unesterified three P–O distances are electroni-

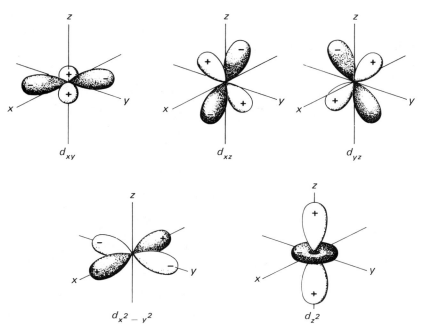

Figure 4-21.(a) Shapes and spatial orientations of the five atomic d orbitals. Three of these d_{xy}, d_{xz}, d_{yz}, are located on diagonals between axes x, y, z of a Cartesian system (top) whereas $d_{x^2-y^2}$ and d_{z^2} are along the axes (bottom). Only the latter two (as $3d_{x^2-y^2}$ and $3d_{z^2}$) are engaged in P–O π bonding. From (400).

cally equivalent. In crystal structures they are nearly equal, 1.51 Å, and similar to the free P–O bonds in phosphodiester groups.

The O–P–O bond angles depend on the oxygen substituents. Angles involving P–O ester bonds are usually around 105° to 109°; those between charged oxygens, $O^{\ldots}P^{\ldots}O^-$, are in the range 115° to 118°; angles HO–P–O$^-$ are between 105° and 114°; and angles HO–P–OC are 100° to 108°. This general angle distribution is found in mono-, di-, and triesters. The spread of O–P–O angles from 100° to 120° indicates that charge-charge repulsion in $(O^{\ldots}P^{\ldots}O^-)$ has a dominating influence, but bond length differences also determine the angle distribution. Bond length differences are mainly caused by bond polarization due to metal ion coordination or hydrogen bonding; they are hardly predictable and are of significance only in the solid state. In solution, differences between unesterified and protonated oxygens will disappear on a time average because the hydrogen and metal ions are migrating, but differences involving ester oxygens will remain as found in crystal structures.

What is particular about "energy-rich" phosphates? Structural, quantum chemical, and thermodynamic data. The pyrophosphate group in nucleoside di- and triphosphates was termed "energy rich" (343). From a structural point of view, however, the atomic arrangement within this

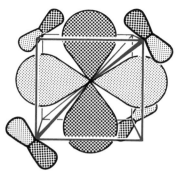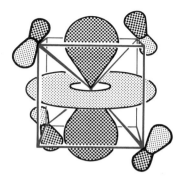

Figure 4-21.(b) Schematic description of orbital overlap of phosphorus $3d$ orbitals with oxygen $2p$ orbitals in the phosphate tetrahedron. (Left) $3d_{x^2-y^2}$ and $2p\pi$; (right) $3d_{z^2}$ and $2p\,\pi'$. After (335).

group is "normal" and corresponds to two fused phosphate groups. The pyrophosphate $P-O_{bridge}$ bond lengths are of the same order as $P-O_{ester}$ bonds (Figure 4-23), and the free $P-O$ bonds carrying the negative charge are as short as in mono- and diesters, 1.50 to 1.55 Å, indicating considerable double-bond character.

The relatively long $P-O_{bridge}$ bonds are associated with an obtuse $P-O-P$ angle of $\sim 130°$ and both geometrical peculiarities together provide for a long $P\cdots P$ distance. This feature allows nearly uninhibited rotation about the pyrophosphate $P-O$ bonds and even eclipsed $O-P-O-P$ torsion angles are possible (344), (Figure 4-24).

As shown by quantum chemical calculations employing extended Hückel methods (345), the $P-O_{bridge}$ bond lengthening and $P-O-P$ angle widening are due to electrostatic repulsion between the two negatively charged phosphate groups and to opposing resonance. The π-bonding of the bridge oxygen is about 35%, which is significantly less than that of terminal (charged) $P-O$ bonds with 42 to 48% π-bonding contribution. These studies discredit earlier quantum chemical calculations [reviewed

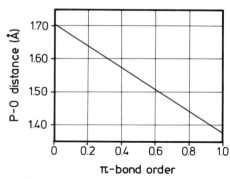

Figure 4-22. Relation between $P-O$ bond distance and π-bond order. After (395).

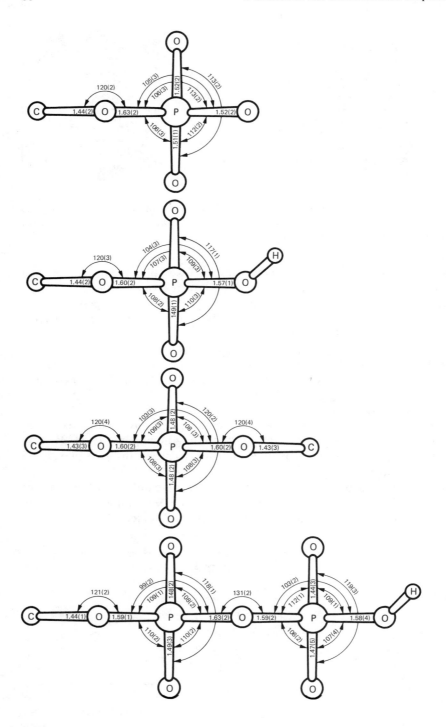

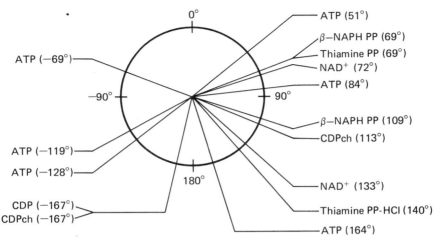

Figure 4-24. Radial lines in conformational wheel describe torsion angles O–P–O–P (°) found in pyrophosphate crystal structures. The almost even distribution of torsion angles indicates nearly free rotation about P–O$_{bridge}$ bonds. (CDPch, CDPcholine; β-Naph PP, di-β-naphthyl-pyrophosphate; thiamine PP, thiamine pyrophosphate; thiamine PP-HCl, thiamine pyrophosphate hydrochloride). From (344).

in (346)] which claimed that the bridge oxygens in ATP should be positively charged. In fact, the phosphorus atoms carry a partial positive charge of $+0.17e$ to $+0.34e$, the bridge oxygens are slightly negative, $-0.12e$ to $-0.17e$, and the terminal oxygens are still more negative, $-0.63e$ to $-0.70e$. All these data, structural and theoretical, show that the term "energy rich" does not mean energy storage in a P–O bond but rather refers to the free energy of hydrolysis near pH 7 (345,347).

If the terminal phosphate or pyrophosphate of ATP is hydrolyzed, inorganic phosphate P$_i$ or pyrophosphate PP$_i$ are released:

$$ATP + H_2O \rightleftharpoons ADP + P_i + H^+,$$

$$ATP + H_2O \rightleftharpoons AMP + PP_i + H^+.$$

In both processes, the same standard free energy $\Delta G°$ of -7.3 kcal/mole is gained, considerably more than for hydrolysis of an ester bond such as that of glycerol 3'-phosphate, $\Delta G° = -2.2$ kcal/mole (347,348):

$$HOCH_2CHOHCH_2OPO_3^- + H_2O \rightleftharpoons HOCH_2CHOHCH_2OH + P_i.$$

Figure 4-23. Geometrical data for phosphate and pyrophosphate mono- and diesters. Data for monoesters from (191); those for diesters and pyrophosphate ester from (163). Standard deviations σ obtained with formula given in legend to Figure 4-1 are presented in parentheses. Numbers of observations, N, are 11 and 15 for monoester di- and monoanion, 33 for diester, and 4 for pyrophosphate ester.

The ATP hydrolysis is largely dependent on pH and Mg^{2+} concentration (349,350) and above values are for physiological pH. The higher "phosphate-transfer potential" of ATP relative to glycerol- 3'-phosphate can be explained with electrostatic repulsion of phosphate moieties in the pyrophosphate, with favorable resonance stabilization of the resulting products (ADP and P_i, or AMP and PP_i) relative to ATP, and with a gain in solvation energy if ATP is broken into fragments.

4.12 Orientation About the C–O and P–O Ester Bonds

Continuing with the description of the conformational properties of nucleotides, we consider torsion angles involving the C–O and P–O ester bonds. As we shall see, the rotations about C–O ester bonds are restricted but rotations about P–O ester bonds are less so. Thus P–O bonds are the major pivots affecting polynucleotide structure.

Rotations about C–O and P–O bonds are restricted. In crystals of mono-, oligo-, and polynucleotides, the torsion angle β defining rotation about the $C_{5'}–O_{5'}$ bond (Figure 2-3) is largely limited to the *ap* range, with only few deviations in $+ac$ or $-ac$. The rotation about the $C_{3'}–O_{3'}$ bond, denoted by ϵ, follows a similar trend yet the main clustering is not *ap* but is shifted slightly to 220° in the $-ac$ range (166,169,351) (Figure 4-25).

These observations are corroborated by NMR-spectroscopic studies evaluating $^{31}P–^{1}H$ and $^{31}P–^{13}C$ heteronuclear coupling (130–132,143–147, 175,176,178,218,293,352–354) as well as lanthanide shift methods (330,354,355) on mono-, oligo-, and polynucleotides. Irrespective of sugar puckering, the atomic chain $H_{4'}–C_{4'}–C_{5'}–O_{5'}–P$ is predominantly extended, corresponding to $+sc$ orientation for γ and *ap* for β torsion angles. This is also in agreement with theoretical studies (Figure 4-19). The torsion angle ϵ about the $C_{3'}–O_{3'}$ bond cannot be determined from $^{31}P–^{1}H$ coupling because only one coupling constant with $H_{3'}$ is accessible and it is insufficient for defining ϵ unambiguously (131). Comparative NMR studies of several dinucleoside phosphates suggest that ϵ is in the *ap* or $-sc$ range, depending on $C_{3'}$-*endo* or $C_{3'}$-*exo* puckering mode (353). A more direct analysis (143) used $^{31}P–^{13}C$ coupling between P and $C_{2'}$ as well as $C_{4'}$ atoms. This leads to the conclusion that torsion angle ϵ is confined to ranges in the (–)hemisphere of the Klyne-Prelog cycle (Figure 2-5) and ϵ depends on temperature and nature of base. Intramolecular base stacking appears to be the major force in determining structures of dinucleotides and higher oligonucleotides and this interaction also directly influences ϵ.

Theoretical considerations greatly agree with experimental data, showing that severe steric hindrance between the phosphate group and sugar moiety restricts C–O torsion angles β essentially to *ap* and ϵ to the *ap* and $-ac$ ranges (60,182,314,316,356,357).

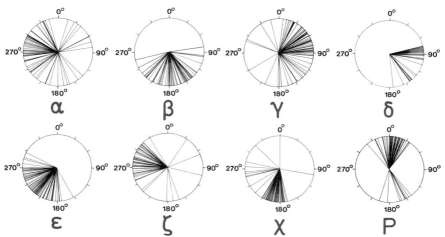

Figure 4-25. Radial lines in conformational wheels show crystallographically determined torsion angles (°) and pseudorotation phase angle P (°) for polyribonucleotide tRNA. Note that P displays bimodal distribution with $C_{3'}$-*endo* more populated than $C_{2'}$-*endo*, also reflected in the $C_{3'}$–$C_{4'}$ torsion angle δ with $\delta \sim 90°$ $C_{3'}$-*endo* and $\delta \sim 140°$ $C_{2'}$-*endo*. Torsion angle $\alpha(P$–$O_{5'})$ is less restricted than $\zeta(P$–$O_{3'})$, which prefers $-sc$ range, and adjacent torsion angles $\beta(O_{5'}$–$C_{5'})$ and $\epsilon(O_{3'}$–$C_{3'})$ are in the *ap* range, the latter shifted toward $-ap/-ac$. Finally, γ greatly prefers the $+sc$ range and χ is predominantly *anti*, in some cases *high anti* (-*sc*) and in one case *syn*. After (396).

The gauche or anomeric effect limits orientations about P–O ester bonds to $-sc$ for α and $-sc$ or $-ap$ for ζ. Rotations about the P–O ester bonds P–$O_{5'}$, α, and P–$O_{3'}$, ζ, in 5'- and 3'-mononucleotides correspond to the classical ethane-like staggered scheme because all three nonesterified P–O bonds are equivalent. However, with a dinucleoside monophosphate or a higher oligonucleotide, steric hindrance comes into play and dictates rotational limits for α and ζ.

X-ray crystal studies of oligonucleotides and tRNA show that the two main geometrical arrangements of a dinucleoside monophosphate moiety are helical and nonhelical. The right-handed helical conformation occurs only if both α and ζ are in the $-sc$ range, around 270° (Sets I and IV in Table 4-3). For nonhelical conformations, several possibilities exist (Figure 4-26 and Tables 4-3,4-4), corresponding to extended, looped, and left-handed configurations like those found in the oligonucleotides UpA (317, 318), ApApA (160), pTpT (358), and dApTpdApT (359) (Sets II and III in Table 4-3). In tRNA (Figure 4-25), ζ is more confined to $-sc$ than α. which spreads over $-sc$ and $-ap$, and α also populates the $+sc$, $+ap$ ranges more frequently than ζ does. These differences in rotational mobility about the two P–O ester linkages are due to the attachment of P–$O_{5'}$ to the primary $C_{5'}$ which is sterically less restricting than the secondary $C_{3'}$ bound to P–$O_{3'}$.

In solution studies of oligoribonucleotides, NMR methods have deter-

Table 4-3. Backbone Torsion Angles for Dinucleoside Monophosphate Fragments from Single-Crystal Studies and for Double-Helical Polynucleotides Derived from Fiber Diffraction Patterns[a]

		δ(5'-end) $C_{4'}-C_{3'}$	ϵ $C_{3'}-O_{3'}$	ζ $O_{3'}-P$	α $P-O_{5'}$	β $O_{5'}-C_{5'}$	γ $C_{5'}-C_{4'}$	δ(3'-end) $C_{4'}-C_{3'}$
Set I	GpC	89	211	292	285	184	50	77
	GpC(1)	79	222	294	291	181	47	79
	GpC(2)	73	217	291	293	172	57	80
	GpC(3)	96	224	290	286	167	63	74
	GpC(4)	88	216	288	283	181	52	87
	ApU(1)	84	213	293	288	177	57	74
	ApU(2)	78	221	284	295	168	58	77
	ApA+*	82	223	283	297	160	53	81
	ApT**(1)	90	213	294	293	176	68	134
	ApT**(2)	83	212	284	302	171	64	139
Set II	UpA(2)	77	224	164	271	192	53	93
	TpT**	157	252	163	288	187	41	158
	TpA**	134	204	168	286	186	49	83

Set III	UpA(1)	86	206	81	82	203	55	85
	A⁺pA⁺*	81	207	76	92	186	56	79
	ApA***	147	219	61	66	186	46	77
Set IV	A RNA	95	202	294	294	186	49	95
	A DNA	83	178	313	285	208	45	83
	B DNA	156	155	264	314	214	36	156
	C DNA	141	211	212	315	143	48	141
	D DNA	156	141	260	298	208	69	156
	B DNA (left)	140	200	212	282	165	40	140

^{a} Sugar puckering mode is indicated by δ, 74° to 95° corresponding to C$_{3'}$-endo and 134° to 158° to C$_{2'}$-endo. Data have been grouped into sets I to IV. Sets I to III have torsion angles α/ζ in the ranges $-sc/-sc$, $-sc/ap$, $+sc/+sc$ and in set IV are polynucleotides with $-sc/-sc$. Molecular fragments in sets I and II belong to right- and left-handed helices, those in set III to looped structures, set IV to right-handed double helices, with parameters for B-DNA (left-handed) obtained not experimentally but by theoretical considerations, see Table 4-6. Numbers in parentheses after compound names refer to different molecules in the crystal or to independent X-ray analyses. [From (383)].

* Part of ApA⁺pA⁺ (160).

** Deoxy sugar.

*** As Ca²⁺-salt (383a).

mined essentially the same $-sc$ range for both α and ζ as derived from crystallographic data. Because the $-sc$ orientation about both P–O ester bonds is, for a "rigid" nucleotide, synonymous with formation of a right-handed helix (360), stacking of adjacent bases occurs and this stabilizes the structure (see Chapter 6). This helix formation influences the conformational flexibility of the nucleotides and pushes them toward only one sugar pucker, $C_{3'}$-endo, leading to greater conformational purity or "persistence" as the chain grows from monomer to dimer to trimer, etc. (130–132,138–140,143–148,331) (see Table 4-5).

Theoretical studies show that the *gauche* effect dictates orientation about P–O bonds. The orientations about the P–O ester bonds are crucial for understanding polynucleotide structure. They have been the topic of several theoretical studies, starting with the simple models dimethyl phosphate (361–363) and ethyl methyl phosphate (364). *Ab initio* quantum chemical and extended Hückel calculations have resulted in energy contour maps (Figure 4-27). The map for dimethyl phosphate (Figure 4-27a) shows the twofold symmetry characteristic of the molecule and exhibits energy minima at $C–O–P–O_{ester}$ torsion angles in the sc,sc and $-sc,-sc$ ranges. The map does not show the usual threefold staggered potential with ap contribution because the "anomeric" or "*gauche*" effect (365–368) favors $+sc$ or $-sc$ orientations. This "*gauche*" effect is operative in a system $C–\ddot{O}–P–O_{ester}$ with lone-pair electrons at $\ddot{O}$ and an adjacent polar bond $P^{\delta+}–O^{\delta-}{}_{ester}$. This leads to partial donation of the lone pair electrons into the polar bond when the lone-pair orbital and $P–O_{ester}$ bond are antiparallel, i.e., if the $C–O–P–O_{ester}$ torsion angle is in the $+sc$ or $-sc$ range (Figure 4-28). In a fully extended, zig-zagged ap orientation of the

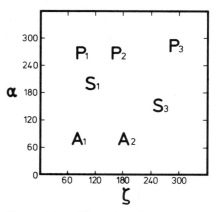

Figure 4-26. Allowed conformations for dinucleoside monophosphates, calculated with classical potential energy methods. Symbols P, S, and A stand for parallel, skewed, and antiparallel orientations of sugar units relative to each other; the orientations are given by vectors bisecting the $C_{2'}–C_{3'}$ bond and running through $O_{4'}$. Both $C_{2'}$-endo and $C_{3'}$-endo sugar puckers give essentially the same map.

Table 4-4. Properties of Single Stranded Polynucleotides Generated from the Seven Basic α and ζ Conformations Shown in Figure 4-26 From [(360)].

	Sugar oxygen orientation[b]	Feasibility of base parallelism	Repeatability	α/ζ	Relative position of base with respect to propagation axis
P_1	Same direction	Yes	Yes	270/90	Away
P_2	Same direction	Yes	Yes	270/170	Away
P_3	Same direction	Yes	Yes	280/290	Toward[a]
S_1	Skew direction	No	Yes	200/110	Away
S_3	Skew direction	No	Yes	160/250	Away
A_1	Opposite direction	Yes	No	80/80	—
A_2	Opposite direction	Yes	Yes	80/180	Away

[a] Only the conformation P_3 leads to Watson–Crick type double helices with parallel bases oriented toward the helix axis.

[b] The orientation is defined by the vector $C_{3'}-C_{2'}\to O_{4'}$ (see legend to Figure 4-26). "Same direction" means that all sugars of the polynucleotide are in parallel arrangement.

C–O–P–O–C chain, the lone-pair orbitals would not be antiparallel to the polar P–O_{ester} bonds, a situation corresponding to 7 kcal/mole above the global energy minimum in the map. It should be added that inclusion of phosphorus $3d$ orbitals in the calculations did not significantly influence the energy distribution with varying α or ζ. Further, a subtle interrelation between distortion of the phosphate tetrahedron and left- or right-handed arrangement of the two ester groups was noted (362).

If the model is extended and calculations are performed with sugars or nucleosides attached to the pivotal phosphate group, the situation

Table 4-5. Effect of Phosphodiester Bond Formation on Nucleotide Conformation[a]

Conformational bond (group)	Monomer	Dimer
χ C–N (glycosyl bond)	> 225°	→ 180°
Ribose ring (population ratios)	$C_{2'}$-endo/$C_{3'}$-endo (65:35)	$C_{2'}$-endo/$C_{3'}$-endo (40:60)
γ $C_{4'}-C_{5'}$	+sc (74%)	+sc (85%)
β $C_{5'}-O_{5'}$	ap (75%)	ap (86%)
ϵ $C_{3'}-O_{3'}$	+sc ⇌ −sc	−sc
α P–$O_{5'}$	Free rotation	(320° and 80°)
ζ P–$O_{3'}$	+sc, −sc, and/or free rotation	(330° and 50°)

[a] Table summarizes changes in going from pA and Ap to ApA. Note that conformation ranges are frozen by oligomer formation, leading to greater "conformational purity" or "persistence." [After (131).]

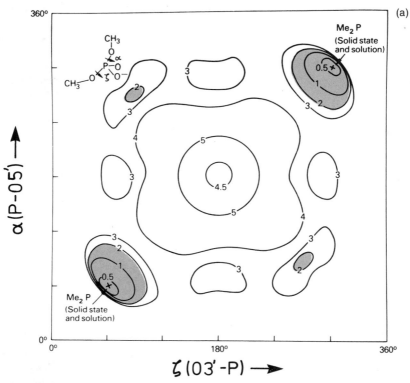

Figure 4-27. Conformational maps for P–O bond rotations calculated with classical potential energy functions including an extra term to account for *gauche* effect (Figure 4-28). Shown are isoenergy contour maps for (a) dimethyl phosphate, (b, facing page) diribose phosphate, (c, facing page) diribose triphosphate, with riboses in $C_{3'}$-*endo* puckering form. Allowed conformations in bottom map are rather restricted and show that polynucleotides can adopt only certain α and ζ values. For other, similar representations, see (361,369,373). From (375).

changes. As summarized and critically reviewed (182,370), applications of classical potential energy (58–61,369–371) and extended Hückel methods (316,372) are unsatisfactory because they neglect the *gauche* effect and thus incorrectly predict a favorable energy minimum with α and ζ in the *ap* range. This is avoided with the more refined CNDO (373) and PCILO (182,357,374) calculations or if an extra term is included in the classical potential energy formalism (375). As shown in Figure 4-27b, computed minima then correspond to observed values of α and ζ, with main populations in the $+sc, +sc$ and $-sc, -sc$ ranges. The $-sc, -sc$ region extends into *ap* for ζ because, as pointed out earlier, the primary $C_{5'}$ is less restricted than the secondary $C_{3'}$.

Second neighbor phosphates considerably restrict conformational space. The picture is further modified if, instead of a dinucleoside monophosphate, dinucleoside di- or triphosphates are utilized as model compounds

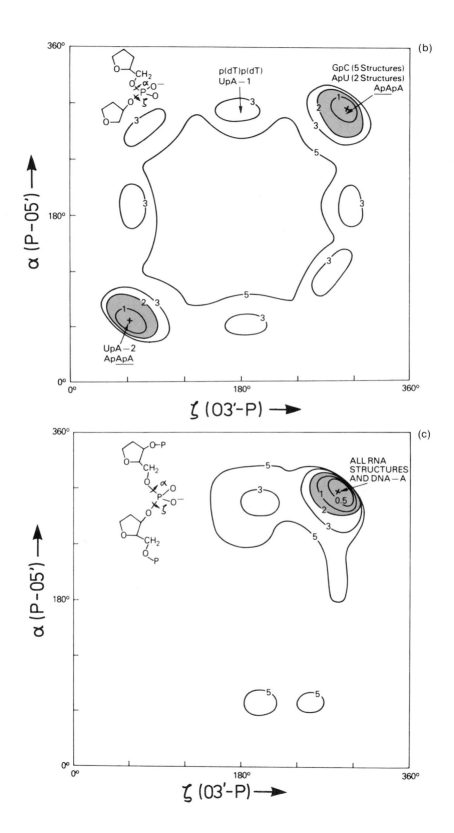

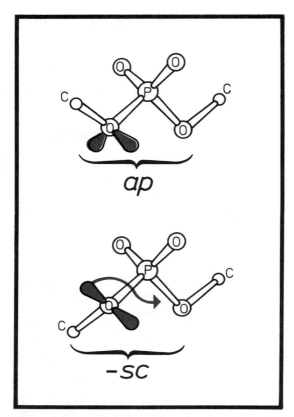

Figure 4-28. Description of the *gauche* effect. The *gauche* (+ or −*sc*) conformation of the C–O–P–O group is favorable because an oxygen electron lone pair is *ap* to the adjacent, polarized P–O bond and can donate electrons (bottom). In the all-*trans* (*ap*) orientation (top), however, orbitals and adjacent P–O bonds are in +*sc* and −*sc* positions and electron transfer is diminished.

(369,375,376) (Figure 4-27c). In that case the allowed conformation space narrows considerably and the +*sc*, +*sc* orientations for α, ζ are now forbidden because of close contacts between second neighbor phosphate groups. This result suggests that oligonucleotides of increasing length become conformationally more limited and "pure," as indeed observed by NMR-spectroscopic methods (Table 4-5).

4.13 Correlated Rotations of Torsion Angles in Nucleotides and in Nucleic Acids

Are torsion angle rotations independent of each other? We already have seen that rotations about single bonds in nucleosides and nucleotides are not independent of other structural features. Because the nucleotide in itself is a rather compact molecule with several interactions between non-bonded atoms, this is not surprising and the question arises as to whether some systematic correlation exists among torsion angles.

Only γ is independent. In a detailed study of 127 nucleoside/-tide crystal structures, Tomita and co-workers showed by application of circular correlation and regression analysis that almost all of the torsion angles are linearly related with correlation coefficients between 0.78 and 0.89 (perfect correlation = 1.0) (377). A scheme can be drawn where arrows indicate strong correlation and dots designate weak or no interdependence; torsion angle α has been omitted because it is not relevant in nucleosides/-tides:

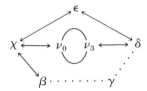

It is obvious that δ and ν_3 have to be related since they describe torsions about the same $(C_{3'}-C_{4'})$ bond and define sugar puckering which again involves ν_0. Surprisingly, torsion angle γ is not correlated at all with any other angle and probably adopts the three staggered conformations (Section 4.8) regardless of the overall structure of the nucleoside. For the nucleoside as a whole, however, one could in principle predict the other torsion angles because they depend upon one another.

In helical polynucleotides, all backbone torsion angles are correlated. Extension to nucleic acid helices has given additional insight into the relationship between structural parameters (378,379). Along the backbone, correlations are observed for

$$\alpha \leftrightarrow \beta, \ \alpha \leftrightarrow \gamma, \ \beta \leftrightarrow \epsilon, \ \epsilon \leftrightarrow \zeta.$$

We now find that γ is also no longer independent. Although correlation coefficients in the range 0.65 to 0.71 are smaller than with monomeric units, the statement can be made that all torsion angles of the sugar-phosphate backbone and of the nucleoside moiety are interrelated. Therefore, conformational changes in helical polynucleotides are associated with concerted motions of all torsion angles. (See also Ref. 894a.)

4.14 Helical or Not Helical—and if, What Sense?

Potential energy calculations can probe polynucleotide helicity. Conformations of Watson-Crick type double-helical oligo- or polynucleotides are largely restricted to torsion angles α, ζ both in the $-sc$ range, with nucleotides in their most preferred "rigid" forms, *viz.* $C_{2'}$-*endo* or $C_{3'}$-*endo*

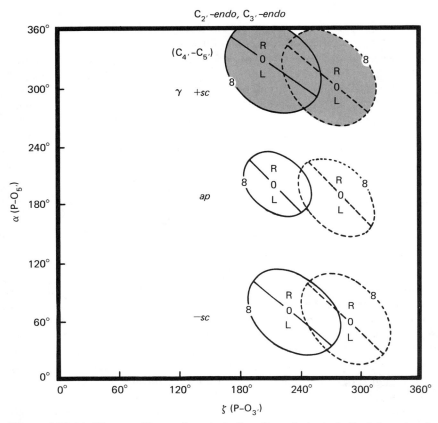

Figure 4-29.(a) Diagram illustrating sterically allowed single-helical domains for polynucleotides. Sugar puckerings are C$_{2'}$-*endo* (solid lines) and C$_{3'}$-*endo* (dashed lines); torsion angle γ(C$_{4'}$–C$_{5'}$) is assumed in the three staggered conformations $+sc$, $-sc$, and ap, and α and ζ are varied from 0° to 360°. For the sake of clarity, only contours $n = 8$ and $h = 0$ are drawn, with n indicating number of nucleotides per helical turn, and h the rise per residue (Å) along the helix axis. Parameters leading to Watson-Crick-type double helices are found in the region with α and ζ both at $-60°$ (300°), marked in color. From (383).

sugar pucker, *anti* about glycosyl link χ, $+sc$ about C$_{4'}$–C$_{5'}$ bond γ, and *ap* about C–O bonds β and ϵ. Classical potential energy calculations have been applied to systematically investigate the limitations which nucleotide geometries pose for single- and double-helical arrangements; that is, what are the possible values for n, the number of nucleotides per helix turn, and h, the axial rise per residue (380–387)?

Theoretically, right- and left-handed single and double helices are feasible. Assuming sugar puckers C$_{2'}$-*endo* and C$_{3'}$-*endo*, calculations of n-h plots were carried out with γ in the three staggered conformations $+sc$, $-sc$, and *ap*, and for variation of the P–O torsion angles α and ζ. In all

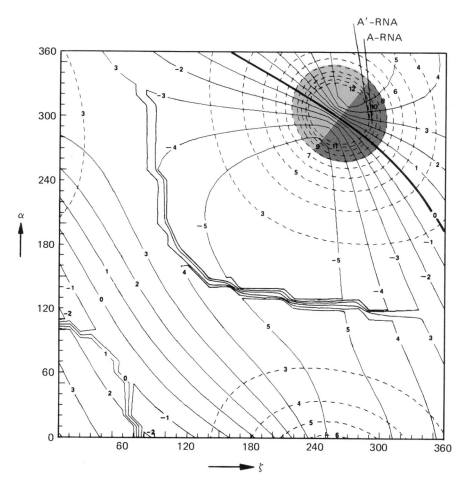

Figure 4.29.(b) A more detailed plot than in Figure 4-29a, indicating $n-h$ dependence for single-stranded polynucleotide with $C_{3'}$-*endo* sugar puckering; γ in $+sc$, β and ϵ in ap, and χ in *anti* orientations. α and ζ are varied from 0 to 360°. On curves of iso-n (dashed) and iso-h (solid), figures give number of nucleotides per helix turn (n) and axial rise per residue (h). $h = 0$ Å indicates a helix degenerated into a straight line. h is positive for right-handed and negative for left-handed screw sense. Each curve h is cut twice by curves n, leading to regions R_I, R_{II} for right-handed and L_I, L_{II} for left-handed helices. Among these, R_I is energetically preferred because of favorable base-base and sugar-base interactions between adjacent nucleotides. R_I, dark, top; R_{II}, light, above line $h = 0$. L_I, L_{II}, below line $h = 0$. From (381).

cases, only *one* domain for stereochemically allowed helix formation was found, as summarized in Figure 4-29. The location of this domain within the α, ζ plot indicates that sugar puckering correlates primarily with orientation ζ about the $O_{3'}-P$ bond and conformation about $C_{4'}-C_{5'}$ influences the torsion angle α about the $P-O_{5'}$ bond. The most striking feature of these studies is, that a line of $h = 0$ Å corresponding to a single (or double) straight, ladder-like "helix" passes through all helical domains and divides them into two sections for right- and left-handed screw sense. This finding demonstrates that minor changes in α, ζ occurring without major alterations in other torsion angles can reverse the handedness of a helix. Turning the right-handed B-DNA helix into a stereochemically acceptable smooth, left-handed form requires changes in torsion angles of less than 30° (except for χ, which is in *high anti* ($-sc$) range in right-handed form and *low anti* if left-handed); see Table 4-6.

Although stereochemically possible, left-handed helices have, thus far, not been found by crystallographic methods: all oligonucleotide crystal structures of DNA or RNA fragments or of tRNA display right-handed helical sense. While model building studies have indicated that optimum strand interactions between adjacent nucleotides favor a right-handed helical sense, this does not mean that a left-handed double helix could not exist under certain circumstances.

These circumstances could be protein complexation and, in fact, the crystal structure of catabolite activator protein (CAP) suggested that it could bind to smooth, left-handed double-helical DNA (however, see Chapter 18). Also, a smooth left-handed duplex was proposed earlier for

Table 4-6. Conformational Parameters for B-DNA-Type Double Helices in Right- and Left-Handed Screw Sense, as Obtained from Potential Energy Calculations[a]

		Right-handed	Left-handed
Backbone and glysosyl torsion angles	α	302° (314°)	270°
	β	144° (214°)	135°
	γ	41° (36°)	36°
	δ	149° (156°)	137°
	ϵ	228° (155°)	241°
	ζ	202° (264°)	204°
	χ	−106° (−100°)	177°
Furanose conformation angles	ν_0	−33.7° (−34.8°)	−38.6°
	ν_1	18.7° (33.3°)	18.1°
	ν_2	6.3° (−18.6°)	11.2°
	ν_3	−30.0° (−4.0°)	−36.1°
	ν_4	40.5° (24.7°)	35.2°
Base parameters as defined in Figure 2-14	θ_R	−4.0° (−1.0°)	0.0°
	θ_T	−7.5° (−6.0°)	−3.0°
	D	−0.30(−0.14)Å	−1.22Å

[a] Sugar conformation is assumed to be $C_{2'}$-*endo* and γ in the $+sc$ range. Numbers in parentheses give data for B-DNA from fiber diffraction. [From (391).]

poly [d(I−C)] · poly [d(I−C)] on the basis of X-ray fiber diffraction and circular dichroism studies in connection with model building (388). In this context, it should be stressed that the left-handed double helices described here are quite distinct from Z-DNA with its zig-zag appearance. In that case, the motif is a dinucleotide d(G−C) with dG and dC in very different conformations (Chapter 12), whereas in the smooth left handed helices all nucleotides display very similar structural characteristics.

The virtual bond concept: A nucleotide blocked into two units. The calculations can be greatly simplified and speeded up if backbone torsion angles are not treated individually but in terms of two blocks or "virtual bonds." These blocks consist of chains of atoms $P_{5'}-O_{5'}-C_{5'}-C_{4'}$ and $C_{4'}-C_{3'}-O_{3'}-P_{3'}$, which in nucleotides are generally in the *trans* (or *antiperiplanar*) configuration and therefore behave as planar systems. The virtual bonds $P_{5'}\cdots C_{4'}$, and $C_{4'}\cdots P_{3'}$, are linked at phosphorus and $C_{4'}$ atoms; they form virtual bond angles and torsion angles and thus enable us to construct polymeric structures, be it in helical or in random coil organization which can be tested with respect to macroscopically measured experimental data (380,389).

The unusual Z-DNA: Can it be "predicted" on a theoretical basis? Faced with the left-handed double-helical structure of Z-DNA which is adopted by the alternating polynucleotide poly [d(G−C)], an additional note is needed. Under the assumption that a mononucleotide is the repeating motif in a helical structure, Z-DNA could never have been predicted because its basic unit is a dinucleotide (Chapter 12). Such situations, however, can also be simulated theoretically if the size of the motif is changed from mono- to dinucleotide or even to longer units (390,391). One could even devise sequence-dependent motifs which find their justification in the fine details of the DNA-dodecamer crystal structure to be discussed in Section 11.2.

Summary

The main lessons learned from Chapter 4 can be summarized in a few statements:

1. Bases are planar. They occur in keto and amino tautomer forms. The amino groups are coplanar with the heterocycles due to electron resonance, with a barrier to rotation of ∼ 15 kcal/mole.
2. The sugar, ribose or deoxyribose, is puckered $C_{2'}$-*endo*(S) or $C_{3'}$-*endo*(N). In solution, both states are in equilibrium and rapid transition from one to the other follows the path of pseudorotation with $O_{4'}$-*endo* as the transition state. The energy barrier between states is low: 0.6 to 3.8 kcal/mole is calculated and 4.7 kcal/mole is experimentally observed. The $C_{2'}$-*endo* $\rightleftharpoons C_{3'}$-*endo* equilibrium is determined by the

Table 4-7. Summary of Preferred Nucleoside and Nucleotide Conformations[a]

Type of nucleoside/-tide	Sugar pucker (δ)	χ $C_{1'}$–N	γ $C_{4'}$–$C_{5'}$	β $C_{5'}$–$O_{5'}$	ϵ $C_{3'}$–$O_{3'}$	α P–$O_{5'}$	ζ P–$O_{3'}$
Nucleosides							
Pyrimidine							
Ribo	$C_{3'}$-_endo_/$C_{2'}$-_endo_	_anti/syn_	_sc/−sc/ap_				
Deoxyribo	$C_{3'}$-_endo_/<u>$C_{2'}$-_endo_</u>	_anti/<u>syn</u>_	<u>_sc_</u>/−_sc/ap_				
Purine							
Ribo	$C_{3'}$-_endo_/$C_{2'}$-_endo_	_anti/syn_	_sc/−sc/ap_				
Deoxyribo	$C_{3'}$-_endo_/<u>$C_{2'}$-_endo_</u>	_anti/syn_	_sc/−sc/ap_				
5'-Nucleotides							
Pyrimidine							
Ribo	$C_{3'}$-_endo_/$C_{2'}$-_endo_	_anti_	_sc_	_ap_ (180°)	−_ac_ (220°)		
Deoxyribo	$C_{3'}$-_endo_/$C_{2'}$-_endo_	_anti_	_sc_	_ap_ (180°)	−_ac_ (220°)		
Purine							
Ribo	$C_{3'}$-_endo_/$C_{2'}$-_endo_	_anti_	_sc_	_ap_ (180°)	−_ac_ (220°)		
Deoxyribo	$C_{3'}$-_endo_/$C_{2'}$-_endo_	_anti_	_sc_	_ap_ (180°)	−_ac_ (220°)		
Helical polynucleotides							
Ribo	$C_{3'}$-_endo_	_anti_	_sc_	_ap_ (180°)	−_ac_ (220°)	−_sc_	−_sc_
Deoxyribo	$C_{3'}$-_endo_/$C_{2'}$-_endo_	_anti_	_sc_	_ap_ (180°)	−_ac_ (220°)	−_sc_	−_sc_

[a] Dominating conformations are underlined. In previously used nomenclatures, _sc_ is +_gauche_, −_sc_ is −_gauche_, _ap_ is _trans_.

electronegativity of substituents at $C_{2'}$ and $C_{3'}$ (a more electronegative substituent preferring axial orientation due to the *gauche* effect) and by the orientation of the base (*syn* correlating largely with $C_{2'}$-*endo*). In helical polynucleotides, the equilibrium is "frozen" in either $C_{3'}$-*endo* or $C_{2'}$-*endo* form (Table 4-7).

3. The orientation about the glycosyl $C_{1'}$-N bond, χ, is predominantly *anti* for pyrimidine nucleosides, but in the case of purine nucleosides *syn* and *anti* are about equally common and in solution the two states are in rapid equilibrium with each other. *Syn* is stabilized by $O_{5'}$-H$\cdots$N$_3$ (purine) or $O_{5'}$-H$\cdots$O$_2$ (pyrimidine) interactions and by bulky substituents in C_8 (purine) or C_6 (pyrimidine). In some crystal structures of (modified) nucleosides, the *high anti*(*-sc*) form with χ in the range $-100°$ is found. 5'-Nucleotides and polynucleotides adopt the *anti* conformation, but in case of guanine *syn* is equally possible.

4. In nucleosides, the orientation about the $C_{4'}$-$C_{5'}$ bond, described by γ, is preferentially $+sc$, but both $-sc$ and ap are sometimes found. In 5'-nucleotides, $+sc$ is by far predominant; it is stabilized by (base)C-H$\cdots$O$_{5'}$ hydrogen bonding which also influences the *syn/anti* orientation (Table 4-7).

5. The phosphate group displays a tetrahedral configuration with P-O bond lengths varying from 1.5 Å for P-O$^-$ to 1.6 Å for P-O$_{ester}$ because of different contributions of π bonding. This is due to phosphorus $3d_{x^2-y^2}$ and $3d_{z^2}$ orbital overlap with $2p$ orbitals of oxygen which also influences O-P-O angles, from 105° to 120°. In pyrophosphates, the P-O-P bonds are 1.6 Å and the angle is $\sim130°$, but bonding geometry is largely "normal" and the extra energy of the "energy-rich" pyrophosphates is released through phosphate-phosphate repulsion and favorable hydration of fragments.

6. Orientation of phosphate groups about C-O bonds is *ap* for $C_{5'}$-$O_{5'}$ (β) and slightly shifted *ap* to $-ac$ (about 220°) for $C_{3'}$-$O_{3'}$ (ϵ).

7. In nucleosides, all torsion angles except γ are correlated, and in polynucleotides γ is also involved. This means that structural changes follow a concerted motion.

8. Torsion angles α and ζ defining orientations about P-$O_{5'}$ and $O_{3'}$-P ester bonds are preferentially $+sc$ or $-sc$ due to the *gauche* effect. In single-stranded polynucleotides, P-O rotations α and ζ are greatly restricted to $-sc$ or $ap/-sc$ or to $-sc/ap$, and in base-paired helical structures only $-sc$ is allowed for both α and ζ.

9. Taking a nucleotide in standard ("rigid") conformation χ (*anti*), γ ($+sc$), β (*ap*, 180°), ϵ ($-ac$, 220°), and $C_{3'}$-*endo* or $C_{2'}$-*endo* sugar puckering, helical parameters n (number of nucleotides per turn) and h (axial rise per nucleotide) were calculated for varying α and ζ. The resulting $n-h$ plots indicate that only *one* helical domain exists with smooth left- and right-handed helices separated by a line $h = 0$ Å corresponding to a ladder-type structure. Small variations in α or ζ reverse

helical sense. Right-handed helices are preferred because of close intra- and interstrand contacts between neighboring nucleotides.

10. A virtual bond concept, considering polynucleotides as successive $P_{5'}\cdots C_{4'}$ and $C_{4'}\cdots P_{3'}$ units, can be used to predict structural properties of helical and randomly coiled polymers.

Chapter 5

Physical Properties of Nucleotides: Charge Densities, pK Values, Spectra, and Tautomerism

The physical features of nucleotides discussed in this chapter will lead to an understanding of some of the characteristic properties of these molecules, especially if base–base recognition through base-pairing and base–metal interactions are concerned. Both of these processes depend on charge densities, pK values, and tautomeric states of nucleotides and are of fundamental, functional importance in biological systems.

5.1 Charge Densities

Hydrogen bonding affinities are indicated by charge densities. Charge densities have been calculated for a number of nucleotides and their derivatives applying either Del Re's and Hückel's methods to obtain partitioned σ- and π-charge contributions (72) (Figure 5-1) or PCILO to produce total charges (70,71,180).

The charge density distributions indicated in Figure 5-1 suggest that in the bases, hydrogen atoms attached to amino groups in adenine, guanine, and cytosine have nearly equivalent positive σ charges of $+0.22e$ and therefore exhibit a weak acidity favoring them as donors in hydrogen bonds of the form

$$\overset{\delta-}{N}\!-\!\overset{\delta+}{H} \cdots \overset{\delta-}{O}$$

Donor   Acceptor

Similarly, the "amide-like" hydrogen atoms at uracil N_3 and at guanine N_1 have a positive σ charge of $+0.19e$, nearly equal to each other but significantly less than the charges at amino group hydrogens. On the other hand, bases can act as hydrogen bond acceptors, viz., N_1, N_3, and N_7 in adenine; N_3, N_7, and O_6 in guanine; O_2 and O_4 in uracil/thymine; N_3 and O_2 in cytosine. All these atoms display negative charges in the range

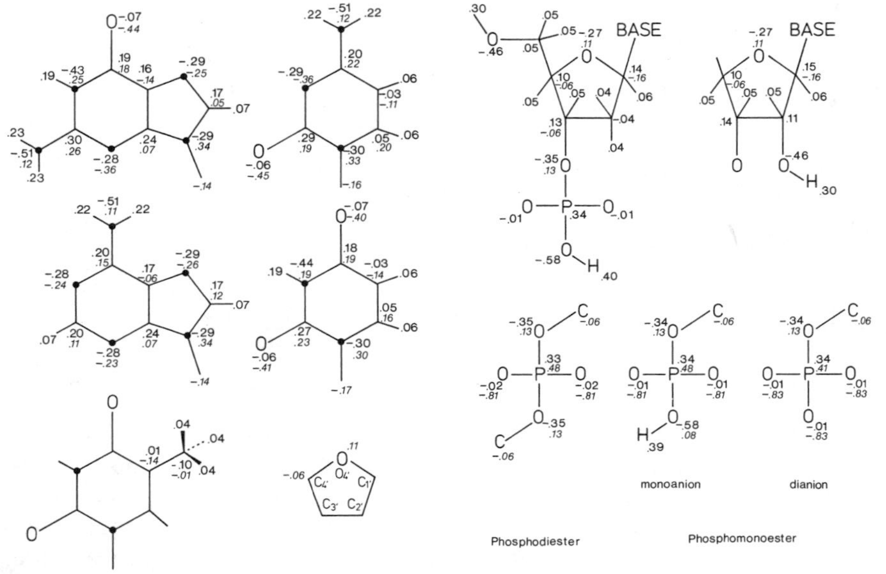

Figure 5-1. Charge density distributions in nucleotides, partitioned into Del Re's σ charges (Roman) and Hückel's π charges (italic). After (72).

$-0.47e$ (O$_2$ of thymine) to $-0.65e$ (N$_3$ of cytosine) which render them good acceptors for hydrogen bonds.

Figure 5-1 indicates that the π-charge contributions in the heterocyclic systems are also conducted through the adjacent furanose atoms $C_{1'}$, $O_{4'}$, $C_{4'}$, explaining in part the shortening of the $C_{1'}$–$O_{4'}$ bond relative to $C_{4'}$–$O_{4'}$ which is due to resonance of the lone electron pairs at $O_{4'}$ with the electronic system of the heterocycle. The $O_{4'}$ oxygen carries a total charge of only $-0.16e$, suggesting that it is a weak hydrogen bond acceptor and in fact hydrogen bonds to this atom have been observed in only a few nucleoside crystal structures. However, in case of favorable stereochemical juxtaposition of $O_{4'}$ and H–O groups, such hydrogen bonds appear feasible (see Section 4.3) (Figure 4-11).

The charge distribution within a phosphate group suggests that esterified oxygens with total charges around $-0.22e$ are only weak hydrogen bond acceptors whereas terminal oxygens carrying $-0.83e$ are good ones, in agreement with crystallographic observations.

The affinity of negatively charged N and O atoms of bases and of phosphate groups for hydrogen bond formation is largely paralleled by their ability to act as ligands for a great variety of metal ions. Because these complexes are not only of biological importance but also of stereochemical interest, they are discussed more fully in Chapter 8.

Charge densities give dipole moments. The detailed information on charge density distribution contained in Figure 5-1 can be cast into a sin-

Table 5-1. Calculated and Experimentally Determined Dipole Moments of Bases[a]

Compound	Dipole moment		Measured dipole (D)[b]
	Calculated		
	Magnitude (D)	Direction (°)	
Adenine	2.9	64°	3.0
Uracil	4.6	36°	3.9
Guanine	7.5	−31°	—
Cytosine	7.6	102°	—

[a] Directions of calculated dipoles are from + to −, with the angle to axis N_1–C_4 in pyrimidines and C_4–C_5 in purines counted counterclockwise if structural formulas are drawn such that atom C_2 is left with regard to the glycosyl link. [After (180).]
[b] For 9-methyladenine and 1,3-dimethyluracil.

gle vector for each individual base. The orientation and magnitude of this vector give the respective dipole moment μ. In general, the dipole moment lies within the molecular plane of the base and its orientation is indicated by an angle relative to a reference line in the molecule (Table 5-1). The theoretical data can be tested by measurements which yield the magnitude of the moment, but its orientation can only be derived if moments of a series of differently substituted compounds of the same molecule are at hand.

5.2 pK Values of Base, Sugar, and Phosphate Groups: Sites for Nucleophilic Attack

Protonation and deprotonation of bases occur within 3.5 pH units from neutrality, i.e., between pH 3.5 and 10.5 (Table 5-2). This indicates that amidelike hydrogens (at N_3 of uracil and N_1 of guanine) are rather acidic in character and ring nitrogens are basic, as might be inferred from the charge densities. In nucleotides, due to the electrostatic attraction of the negatively charged phosphate groups at different distances from the bases, the pK values of bases increase by about 0.5 unit when going from nucleoside to 3'-nucelotide to 5'-nucleotide (404).

At alkaline pH, hydrogens at N_3 of uracil/thymine and at N_1 of guanine are removed. At slightly acidic pH > 3, sites of protonation are N_1 of adenine and N_3 of cytosine. More acidic conditions, pH < 2, also protonate N_7 of guanine and of adenine (Figure 5-2) and O_4 of uracil (401–403,405–412). Crystallographic studies have shown that at sufficiently low pH (<2), N_9-substituted guanine is protonated at N_7 (413,414), adenine is double-protonated at N_1 and N_7 (415), and 1-methyluracil is protonated at

Table 5-2. pK_a Values for Bases in Nucleosides and in Nucleotides, Extrapolated to Zero Ionic Strength[a]

Compound/site of protonation or deprotonation	Nucleoside	3'-Phosphate	5'-Phosphate
Adenosine/N_1	3.52	3.70	3.88
Cytidine/N_3	4.17	4.43	4.54
Guanosine/N_1	9.42	9.84	10.00
Uridine/N_3	9.38	9.96	10.06
2'-Deoxythymidine/N_3	9.93	—	10.47

[a] Concentration of compounds, $5-10^{-5}$ M, 20°C. Because it is closer to the base, 5'-phosphate has more influence than 3'-phosphate. [From (401). For compilation of more data, see (402, 403).]

O_4 (416). In all these cationic molecules, bonding geometry is changed according to canonical resonance structures: if ring nitrogens are protonated, the corresponding C–N–C angles widen by about 5° (191) (see Section 4.5).

In all above mentioned cases, ring nitrogens are protonated rather than amino nitrogens (417). This is in agreement with the charge distributions given in Figure 5-1 and, moreover, quantum chemists have given an explanation for this behaviour. Thus, charge densities are calculated for isolated molecules in the ground state (418) while chemical reactions as protonation depend on charge distribution in the transition state, which is markedly different. A picture of the electrostatic isopotential curves of the four bases, calculated for the approach of a proton or of a nucleophilic alkylating agent, clearly demonstrates that the attack is primarily at ring nitrogens and at keto oxygens, not at amino groups (418).

Under the influence of stronger alkali, the secondary hydroxyls in riboses are deprotonated. The corresponding pK_a values of ~ 12.4 are lower than pK_a's of either primary or single, isolated secondary hydroxyls because the negatively charged C–O⁻ groups are stabilized by the presence of the vicinal ribose hydroxyl working via a combined inductive effect and hydrogen bonding (420,421) (Figure 5-2) [for a comparable observation in glucose hydroxyls, see (422,423)].

The phosphate groups in nucleoside-mono-, di-, and triphosphates display two main pK values (424,425). At low pH, around 1–2, either all negative charges at phosphate groups are neutralized by protonation or, in di- and triphosphates, they are neutralized in part, resulting in average charge/phosphate ratios of ½ in diphosphates and ⅓ or ⅔ in triphosphates. In the primary ionization state, one proton per each phosphate is removed, resulting in a single negative charge per phosphate group (a charge/phosphate ratio of 1.0). This ionization state prevails up to a pH around 7 where the last proton is removed from the terminal phosphate group in the secondary ionization step (Table 5-3). This protonation at

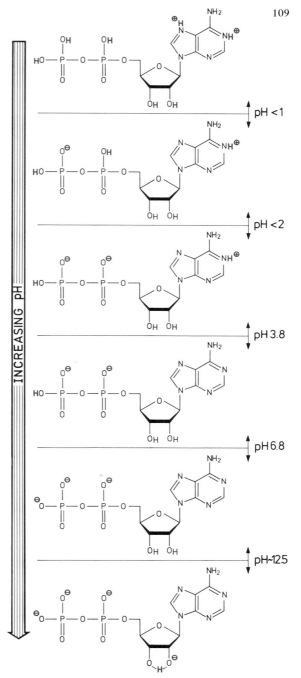

Figure 5-2. Sites of protonation in adenosine diphosphate (ADP). The adenine amino group is not protonated, even at low pH. Primary ionization of the phosphate groups is complete around pH 2, and above pH 3.8, adenine becomes neutral. At pH 6.8, secondary ionization of phosphates takes place and at pH > 12.5 secondary ribose hydroxyls are deprotonated.

Table 5-3. pK_a Values for Secondary Phosphate Ionization in Nucleoside Mono-, Di-, Triphosphates[a]

	Phosphate		
Compound	Mono	Di	Tri
Adenosine	6.67	7.20	7.68
Cytidine	6.62	7.18	7.65
Guanosine	6.66	7.19	7.65
Uridine	6.63	7.16	7.58

[a] The primary ionization occurs at pH 1–2 (Figure 5-2). [From (424).]

physiological pH does not occur with diesters like DNA and RNA or in the coenzymes NAD$^+$ and CDP-choline where only the primary ionization state is of interest.

pK_a values of bases can be directly influenced by chemical modification, as shown for uridine derivatives in Table 5-4 (402,410,426). The pK_a for 5-methyluridine, 9.7, is higher than that for uridine, 9.3. Replacement of keto- by thioketo groups reduces the pK_a to 7.4 in the case of 2,4-dithiouridine, and a similar effect is observed for 5-bromo substitution, pK_a 7.8. Because the pK_a values of some of these modified nucleosides are close to physiological pH, they can occur as ionized species and several tautomeric forms are feasible, as discussed in the next paragraph.

5.3 Tautomerism of Bases

Heterocyclic molecules in solution frequently yield a mixed population of species in rapid equilibrium when hydrogen atoms attached to nitrogens are able to migrate to other, free nitrogens or to keto oxygens within the same molecule (198,427). This kind of tautomerism involving prototropic change depends largely on the dielectric constant of the solvent and

Table 5-4. Some Substituted Uracil Derivatives and their pK-Values.

Compound name	Chemical structure	pK
5,6-Dihydrouracil	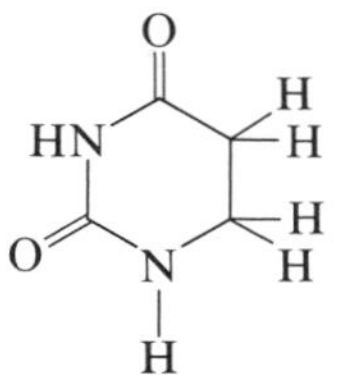	>11

Table 5-4. (*Continued*)

Compound name	Chemical structure	pK
5-Methyluridine		9.7
Uridine		9.3
2-Thiouridine		8.8
4-Thiouridine		8.2
1-Methyl-5-bromouracil		7.8
2,4-Dithiouridine		7.4
1-Methyl-6-azauracil		6.9

R = Ribose or methyl.

on the pK of the respective hetero atoms and has been the subject of numerous investigations (428,429).

The heterocyclic systems in nucleosides are also liable to tautomeric changes. In unsubstituted bases, a prototropic migration from one ring nitrogen to another can take place, involving $N_7-H \rightleftharpoons N_9-H$ in purine and $N_1-H \rightleftharpoons N_3-H$ in pyrimidine residues (Figure 5-3). This type of isomerization will not be treated here further because it requires unsubstituted purine N_9 or pyrimidine N_1 and therefore cannot occur in nucleosides. With nucleosides we deal with the keto $\rightleftharpoons$ enol and amino $\rightleftharpoons$ imino transitions (Figure 5-4). It is striking to see that a change in hydrogen atom position leads to new hydrogen-bonding characteristics of the base: a keto group with acceptor properties transforms as an enol into a donor, and a donating amino group becomes an accepting imino substituent. However, if the hydrogen of the latter is rotated, the imino group can also act as hydrogen bond donor—a remarkable situation.

By tautomeric changes uracil and guanine in the enol form simulate cytosine and adenine, and cytosine and adenine in the imino form may substitute for uracil and guanine (Figure 5-4). Since this kind of metamorphosis of base properties would be disastrous for a self-replicating system relying on the base–base recognition proposed by Watson and Crick (7,8) (Chapter 6), tautomerism has been scrutinized by UV, IR, NMR, crystallo-

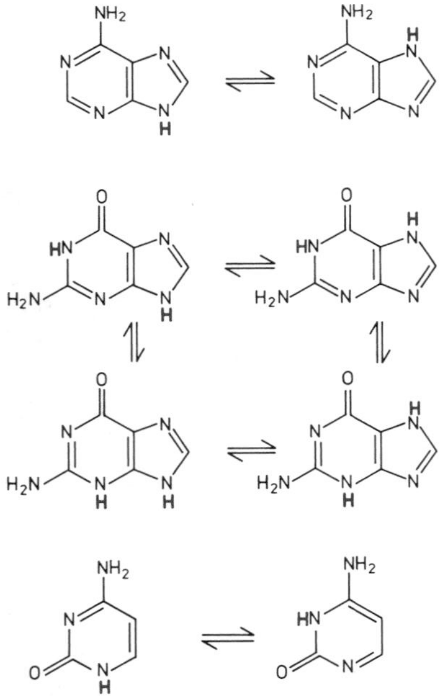

Figure 5-3. Prototropic tautomerism in unsubstituted adenine, guanine, and cytosine.

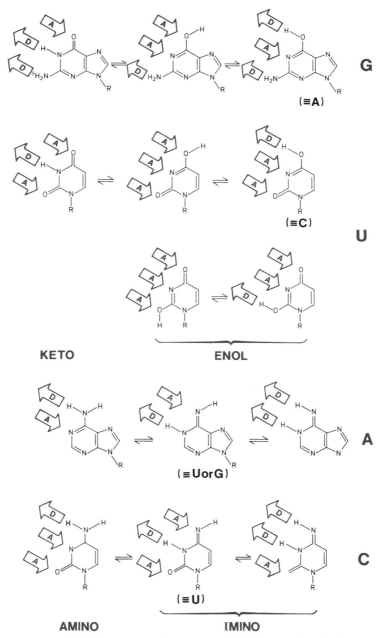

Figure 5-4. Keto⇌enol and amino⇌imino tautomerism in nucleoside bases. Arrows denoted A and D symbolize acceptor and donor sites for hydrogen bond formation. Note that in the enol forms, G becomes equivalent to A, and U to C; in the imino forms, A is equaivalent to U or G, and C to U. The situation changes, however, if the imino (=N—H) or enol (O-H) groups rotate, giving rise to a diversity of hydrogen-bonding possibilities. As discussed in the text, it is comforting to know that only keto and amino forms occur to >99.99% under physiological conditions.

graphic methods [summarized in (429); for newer work see (430–433)] and quantum chemical methods (179,180).

The essential results can be summarized with a few sentences. The main tautomeric forms of the naturally occurring bases are as displayed in Figures 2-1 and 2-2; i.e., amino and keto configurations dominate, with less than a 0.01% population estimated for the imino and enol tautomers in adenosine (434) and in 1-methyluracil (410). This holds for a variety of solvents with different dielectric constants and even in the vapor state (435). Earlier NMR investigations on cytosine and guanine derivatives which traced about 15% of the rare imino and enol forms later proved erroneous (436,437). Quantum chemical calculations are not adequate to furnish details on tautomeric states because they are too sensitive to uncertain O–H bond lengths (438). Also, crystallographic studies interpreted in terms of tautomerism should be met with reservation because C–O bond length differences (in uracil) should exceed the 3σ level to be significant (439). In addition, direct coordination of O_2 keto oxygen to cations like K^+ might influence the C–O bond length and artificially induce a trend toward enolization (440).

In general, we can state safely that the naturally occurring bases display predominantly keto and amino tautomeric forms and are therefore suitable for their task of recognizing the respective base-pair partner, in harmony with the requirements demanded for a reproductive genetic apparatus. A few remarks on mutation rates caused by mismatching due to tautomerism are given in Chapter 6.

The picture changes somewhat if chemically modified bases are considered and it becomes clear why nature has selected the four normal bases as primary tools. Naturally occurring isoguanosine, a structural isomer of guanosine (Figure 5-5) (441,442), is not incorporated into DNA or RNA although it could, on a purely functional and structural basis, form a base-pair with the synthetically available isocytidine and replace the common analog. The reason for the discrimination is found in the extraordinary keto–enol tautomerism of isoguanosine or correspondingly substituted isoguanine which is largely dependent on solvent polarity and on temperature (443).

Other chemical modifications such as replacement of keto by thioketo groups in position 6 of purine nucleosides increase the population of enol tautomeric forms to about 7% (444). In contrast, in 1-substituted 4-thiouracil, only the keto tautomeric form could be detected (445). In 5-halogenated uridine derivatives, introduction of fluorine apparently does not reduce the dominating population of the keto form although the pK_a value, 7.4, is close to physiological conditions (446). It should be anticipated that bromo substitution, with a similar effect on pK_a (7.8) would also not influence the keto–enol tautomerism but in fact a tenfold increase in the enol form is found, still a relatively low amount, about 0.1 to 1% in comparison with the keto form (410). However, it is not necessary to modify the base in order to increase the amount of imino or enol tauto-

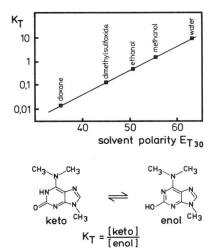

Figure 5-5. Dependence of keto⇌enol equilibrium, expressed by constant K_T, on the solvent in N^6,N^6,N^9-trimethylisoguanine. Polarity parameters $E_{T_{30}}$ as defined in (447), K_T on a logarithmic scale. The strong influence of the solvent on the keto⇌enol equilibrium prohibits the use of isoguanine as a suitable base in nucleic acid chemistry. After (443).

meric forms. As described in Section 6.1, NMR data on tRNA suggest that upon base-pair formation, about 5 to 15% of the bases occur as rare tautomers because hydrogens involved in hydrogen bonding jump in concerted motion.

Summary

Charge densities of atoms in nucleotides explain most of their interactive behavior with other molecules. Thus, base amino groups bear hydrogen atoms with partial positive charge and are good hydrogen bond donors, whereas base keto oxygens and nitrogens are good acceptors and the sugar ring oxygen accepts only weakly. Similarly, phosphate ester oxygens as hydrogen bond acceptors are less important and excelled by free oxygens with partial negative charge. pK values of bases and phosphate groups reflect only in some cases charge densities because basicity rather than electronegativity comes into play. Thus, at low pH ring nitrogens of bases become protonated rather than amino groups and phosphate-monoesters display two main pK values, one around pH 1–2, the other around pH 7.

In principle, bases in solution could occur in several tautomeric forms. The normally accepted keto and amino forms, however, predominate and enol or imino forms exist in less than 0.1% range, a figure that can change drastically if chemically modified bases are considered or if base-pairs are formed.

Chapter 6

Forces Stabilizing Associations Between Bases: Hydrogen Bonding and Base Stacking

Before describing structural features of nucleotides and their oligo- and polymeric complexes, a few remarks about base–base interactions are in order. These interactions are of two kinds: (a) those in the plane of the bases (horizontal) due to hydrogen bonding and (b) those perpendicular to the base planes (base stacking) stabilized mainly by London dispersion forces and hydrophobic effects. Hydrogen bonding is most pronounced in nonpolar solvents where base stacking is negligible, and base stacking dominates in water where base–base hydrogen bonding is greatly suppressed due to competition of binding sites by water molecules. Both are individually accessible to measurement and have been investigated in detail, especially hydrogen bonding because it is fundamental to the genetic code. For reviews see Refs. (448,449).

6.1 Characterization of Hydrogen Bonds

Hydrogen bonds are mainly electrostatic in character. They play a key role in the stabilization of protein and nucleic acid secondary structure and have been the topic of several monographs and review articles (192,198,450–452,455,456). Therefore, in this discussion only some characteristics relevant to base–base hydrogen bonding will be described.

In general, a hydrogen bond

$$X{-}H \cdots Y$$

is formed if a hydrogen atom H connects two atoms X,Y of higher electronegativity. As hydrogen bonds are largely electrostatic in nature, their strength (reflected in the length of the H $\cdots$ Y distance) depends on (partial) charges located on X,H,Y. Hydrogen bonds with X = carbon and Y = oxygen have already been alluded to in Section 4.9, for charge densities see Figure 5-1.

Hydrogen-bonding interactions between bases are of the type N–

Table 6-1. Comparison of Some Energy Values in Covalent and Hydrogen Bonds

Bond type	Bond length Å	Bond energy kcal/mole	Energy required for lengthening by 0.1 Å kcal/mole
Covalent			
C–C	1.54 ± 0.02	83.1	3.25
C–H (in ethane)	1.09 ± 0.02[a]	98.8[b]	3.60[c]
Hydrogen bond			
O–H $\cdots$ O	2.75 ± 0.2[d]	3 to 6[c]	0.1[b]
	(O $\cdots$ O distance)		

[a] From (197).
[b] From (198).
[c] From (455).
[d] From (456).

H $\cdots$ N and N–H $\cdots$ O with the donor N–H group of either the amino or imino type. In some modified bases containing thioketo groups, N–H $\cdots$ S hydrogen bonds also occur, although in general sulfur is thought to be a weaker hydrogen bond acceptor than oxygen (453,454).

Hydrogen bonds are "soft" and only weakly directional. Compared with covalent bonds with well-defined length, strength, and orientation, hydrogen bonds are about 20 to 30 times weaker (Table 6-1). Therefore they are more susceptible to bending and stretching, and this results in variable geometries for the X–H $\cdots$ Y system (Table 6-2). In some extreme cases with long H $\cdots$ Y distances, the usual criterion that H $\cdots$ Y should be shorter than the sum of the van der Waals radii might not be sufficient. In these instances, Allinger's van der Waals' potential minimum contact radii (Table 3-1) can instead be used to relax the criterion considerably (158, 451). Relatively "long," weak hydrogen bonds have been observed when two acceptors Y_1, Y_2 compete for the same hydrogen atom in bifurcated or "three-centered" systems (321,451):

Additivity and cooperativity of hydrogen bonds. Under the influence of a hydrogen bond, the charges on the atoms involved are modified due to polarization, H becoming more electropositive and X,Y more negative. This effect leads to increased affinity of X,Y for accepting further hydrogen bonds. If Y is the oxygen atom of a hydroxyl group, the hydrogen attached to it will also be affected by polarization and becomes a better donor. Since this cooperativity (451) holds in general for bifunctional X–H $\cdots$ X–H $\cdots$ X–H $\cdots$ systems where each donor plays simultaneously the role of an acceptor and vice versa, it can involve nucleoside

Table 6-2. Some Geometrical Characteristics of Hydrogen Bonds Involved in Base-Base Interactions [From (192, 321, 450, 453, 454)]

System	Distances or angles		
	Minimum	Maximum	Mean
N–H · · · N			
N · · · N	2.75	3.15	2.90
N–H	0.86	1.02	0.95
H · · · N	1.78	2.02	1.99
N⟨H⟩N	2°	17°	9°
N–H · · · O			
N · · · O	2.74	3.07	2.95
N–H	0.83	1.06	0.95
H · · · O	1.83	2.17	1.95
N⟨H⟩O	3°	23°	9°
O–H · · · O			
O · · · O	2.60	3.05	2.73
O–H	0.68	1.16	0.86
H · · · O	1.74	2.20	1.95
O⟨H⟩O	1°	20°	8°
N–N · · · S			
N · · · S	3.25	3.55	3.32
N–H	0.84	1.04	0.95
H · · · S	2.27	2.57	2.40
N⟨H⟩S	1°	25°	15°

hydroxyls as well as bases, because these display a formally analogous electronic structure:

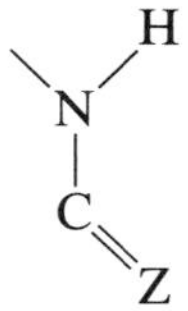

Here, Z represents (unsubstituted) ring nitrogen or keto oxygen. The equivalence to a simple X–H system becomes clear if electrons are delocalized:

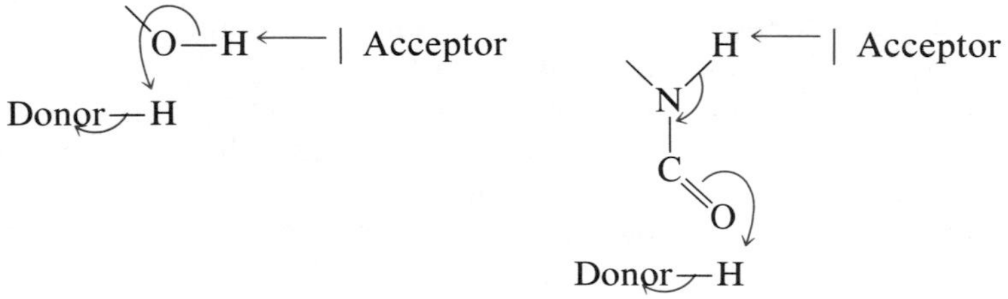

Besides geometrical factors involved, this scheme suggests that bases form preferentially complexes with at least two (cyclic) hydrogen bonds. We should even expect that under the influence of the cooperative effect, hydrogens in base-pairs can jump in concerted mechanism from the donor in one base to the acceptor on the partner base:

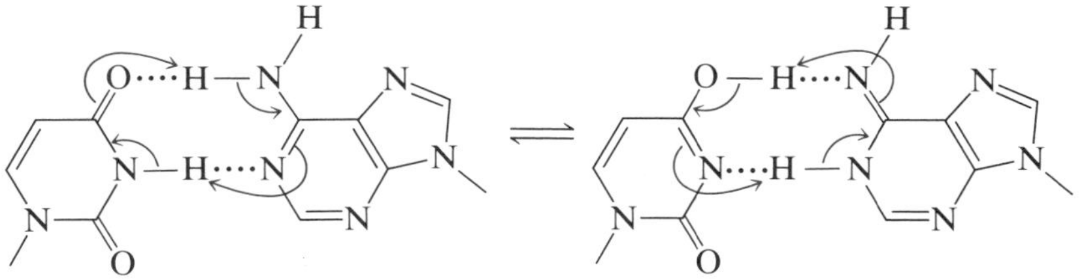

The hydrogen transfer would be only of minor importance on base-pair geometry and, above all, it would not disturb Watson-Crick type recognition because complementarity within the base-pair is obeyed at all times. A recent NMR study on tRNA in aqueous solution has in fact estimated that 5 to 15% of all base-pairs are in such imino/enol form at a given time, with rate constants as slow as 100 to 300 per second (456a). However, the spectra can also be explained in other ways (anisotropy effects) and it appears more likely that hydrogen atoms remain in their amino/keto positions and do not jump.

6.2 Patterns of Base–Base Hydrogen Bonding: The Symmetry of a Polynucleotide Complex

Interactions between like (homo) and different (hetero) bases have been observed in crystal structure analyses of individual bases, nucleosides, or nucleotides and of complexes formed by two or more different compounds of this type [reviewed in (192)]. Under the assumption that at least two "cyclic" N–H $\cdots$ O or N–H $\cdots$ N hydrogen bonds must form in order to produce a stable base-pair, the four bases substituted at the glycosyl nitrogens (N_1 in pyrimidine and N_9 in purine) can be arranged in 28 different ways (Figure 6-1) (33,457).

Twenty-eight base-pairs with dyad, pseudodyad, and no symmetry. The 28 base-pairs are grouped in Figure 6-1 according to interactions between like and different bases in the purine–purine and pyrimidine–pyrimidine series, followed by purine–pyrimidine pairs. In each group, the orientations of glycosyl $C_{1'}$–N linkages can be either unrelated by symmetry elements (asymmetric) or related by dyads (twofold axes) located perpendicular to or within the base-pair planes as indicated by symbols ● and ↑ (Figure 6-2). Note that the dyads in the homopurine and homopyrimidine base-pairs, I to IV and XII to XV, transform one base exactly into the

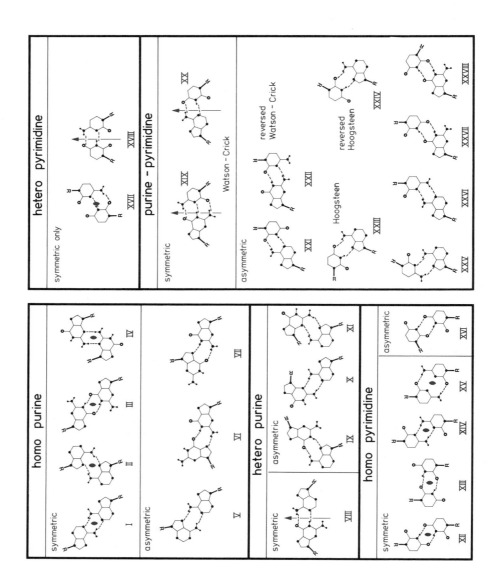

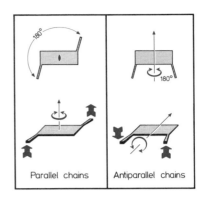

| Parallel chains | Antiparallel chains |

Figure 6-2. Base-pairs can display two kinds of twofold symmetry, depending on the orientations of the glycosyl $C_{1'}$–N bonds. In this illustration, the latter are indicated as sticks protruding from rectangular plates representing base-pair planes. Twofold rotation axes (dyads) can be arranged either perpendicular to (left) or within the base-pair planes (right); this determines the orientation of the attached sugar–phosphate groups or backbones. In a double helix, the two strands are parallel to each other if the dyad is perpendicular to the base-pair (and therefore coinciding with the helix axis; left) and they are running in opposite direction, antiparallel, if dyads are within the base-pair plane and consequently perpendicular to the axis of the double helix (right). The Watson–Crick type DNA or RNA corresponds to this latter arrangement; see also Figures 6-3(a) and (b).

other. However, in heteropurine and heteropyrimidine (VIII and XVII, XVIII) and in the purine–pyrimidine Watson–Crick base-pairs (XIX and XX), the operation of the dyad is restricted to the glycosyl $C_{1'}$–N linkage and not applicable to the bases. It is therefore called pseudo-dyad.

Symmetry elements are also found in polynucleotide complexes. The symmetry elements are of special interest because in polynucleotide complexes (Chapters 9–13) they relate not only the glycosyl bonds but also the attached sugar–phosphate backbones. As a consequence, base-pairs with dyads *perpendicular* to the base planes (symbol ●) direct the $3' \rightarrow 5'$ orientations of the backbones parallel and identical to each other, whereas a dyad *within* a base-pair (symbol ↑) gives rise to antiparallel

Figure 6-1. The 28 possible base-pairs for A, G, U(T), and C involving at least two (cyclic) hydrogen bonds. Hydrogen and nitrogen atoms displayed as small and large filled circles, oxygen atoms as open circles, glycosyl bonds as thick lines with R indicating ribose $C_{1'}$ atom. Base-pairs are boxed according to composition and symmetry, consisting of only purine, only pyrimidine, or mixed purine/pyrimidine pairs and asymmetric or symmetric base-pairs. Symmetry elements ● and ↑ are twofold rotation axes vertical to and within the plane of the paper (see Figure 6-2). In the Watson–Crick base-pairs XIX and XX and in base-pairs VIII and XVIII, pseudosymmetry relating only glycosyl links but not individual base atoms is observed. Drawn after compilations in (33,457).

orientations of identical backbone structures; see hands in Figure 6-2. An example of the parallel case is poly (AH$^+$)]$_2$ with base-pair II of Figure 6-1 (458) and examples of antiparallel structures are DNA, RNA, and poly (A) · poly(U) all of which display Watson–Crick-type base-pairs (8). A special case is encountered with poly(U) and the 2-thioketo derivative poly(s^{2}U) both of which form antiparallel double helices with the asymmetric base-pair XVI. As a result, the two sugar–phosphate backbones are not identical to each other but have different conformations (459–461) (Chapter 13).

The term "symmetry related" applies in a strict sense only to crystalline polynucleotide complexes. In these, the space group dyads coincide with dyads relating glycosyl links (Figure 6-2). Therefore, space group constraints must be obeyed and the molecule is subject to crystallographic law and order. However, if the material is dissolved and escapes the limiting lattice, the overall, gross symmetry properties will be largely retained but they break down in crystallographic terms. A good example is presented by crystalline fibrous DNA which in the B form is pierced by a space group dyad and repeats exactly after 10 base-pairs (= one turn or one c-axial length). In solution and when DNA is adsorbed on a flat surface, however, repeat lengths of 10.4 ± 0.1 and 10.6 ± 0.1 base-pairs are observed (462,463). Moreover in a crystalline, double-helical DNA dodecanucleotide no space group dyad is colinear with a base-pair dyad; the overall repeat is 10.3 to 10.4 base-pairs (464), and for the central 6 base-pairs repeat lengths of about 9.8 nucleotides are observed (Chapter 10).

6.3 Detailed Geometries of Watson–Crick and Hoogsteen Base-Pairs

It is worthwhile to look at the Watson–Crick and Hoogsteen base-pairs more closely (Figures 6-3 and 6-4). The data entered in Figure 6-3(a) were obtained from the X-ray crystal structure analyses of ApU (465) and GpC (466) which located all second row atoms of the base-pairs but not hydrogens; the hydrogen atoms are shown in positions calculated from the C,

---

Figure 6-3. (a) Watson–Crick base-pairs observed in crystal structures of GpC (top) and ApU (bottom). Hydrogen atoms were not located experimentally but are calculated from the positions of the other atoms. Note differences in hydrogen bond lengths N–H $\cdots$ O, from 2.86 to 2.95 Å, and N–H $\cdots$ N, from 2.82 to 2.95 Å, consistent with the spread given in Table 6-1 and reflecting the "softness" of this type of interaction. The distances between glycosyl $C_{1'}$ atoms, 10.46 Å and 10.67 Å, are remarkably similar and are the basis, together with the almost coinciding angles, $C_{1'} \cdots C_{1'}$–N, around 53°, of the observed geometrical isomorphism. From (465,466). **(b)** Schematic description of isomorphism and pseudo-symmetry in Watson-Crick base-pairs.

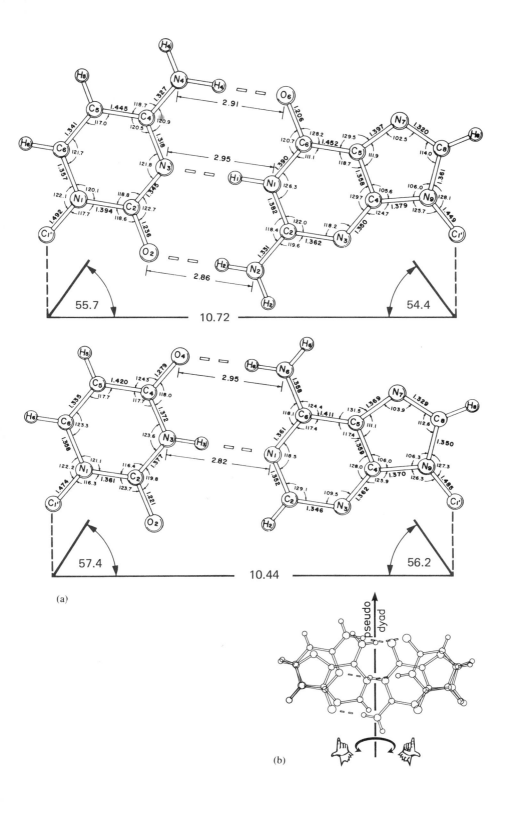

(a)

(b)

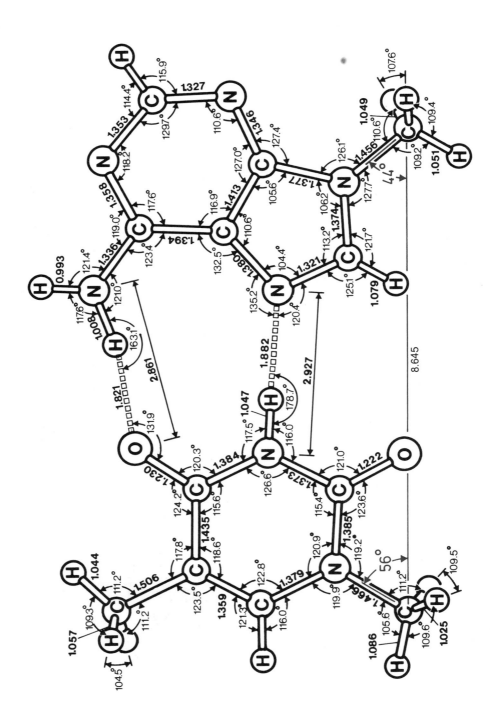

N, and O atoms. In Figure 6-4 the 9-ethyladenine:1-methylthymine Hoogsteen pair as derived from an accurate neutron diffraction study is displayed, with all hydrogen positions experimentally determined.

Geometrical isomorphism and propeller twist. In the A–U and G–C Watson–Crick base-pairs, the structural isomorphism is noteworthy; i.e., the overall shapes and dimensions are comparable. The $C_{1'} \cdots C_{1'}$ distances are ~0.3 Å smaller for A:U than for G:C and the angles which these lines form with the glycosyl $C_{1'}-N$ bonds are between 54.4° and 57.4°. Neglecting the small difference of 3°, we can say that the glycosyl links are related by a pseudodyad (Figure 6-3(b)).

The two bases in the A:U and G:C pairs are not coplanar. They are twisted about the hydrogen bonds like the blades of a propeller. This twist is about 12° in A:U and 7° in G:C. Comparable twists have also been observed in other base-pairs (192).

In contrast, the bases in the Hoogsteen pair (Figure 6-4) (467) are perfectly coplanar because they are located on a crystallographic mirror plane. In other crystal structures of Hoogsteen A:U base-pairs, however, where such crystallographical constraint is absent, a propeller-like twist similar to the Watson–Crick base-pairs has been observed, with 9° in 9-ethyladenine:1-methyl-5-iodouracil (468). Compared with the Watson–Crick base-pair, in the Hoogsteen pair the $C_{1'} \cdots C_{1'}$ distance, 8.645 Å, is reduced by about 2 Å and the angles between this line and the glycosyl $C_{1'}-N$ bonds differ by more than 10° between purine and pyrimidine bases; i.e., a pseudodyad is not present.

A structural disorder: Hoogsteen and reversed Hoogsteen base-pairs. As discovered in the neutron study (467) and mentioned by Hoogsteen in his original paper (469), there is some disorder in the 1-methylthymine:9-methyladenine crystal structure, with 10–13% of the 1-methylthymine molecules rotated 180° about the C_6-N_3 axis. This simultaneous occurrence of Hoogsteen and reversed Hoogsteen base-pairs (XXIII and XXIV in Figure 6-1) can be explained by the symmetrical shape of 1-methylthymine. The rotation leads to substitution of O_4 by O_2, maintaining the hydrogen-bonding scheme. A reversal of this behavior is found in the complex 1-methyl-5-bromouracil:9-ethyladenine (with the 5-bromo group substituting for the 5-methyl in thymine) where only 6% of the 1-methyl-5-bromouracil molecules are in the Hoogsteen base-pairing scheme and 94% are in the reversed Hoogsteen mode (470). It could be argued that the altered electronic structure in 5-bromouracil relative to thymine determines the configuration of these base-pairs. However, the crystal struc-

Figure 6-4. Hoogsteen base-pair formed by 9-ethyladenine:1-methylthymine. Data obtained from a neutron study which experimentally located hydrogen atoms (467). Note differences in $C_{1'} \cdots C_{1'}-N$ angles, 56° and 44°, and the relatively short $C_{1'} \cdots C_{1'}$ separation of 8.465 Å. NH $\cdots$ O and NH $\cdots$ N hydrogen bonds are indicated by broken lines.

ture of 1-methyl-5-fluorouracil:9-ethyladenine which features a Hoogsteen base-pair suggests that crystal packing and base stacking effects may be the dominant factors (471).

6.4 The Stability and Formation of Base Pairs as Determined by Thermodynamic, Kinetic, and Quantum Chemical Methods: Electronic Complementarity

Base–base interaction through hydrogen bonding has been investigated by thermodynamic methods in order to derive association constants as well as enthalpies and entropies for this process. Complex formation in water is relatively weak due to competition of water molecules for hydrogen bond donor and acceptor sites, and therefore, only qualitative data can be obtained. For the 5'-nucleotides, NMR downfield shifts of amino group signals gave the following sequence of decreasing association (base-pair formation) tendencies (204):

GMP with CMP > UMP > IMP >> AMP,
AMP with UMP ~ CMP >> IMP, GMP,
CMP with GMP ~ CMP > UMP > XMP, AMP, IMP.

For more quantitative data, apolar solvents must be used where hydrogen bonding is pronounced and where stacking interactions between bases are negligible (204). Several studies were carried out using bases substituted at the glycosyl nitrogen and solvents such as tetrachloromethane, chloroform, and dimethylsulfoxide. Base–base association has been monitored as change in amino group signals by both NMR and IR methods (Figure 6-5) (472,473), and has been studied by calorimetric and osmometric methods.

In solution, a mixture of Watson–Crick and Hoogsteen base-pairs are formed with at least two hydrogen bonds and involving all potential binding sites. The data summarized in Figure 6-6 show that the association constants depend greatly on the chemical nature of the two partners. In the 9-ethyladenine:1-cyclohexyluracil series, the 3-methyluracil derivative does not form a complex with adenine. This indicates that, for dimer association, one hydrogen bond between the adenine amino group and uracil carbonyl oxygens is not sufficient; at least two hydrogen bonds (a cyclic dimer) are required. It is obvious that the association constants depend on the acidity of the uracil imino hydrogen (Figure 5-3), low acidity as in 5,6-dihydro-1-cyclohexyluracil ($pK = 11$) correlating with weak association, and high acidity as in the 5-bromo- derivative ($pK = 7.8$) with strong association. In the 4-thiouracil compound ($pK = 7.4$), the acidity effect is counterbalanced by the weak hydrogen bond acceptor properties of the sulfur atom.

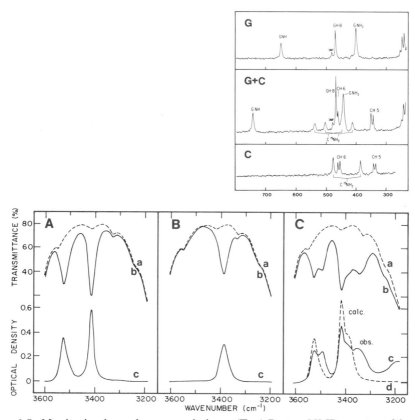

Figure 6-5. Monitoring base–base associations. (Top) Proton NMR spectra of 9-methylguanine (G), 1-[7-¹⁵N]methylcytosine (C), and their 1:1 mixture (G + C) in deuterated dimethylsulfoxide/dimethylformamide. ¹⁵N-Substituted cytosine was used in order to display the C–¹⁵NH₂ doublet which otherwise would be obscured by signals from aromatic protons. Recorded with a Varian A-60 spectrometer; abscissa given in cps (cycles per second) as downfield shifts from internal TMS (tetramethylsilane) standard. From (472). (Bottom) Infrared absorption spectra of 2,6-diamino-9-ethylpurine (A), 1-cyclohexyl-5-bromouracil (B), and their 1:1 mixture (C), all at 0.002 M in deuterochloroform. In the upper part of this picture, absorption spectra of pure solvent (a) and of solutions (b) are recorded. In the lower part difference spectra (a − b) are plotted as optical density (c). Dashed curve (d) gives calculated sum of the optical density curves in A(c) and B(c); difference with measured curve C(c) directly indicates presence of hydrogen-bonded complexes. From (473).

In the series with 9-ethyladenine derivatives (Figure 6-6), compare the association constants for complex formation of 1-cyclohexyluracil with those for 9-ethyl-6-methylaminopurine, 9-ethyl-2-aminopurine, and 9-ethyladenine. The latter (100 liters/mole) is twice that of the former two (50 and 45 liters/mole), suggesting that adenine binds uracil derivatives in

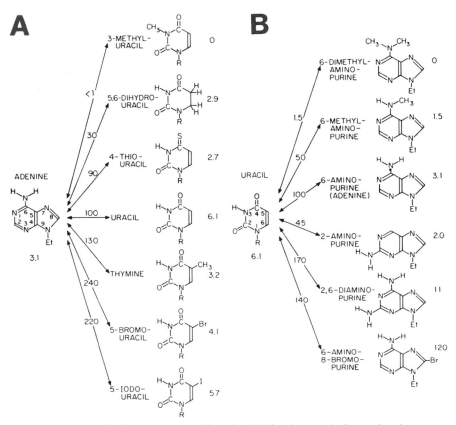

Figure 6-6. Association constants (liters/mole) for base-pair formation between various substituted 1-cyclohexyluracil and 9-ethyladenine derivatives, measured in deuterochloroform solution at 25°C using infrared methods displayed in Figure 6-5. Numbers near arrows are constants for association between adenine and uracil derivatives whereas figures shown close to structural formulas are self-association constants. (A) Association of 9-ethyladenine with uracil derivatives. (B) Association of 1-cyclohexyluracil with adenine derivatives. From (473).

Watson–Crick and Hoogsteen mode (base-pairs XX and XXIII in Figure 6-1) and thus has a statistical advantage of two binding sites with respect to the other two compounds. In all cases, Watson–Crick and reversed Watson–Crick, Hoogsteen and reversed Hoogsteen base-pairs cannot be differentiated so that all data for A : U association refer to a combination of base-pairs.

Hydrogen bonding to O_2/O_4 in uracil is determined by electronegativity. The mixture of A : U base-pairs has been studied by ^{13}C NMR methods looking at signals assigned to C_2- and C_4-carbonyl carbons in 1-cyclo-hexyluracil derivatives when complexed with 9-ethyladenine in chloroform, i.e., in a system identical to that above (Figure 6-6). For A : U association, the frequency of O_4 binding decreases and that of O_2 binding increases in the order (474)

thymine > uracil > 5-bromouracil > 4-thiouracil.

This sequence parallels the electronegativity of the O_4 oxygen in these compounds; the 5-methyl group in thymine pushes electrons and provides more electronegativity at O_4 whereas the 5-bromo group withdraws electrons and has an opposite effect. This latter observation could explain why incorporation of 5-bromouridine into DNA leads to mutations via mispairing. If O_2 is more likely to be involved in hydrogen bonding than O_4 (in contrast to thymidine), then it is quite obvious that G:5-BrU pairing (base-pair XXVIII in Figure 6-1) can form without involving improbable tautomeric forms of 5-bromouracil to simulate a G:C geometry (475).

At high concentrations, oligomeric base multiplets can form. Under usual experimental conditions, A:U complexes are dimers and not trimers or higher oligomers and the same holds for the self-association, U:U and A:A. This is clear from an IR band observed at 3490 cm^{-1} which is attributed to a free amino N–H hydrogen in adenine and which should disappear if both Watson–Crick and Hoogsteen binding sites are simultaneously occupied (Figure 6-7). Such higher complexes do indeed occur at elevated concentrations (476) and they are of importance in trimer formation of one poly(A) with two poly(U) chains, poly(A) · 2poly(U); see Chapter 10.

Thermodynamic data describing formation of homo- and heterocomplexes of the four bases are summarized in Table 6-3 (476–480). It is evident that solvent effects have a strong influence on binding constants and enthalpies, and that in general the self-associates are less stable than the complementary A:U and G:C base–pairs. The loss in entropy (− 11 to − 16 e.u.), however, is comparable in all cases, and this suggests that the overall structures of the associates are similar, dimeric complexes stabilized by at least two (cyclic) hydrogen bonds. If the enthalpy of base-pair-

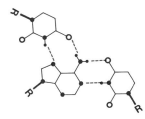

Figure 6-7. In the base trimer A:2U, both Watson–Crick and Hoogsteen base-pairing sites of A are engaged simultaneously. ●, nitrogen; ○, oxygen; the glycosyl link is indicated by the solid line. In this illustration, only normal Watson–Crick and Hoogsteen pairs are drawn as formed in the poly(A):2 poly(U) triple helix, (Section 10.2). If these as well as reversed base-pairs are utilized four different schemes emerge.

Table 6-3. Thermodynamic Data for Complementary and Self-pairing of Bases via Hydrogen Bonding

Compounds, solvents, and methods	Reaction	$K_{assoc.}$ [l/mole]	$-\Delta H$ [kcal/mole]	$-\Delta S$ [e.u.]
Guanosine, Cytidine in	$G+G \rightleftharpoons G_2$	0.18	1.0 ± 1	
dimethylsulfoxide	$C+C \rightleftharpoons C_2$	0.1	1.7 ± 1.5	
NMR, 32°C[a]	$G+C \rightleftharpoons GC$	3.7 ± 0.6	5.8	16
9-Ethyladenine,	$A+A \rightleftharpoons A_2$	3.1 ± 0.3	4.0 ± 0.8	11.4 ± 2
1-Cyclohexyluracil	$U+U \rightleftharpoons U_2$	6.1 ± 0.6	4.3 ± 0.4	11.0 ± 1
in chloroform IR, 25°C[b]	$A+U \rightleftharpoons AU$	103 ± 36	6.2 ± 0.6	11.8 ± 1.2
1-Methylpyrimidine,	$A+U \rightleftharpoons AU$		14.5	
9-Methylpurine in	$G+C \rightleftharpoons GC$		21.0	
vacuum mass spectrum[c]	$U+U \rightleftharpoons U_2$		9.5	
	$C+C \rightleftharpoons C_2$		16.0	

[a] From (477).
[b] From (478).
[c] From (480).

ing, around 6 kcal/mole dimer, can be attributed to only hydrogen-bonding effects, then 2 to 3 kcal/mole is released per mole hydrogen bond formed, a value within the normally accepted range for such interactions (Table 6-1). The "vacuum" data, however, obtained for the gaseous state indicate enthalpies of around 7 kcal/mole per hydrogen bond (480).

Rates of base–base association are diffusion controlled. Using ultrasonic attenuation measurements, the kinetics of base–base association have been determined for the system 9-ethyladenine:1-cyclohexyluracil in chloroform. Rates of duplex formation for both homo- and heterocomplexes are practically diffusion controlled and are similar in all three cases (481) (Table 6-4). The dissociation rates, however, are slower for A:U than for A:A and U:U, indicating that the relative strengths of the base-pair hydrogen bonds decrease A:U > A:A ~ U:U, as already suggested by the respective association constants.

Electronic complementarity is of major importance for specific base–

Table 6-4. Rate Constants for Association and Dissociation of 9-Ethyladenine and 1-Cyclohexyluracil in Chloroform [From (481)]

Reaction	Rate constant		Temperature (°C)	
	Association $(mole^{-1} \cdot sec^{-1})$	Dissociation (sec^{-1})		
$U+U \rightleftharpoons U_2$	1.5×10^9	25×10^7	25	
$A+A \rightleftharpoons A_2$	$\geq 2 \times 10^9$	$\geq 60 \times 10^7$	25	
$A+U \rightleftharpoons AU$	4.0×10^9	3.2×10^7	20	$\Delta H = -6.1$ kcal/mole

base interactions. This has been demonstrated by quantum chemical studies (482–485) which are summarized for a few selected base-pair combinations in Table 6-5. The separation of the total hydrogen-bonding interaction energy E_{tot} into contributions from dispersion (E_d), polarization (E_p), and electrostatic (E_{el}) effects clearly shows that the latter by far dominate and account for about 80% of the total energy; i.e., hydrogen bonds are mainly electrostatic in nature. From the variation of total energies (E_{tot}) with base-pair constitutents, it becomes clear that the stability of such a complex is not merely defined by the number of hydrogen bonds. Rather, it is necessary to consider the intrinsic electronic structures of associated bases, a fact previously deduced from measurements of binding constants and termed *"electronic complementarity"* (473).

The relative E_{tot} values for different base-pairs suggest that complementary pairs in the Watson–Crick sense are more stable than the self-associates of the individual components. All noncomplementary base-pairs such as A:G and G:U are less stable than the corresponding self-associate pairs. This finding explains why noncomplementary pairs have not been crystallized except for substituted bases like 5-fluorouracil complexing with cytosine (486,487) which have an altered electronic structure and therefore exhibit binding energies different from those of the natural analogs. A special role is assigned to some "wobble" base-pairs discussed in Section 6.9.

Table 6-5. Interaction Energies (E_{total}) in Some Selected Base–Pairs Calculated by Quantum Chemical Methods [From (484)]

Base–pair[a]	Energy contributions [kcal/mole]			
	E electrostatic	E polarization	E dispersion	E total
Complementary Watson-Crick and Hoogsteen Base–Pairs				
G–C (XIX)	−14.10	−1.90	−0.79	−16.79
A–T (XX)	− 5.7	−0.57	−0.73	− 7.00
A–T (XXI)	− 5.47	−0.64	−0.86	− 6.97
A–U (XX)	− 5.68	−0.66	−0.87	− 7.21
A–U (XXIII)	− 4.93	−0.70	−0.98	− 6.61
A–U (XXIV)	− 5.26	−0.64	−0.95	− 6.85
Self-Complementary Base–Pairs				
A–A (V)	− 4.40	−0.49	−0.71	− 5.60
U–U (XII)	− 4.13	−0.61	−0.68	− 5.42
G–G (III)	−13.71	−1.72	−0.61	−16.04
C–C (XIV)	− 8.61	−1.41	−0.71	−10.73
A–G (VIII)	− 7.68	−1.02	−0.70	− 9.40
G–U (XXVIII)	− 6.08	−1.12	−0.67	− 7.78

[a] Roman numerals in parentheses refer to base–pairs in Figure 6-1.

6.5 Patterns of Vertical Base–Base Interactions

Bases pile up in long stacks like coins in a roll. Horizontal base–base association via hydrogen bonding is observed in nonaqueous solvents and in the gaseous and crystalline states. Additionally, in the solid state bases are found almost exclusively stacked such that one base plane is at the van der Waals distance, ~3.4 Å, and parallel to the adjacent one, an arrangement due to vertical rather than horizontal interactions. In aqueous solution, such base stacks form as well. Since stacking is important for the stabilization of nucleic acid helices (488,489), the principal geometric and thermodynamic features and the main forces responsible for the interactions will be discussed here. For a review, see (448).

In two comparative studies (490,491), packing patterns of bases, of charge–transfer complexes, and of nonpolar aromatics such as benzene, naphthalene, anthracene, and phenanthrene were described. The nonpolar aromatics have no dipole moments and crystallize in a herringbone-type arrangement with adjacent molecules perpendicular rather than parallel to each other. If, however, as in acridine and ethidium, a nitrogen is introduced into anthracene or phenanthrene to create a dipole moment, the molecules orient parallel to each other in the crystal lattice and form long stacks.

Recurring stacking patterns suggest specificity. Stacked arrangements are also dominant in crystal structures of bases (490,491). It is striking to find that the stacking patterns in these bases are rather specific, with polar substituents $-NH_2$, $=N-$, $=O$ or halogen of one base superimposed over the aromatic system of the adjacent base (Figure 6-8). This kind of stacking specificity even overrides hydrogen-bonding effects. In the series deoxyadenosine monohydrate (492), adenosine 5'-phosphate (493), adenosine : 5-bromouridine (494), deoxyguanosine : 5-bromodeoxyuridine (398), the purine bases all stack in the pattern described in Figure 6-8a although crystal space groups, cell dimensions, molecular packing, and hydrogen-bonding schemes differ from each other. Similarly, the stacking specificity is pronounced in halogenated bases. Halogen atoms are located over the adjacent heterocycle, with even less than a van der Waals distance between halogen atoms and atoms of the heterocycle (Figure 6-8i).

Several nucleoside and nucleotide crystal structures display close intermolecular contacts between sugar $O_{4'}$ atoms and adjacent heterocycles (Figures 6-8a,j). This kind of interaction is probably due to the relatively low electronegativity of $O_{4'}$ (Section 5.1.), which encourages close approach to a π-electronic system.

No stacking at all is found in some crystal structures of protonated pyrimidine bases and nucleotides, suggesting that charged pyrimidines tend to unstack. In the purine series, however, stacking is observed but

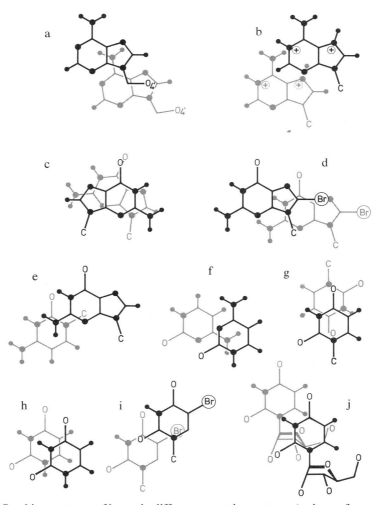

Figure 6-8. Stacking patterns of bases in different crystal structures (redrawn from 491). **(a)** The same adenine stack occurs in different crystal structures: deoxyadenosine monohydrate (492), adenosine 5'-phosphate (493) and adenosine:5-bromouridine complex (494); **(b)** stacking of protonated purines, 9-methyladenine dihydrobromide (415); **(c)** stacking pattern in guanosine and inosine crystal structures (495); **(d)** influence of halogen substituent on stacking pattern in 8-bromoguanosine (496), compare with (c); **(e–g)** interaction between amino or keto substituents and purine or pyrimidine bases: (e) 9-ethylguanine:1-methyl-5-fluorocytosine (497); (f) cytosine monohydrate (498); (g) 9-ethyladenine:1-methyluracil (499); **(h)** pyrimidine overlap involving ring nitrogen atom in uracil (500); **(i)** influence of halogen substituent on pyrimidine stacking in 5-bromouridine (501) and in 5-fluoro-2'-deoxyuridine (502); **(j)** interactions of ribose ring with pyrimidine in cytidine (503).

the stacking pattern is modified relative to neutral purine bases, with atoms N_3 and N_7 of adjacent protonated bases overlapped (Figure 6-8b).

A survey of stacking patterns in base, nucleoside and nucleotide crystal structures suggested that forces between permanent dipoles are only of minor importance for the stabilization of base stacks. Rather it appears that dipole-induced dipole interactions play the major role, with the permanent dipole, predominantly in C=O or C—NH$_2$ groups, superposed over the π-electronic system of the adjacent base (491).

6.6 Thermodynamic Description of Stacking Interactions

Base stack formation is additive, diffusion controlled, and stabilized by weak interactions. Studies on association of bases and nucleosides in aqueous solution using osmometric techniques (448,504) led early to the conclusion that vertical base stacking occurs and goes beyond the dimeric state (Figure 6-9). Sedimentation equilibrium experiments indicated that the process is reversible with a constant free energy increment for each step, suggesting that addition of a base to another one or to an existing stack is additive and not cooperative and thus follows *isodesmic* behavior (505). The data for purine and pyrimidine nucleosides in aqueous solution in Table 6-6 indicate: (i) that association constants K are characteristic of

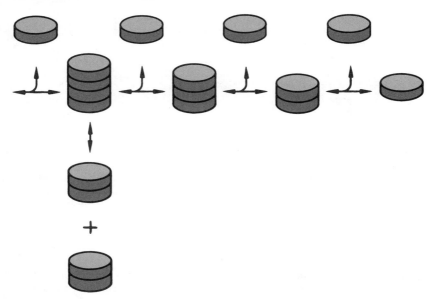

Figure 6-9. Stacking of bases in aqueous medium goes beyond the dimeric state and follows in a broad sense *isodesmic* behavior; i.e., each step is independent and displays the same thermodynamic and kinetic parameters. From (544).

Table 6-6. Thermodynamic Parameters of Self-association (Stacking) of Nucleosides and Bases in Water [From (448)]

Compound	K [l/mole]	ΔH [kcal/mole]	ΔS [e.u.]	ΔG [kcal/mole]
6-Methylpurine	6.7	-6.0 ± 0.4	-16	-1.12
Purine	2.1	-4.2 ± 0.2	-13	-0.44
Ribosylpurine	1.9	-2.5 ± 0.1	-7	-0.38
Deoxyadenosine	4.7 to 7.5	-6.5 ± 1.0	-18	-1.00
Cytidine	0.87	-2.8 ± 0.1	-10	0.08
Uridine	0.61	-2.7 ± 0.1	-10	0.29
Thymidine	0.91	-2.4 ± 0.3	-9	0.06

weak interactions, (ii) that both enthalpies ΔH and entropies ΔS are negative, (iii) that the standard free energy change ΔG is in the order of the thermal energy kT (0.6 kcal/mole), and (iv) that, in general, methylation of bases leads to moderately increased stacking interactions.

Information regarding the structure of the stacks in aqueous solution has also been provided by NMR measurements (506,507). If two heterocycles aggregate by stacking, the magnetic anisotropy associated with the aromatic ring current in one molecule has a deshielding effect on the protons of the adjacent heterocycle and shifts their resonance signals to higher fields with increasing concentration (Figure 6-10). Relative shifts of base and sugar protons in purine nucleosides indicated that the six-membered pyrimidine ring participates preferentially in stacking rather than the five-membered imidazole fragment, and that the orientation of the heterocycles in a stack depends on the nature of bases, i.e., their amino and keto substituents (448). Because there is no obvious line broadening observed with stack formation, the aggregates build up and break down rapidly on the NMR time scale. Sound absorption techniques indicate that the processes displayed in Figure 6-9 are actually so rapid as to be diffusion controlled (508).

Purine–purine stacks are most stable. Solubility experiments in biphasic systems and NMR data show that stacking interactions between purine and pyrimidine bases follow the trend (509–513)

purine–purine > pyrimidine–purine > pyrimidine–pyrimidine.

If bases are linked to each other in oligo- and polynucleotides, stacking interactions between adjacent bases occur and give rise to stable, single-stranded, helical structures (514). The stabilities of these helices reflect the same trend as given above, with poly(A) mainly helical and poly(U) predominantly random coil at room temperature.

Single-stranded helix formation follows a two-state mechanism. In most cases a simple two-state process (514–522)

helical (stacked)⇌random coil (unstacked),

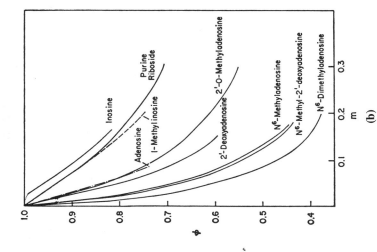

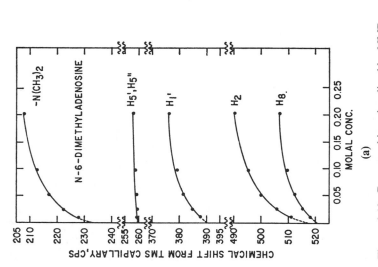

Figure 6-10. Base-stacking visualized by NMR spectroscopy and by vapor pressure osmometry. (a) Dependence of chemical shift of base protons on base concentration yields indication of association constants and structure of formed stacks. N_6-Dimethyladenosine in D_2O, all spectra taken at 10°C. (b) Osmotic coefficient ϕ for various nucleosides in aqueous solution plotted versus molal concentrations (m). Curves allow the derivation of association constants and standard free energy of complex formation. Note strong dependence of ϕ on methylation of bases. From (506).

Table 6-7. Thermodynamic Parameters of Stacking in Some Dinucleoside Phosphates XpY

Molecule	ΔH [kcal/mole]	S [cal $K^{-1}mol^{-1}$]	T_m [°C]
ApA[a]	−8.5 ± 0.5	−28.5 ± 1.5	26.2 ± 3
CpC	−8.5 ± 0.5	−30.0 ± 1.5	13.3 ± 5
ApU	−7.3 ± 0.5	−24.7 ± 1.7	22.0 ± 1.1
m^6ApU	−7.4 ± 0.3	−24.8 ± 0.9	26.9 ± 0.9
m_2^6ApU[b]	−6.5 ± 0.2	−20.9 ± 0.7	36.7 ± 0.8
ApA	−7.2 ± 0.3	−24.5 ± 0.9	21.9 ± 0.9
m_2^6Apm_2^6A	−6.7 ± 0.2	−21.1 ± 0.7	44.5 ± 0.8

[a] From (522).
[b] From (518).

is sufficient to explain the helix formation behavior of single-stranded oligomers. Thermodynamic parameters for some oligonucleotides are given in Table 6-7, ΔH, ΔS, and T_m for ApU and ApA are virtually identical, and apparently the above given sequence-stacking rule does not apply strictly to oligonucleotides that short. However, the trend is clear for CpC which displays largely reduced stacking properties. As already noted for individual bases in Table 6-6, alkylation increases stacking interactions, indicated by rising T_m values with degree of methylation in Table 6-7.

The earlier proposal of a more complicated mechanism suggested by NMR (516) and kinetic experiments (517) for ApA has been criticized (520). If the more recent circular dichroism and NMR data are considered, then the two-state mechanism appears to be sufficient. These data also show that the overall conformation of sugar moieties in helical oligoribonucleotides is $C_{3'}$-endo. In the helical deoxyribo series, however, $C_{2'}$-endo prevails, with the 3′-terminal nucleotide in rapid $C_{2'}$-endo ⇌ $C_{3'}$-endo equilibrium (519). This floppy end of a helical oligonucleotide might give rise to a more complex spectrum of conformational states, and the NMR and kinetic data (516,617) should be reconsidered in the light of this new information.

6.7 Forces Stabilizing Base Stacking: Hydrophobic Bonding and London Dispersion

Dipoles, π-electron systems, and dipole-induced dipole moments appear to be important in vertical base stacking. In addition, evidence has been presented for the contribution of London dispersion forces (488,489) and for base stacking in aqueous solution, hydrophobic forces also are involved (523,524). How do these forces operate?

Base stacking is stabilized by entropic contributions: Iceberg water and water cavities. If a base is dissolved in water, the individual solute molecules tend to aggregate and reduce the number of solute–solvent contacts, yielding in extreme cases a two-phase system. This hydrophobic bonding is of particular interest in biological macromolecules because it encourages interactions between nonpolar amino acid side chains or between nucleic acid bases and significantly contributes to secondary and tertiary structure stabilization (525–528). Although widely recognized in a broad sense, hydrophobic forces are still a matter of debate (529).

The aggregation of solute molecules in water has been explained in two ways.

(A) If a (hydrophobic) or nonpolar molecule is dissolved in water, water molecules cluster around its surface. They adopt an "iceberg-like" structure reminiscent of ice clathrates (530) and lose entropy—an unfavorable situation. If the solute molecules aggregate, their surface exposed to water is reduced and structured water molecules are released, resulting in an overall gain in entropy for water which stabilizes the solute complexes (531–533).

(B) A solute molecule entering the water must create a cavity against the surface tension of water. If two or several dissolved solute molecules aggregate, larger cavities are formed and the surface is diminished, leading to a reduction in surface tension which promotes this process (534,535).

Is it possible to distinguish between the two proposals in the case of base stacking? Table 6-6 demonstrates that base stacking in aqueous solution is exothermic and accompanied by negative entropy changes and thus mechanism A seems to be ruled out. However, studies on the system actinomycin:deoxyguanosine, involving the addition of methanol to the aqueous solution, and thermal denaturation experiments on dinucleoside phosphates have indicated that, in fact, a "hidden" positive entropy change occurs with stack formation and that this is, in general, masked by an overall negative entropy change (507,536,537). It appears that mechanism A contributes more significantly to the stacking process than mechanism B, and it is clear from a number of additional experiments that hydrophobic forces are of importance (512,538,539).

Dipolar and London dispersion forces convey stacking specificity. The hydrophobic interactions cannot explain some effects related to specific bases; e.g., purines stack better than pyrimidines and methylation enhances stacking. These individual properties are related to the electronic systems of the bases and are mainly due to London dispersion forces and to interactions between dipoles.

As recognized by London (540), the principal contribution to the ever-effective van der Waals attractive forces between atoms in close proximity resides in electrokinetic interactions between the systems. At any instant, the electronic charge distribution within atomic groups is asymmetric due to electron fluctuations. Therefore, dipoles created in one group of atoms polarize the electronic system of neighboring atoms or molecules, thus inducing parallel dipoles which attract each other. These forces are additive and decrease with the sixth power of the distance. They are independent of temperature and increase with the product of the polarizabilities of the partner molecules (541). As the bases possess in addition a permanent dipole moment, the two electronic effects, London dispersion and permanent dipoles, combine and lead to appreciable effects which are more pronounced in purine than in pyrimidine bases.

Base-pair hydrogen bonding depends on composition whereas stacking is

Table 6-8. Total Stacking Energies [kcal/mole dimer] for the Ten Possible Dimers in B-DNA Type Arrangement Obtained by Quantum Chemical Calculations[a] [From (542)]

Stacked dimers		Stacking energies [kcal/mole dimer]
↑C·G↓ / ↑G·C↓		−14.59
↑C·G↓ / ↑A·T↓	↑T·A↓ / ↑G·C↓	−10.51
↑C·G↓ / ↑T·A↓	↑A·T↓ / ↑G·C↓	− 9.81
↑G·C↓ / ↑C·G↓		− 9.69
↑G·C↓ / ↑G·C↓	↑C·G↓ / ↑C·G↓	− 8.26
↑T·A↓ / ↑A·T↓		− 6.57
↑G·C↓ / ↑T·A↓	↑A·T↓ / ↑C·G↓	− 6.57
↑G·C↓ / ↑T·A↓	↑T·A↓ / ↑C·G↓	− 6.78
↑A·T↓ / ↑A·T↓	↑T·A↓ / ↑T·A↓	− 5.37
↑A·T↓ / ↑T·A↓		− 3.82

[a] Arrows designate direction of sugar-phosphate chain and point from $C_{3'}$ of one sugar unit to $C_{5'}$ of the next, both carbons attached to the same phosphodiester link.

influenced by composition and sequence: Results of quantum chemical calculations. Electrostatic forces between stacked bases and base-pairs are intrinsically related to charge distributions and are therefore accessible to quantum chemical calculations. The first rather crude estimations using dipole–dipole approximations were later put on a more refined level employing monopole–monopole interactions, polarization and dispersion interactions and repulsions due to the $1/r^{12}$ term in the Lennard-Jones approach [Eq. (3-5)] (483,484,489,537,542,543).

The data in Table 6-8 for the ten possible combinations of stacks formed by G·C and/or A·T base-pairs were calculated within the frame of B-DNA geometry (Chapter 11). From these (542) and related studies (484) partitioning the total stabilizing energy of base-paired dimers into horizontal (base-pairing) and vertical (base-stacking) components, it becomes clear that horizontal, complementary hydrogen bonding is only composition dependent and more important than vertical stacking which is influenced by both composition and sequence.

As G·C base-pairs are more stable than A·T pairs (Table 6-5), stacked dimers with high G·C content are energetically preferred to those rich in

A·T; compare $\left|{}^{\text{C·G}}_{\text{G·C}}\right|$ and $\left|{}^{\text{T·A}}_{\text{A·T}}\right|$ which differ by 8 kcal/mole dimer. The influence of stacking energy is expressed in the general sequence dependences showing that

$$\left|{}^{5'\text{pyrimidine·purine}_{3'}}_{3'\text{purine·pyrimidine}_{5'}}\right| \text{ is more stable than } \left|{}^{5'\text{purine·pyrimidine}_{3'}}_{3'\text{pyrimidine·purine}_{5'}}\right|$$

See, for instance,

$$\left|{}^{\text{C·G}}_{\text{G·C}}\right| \text{ and } \left|{}^{\text{G·C}}_{\text{C·G}}\right|$$

which differ by about 5 kcal/mole dimer (Table 6-8). The origin of this energy–sequence correlation becomes clear if the geometries of the stacking overlaps displayed in Figure 9-6 are considered. In the alternating purine, pyrimidine sequences, overlap between adjacent bases in a stack is governed by the sequence

$$\left|{}^{5'\text{pyrimidine}}_{3'\text{purine}}\right| \text{ or } \left|{}^{3'\text{purine}}_{3'\text{pyrimidine}}\right|$$

and is very different from the overlap observed in base-paired dimers containing only purine or only pyrimidine bases in one strand.

It should be stressed that the data in Table 6-8 consider only molecules *in vacuo* and do not take into account hydrophobic interactions which, however, contribute significantly to stacking interactions. In experiments discussed in the next section, base-pairing and stacking interactions in aqueous solutions are obtained as total contributions to oligo- and polynucleotide helix stability.

6.8 Formation and Breakdown of Double-Helix Structure Show Cooperative Behavior

A helix with the pitch as basic repeating motif may be considered as if it were a one-dimensional, crystalline lattice. A formal analogy of helical and crystalline order is reflected in physical properties: both break down suddenly at a certain (melting) temperature and both grow and propagate after a nucleus (seed) has been formed.

Cooperative zipper mechanism of helix formation requires three-base-pair nucleus. As shown schematically in Figure 6-11 for double-helix formation between poly(U) and poly(A), the first step is association of a single A·U base-pair with a stability constant expressed as the product of nucleation parameter β (10^{-3} liters/mole), and chain growth parameter s (around 10 at 0°C and 1 at melting temperature) which were both determined by relaxation kinetics. This first, isolated base-pair is rather unstable due to the influence of β but addition of another, neighboring and stacked base-pair follows with stability constant s, not diminished by this parameter β. A third base-pair is stacked on top of the first two and creates a suitable nucleus from which further addition of stacked base-pairs leads to stepwise construction of a helix just as a zipper is closed (544).

The energy profile of this process is schematically illustrated in Figure 6-12. Formation of the initial, isolated base-pair is unfavorable and gives

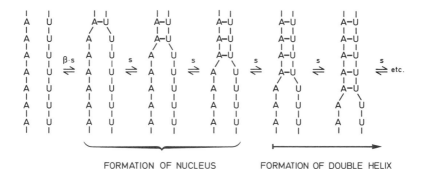

FORMATION OF NUCLEUS   FORMATION OF DOUBLE HELIX

Figure 6-11. Schematic description of double-helix formation in the case of oligo(A)·oligo(U). In this system, helix growth parameter s is about 10 at 0°C and 1 at the melting temperature. Nucleation parameter β, 10^{-3} liters/mole, diminishes stability constant $K = \beta \cdot s$ of primary base-pair formation but does not influence formation of additional, stacked base-pairs which form cooperatively with $K = s$ according to a *linear Ising model*. In contrast to the *isodesmic model* for base stacking (Figure 6-9), where each step is independent of the other, in the *cooperative process* described by the Ising model, base-pair formation and stacking are influenced by the next neighbors, except for the very first base–base association. After (544).

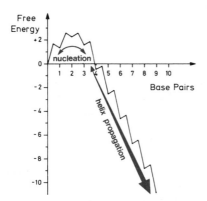

Figure 6-12. After the unfavorable positive free energy contribution in the nucleation process is overcome, the free energy for additional steps becomes negative and the helix grows spontaneously. Relative total free energy (ΔG) of helix formation in arbitrary units is plotted as a function of the number of consecutive, stacked base-pairs assembled into a helical array. From (544).

positive contributions to free energy. After three consecutive, stacked base-pairs have created a nucleus which can still dissociate easily into its components, addition of new base-pairs stacked to the nucleus leads to favorable, negative contributions to free energy. From then on, growth of the double helix is spontaneous, due mainly to geometrical constraints of

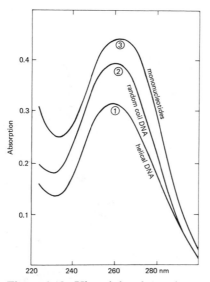

Figure 6-13. Ultraviolet absorption spectra of DNA in native double helical (1) and denatured random coil (2) states and spectrum of individual, monomeric, unstacked nucleotides of the same concentration as in native DNA (3). In denatured state (2), bases are still considerably stacked. From (185).

the sugar–phosphate backbone which are implied by the stereochemistry
of the nucleotide unit which, in its preferred conformation, is preformed
to suit this purpose. Beyond that, base-pair stacking and hydrogen-bonding interactions act along the same line and these effects taken together
are responsible for the cooperative behavior of nucleic acid helix formation and break down (melting).

Melting temperature T_m characterizes a double helix. Base stacking is
accompanied by reduction in UV absorption (hypochromicity, Figure 6-
13), so the UV spectrum is a convenient monitor of the formation and
breakdown of double helices. If the temperature of a solution containing
double-helical DNA (or RNA) is slowly raised, UV absorption increases
suddenly at a certain temperature because ordered double helices dissociate. The midpoint of transition is called the "melting temperature" or T_m.
This process is pictured for oligo(A)·oligo(U) double helices of different
chain lengths in Figure 6-14. It is obvious that, with increasing chain

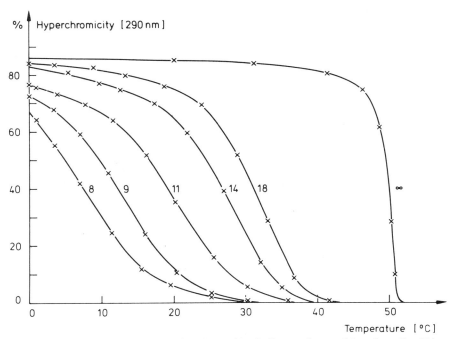

Figure 6-14. Melting profiles of cooperative helix⇌coil transition for oligo(A)
oligo(U) double helices of various chain lengths, from 8 to ∞ nucleotides.
Melting points T_m correspond to 50% transition and are around 49°C for long
poly(A) · poly(U) duplexes. Plotted are changes in ultraviolet absorption at
290 nm in the form of hyperchromic effects defined as $100 \times (E_T - E_C) \cdot E_C$, where
E_T = absorption at temperature T, E_C = absorption at 100% coil form. Concentrations are 5 to 11.3 mM oligonucleotides, 50 mM sodium cacodylate buffer at pH 6.9.
From (545).

length, T_m increases and the slope at the point of inflection (T_m) becomes steeper, synonymous with enhanced cooperativity (545–547).

Helix–coil transitions for double-helical structures are approximately an all-or-none process: Helix⇌Coil. As shown in Figure 6-15, a population analysis of the system $A(pA)_{17} \cdot U(pU)_{17}$ at various temperatures yields exclusively either dissociated, coiled monomers or fully integrated double helices with only the terminal base-pairs in a rapid dissociation–recombination process.

The melting temperatures of double-helical nucleic acids increase not

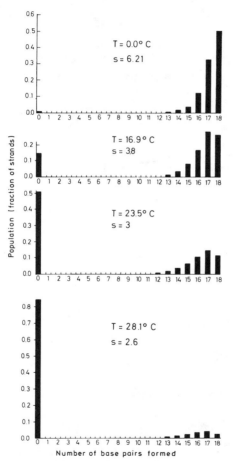

Figure 6-15. Population analysis for the system $A(pA)_{17} \cdot U(pU)_{17}$ at different temperatures. Product of [(nucleation parameter β) × (concentration c)] used in this analysis is $\beta \cdot c = 10^{-7}$. Note that base-pairs at ends of the helix open and close easily, giving rise to a distribution of helices with 13 to 18 base-pairs but helices with 1 to 12 base-pairs are practically not observed; they either dissociate or form helices with 13 to 18 base-pairs. From (548).

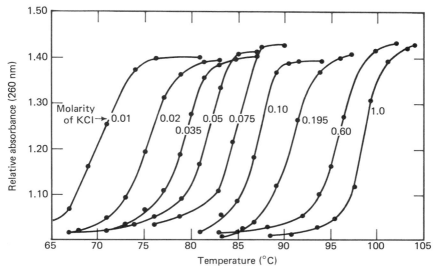

Figure 6-16. Dependence of thermal denaturation of *D. pneumoniae* (R-36A) DNA on ionic strength. Various KCl concentrations in citrate buffer, pH 7.0. From (549).

only with their lengths but also with the ionic strength of the medium and with the GC/AT ratio of the polynucleotide (Figures 6-16,6-17). As the composition linearly influences the melting temperature to a good approximation, it can be evaluated from the T_m value. Because of this dependence of melting behavior on nucleotide composition, it follows that in a double-helical DNA or RNA with random sequence, A–T(U)-rich regions should melt at lower temperatures than G–C-rich regions (Figure 6-18). This gives rise to local breakdown of helical order (550). It leads to the observed broad relaxation spectrum of the melting process (544,550). Differential melting curves, in which the differential change of absorption with temperature, $\Delta A / \Delta T$, is plotted versus temperature, T, display a rugged profile with a multitude of maxima which can be deconvoluted to yield individual double-helical domains melting at similar, yet significantly distinct temperatures and indicative of different GC/AT(U) content (551–554) (Figure 6-19).

Analysis of a number of such profiles produced the "stability matrix" for stacked, base-paired dinucleotides in B-DNA geometry which gives T_m values for base-paired doublets at certain conditions, Table 6-9a. From these data, the melting temperature T_m of a DNA double helix of any sequence can be estimated. Some examples are entered in Table 6-9b. If melting temperatures of duplexes calculated on the basis of Table 6-9 are plotted against stacking energies obtained by quantum chemical calculations (Table 6-8), it is striking to find that a linear correlation emerges (Figure 6-20). Experimental and theoretical approaches form a consistent picture although the theoretical work neglects contributions due to hydro-

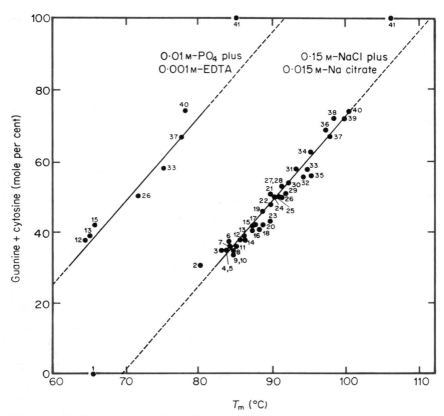

Figure 6-17. Dependence of melting temperature T_m on guanine + cytosine (G + C) content of various samples of DNA obtained from different sources. DNA was dissolved in 0.15 M NaCl + 0.015 M Na-citrate, pH 7.0. Points 1 and 41, for poly(dA-dT) and poly(dG-dC), fall off the least-squares line which is described analytically by $T_m = 69.3 + 0.41$ (%C). From (549).

phobic effects, steric constraints, positive ions around phosphate groups, hydration, etc., which, to a certain extent, seem to counterbalance each other.

Using oligo-ribonucleotides with defined sequence as model compounds, thermodynamic parameters for adding or subtracting one base-pair to or from an existing helix were derived (554–557). Table 6-10 predicts that depending on nucleotide sequence and composition, RNA duplexes exhibit the following trend in helix stability: poly(G)·poly(C) > poly(G-C)· poly(G-C) > (double helix with random sequence) > poly(A-U)·poly(A-U) > poly(A)·poly(U). The A/U case is surprising because in the DNA series, poly(dA)·poly(dT) is more stable than the alternating copolymer poly(dA-dT)·poly(dA-dT) (558). For the latter, formation of short hair-

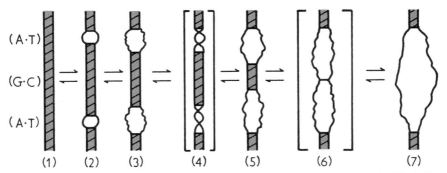

Figure 6-18. Scheme describing processes involved in DNA or RNA double-helix de- and renaturation. A–T-rich regions melt first, giving rise to states (2) and (3). In (4), additional base-pairs are opened and the twist is taken up in coil regions. From (550).

pin loops with reduced thermal stability has been invoked to explain the overall low T_m. Since the RNA analog is also able to produce such hairpins (as is known from tRNA), it appears that the differences in stability are correlated not with composition but rather with helix structure, the deoxy analogs preferring B- and D-type helices whereas all ribo-polynucleotides can only occur in the A-form (Chapter 9).

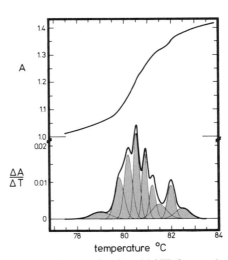

Figure 6-19. Melting profile of DNA (top) and its first derivative dA/dT (bottom). The latter curve is deconvoluted into nine individual peaks characterized by temperature, amplitude, and breadth. A indicates UV absorption at 260 nm; dA/dT or $\Delta A/\Delta T$ are first derivatives with respect to temperature T. These curves are simulated; for some realistic data see Ref. (557).

Table 6-9. Prediction of DNA Double Helix Stability from Base Sequence [From (555)]

(A) Stability Matrix for Nearest-Neighbor Stacking in Base–Paired Dinucleotides in B-DNA Geometry[a]:

5'	3'			
	A	T	G	C
T	36.73	54.50	54.71	86.44
A	54.50	57.02	58.42	97.73
C	54.71	58.42	72.55	85.97
G	86.44	97.73	85.97	136.12

[a] Numbers give T_m values in °C at 19.5 mM Na$^+$.

(B) T_m Values Predicted with This Matrix for a Collection of Synthetic DNA Polymers with Defined Sequence:

Polynucleotide	T_m(°C)		
	Experimental[a]	Calculated[b]	Difference[c]
Poly(dA-dT)·poly(dA-dT)	45.0	46.9	−1.9
Poly(dA-dA-dT)·poly(dA-dT-dT)	49.2	49.4	−0.2
Poly(dA)·poly(dT)	53.0	54.5	−1.5
Poly(dG-dA-dA)·poly(dT-dT-dC)	64.5	66.5	−2.0
Poly(dG-dT-dA)·poly(dT-dA-dC)	66.8	64.3	2.5
Poly(dA-dA-dC)·poly(dG-dT-dT)	70.2	69.0	1.2
Poly(dG-dA)·poly(dT-dC)	71.3	72.4	−1.1
Poly(dG-dA-dT)·poly(dA-dT-dC)	72.0	66.1	5.9
Poly(dG-dG-dA)·poly(dT-dC-dC)	76.3	76.9	−0.6
Poly(dG-dT)·poly(dA-dC)	77.4	76.2	1.2
Poly(dG)·poly(dC)	87.8	86.0	1.8
Poly(dG-dC)·poly(dG-dC)	99.2	104.3	−5.1

[a] Experimental melting temperatures at various ionic strengths are interpolated to 19.5 mM Na$^+$.
[b] Calculated from values in Table 6-9(A) and nearest-neighbor frequencies in each polymer.
[c] T_m (experimental) − T_m (calculated).

Hairpins, bulges, and loops. In addition to next-neighbor stabilization in RNA double helices, contributions due to formation of hairpin loops, bulges, and interior loops have been evaluated (557–561) (Table 6-10, Figure 6-21). The data suggest that interior loops and bulges are more stable than hairpin loops which are optimum for six nonbonded bases.

It should be stressed that in a double-helical DNA or RNA structure, base-pairs at the termini and base-pairs within the helical stem can both potentially open. The expected frequency for the latter process is rather

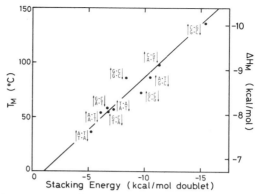

Figure 6-20. Correlation diagram between experimental melting temperatures T_m (Table 6-9) and theoretically evaluated stacking energies (Table 6-8) of base-paired dinucleotides in B-DNA geometry. Note that two different approaches to the same problem yield a satisfactory correlation with a correlation coefficient of 0.97. Stacking energies for dinucleotide duplexes entered in this plot differ slightly from data in Table 6-8 because different results given in Ref. (536) were used. From (555).

low and has been evaluated from measurements of oligoribonucleotide double-helical complexes (563,564) (Table 6-11). The G–C base-pair is, in general, more stable by a factor 100 than an A–U pair, and the flanking base-pairs play an important, additional role. On the basis of these data, one can estimate that in a double-helical RNA of random sequence and one million base-pairs long, at 25°C about 10 G–C and 500 A–U pairs are in non-hydrogen-bonded, unstacked configuration. This description suggests that double-helical nucleic acids should not be considered as static columns but rather as flexible, "breathing" entities, in harmony with chemical modification experiments (563).

6.9 Base-Pair Tautomerism and Wobbling: Structural Aspects of Spontaneous Mutation and the Genetic Code

If, in living organisms, base-pairing always occurred in a strict Watson–Crick sense, life on this planet would not display the great variety of species in fauna and flora which are due to mutations. Altered genetic information is given to daughter generations via replication and is passed on to protein biosynthesis via transcription and translation as described schematically in Figure 6-22.

Replication, transcription, translation, and the genetic code. In all three processes, base–base recognition by means of Watson–Crick pairing is

Table 6-10. Experimentally Determined Free Energies at 25°C for RNA Double Helical and Looped Structures[a] [From (557)]

Base paired regions	ΔG(kcalorie)$\pm 10\%$
$-\overrightarrow{A-A}-$ $-\dot{U}-\dot{U}-$	-1.2
$-\overrightarrow{A-U}-$ $\overrightarrow{-U-A}-$ $-\dot{U}-\dot{A}$ ' $-\dot{A}-\dot{U}-$	-1.8
$\overrightarrow{-A-C}-$ $\overrightarrow{-C-A}-$ $-\overrightarrow{A-G}-$ $-\overrightarrow{G-A}-$ $-\dot{U}-\dot{G}-$ ' $-\dot{G}-\dot{U}-$ ' $-\dot{U}-\dot{C}-$ ' $-\dot{C}-\dot{U}-$	-2.2
$-\overrightarrow{C-G}-$ $-\dot{G}-\dot{C}-$	-3.2
$-\overrightarrow{G-C}$ $\overrightarrow{G-G}-$ $-\dot{C}-\dot{G}-$ ' $\dot{C}$ $\dot{C}-$	-5.0
$-\overrightarrow{G-U}-$ $-\dot{U}-\dot{G}-$	-0.3
$-\overrightarrow{G-X}-$ $-\overrightarrow{X-G}-$ $-\dot{U}-\dot{Y}-$ ' $-\dot{U}-\dot{Y}-$	0

Unbonded regions	ΔG(kcalorie)± 1 kcalorie
Number of bases unbonded	Interior loops
2–6	+2
7–20	+3
m (>20)	1+2 log m
	Bulge loops
1	+3
2–3	+4
4–7	+5
8–20	+6
m (>20)	4+2 log m

	Hairpin loops	
	Closed by G·C	Closed by A·U
3	+8	>8
4–5	+5	+7
6–7	+4	+6
8–9	+5	+7
10–30	+6	+8
m (>30)	3.5+2 log m	5.5+2 log m

[a] The free energies (ΔG) for the base-paired regions refer to the free energy of adding a base-pair to a pre-existing helix; therefore the magnitude depends on the sequence of two base-pairs.

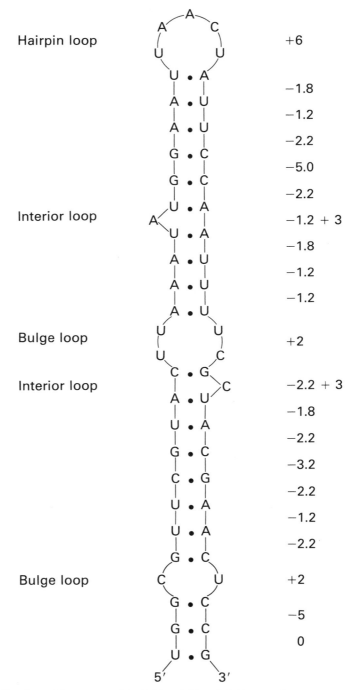

Hairpin loop	+6
	−1.8
	−1.2
	−2.2
	−5.0
	−2.2
Interior loop	−1.2 + 3
	−1.8
	−1.2
	−1.2
Bulge loop	+2
Interior loop	−2.2 + 3
	−1.8
	−2.2
	−3.2
	−2.2
	−1.2
	−2.2
Bulge loop	+2
	−5
	0

Figure 6-21. Application of data entered in Tables 6-10 and 6-11 to evaluate a possible secondary structure for a 55-nucleotide fragment from R17 virus (557). Overall ΔG at 25°C is −21.8 kcal/mole. For slightly different, revised data see (556).

Table 6-11. Transient Opening Probabilities for
A·U and G·C Base–Pairs in Double Helical RNA
[From (561)]

Arrangement	ΔG for opening central pair (25°C, kcal/mole)	Probability of opening
G–G–G Ċ–Ċ–Ċ	7.5	0.3×10^{-5}
G–G–A Ċ–Ċ–U̇	6.75	1.2×10^{-5}
A–G–G U̇–Ċ–Ċ	6.75	1.2×10^{-5}
A–G–A U̇–Ċ–U̇	6.0	4.2×10^{-5}
A–A–A U̇–U̇–U̇	4.0	120×10^{-5}
A–A–G U̇–U̇–Ċ	4.15	100×10^{-5}
G–A–A Ċ–U̇–U̇	4.15	100×10^{-5}
G–A–G Ċ–U̇–Ċ	4.3	70×10^{-5}

crucial and leads to direct copying of DNA (*replication*) and to transformation of sequence in DNA into corresponding sequence (*transcription*) in messenger RNA (mRNA). In *translation,* the information contained in the mRNA nucleotide sequence is translated, at the ribosomal machinery, into polypeptide amino acid sequence with the amino acid carrying tRNA serving as mediator (564). According to its amino acid specificity, the tRNA molecule consists of an anticodon of three nucleotides (triplet) which recognizes a complementary codon on the mRNA and, following the rules implied by the genetic code (Figure 6-23), incorporates amino acids sequentially into a growing polypeptide chain, with mRNA as guideline read off from the 5'- to 3'-terminus.

Mutations can occur if non-Watson-Crick base-pairs are formed. Let us look first at DNA replication. In this process two types of simple spontaneous substitution mutations can occur: (i) *transitions,* where a purine is replaced by another purine or a pyrimidine is replaced by another pyrimidine; and (ii) *transversions,* where a purine is exchanged for a pyrimidine or vice versa (565) (Figure 6-24). Experimentally observed mutation rates are of the order 10^{-8} to 10^{-11} per base-pair synthesized (566) and represent those cases in which noncomplementary base-pairs in the strict Watson-Crick sense have escaped two DNA polymerase checking processes. This

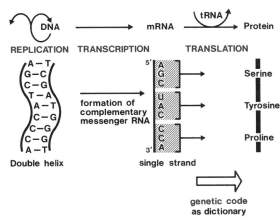

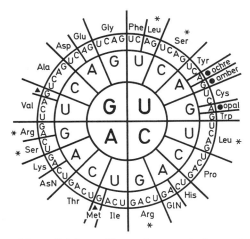

Figure 6-22. Schematic, simplified description of the central dogma in molecular genetics (584). Double-helical DNA is replicated and passed on from one generation to the next. In the DNA nucleotide sequence, information on amino acid sequence in protein is encoded. In order to synthesize protein, DNA is transcribed into complementary messenger RNA (mRNA) the sequence of which, at the ribosome, is translated into a protein sequence, with one nucleotide triplet coding for one amino acid according to the rules given by the genetic code (Figure 6-23). Transfer RNA, as adapter molecule, carries an anticodon complementary to the mRNA codon and an amino acid specific for that anticodon.

Figure 6-23. The genetic code displayed in radial form. The codons are to be read from the center (5′) outward (3′) and represent mRNA triplets coding for amino acids entered at the periphery. The last (3′-terminal) base is redundant (degenerate) except for amino acids Trp and Met whereas the first two bases are specific. Amino acids marked * appear twice, dots ● and triangles ▲ indicate terminator and starting codons, respectively. From (585).

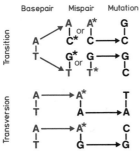

Figure 6-24. Pathways for spontaneous substitution mutations of A–T base-pairs caused by *transitions* and *transversions* (566). A comparable scheme can be drawn for G–C pairs. Both mutation processes are explained with rare tautomeric imino and enol forms of bases indicated by *. In transversion, purine–purine base-pairs with one base in *syn* conformation, as described in Figure 6-25, are involved. Pyrimidine–pyrimidine pairs are unlikely to pass enzymatic checking mechanisms because they have too short geometry. After (570).

enzyme travels along DNA in 5'→3' direction, takes up nucleoside triphosphates, and incorporates them into a newly synthesized DNA, complementary to the original DNA (567). Each step in synthesis is tested for correctness in a subsequent proofreading step, and the newly added nucleotide is excised by a counterrunning 3'→5' exonuclease if found erroneous (568,569).

Recalling Section 5.4, we find that bases can occur in rare tautomeric enol and imino forms at concentrations of 10^{-4} to 10^{-5} moles/liter. This can give rise to the non-Watson–Crick mispairs displayed in Figure 6-25. If in both synthesizing and proofreading steps, which are independent of each other, rare tautomeric forms do occur and do mimic Watson–Crick geometry, they are not detected and pass the enzymatic tests unhindered.

For an assessment of overall mutation rate, therefore, the two individual tautomer concentrations are multiplied, resulting in $10^{-4} \times 10^{-4}$ to $10^{-5} \times 10^{-5}$ or 10^{-8} to 10^{-10} mutations per newly synthesized DNA, in agreement with experiment (570).

Transversions follow a different scheme than transitions. For transversions, a different mechanism has been proposed involving purine–purine instead of purine–pyrimidine mispairs in order to explain the mutations described in Figure 6-24. Because "standard" purine–purine base–pairs are longer than purine–pyrimidine pairs and would not be tolerated by the synthesizing and proofreading apparatus, an *anti*→*syn* conversion of the new, incoming nucleoside triphosphate was postulated to allow for a "normal-size" purine–purine pair which additionally requires rare tautomeric forms of the bases (Figure 6-25). The *anti* →*syn* rotation will decrease the probability of mutations by another factor of 10^{-1} to 10^{-2}, yielding 10^{-10} to 10^{-12} transversions per newly incorporated nucleotide, again in harmony with experimental findings (570,571).

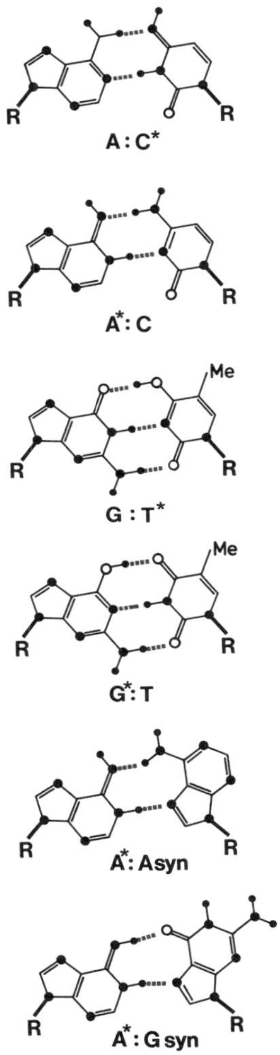

Figure 6-25. Some tautomeric non-Watson–Crick base-pairs likely to be involved in the mutation schemes described in Figure 6-24. For further base-pairs, see (570). Bases in rare tautomeric forms are marked *.

"Wobbling" of bases in mutations and in tRNA–mRNA recognition. The explanation of spontaneous substitution mutations of DNA with rare tautomeric forms of the bases is attractive and encompasses even chemical mutations due to incorporation of 5-bromouracil and 2-aminopurine. On the other hand, mutation can, at least in some cases like G–T mispairing, be accounted for by misadjustment of juxtaposed bases without invoking rare tautomeric forms. This "wobbling" due to steric misalignment of bases in a base-pair was proposed by Crick for the translation process (572) and was later verified experimentally (573–576).

The "wobble" theory was put forward in order to explain redundancies in the genetic code which relates amino acid sequence of newly synthesized polypeptide with nucleotide sequence in mRNA. In a series of experiments (577), it was demonstrated that a nucleotide triplet in mRNA (a codon) is recognized by a complementary triplet in tRNA (an anticodon) carrying, attached to its 3' terminus, the anticodon-specific amino acid to be incorporated into the growing polypeptide chain. Because 20 amino acids are coded by triplets of the four different nucleotides A,U,G,C, i.e., by $4^3 = 64$ possible words, a degeneracy is programmed in this translation code. If regarded in the reading direction from the 5' to the 3' terminus in mRNA, the first two letters of a codon are fixed according to the amino acid incorporated but the third one is variable (degenerate) and can include from one to four different bases (Figure 6-23).

The question whether each of the 64 possible codons has one partner tRNA or whether one particular tRNA can recognize several codons with the same first two letters but different letters in the third place was resolved in favor of the latter, and supported the wobble hypothesis.

According to this hypothesis, three base-pairs are formed between mRNA codon and tRNA anticodon. The first two of these are of the standard Watson–Crick type but in the third position, deviations are allowed to occur and consist of pairing between:

Base on the tRNA anticodon in third (3') position	Bases recognized on the mRNA codon	
	Standard Watson–Crick	Wobble
U	A	G
C	G	
G	C	U
I	C	U,A

It is remarkable that thus far no tRNA could be found with an A in the third (wobble) position of the anticodon (578), probably due to deamination after synthesis.

Some "wobble" base-pairs are more likely to occur than others. In the wobble hypothesis, pairings between pyrimidine bases like U–U and U–C (base-pairs XVI and XVIII in Figure 6-1) were considered improbable because the distance between glycosyl links in these base-pairs, ~8.5 Å, is about 2 Å shorter than the 10.6 Å distance in a standard Watson–Crick base-pair. The overall Watson–Crick geometry is probably essential because base-paired codon–anticodon triplets will stack in a helical arrangement and large deviations from Watson–Crick geometry could not be tolerated (579) (Section 15.7). This raises questions about the long A–I "wobble" base-pair because of the 12.8 Å separation between glycosyl $C_{1'}$ atoms. In a more recent proposal (580), an A–I$_{syn}$ base-pair as found in the 9-ethyl-8-bromoadenine·9-ethyl-8-bromohypoxanthine complex (581) was advocated; its overall dimensions imitate Watson–Crick geometry.

The wobble base-pairs proposed in (572) are displayed and compared with standard A–U Watson–Crick base-pairs in Figure 6-26. They are all linked by two cyclic hydrogen bonds, but their geometry is different from that of standard pairs because the glycosyl $C_{1'}$–N bonds do not obey dyad symmetry and the isomorphism of A–U and G–C base-pairs is lost.

Evidence for the "wobble" base-pairs comes not only from the translation process: The G–U base-pair was directly visualized in the acceptor stem of $tRNA_{yeast}^{Phe}$ (Chapter 15). It probably accounts for the stability of the poly(G)·poly(U) double helix (582) and was found to approximate, in a complementary double-helical oligonucleotide, an A–U base-pair in stabilizing efficiency (583). As for the long A–I base-pair, it could be verified in the all-purine double-helix poly(A)·poly(I) (Chapter 13).

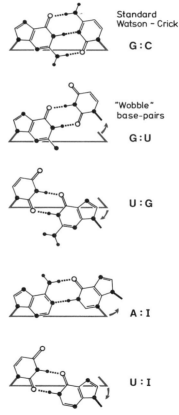

Standard
Watson – Crick

G : C

"Wobble"
base-pairs

G : U

U : G

A : I

U : I

Figure 6-26. Comparison of Watson–Crick and "wobble" base-pairs. Deviations from standard geometry are indicated by colored arrows. For more base-pairs involving not wobbling but enol/imino tautomerism, see (580).

Summary

This chapter covers associative forces between bases which are greatly responsible for the stability of a double-helical array. Most important are hydrogen-bonded base-pairs between like (homo) and different (hetero) bases which can be arranged in 28 different ways. Among the hetero pairs, only those with Watson–Crick geometry lead to regular double-helical structures because they are isomorphous. The A–T(U) and G–C base-pairs are pseudosymmetrical; a twofold rotation axis relates glycosyl links (and attached sugar–phosphate backbones) but not individual base-pair atoms. Hydrogen-bonding interactions between bases in a pair are specific and indicate electronic complementarity, with G–C more stable than A–U(T) and homo or other hetero base-pairs less favored. In crystal structures of nucleic acid constituents and double-helical arrays, bases form vertical stacks which show rather specific patterns, exocyclic groups (C=O and C—NH$_2$), or ring nitrogen located over the heterocycle of the adjacent base. Stacking interactions are mainly due to dispersion and dipole-induced dipole forces and hydrophobic bonding. In double-helical structures, formation and breakdown display cooperative characteristics, associated with geometrical constraints due to sugar–phosphate backbone and interbase interactions due to stacking and base-pairing.

The strict Watson–Crick base-pairing scheme is systematically violated in translation processes in tRNA anticodon–codon recognition where "wobble" base-pairs are allowed. This natural disregard of Watson–Crick base-pairing is not only found in translation; in DNA replication it leads to mutation, with mutation rates explicable by the occurrence of tautomeric imino and enol forms and by *syn/anti* conformers of purine nucleotides.

Chapter 7

Modified Nucleosides and Nucleotides; Nucleoside Di- and Triphosphates; Coenzymes and Antibiotics

The modified bases, nucleosides, and nucleotides of prime concern in the present chapter are found as natural substances or are obtained synthetically. They are of importance in biochemical regulation and are used extensively in chemistry, biochemistry, and pharmacology as probes to study biological mechanisms. They have also found successful applications as antibiotics or antineoplastic agents. A full account of the structure and function of modified nucleosides and nucleotides would by far exceed the scope of this presentation; only a selection of the more widely known and, from a structural point of view, important molecules will be discussed here. For specialized literature, see Refs. (21,586–590).

7.1 Covalent Bonds Bridging Base and Sugar in Fixed Conformations: Calipers for Spectroscopic Methods

The intrinsic structural flexibility of the nucleoside can be suppressed if covalent bonds are formed which intramolecularly join base and sugar or sugar hydroxyls. Using this cross-link as a tool, nucleosides locked in *syn* or in *anti* conformation can be designed, depending on whether C_8 (purine) or N_3 (purine), C_6 (pyrimidine), or C_2 (pyrimidine) is utilized as a bridgehead atom. The glycosyl torsion angle χ can be further tuned by choosing sugar carbon atoms $C_{2'}$, $C_{3'}$, or $C_{5'}$ as connectors (Table 7-1).

Base–sugar bridge $C_{2'}$–X–C(base). If the base is fixed relative to the sugar by bridges of the type $C_{2'}$–X–C(base), the heterocycle and the newly formed five-membered ring containing atoms $C_{1'}$–$C_{2'}$–X are coplanar and rigidly connected. This locks the nucleoside in a *high-anti* ($-sc$) or in a *syn* conformation according to the particular base atom that

Table 7-1. Structural Characteristics of a Selection of Nucleosides with Fixed Conformation Due to Base–Sugar Cyclization

Chemical structure	Compound name[a]	Sugar pucker	Pseudorotation parameters P (°)	ν_{max}(°)	Glycosyl torsion χ Range	χ (°)[d]	Orientation γ about $C_{4'}$–$C_{5'}$ Range	γ (°)	Reference
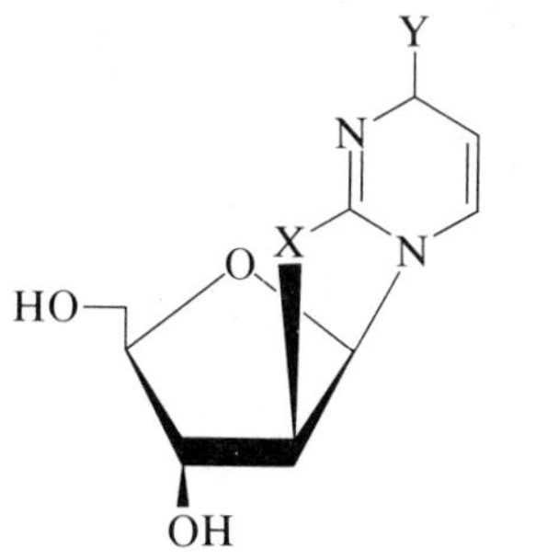	2,2'-Anhydro-1-β-D-arabinofuranosyl-uracil (X = Y = O)								
	A	$C_{4'}$-endo–$C_{3'}$-exo, $^4T_3$	227	29	syn	115	ap	175	(591)
	B	$C_{4'}$-endo–$C_{3'}$-exo, $^3T_4$	214	34	syn	110	ap	174	(592)
	-2-thiouracil (X = S, Y = O)	$C_{4'}$-endo, ^4E	232	34	syn	116	ap	187	(593)
	-cytosine (X = O, Y = NH₂)[b]	$C_{4'}$-endo–$O_{4'}$-exo, $^4T_0$	247	22	syn	119	sc	49	(594)
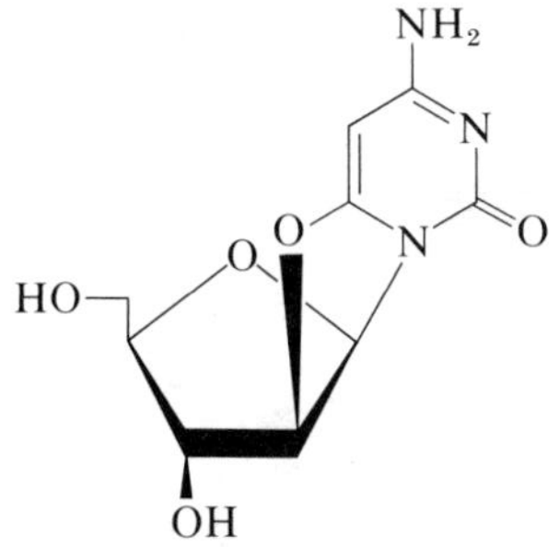	6,2'-Anhydro-1-β-D-arabinofuranosyl-6-hydroxycytosine	$C_{4'}$-exo–$O_{4'}$-endo, $_4T^0$	63	44	high-anti	291	sc	57	(595)

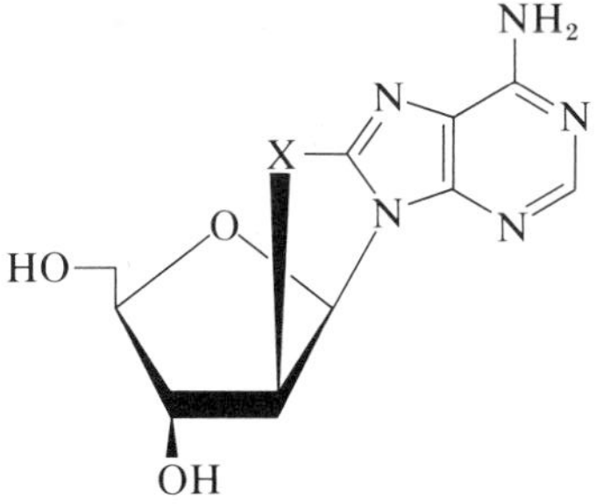

Compound	Sugar conformation							Ref.
8,2'-Anhydro-9-β-D-arabinofuranosyl-adenine (X = O)	$C_{4'}$-endo–$C_{3'}$-exo, ^4_3T	218	19	high-anti	287	sc	59	(596)
-8-mercapto-adenine (X = S)[c]	$C_{4'}$-endo, ^4E	232	34	high-anti	299	ap	163	(597)
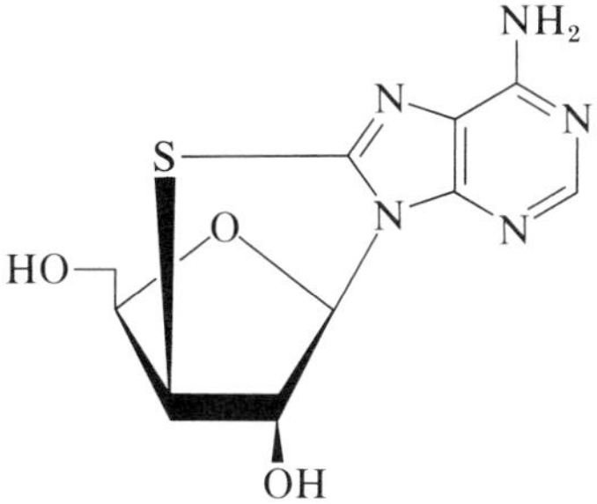 8,3'-Anhydro-8-mercapto-9-β-D-furanosyladenine	$C_{2'}$-exo, $_2E$	345	45	high-anti	256	sc	74	(598)
2,5'-Anhydro-1-(2',3'-O-isopropylidene-β-D-ribofuranosyl)-uracil (X = O)	$C_{4'}$-endo–$O_{4'}$-exo, ^4_0T	254	42	syn	71	sc	42	(599, 600)
-2-thiouracil (X = S)	$C_{4'}$-endo–$O_{4'}$-exo, ^4_0T	250	35	syn	64	sc	45	(601)

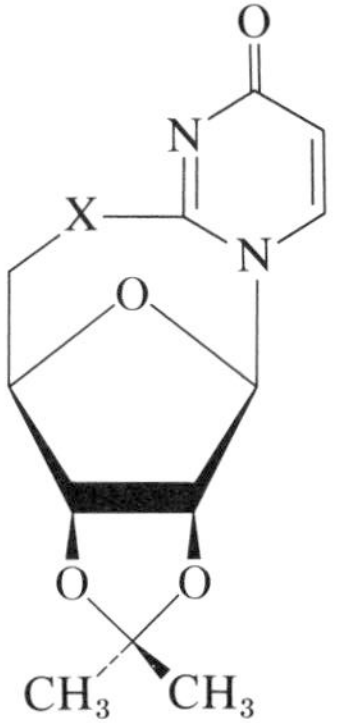

Table 7-1. (*Continued*)

Chemical structure	Compound name[a]	Sugar pucker	Pseudorotation parameters		Glycosyl torsion χ		Orientation γ about $C_{4'}$–$C_{5'}$		Reference
			P (°)	ν_{max} (°)	Range	(°)[a]	Range	(°)	
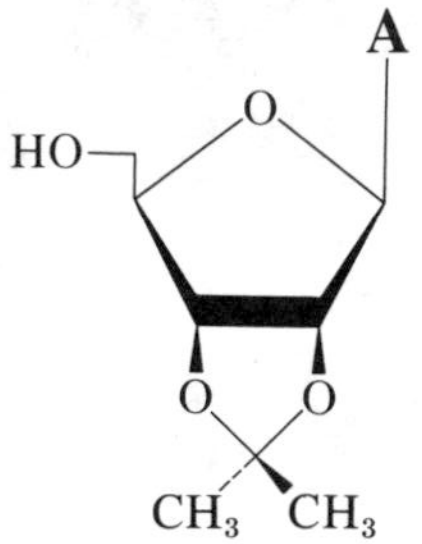	2′,3′-*O*-Isopropyli-deneadenosine								
	A	Planar	—	—	*anti*	−170	*sc*	54	(602)
	B	C_{4}-*endo*–$C_{3'}$-*exo*, $^{4}_{3}T$	215	32	*anti*	−164	*ap*	175	
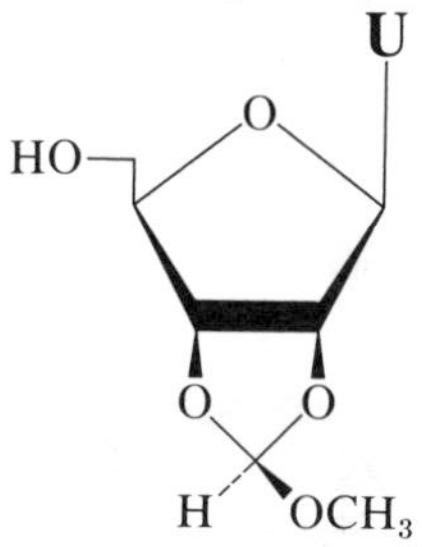	2′,3′-*O*-Methoxy-methyleneuridine	$C_{2'}$-*endo*, ^{2}E	163	23	*anti*	−124	*sc*	53	(603)

$^{a-c}$ A and B refer to two independent molecules within the asymmetric unit: ^{b} as 3′,5′-diphosphate; ^{c} as 5′-monophosphate.

^{d} Published values for χ refer to $O_{4'}$–$C_{1'}$–N_{9}–C_{8} (purines) and $O_{4'}$–$C_{1'}$–N_{1}–C_{6} (pyrimidines). They have been converted to the now-used nomenclature (Chapter 2) by adding or subtracting 180°. The χ data given here are therefore only approximate, ±3°.

has been used for cyclization (Figure 7-1a). Because this link freezes rotations about the glycosyl $C_{1'}-N$ bond as well as about $C_{1'}-C_{2'}$ and rigidly adjusts the endocyclic torsion angle ν_1 ($O_{4'}-C_{1'}-C_{2'}-C_{3'}$) to near 0°, only puckerings of $C_{4'}$ are possible. This yields $C_{4'}$-*exo* or $C_{4'}$-*endo* envelope sugar conformations with pseudorotation phase angle, P, around 240° or 60° (Table 7-1).

It goes without saying that the base–sugar cyclization involving atom $C_{2'}$ has some drastic effects on bonding geometry, especially on valence angles around the glycosyl link and at atom $C_{2'}$. If the oxygen O_2 in pyrimidine nucleosides becomes bridge atom, the whole electronic system (and with it hydrogen-bonding donor and acceptor sites of the base) is altered because the double-bonded keto oxygen O_2 is changed to an oxygen involved in two single bonds.

Structure (I) in Figure 7-2 is predominant among the possible canonical electronic structures for cytosine derivatives (594). It is of interest that in no crystallographic study has an imino form of type (V) been observed even in bridged cytosine derivatives. Rather, the mesomeric forms (II and III) with the positive charge located on N_1 appear to prevail. For this reason, in the bridged cytosine derivatives the orientation about $C_{4'}-C_{5'}$ is predominantly $+sc$, positioning hydroxyl $O_{5'}-H$ in favorable contact to positively charged N_1. In uracil derivatives, the situation is quite different because the heterocycle does not bear a significant charge and $O_{5'}$ is therefore more frequently directed away, in *ap* position (604).

Base–sugar bridge $C_{3'}-X-C$(base). The angular distortions around atoms of the glycosyl bond and at the sugar bridgehead atom are less dramatic when $C_{3'}$ rather than $C_{2'}$ is involved in cyclization. For geometrical reasons, the resulting cyclonucleoside will adopt preferentially the $C_{2'}$-*exo* (or the equivalent $C_{3'}$-*endo*) pucker. The glycosyl torsion angle χ, although still in the *high-anti* ($-sc$) or *syn* range depending on the attachment at the base, differs by about 40° from $C_{2'}$ bridged compounds. It is remarkable that in the only known crystal structure of such a molecule, 8,3'-anhydro-8-mercapto-9-β-D-xylofuranosyladenine (598) (Figure 7-1b), atom $O_{5'}$ is still located over the ribose although steric interference with the bridge sulfur atom pushes $O_{5'}$ more toward atom $O_{4'}$, as indicated by an increased γ angle, 74.3°, the largest value in the sc range in Table 7-1.

Base–sugar bridge $C_{5'}-X-C$(base). The last possibility to bridge base and sugar is $C_{5'}-X-C$(base). The two molecules of this kind crystallographically investigated are 2',3'-O-isopropylidene derivatives of 5',2-O-cyclouridine, with oxygen and sulfur as bridge atoms (Table 7-1, Figure 7-1c). Because in this particular base–sugar link, $C_{5'}$ and base approach each other closer than allowed in normal nucleosides, the usually forbidden $O_{4'}$-*exo* sugar puckering now becomes a standard structural feature.

Model building studies suggest that in bridges of the type $C_{5'}-X-C$(base), torsion angles χ can adopt two values, around 0° (*sp*) and 60° (*sc*), which position atom X approximately over sugar atom $O_{4'}$ or over

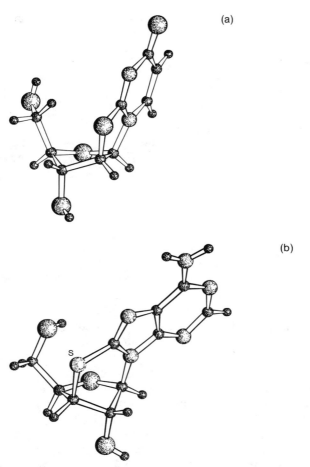

Figure 7-1. Molecular structures of some nucleosides with fixed conformations selected from Table 7-1. **(a)** 2,2'-Anhydro-1-β-D-arabinofuranosyluracil (591,592) with sugar constrained to C$_{4'}$-*endo* pucker and base in *high-anti* ($-sc$) orientation. **(b)** 8,3'-Anhydro-8-mercapto-β-D-xylofuranosyladenine (598), with sugar fixed in C$_{2'}$-*exo* pucker $_2$E and base oriented *high anti* in the ($-sc$) range.

the furanose ring. As the former orientation would lead to stereochemically unfavorable short contacts between atoms X and O$_{4'}$, it is avoided and thus far the only observed conformation of this kind is with atom X located over the sugar ring.

Bridge between vicinal ribose O$_{2'}$, O$_{3'}$ hydroxyls. A frequently used modification of ribonucleosides is blocking and bridging of the vicinal *cis*-oriented O$_{2'}$, O$_{3'}$ hydroxyls by isopropyl or other similar groups in order to protect them from further chemical attack. Such cyclized nucleosides

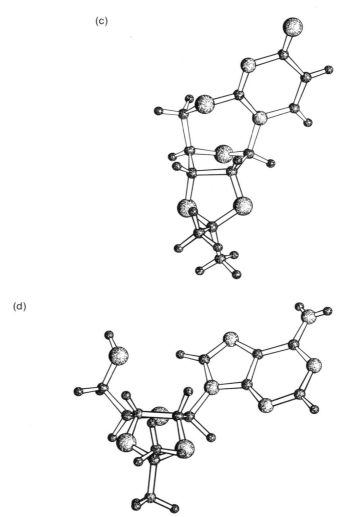

(c)

(d)

(c)   2,5'-Anhydro-1-(2',3'-*O*-isopropylidene)-β-D-ribofuranosyluracil   (599,600),
with base in *syn* position and sugar in the otherwise forbidden $O_{4'}$-*exo* puckering
mode $_{o}E$. (d) 2',3'-*O*-Isopropylideneadenosine (602), a nucleoside with planar
ribose. This situation never occurs in unmodified nucleosides.

still exhibit full flexibility about glycosyl and $C_{4'}-C_{5'}$ bonds but, as Table
7-1 indicates, the sugar puckering amplitude ν_{m} is reduced and even
planar ribose moieties are observed, a case not yet encountered in crystallographic studies of unmodified nucleosides (Figure 7-1d) [see Ref. (172)].
In this context it is worth mentioning that a similar planar ribose is found

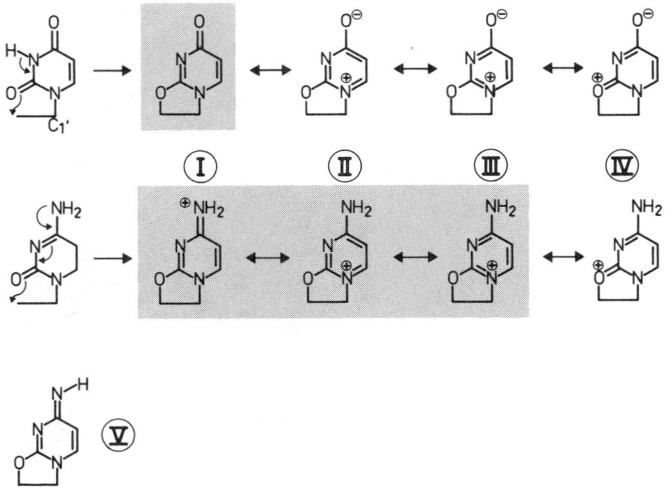

Figure 7-2. Fusion of base to sugar via pyrimidine O_2 changes the electronic system of the heterocycle and, in cytosine derivatives, a positive charge is introduced. The imino form (V) was never observed in crystallographic studies; predominant mesomeric structures are colored in rectangles.

in cytidine $2',3'$-O-cyclophosphate (Table 7-2), suggesting that in general, $2',3'$-O-cyclization of riboses appears to stablize the planar form as intermediate between relatively flat sugar puckering modes $C_{2'}$-$endo \rightleftharpoons C_{3'}$-$endo$. As in unmodified nucleosides, these are in rapid equilibrium. The $syn \rightleftharpoons anti$ distribution, however, is shifted in favor of syn according to nuclear magnetic resonance data (603,605).

Conformationally locked nucleosides offer a unique system to calibrate spectroscopic methods and to test biological systems for their specificity. Owing to the inherent flexibility of nucleosides, spectral assignment of structural features is difficult because the dynamic sugar puckering equilibrium occurs simultaneously with correlated rotational movements about the exocyclic $C_{4'}$–$C_{5'}$ and glycosyl $C_{1'}$–N bonds (Section 4.13). If one or all of these conformational freedoms are blocked, interpretation of ORD, CD, and NMR spectroscopic data can be put on safer grounds (599, 603,606–610). One must keep in mind, however, that the chemical modification used to bridge sugar and base could well alter the spectroscopic signals relative to unmodified nucleosides.

Conformationally locked nucleosides are excellent tools to study the specificity of enzymes which require nucleosides and nucleotides as substrates, inhibitors, or coenzymes. They have also found application in pharmacological investigations (21,588), and it has been demonstrated that cyclic nucleosides can exert strong antitumor and antileukemic activity (611).

7.2 Cyclic Nucleotides

In nature, two kinds of cyclic nucleotides are of importance, with the phosphodiester group bridging 2',3' and 3',5' oxygens (Figures 2-2, 7-3).

The nucleoside 2',3'-O-cyclic phosphates are formed as intermediates during hydrolysis of RNA by ribonucleases (RNases). These enzymes hydrolyze RNA in two steps. In the first step the bond between P and $O_{5'}$ atoms is broken via transesterification to produce RNA fragments with terminal 2',3'-O-cyclic phosphate. In a subsequent reaction step, the cyclic phosphodiester is opened to yield only the terminal 3'-O-phosphate (612). This mechanism is rather well understood for pancreatic RNase and will be described in Section 7.5.

The nucleoside 3',5'-cyclic phosphates, especially the purine derivatives, display complex biochemical action. Adenosine 3',5'-cyclic phosphate (cyclic AMP, cAMP) serves as link between a large group of hormones and a group of regulatory enzymes (protein kinases), and is therefore called a second hormonal messenger (613–615). In bacteria, cyclic AMP also plays a role in the regulation of gene expression (616). As we will see in Chapter 18, it forms a complex with the catabolite gene activator protein (CAP) and exerts allosteric control. The guanosine analog differs in its biological specificity from cyclic AMP and appears to have an opposing effect in cell proliferation and in other cellular events (617).

The overall geometrical features of nucleotides with 2',3'-,3',5'-, and synthetically obtained 2',5'-cyclic phosphodiester groups are summarized in Table 7-2 (see Figure 7-3). As a general rule, these molecular systems are conformationally constrained and similar to the cyclic nucleosides already discussed.

Nucleoside 2',3'-O-cyclic phosphates have flattened ribose pucker and show tendency for *syn* conformation. Crystallographic information for nucleotides with vicinal, *cis*-oriented $O_{2'}$,$O_{3'}$ hydroxyls cyclized by a phosphodiester group is contained in Table 7-2 and Figure 7-3a. Because the five-membered cyclic phosphate ring forces the $O_{2'}-C_{2'}-C_{3'}-O_{3'}$ group into near-planar arrangement, the fused ribose moiety in these systems is restricted in its conformational flexibility. The puckering amplitude is flattened as indicated by the occurrence of a planar ribose in the cytidine derivative (Table 7-2) and by ν_{max} values between 22° and 36°, significantly less than the average, 39°, for unmodified furanose rings (172). Only one of the nucleoside 2',3'-O-cyclic phosphates displayed in Table 7-2 adopts a $C_{2'}$-*endo* pucker whereas for the others, $O_{4'}$-*endo* and even the otherwise unfavorable $O_{4'}$-*exo*-type envelope conformations are found, all considerably flattened and similar to 2',3'-O-isopropylideneadenosine (Table 7-1). If we recall from Section 4.2 that planar and $O_{4'}$-*exo* puckered riboses are forbidden for stereochemical reasons, it becomes clear that these cyclic nucleotides are strained molecular systems.

Table 7-2. Structural Characteristics of a Selection of Cyclic Nucleotides

Chemical structure	Compound name[a]	Sugar pucker	Pseudorotation parameters P (°)	v_{max} (°)	Glycosyl torsion χ Range	(°)[b]	Orientation γ about $C_{4'}-C_{5'}$ Range	(°)	Reference
(11) 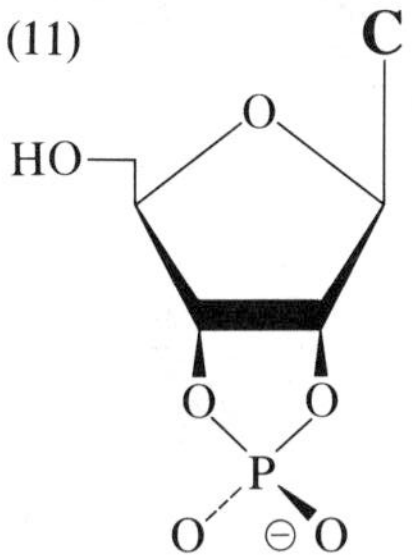	Cytidine 2',3'-O-cyclophosphate, sodium salt								
	A	$O_{4'}$-endo–$C_{4'}$-exo, $^{O}T_4$	82	36	syn	63	sc	56	(618)
	B	Planar	—	—	syn	75	ap	162	(619)
	—, Free acid	$C_{2'}$-endo–$C_{3'}$-exo, $^{2}T_3$	174	29	anti	−118	sc	69	(619)
(12) 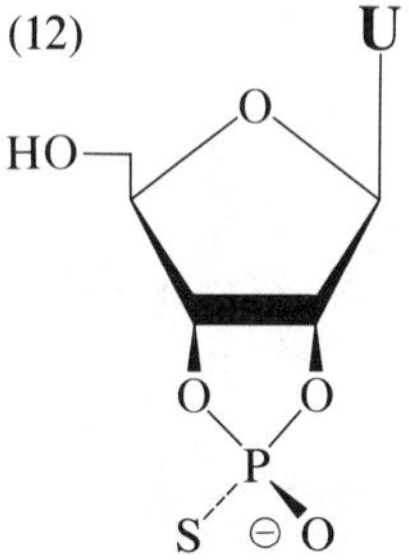	Uridine 2',3'-O-cyclophosphorothioate, triethylammonium salt	$O_{4'}$-exo–$C_{4'}$-endo, $_{0}T^4$	262	23	anti	−167	sc	74	(620)

	Sugar pucker							
1-β-D-Arabinosyl-cytidine 2',5'-O-cyclophosphate, free acid	$C_{2'}$-endo–$C_{3'}$-exo, $^2T_3$	171	42	anti	−154	sc	65	(621)
Guanosine 3',5'-O-cyclophosphate, sodium salt (cyclic GMP)	$C_{4'}$-exo–$C_{3'}$-endo, $_4T^3$	43	44	syn	78	−sc	61	(622)
Uridine 3',5'-O-cyclophosphate, sodium salt								
A	$C_{4'}$-exo–$C_{3'}$-endo, $_4T^3$	42	48	anti	−103	−sc	−61	(623)
B	$C_{4'}$-exo–$C_{3'}$-endo, $_4T^3$	48	47	anti	−122	−sc	−60	
5'-Methyleneadeno-sine 3',5'-cyclo-phosphonate	$C_{3'}$-endo–$C_{4'}$-exo, ^3_4T	37		syn	54	−sc	−64	(624)

^a A and B refer to two independent molecules in the asymmetric unit.

^b Published values, except in the third entry [Ref. (619)], were given for $O_{4'}$–$C_{1'}$–N_1–C_6 (N_9–C_8) and are converted to the now used $O_{4'}$–$C_{1'}$–N_1–C_2 (N_9–C_4) by adding or subtracting 180°. The data given here for χ are therefore only approximate to ±3°.

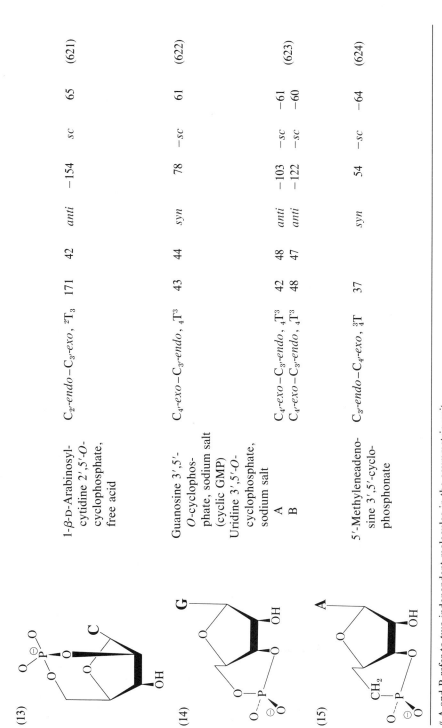

(13) (14) (15)

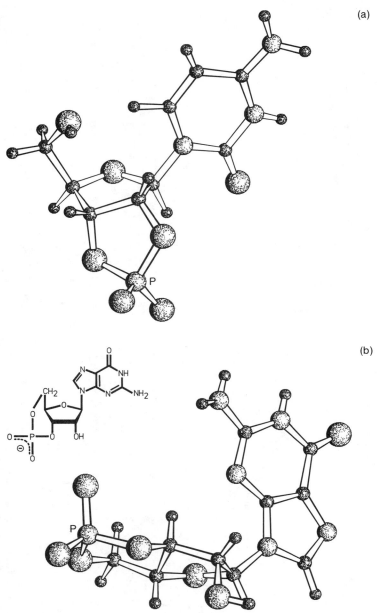

Figure 7-3. Molecular structures of the two naturally occurring cyclic phosphates, with conformational parameters entered in Table 7-2. **(a)** Cytidine 2′,3′-O-cyclophosphate, free acid (619). Base is protonated at N_3 and in *anti* position; sugar puckering is rather shallow due to 2′,3′-cyclization and described as $C_{2'}$-*endo*–$C_{3'}$-*exo*. $^2T_3$. **(b)** Guanosine 3′,5′-O-cyclophosphate, a nucleotide with ribose rigidly fixed in $C_{4'}$-*exo*–$C_{3'}$-*endo* pucker, $_4T^3$. The orientation of the base is *syn* (622).

NMR studies of uridine and cytidine 2′,3′-O-cyclic phosphate. The insensitivity of the H(6) chemical shift of these molecules with increasing temperature contrasts the changes observed in cytidine 3′-phosphate and was interpreted as meaning that pyrimidine nucleoside 2′,3′-O-cyclic phosphates are locked in *syn* conformation, with barriers to rotation estimated to about 25 kcal/mole (625–627). These data, which are in accord with crystallographic studies (Table 7-2), can be rationalized if we assume that the $C_{2'}-H_{2'}$ atomic group has to flex down if the pyrimidine base rotates from *syn* to *anti,* a movement obviously hindered if the vicinal $O_{2'}$, $O_{3'}$ hydroxyls are bridged. This does not hold for purine derivatives in which the sterically less demanding base shows *syn*⇌*anti* interchange (605,627,628).

The conformation about the glycosyl bond is also reflected in the orientation about the exocyclic $C_{4'}-C_{5'}$ counterpart. NMR data suggest that for pyrimidine nucleoside 2′,3′-O-cyclic phosphates which favor *syn,* the $C_{4'}-C_{5'}$ torsion angle γ adopts *ap* and −*sc* ranges, positioning $O_{5'}$ away from the sterically interfering base O_2. For purine derivatives the +*sc* range is preferred and stabilizes the *syn* form via $O_{5'}-H \cdots N_3$ hydrogen bonding (627,628).

NMR data indicate a $C_{2'}$-*endo*⇌$C_{3'}$-*endo* equilibrium for the sugar puckering modes of these cyclic nucleotides. In view of the crystallographic results displaying planar and $C_{2'}$-*endo* puckered nucleoside 2′,3′-O-cyclic phosphates, this equilibrium appears reasonable with the understanding that the sugar puckering amplitude ν_{max} is reduced in comparison to noncyclic nucleotides and that $O_{4'}$-*endo*, $O_{4'}$-*exo*, and planar sugar conformations as transition states in the $C_{2'}$-*endo*⇌$C_{3'}$-*endo* interconversion are stabilized by the cyclization.

Theoretical studies on nucleoside 2′,3′-O-cyclic phosphates using PCILO (629) and semiempirical potential energy methods (630,631) have corroborated the experimental results and have shown that for $O_{4'}$-*endo* furanose pucker, the *syn* conformation is preferred for pyrimidine analogs. In the purine series, *syn* and *anti* are equally probable with the orientation about the $C_{4'}-C_{5'}$ bond preferentially in +*sc* or in *ap* ranges. In $O_{4'}$-*exo* puckered nucleoside 2′,3′-O-cyclic phosphates, however, *anti* is the only possible position for the bases. It is accompanied by a −*sc* or *ap* rotation about the $C_{4'}-C_{5'}$ bond directing the $O_{5'}$ group away from the sugar and from the base which interferes sterically because the $O_{4'}$-*exo* form forces the exocyclic substituents close together.

Nucleoside 3′,5′-cyclic nucleotides have furanose puckering fixed in a high energy conformation. In this class of cyclic nucleotides, the stereochemical situation is very different from that of the 2′,3′-cyclic analogs. In the latter, formation of the five-membered phosphodiester ring constrains the $O_{2'}-C_{2'}-C_{3'}-O_{3'}$ atoms in near-planar arrangement and causes reduced puckering and even planar shape of the attached ribose. In cyclic 3′,5′-nucleotides, however, the phosphodiester group is part of a six-membered ring in fixed chair conformation fused to the ribose through the $C_{3'}$–$C_{4'}$ bond and restricting the ribose pucker to $C_{4'}$-exo–$C_{3'}$-$endo$ ($_4T^3$) Figure 7-3b).

Chair conformations of six-membered rings are associated with endo- and exocyclic torsion angles in the $+sc$ or $-sc$ range, i.e., around $\pm60°$. Hence, rotation about the $C_{3'}$–$C_{4'}$ bond common to both ring systems in cyclic 3',5'-nucleotides tries to adopt this value. It succeeds only in part, however, because the ribose torsion angle ν_3, defined as $C_{2'}$–$C_{3'}$–$C_{4'}$–$O_{4'}$, which in unsubstituted nucleosides adopts values between $+33°$ and $-41°$, can only be increased to about $50°$ and results in extreme ribose puckering amplitudes ν_{max} around $45°$ (172).

In crystal structures of cyclic 3',5'-nucleotides (622,623,632,633), in a modified analog with 8-substituted adenine (634), and in dimethylphosphoroamidate and in benzyl triester derivatives of thymidine and uridine 3',5'-cyclophosphate (635,636), both syn and $anti$ conformations were observed. In one case, cyclic AMP, they coexist even within the same crystal structure (632). These observations are consistent with NMR data demonstrating rapid $syn\rightleftharpoons anti$ interconversion, probably facilitated by the fixed orientation of $O_{5'}$ "away" from the ribose in ap position and by the ribose puckering being restricted to $C_{4'}$-exo–$C_{3'}$-$endo$ (153,637–642). Similar conclusions were also drawn on the basis of theoretical studies (629,630).

Cyclic nucleotides are internally strained as reflected in geometrical features and in enthalpies of hydrolysis. If bond distances in phosphate groups of open-chain diesters and of cyclic diesters with varying ring size are compared, values similar to those in Figure 4-25 are observed, with only small deviations originating mainly from different environments within the crystal lattices. On the other hand, bond angles evidence a strong dependence on ring size, especially if the endocyclic P–O–P and O–P–O angles are considered. As shown in Table 7-3, these angles in the open phosphodiester resemble those in the six-membered cyclic phosphodiester. But they are decreased in the five-membered and increased in the seven-membered cyclic systems, reflecting mechanical strain. Thus, from a distortion-energy correlation of phosphate diesters, cyclic 3',5'-nucleotides should be less strained than their 2',3' and 2',5' analogs. In contrast, enthalpies of hydrolysis for the cyclic 2',3'- and 3',5'-nucleotides display a reversed order (Table 7-3).

This apparent discrepancy can be explained if not only bond angles of phosphate groups but also conformations of the fused furanose rings are taken into account. As Table 7-3 indicates the enthalpy of hydrolysis increases in going from cyclic tetramethylene phosphate to its dimethylene homolog; this difference is due to steric strain in the five-membered circle. Fusion of the ring to a ribose, as in cyclic 2',3'-nucleotides, increases the enthalpy of hydrolysis by another 2.5 kcal/mole. This is an indication of the extra strain introduced by the furanose system which, as pointed out earlier, is flattened (and therefore strained) relative to noncyclized nu-

Table 7-3. Enthalpies of Hydrolysis $-\Delta H$ (kcal/mole) for Cyclic and Noncyclic Phosphodiesters

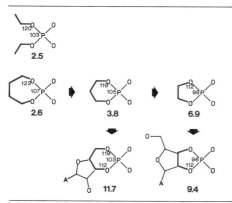

Note. Larger numbers refer to energy values whereas smaller numbers give O–P–O and P–O–C bond angles. If only one P–O–C angle of a phosphate group is indicated, the other one is comparable. Note that the enthalpy of hydrolysis for open-chain diethyl phosphate is the same as for cyclic tetramethylene phosphate (2.5 and 2.6 kcal/mole), and if the size of the cyclic structure is reduced, the enthalpy increases due to increasing steric strain. In cyclic 2′,3′-O- and 3′,5′-O-nucleotides, additional strain is introduced because of geometrical limitations imposed by the fused furanose ring. Data are from (643, 645).

cleotides. On the other hand, if the six-membered cyclic trimethylene phosphate with relatively low enthalpy of hydrolysis, 3.8 kcal/mole, is fused with a ribose to yield a cyclic 3′,5′-nucleotide, it preserves its geometry. The conformation of the ribose, however, is disturbed with torsion angle $C_{2'}-C_{3'}-C_{4'}-O_{4'}$ around 50° being increased by 15° relative to normal ribose. This leads again to a strained system with a 7.9-kcal/mole increase in enthalpy.

In summary, the energy considerations in the fusion of cyclic phosphodiesters with riboses demonstrate that in five-membered cyclic 2′,3′-nucleotides, extra heat of hydrolysis is a result of strain in the two fused five-membered systems. In cyclic 3′,5′-nucleotides, however, it may be traced back to the steric strain of the ribose alone; the six-membered cyclic phosphate group is relatively unaffected (644–646).

7.3 Nucleosides with Modified Sugars: Halogeno-, Arabino-, and α-Nucleosides

Of the great variety of synthetic nucleosides with modified sugars, those with halogen, amino, or azido groups replacing secondary hydroxyls, $O_{2'}H$ or $O_{3'}H$, are of greatest interest. As discussed in Section 4.3, the electronegativity of these substituents appears to directly influence the sugar puckering mode. Moreover, 2′-deoxy-2′-halogenonucleosides, if in-

Box 7-1. On the Mechanism of Pancreatic Ribonuclease (RNase A)

RNase A cleaves single-stranded RNA at the 3' end of pyrimidine nucleosides. An RNA fragment with intermediate terminal 2',3'-*O*-cyclophosphate is formed (Scheme 1) which in a second step is hydrolyzed to produce a 3'-phosphate. In this catalytic process, two histidine residues (His12 and His119) located in the active site region of RNase A are involved.

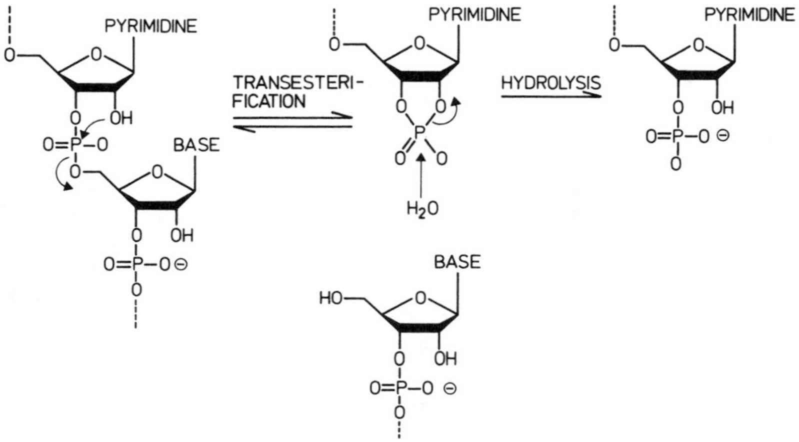

Scheme 1

The question is whether this reaction occurs via an in-line (S_N2) mechanism or whether pyrimidine O_2 is involved and assists in an adjacent mechanism (Scheme 2). In both cases, a pentacovalent phosphorus is formed as intermediate which, in the adjacent mechanism, must undergo pseudorotation in order to bring the leaving group ($O_{2'}$) into apical position (722).

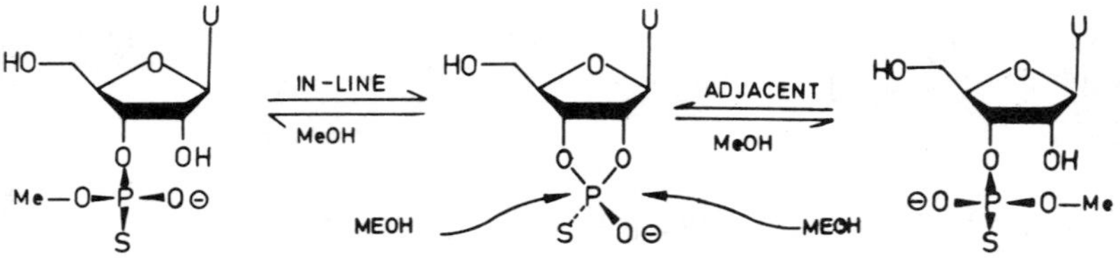

Scheme 2

Since the (cyclic) phosphodiester is not chiral, products of a normal hydrolysis cannot tell us about the reaction path. If, however, the phosphate is made chiral by introduction of a sulfur atom and the

RNase-catalyzed hydrolysis is performed in the presence of methanol, then a chiral product is formed. As Scheme 3 indicates, the stereochemistry of this product allows us to conclude that the RNase-catalyzed reaction proceeded via an in-line mechanism because the alternate adjacent path had brought forth the diastereoisomeric product (699). Based on this principle a number of enzymatic reaction mechanisms involving nucleoside mono-, di-, or triphosphates have been elucidated (698).

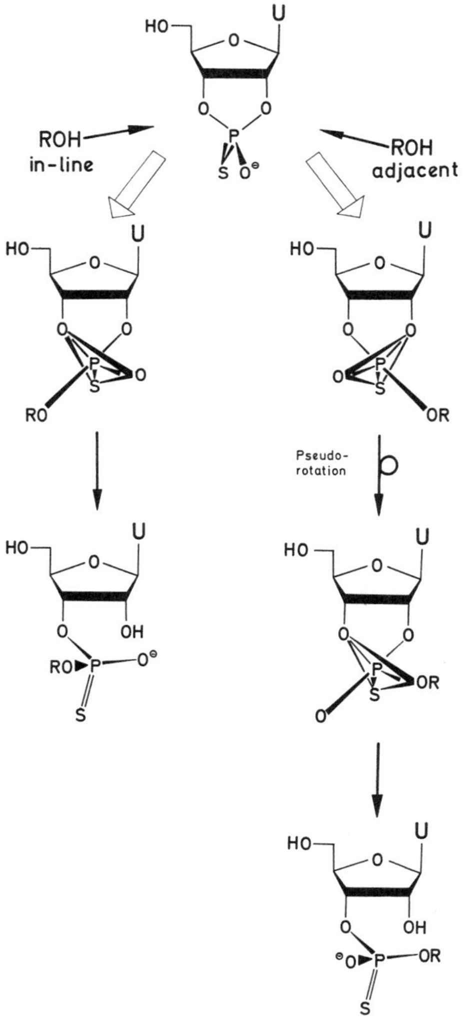

Scheme 3

corporated into polymeric RNA, contribute significantly to its *in vivo* stability because RNases cannot degrade such modified RNA using the mechanism displayed in Box 7-1.

As halogeno-substituted nucleosides are structurally normal they are not treated further here. Rather, we turn to arabino- and α-nucleosides, the first being mainly of biological and the second of structural interest.

Arabinonucleosides have attracted great attention owing to their broad-spectrum antiviral activity against DNA-containing viruses and RNA tumor viruses (oncornaviruses) both *in vivo* and *in vitro*. They occur in natural sources, and 9-β-D-arabinofuranosyladenine (adenine arabinoside, vidarabine, ara-A) has been found especially suitable for pharmacological purposes (647,648). The arabinonucleosides differ from their ribo analogs in the altered configuration at $C_{2'}$. The $O_{2'}$ hydroxyl is *cis*-oriented to the glycosyl $C_{1'}-N$ link (Figure 7-4). This change has slight consequences on furanose bond distances and angles (167). However, it does not influence puckering modes which are still preferentially $C_{2'}$-*endo* or $C_{3'}$-*endo* and it has no obvious effect on the orientations about the exocyclic $C_{4'}-C_{5'}$ and glycosyl $C_{1'}-N$ bonds, which are preserved in the *sc* and *anti* ranges. The *high-anti* ($-sc$) orientation of the base, however, is forbidden, and this conformational characteristic could explain some of the biochemical properties of these nucleosides (649).

In some of the crystal structures described for arabinofuranosyl nucleosides, intramolecular hydrogen bonds joining $O_{2'} \cdots O_{5'}$ are observed (Figure 7-4). As indicated by spectroscopic data, however, they are not a significant solution feature in neutral media (650). Intramolecular $O_{2'} \cdots O_{5'}$ hydrogen bonds are significant only if, in alkaline medium, $O_{2'}H$ is ionized (651) [or, as assumed in theoretical calculations, the molecule exists *in vacuo* (652,653)]. Owing to the particular orientation of $O_{2'}H$ *cis* to the glycosyl $C_{1'}-N$ linkage, one might expect steric hindrance to base rotation. In fact, of the nine arabinofuranosyl nucleosides thus far studied crystallographically, none was found in the *syn* form. But, as evidenced from solution studies, *syn*$\rightleftharpoons$*anti* interconversion still takes place, albeit at a reduced rate due to increased barriers to rotation.

α-Nucleosides are conformational enantiomers of the common ribonucleosides. While for the arabinofuranosyl nucleosides the point of configurational inversion is $C_{2'}$, it is the anomeric carbon $C_{1'}$ for the α-nucleosides (Figure 7-5). They are not found in natural polymeric nucleic acids but do occur as such or as constituents of smaller molecules in living organisms (654–656). The α-nucleoside component of vitamin B-12 is of particular importance (657,658). α-Nucleosides can also act as substrates of 5'-nucleotidase (659,660), and they display pronounced antimetabolic activities in various systems (661,662).

The heterocycle is attached to $C_{1'}$ in α configuration and is *cis*-positioned with respect to the $C_{2'}-O_{2'}$ bond. Thus some steric influence on the conformation of the ribose ring may be anticipated. By and large, the sugar puckering modes in α-nucleosides can be described as the enantio-

(a)

(b)

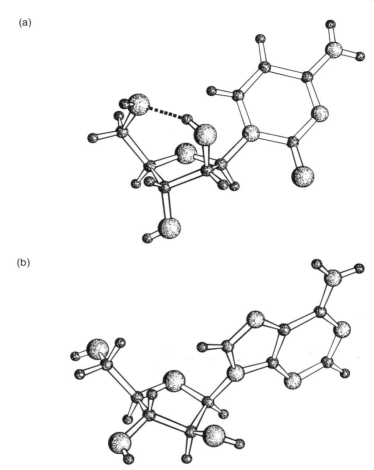

Figure 7-4. Two arabinonucleosides. **(a)** 1-β-D-Arabinosylcytosine, with hydrogen bond $O_{2'}H \cdots O_{5'}$ (714). **(b)** 9-β-D-Arabinosyladenine (vidarabine, ara A), a broad-spectrum antiviral agent (715). The intramolecular hydrogen bond between $O_{2'}$ and $O_{5'}$ is important if ribose hydroxyls are deprotonated (pH > 12.5) but is diminished at physiological pH (651).

meric forms of β-nucleosides; $C_{2'}$-*exo* and $C_{3'}$-*exo* (or $C_{4'}$-*endo*) are now generally preferred (663). This reversal of conformational features demonstrates that the type of glycosyl linkage (α or β) has a direct influence on the conformation of the sugar moiety.

As in β-nucleosides, the preferred orientation about the glycosyl link is *anti*. For the $C_{4'}-C_{5'}$ bond, however, the preponderance of the *sc* conformation is less obvious than in the β-nucleosides, a result indicated by the available crystallographic data [references in (664)] and by solution studies using nuclear magnetic resonance spectroscopy (638,665,666). In terms of intramolecular interactions, these findings suggest that the $O_{5'} \cdots H-C$(base) interactions discussed in Section 4.9 are the main fac-

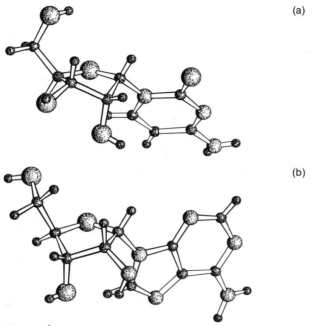

Figure 7-5. Molecular structures of two α-nucleosides. **(a)** 1-α-D-Ribofuranosyl-cytosine (716) **(b)** α-D-2′-amino-2′-deoxyadenosine monohydrate (717a). In both nucleosides, the base is in *anti* orientation. Sugar pucker is $C_{3'}$-*endo* (^3E) for cytosine and $C_{2'}$-*endo* (^2E) for adenine derivative.

tors pulling the $O_{5'}$ hydroxyl into the preferred *sc* orientation. On the other hand, the spectroscopic data on α-nucleosides suggest a subtle influence of the *gauche* effect (Section 4.12): The populations of the three staggered conformations about the $C_{4'}$–$C_{5'}$ bond are distributed such that torsion angles $O_{4'}$–$C_{4'}$–$C_{5'}$–$O_{5'}$ in ± *sc* ranges are slightly more preferred than *ap*. This is similar to results obtained with nucleotides where the base was substituted by H or by O–CH_3 (233a).

7.4 Modified Bases: Alkylation of Amino Groups (Cytokinins) and of Ring Nitrogens, Thioketo Substitution, Dihydrouridine, Thymine Dimers, Azanucleosides

Nature has been more creative with modified bases than with modified sugars. This is probably because the sugar–phosphate backbone in poly-nucleotides acts as inert string holding the functional bases in the right place and any modification on sugars would disturb this scheme.

In tRNA, a great number of base modifications, summarized in Figure

7-6, have been detected (589,590,667–669). Most characteristic are alkylations of adenine, guanine, and cytosine amino groups; alkylations of ring nitrogen atoms such as N_1 of adenine, N_7 of guanine, and N_3 of cytosine; substitution of keto by thioketo groups; saturation of a C=C double bond in 5,6-dihydrouracil; and introduction of a C-glycosyl link in pseudouridine.

Alkylation of adenine amino groups is of particular interest because these substances, better known as cytokinins, are plant hormones and play a major role in directing growth and development of plants (670,671). In all active cytokinins, the amino group of adenine carries an aliphatic or aromatic substituent with a double bond in the β,γ position (Figure 7-7). Crystallographic studies on cytokinins and on comparable minor constituents of tRNA have shown that in all cases, the N_6 substituent is in distal position with respect to the imidazole ring, probably because it would interfere sterically with N_7 if proximally oriented (672–677). This particular conformation, however, blocks N_1 from accepting and N_6 from donating Watson–Crick type hydrogen bonding. It leaves the Hoogsteen site open for base-pair interactions. A further characteristic feature of the cytokinins is that the plane defined by the atoms attached to the double bonds of the side chain is nearly perpendicular to the base (Figure 7-7). If this particular geometry is blocked by bulky substitutents in chemically modified cytokinins, their activity is greatly reduced (673).

A slightly different situation is encountered in the tRNA minor constituent with hypermodified adenine, N-[N-(9-β-D-ribofuranosylpurin-6-yl)carbamoyl]threonine, **t^6A**. Crystallographic data established that the side chain is again distal to N_7 and its amide NH group is involved in a bifurcated hydrogen bond with adenine N_1 as well as the threonyl hydroxyl as acceptors (674) (Figure 7-8). This interaction leads to coplanar arrangement of adenine and side group atoms so that only the terminal carboxyl and methyl groups protrude from this plane.

Extrapolation to N_4-acetylcytidine, a tRNA constituent, predicts that the acetyl group is oriented proximal to N_3 and blocks the Watson–Crick base-pairing site. This suggestion is in line with nuclear magnetic resonance data on N_4-methylcytosine derivatives (205,678). However, a crystallographic study (679) revealed the opposite; that the acetyl group is in distal position, leaving Watson–Crick sites free for base-pairing with guanine. This particular conformation is stabilized because the acetyl substituent pulls electrons out of the heterocycle and acidifies the hydrogen attached to C_5 as evidenced by a downfield shift by 1.21 ppm of the nuclear magnetic resonance signal for H_5. This in turn attracts the carbonyl oxygen in an H $\cdots$ O contact of only 2.15 Å, (Figure 7-9). On the other hand, if the acetyl group were rotated in proximal orientation with respect to N_3, we should expect $N_3 \cdots O_7$ repulsion destabilizing this conformation.

In other modified bases, also occurring in tRNA, Watson–Crick base-pairing sites are free for hydrogen bonding but the Hoogsteen sites are blocked. Such modifications are present in queuosine (Q) (680) and in

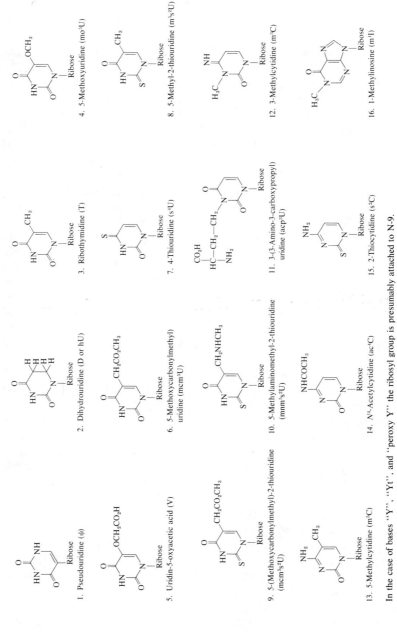

In the case of bases "Y", "Yt", and "peroxy Y" the ribosyl group is presumably attached to N-9.

Figure 7-6. Chemical structures of "rare" or "minor" nucleosides and bases found in tRNAs. From (667).

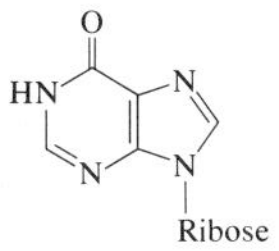

17. Inosine (I)

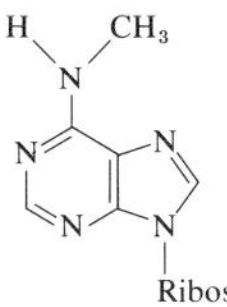

18. N^6-Methyladenosine (m^{6}A)

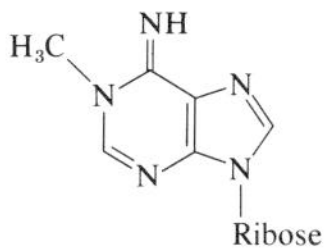

19. 1-Methyladenosine (m^{1}A)

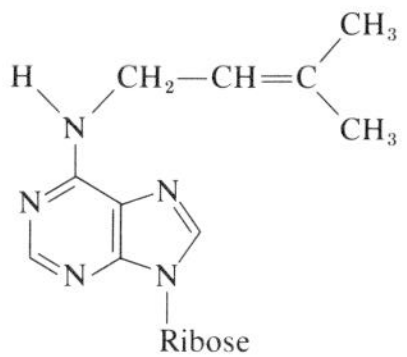

20. N^6-Isopentenyladenosine (i^{6}A)

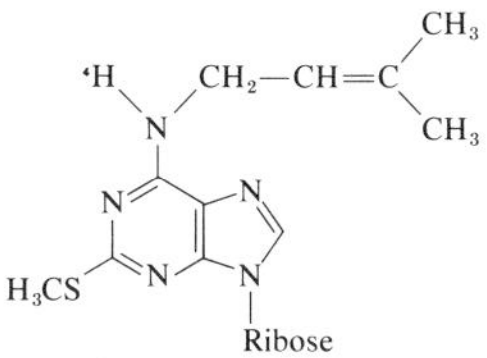

21. 2-Methylthio-N^6-isopentenyladenosine (ms^{2}i^{6}A)

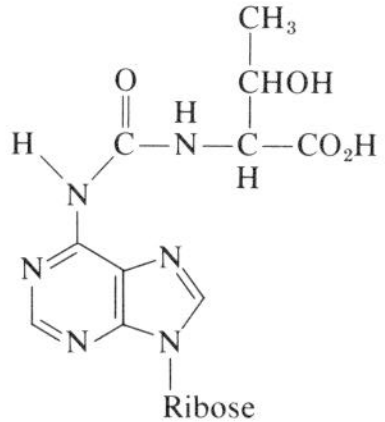

22. N-[N-(9-β-D-Ribofuranosylpurin-6-yl) carbamoyl]threonine (t^{6}A)

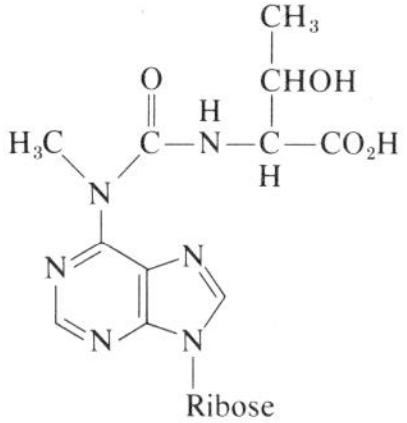

23. N-[N-(9-β-D-Ribofuranosylpurin-6-yl) N-methylcarbamoyl]threonine (mt^{6}A)

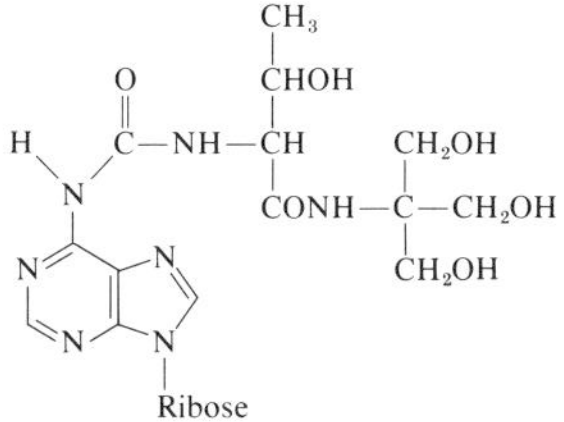

24. N-[N-[(9-β-D-Ribofuranosylpurin-6-yl) carbamoyl] threonyl][2-amido-2-hydroxymethylpropane-1,3-diol

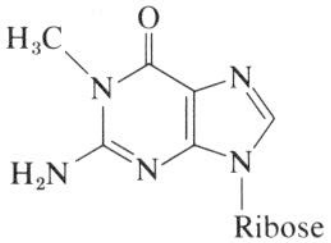

25. 1-Methylguanosine (m^{1}G)

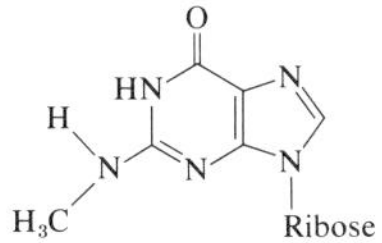

26. N^2-Methylguanosine (m^{2}G)

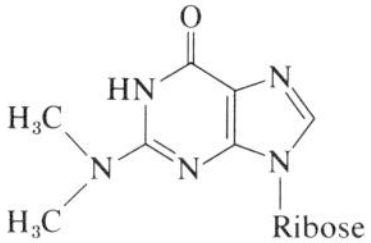

27. N^2, N^2-Dimethylguanosine (m$_2^2$G)

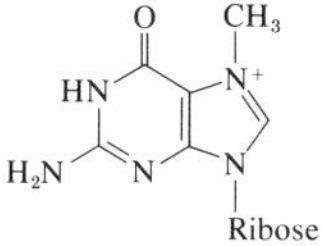

28. 7-Methylguanosine (m^{7}G)

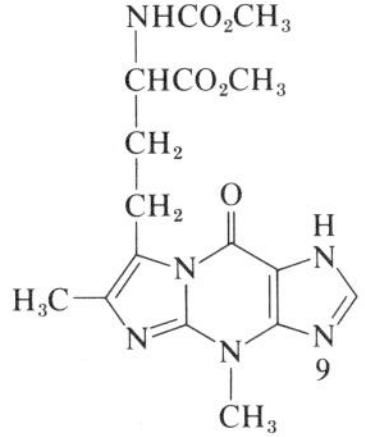

29. Base ''Y'' (yW)

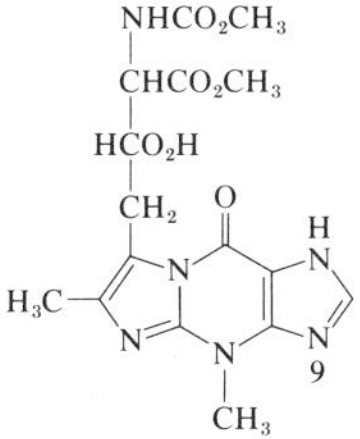

30. Base ''peroxy Y'' (oyW)

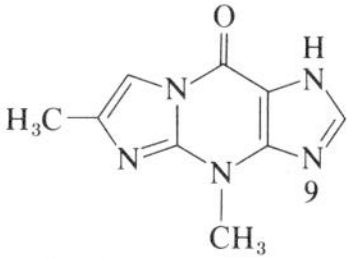

31. Base ''Yt'' (W)

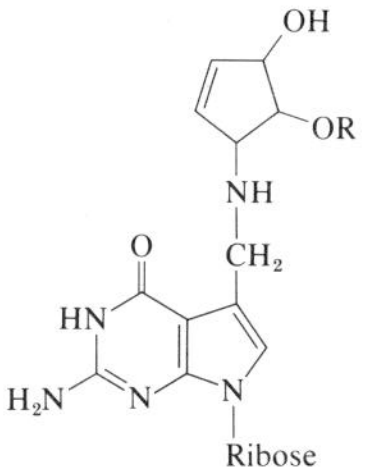

32. R-H Q (queuosine or Quo)[46]
R-β-D-mannosyl.manQ[35]
R-β-D-galactosyl galQ[35]

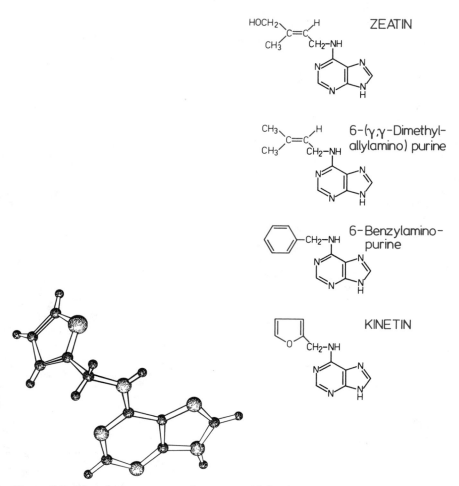

Figure 7-7. Chemical structures of some cytokinins (top). In all cytokinins, a double bond is formed between β and γ atoms of the side chain attached to the adenine amino group. Zeatin and 6-(γ,γ-dimethylallylamino)-purine (2iP) occur in plants; kinetin and 6-benzylaminopurine (BAP) are commonly used synthetic cytokinins. From (670). The molecular structure of kinetin (bottom) shows that the two planes formed by double-bond substituents and by adenine moiety are nearly perpendicular to each other. Note that Watson–Crick-type base-pairing to N_1 is hindered but Hoogsteen binding sites are still available. β–γ and δ–ϵ double bonds are indicated by triple lines. After (673).

7-methylguanosine (m⁷G). With m⁷G, an extra positive charge in the heterocycle is present (Figure 7-6). A variety of 5-substituted pyrimidines are found in tRNAs. If substituents are attached via a 5-oxy group, they are largely coplanar with the heterocycle (Figure 7-10) (681). A methylene group as a link orients the side chains perpendicular to the heterocycle plane and prevents proper stacking of adjacent bases (682,683).

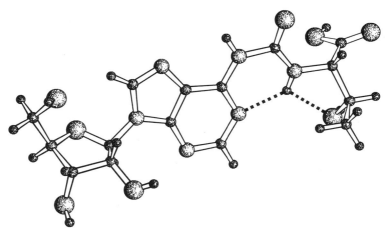

Figure 7-8. Molecular structure of N-[N-(9-β-D-ribofuranosylpurin-6-yl)carba-moyl]threonine, t^6A, a minor constituent of tRNA (for formula see Figure 7-6). A bifurcated hydrogen bond (broken line) fixes the adenine moiety and side chain in a planar arrangement. Watson–Crick-type hydrogen bonding is not possible but Hoogsteen sites are accessible. After (674).

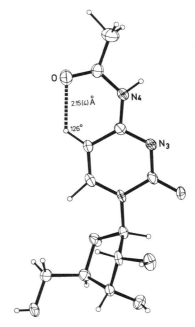

Figure 7-9. Molecular structure of N_4-acetylcytidine. The acetyl group is stabilized by C_5–H $\cdots$ O hydrogen-bonding interaction (broken line) in orientation distal to N_3 and therefore, Watson–Crick-type hydrogen bonding is feasible. C,N,O atoms represented as thermal ellipsoids (Box 3-1). From (679).

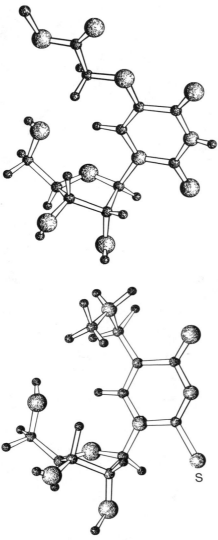

Figure 7-10. Molecular structure of tRNA minor constituent uridine 5-oxoacetic acid (top; for formula see Figure 7-6). Base and side chain are coplanar. After (681). If side chain is coupled not by oxygen but rather by the methylene group, base and side chain are no longer coplanar but vertical to each other as shown for 2-thio-5-methylaminomethyluridine (bottom). After (683).

Nucleoside thioketo derivatives are of interest as anticancer drugs, as tRNA minor constituents, and as spectroscopic probes. From tRNAs, derivatives of pyrimidines with thioketo groups in 2,4, or both 2 and 4 positions could be isolated (see Figure 4-14 for two modifications of 4-thiouridine). In the form of 6-mercaptopurine and of 6-mercaptoguanine, thioketo derivatives have found application in the treatment of cancer (21,684).

In all available crystal structures of thioketo-substituted bases and nucleosides, the C–S bond displays double-bond character with C–S bond lengths slightly longer than expected for an ideal C=S double bond but definitely shorter than for a single bond (Table 7-4). An enolized thioketo group has never been detected in crystallographic studies, although the pK values of amide-like base protons are shifted to lower values if keto oxygen is replaced by sulfur (Figure 5-4).

Sulfur atoms of base thioketo groups act as hydrogen bond acceptors and may replace keto oxygens in this function (686) (See Section 6.1). If keto oxygen is replaced by sulfur, however, there are some geometrical differences involved. First, the C–S distance is about 0.4 Å greater than the C–O distance. Second, the van der Waals radius of sulfur, 1.85 Å, is greater than the 1.4 Å for oxygen (Table 3-1). The hydrogen bond distance, therefore, is 0.4Å to 0.8Å longer if sulfur acts as accepting atom (Table 7-4), and this change might be enough to produce different biological features. In Watson–Crick base-pairs, we may envisage an A–s^{4}U complex with distorted geometry. For s^{6}G–C or G–s^{2}C, however, the situation is more complicated because one of the three G–C hydrogen bonds is widened by ~0.6Å and should result either in weakening of the other two N–H $\cdots$ O and N–H $\cdots$ N interactions or a distinct propeller-like distortion (687).

The large polarizability of sulfur atoms tends to stabilize stacking interactions because it contributes strongly to dispersion forces (Section 6.7). In crystal structures of purine and pyrimidine thioketo derivatives, recurring stacking patterns were traced, with sulfur generally located over the pyrimidine heterocycle (686) especially with sulfur close to ring nitrogens (453). As we shall see in Chapter 11, these stacking interactions have a dramatic effect on the structural stability of double-helical nucleic acids with 2-thioketo sulfur stacking over ring nitrogen N_1 of the adjacent base.

Last, but certainly not least for chemists, thioketo groups exhibit absorption bands in the visible region (4-thiouridine: 330 nm). They can be easily substituted by amino, keto, and other groups and, as they are "soft" bases, they interact readily with "soft" metals like platinum. We will return to this topic in Chapter 8.

Another important nucleoside in tRNA is dihydrouridine, a uridine derivative with a saturated C_5–C_6 bond (589,590,667,668). Comparable analogs are obtained as UV irradiation products of DNA or of frozen thymine or uracil solutions (688,689). In these latter instances, thymine moieties which are located adjacent to each other react to produce dimers (in some cases trimers and tetramers). In the structurally most interesting of the dimers, the monomer units are joined by a cyclobutane ring formed of two C_5–C_6 bonds. As the thymine moieties can associate in different ways, four dimer structures are possible, with *cis/trans* and *syn/anti* orientation of the heterocycles (Figures 7-11, 7-12).

Table 7-4. Geometrical Data for C–S Bond Lengths and Hydrogen-Bonding Interactions Involving Sulfur

C–S bond lengths

Pyrimidine and purine nucleosides	1.645–1.690 Å
Ideal C—S single bond	1.812 Å
Ideal C=S double bond	1.607 Å

Hydrogen-bonding geometry

	Distance (Å) Donor · · · acceptor	Distance (Å) H · · · S	Angle (°) Donor–H · · · S
For O–H · · · S	3.13 to 3.47	2.27 to 2.74	120° to 173°
For N–H · · · S	3.27 to 3.55	2.27 to 2.78	139° to 176°

Note. Data from (685,686).

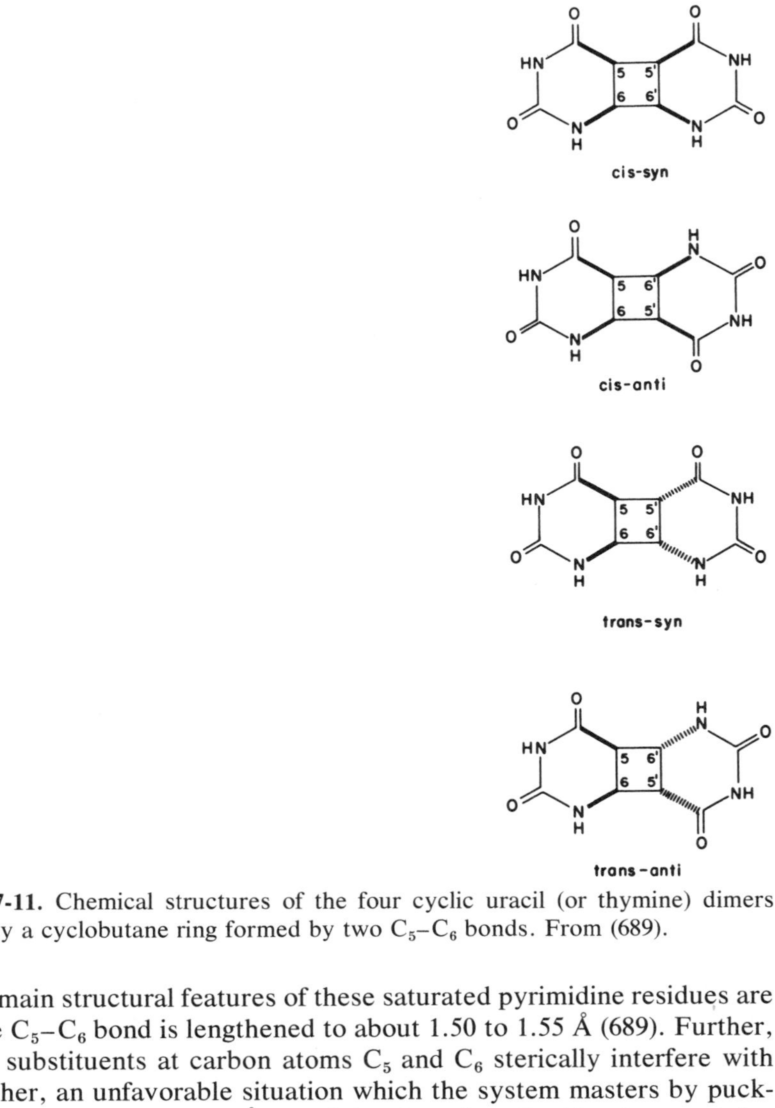

Figure 7-11. Chemical structures of the four cyclic uracil (or thymine) dimers linked by a cyclobutane ring formed by two C_5–C_6 bonds. From (689).

The main structural features of these saturated pyrimidine residues are that the C_5–C_6 bond is lengthened to about 1.50 to 1.55 Å (689). Further, the sp^3 substituents at carbon atoms C_5 and C_6 sterically interfere with each other, an unfavorable situation which the system masters by puckering these atoms 0.3 to 0.7 Å out of the plane described by the remaining four ring atoms (Figure 7-13). In dihydrouridine, the increased volume of the C_6-methylene group influences the sugar puckering which is mainly shifted to $C_{2'}$-endo, with the $+sc$ conformation about the $C_{4'}$–$C_{5'}$ bond slightly destabilized both in the crystalline state and in aqueous solution (351,690).

The base puckering in dihydrouracil derivatives counteracts the parallel stacking of bases essential for double-helix formation. One may conclude, therefore, that these molecules are designed by nature as helix breakers. This finding is corroborated by crystallographic studies on

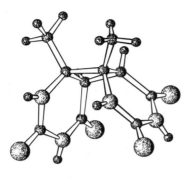

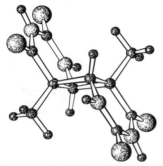

Figure 7-12. Molecular structures of two cyclic thymine dimers shown in Figure 7-11: *cis–syn* [top; after (718)], and *trans–anti* [bottom; after (719)].

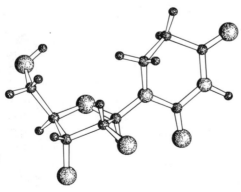

Figure 7-13. Molecular structure of the tRNA minor constituent dihydrouridine. Note that the pyrimidine ring saturated at the C_5–C_6 bond is puckered. The C_6-methylene group is bulkier than C_6–H and displaces $O_{5'}$ "away" from the ribose. In addition, sp^3-CH is less acidic than sp^2-CH, reducing C–H $\cdots$ $O_{5'}$ interactions in dihydrouridine. After (720).

tRNAs which contain two invariant dihydrouridine residues adjacent to each other in the so-called dihydrouracil loop, a region devoid of helically ordered structure.

Azanucleosides occur exclusively in *high-anti* (*−sc*) form and are powerful chemotherapeutic agents. Replacement of C–H groups in bases by aza-N should not, on a purely chemical basis, lead to major structural and functional changes because the modification is isoelectronic and does not alter the π-bonding system. This is indeed observed as far as bond angles and distances in azanucleobases are concerned, except with the understanding that replacement of C–H by the smaller N tends to reduce bond lengths to that atom (289). The charge distributions within the pyrimidine and purine heterocycles, however, change markedly, leading in the free purines to even different tautomeric forms (691) and rendering the aza-substituted bases and their nucleosides effective antineoplastic, cancerostatic agents (586–588,692).

The biochemical particularities of azanucleosides are reflected by characteristic conformational changes if the aza nitrogen is *ortho* to the glycosyl bond as in 6-azapyrimidine or in 8-azapurine nucleosides. In that case, one might have expected a priori formation of a rather strong hydrogen bond $O_{5'}$–H $\cdots$ N. This, however, is never observed because, as quantum chemical calculations have revealed, the aza nitrogens bear virtually no residual charge (289,691,693,694). Rather, as crystallographic (289, 691,695) and solution studies (173) have indicated, repulsion or, at least, nonattraction between aza-N and $O_{5'}$ destabilizes the usually preferred $+sc$ orientation about the $C_{4'}$–$C_{5'}$ bond. Also, the missing pyrimidine C_6–H and purine C_8–H groups allow the bases to rotate into a *high-anti* (*−sc*) position not usually adopted by naturally occurring nucleosides (Figure 4-16). These structural characteristics are also observed in 6-aza-uridine 5'-phosphate which displays *high-anti* (*−sc*) conformation about the glycosyl link and *ap* orientation about the $C_{4'}$–$C_{5'}$ bond, features which have never been observed in unmodified ribonucleotides [except in a transition metal complex of adenosine 5'-phosphate (Chapter 8)]. The conformational differences between 6-azauridine 5'-phosphate and its natural analog are obvious in Figure 7-14 and have led to a proposal concerning the mechanism of the pharmacological activity of 6-azauridine (Box 7-2, 288,696).

7.5 The Chiral Phosphorus in Nucleoside Phosphorothioates

Phosphorothioate analogs of nucleotides, in which one unesterified phosphate oxygen is replaced by sulfur, are powerful tools to investigate enzyme mechanisms which involve reaction at phosphorus (697,698). In contrast to imidodiphosphate analogs of nucleotides, which are dead-end inhibitors for phosphoryl group-transferring enzymes, phosphorothioate

Box 7-2. The Functional Mechanism of the Cytostatic 6-Azauridine Can be Explained with its Unusual Conformation

The cytostatic drug 6-azauridine has been used successfully in the treatment of leukemia. In the body it is converted by uridine kinase to 6-azauridine-5'-phosphate which inhibits the enzyme orotidylic acid decarboxylase. As a consequence, the *de novo* synthesis of uridine-5'-phosphate from orotidylic acid is blocked (Scheme 4). What is the reason for this particular behaviour of 6-azauridine-5'-phosphate?

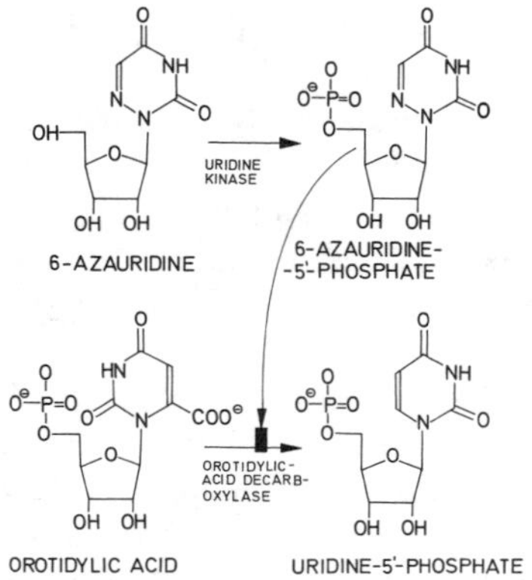

Scheme 4. Mechanism of inhibition of orotidylic acid decarboxylase by 6-azauridine (from Ref. 696)

Orotidylic acid will occur in the *syn*-conformation because the carboxylate group attached to C_6 of the uracil moiety is too bulky to be located over the ribose in a standard *anti* form. In *syn,* however, the orientation about the $C_{4'}$-$C_{5'}$ bond cannot be the normal + *sc* owing to steric hindrance between $O_{5'}$ and O_2 atoms. Therefore, γ will adopt either *ap* or − *sc* orientation. A comparable γ-range, *ap*, is observed in the crystal structure of 6-azauridine-5'-phosphate, whereas *all* normal 5'-ribonucleotides occur in the standard + *sc*, see Figure 7-14.

On a structural basis, therefore, 6-azauridine-5'-phosphate mimics orotidylic acid and binds to the active site of orotidylic acid decarboxylase. It is tempting to predict that all 5'-nucleotides with the base either in *syn* orientation or with an aza group *ortho* to the glycosyl link would act in a similar manner. This in fact is observed and a series of nucleotide analogs with oxipurinol, xanthine, and allo-

purinol substituting for the 6-azauracil base display inhibition of orotidylic acid decarboxylase (Scheme 5).

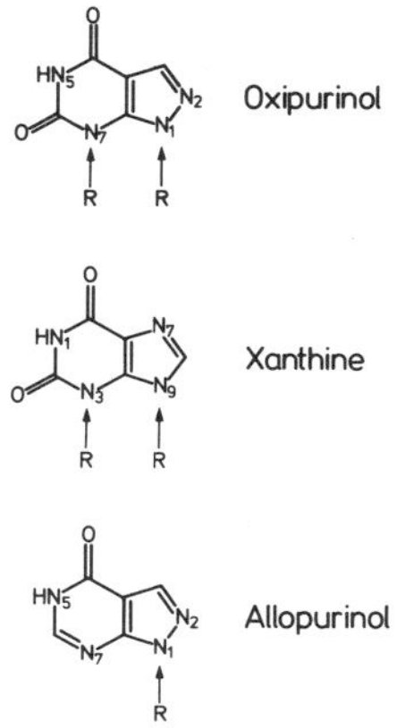

Oxipurinol

Xanthine

Allopurinol

Scheme 5. Inhibitors which act upon orotidylic acid decarboxylase. R
marks the atom to which ribosyl-5'-phosphate residues are attached. The
best inhibitor was 1-ribosyloxipurinol-5'-phosphate (N in *ortho*) followed in
decreasing order of activity by 7-ribosyloxipurinol-5'-phosphate (*syn*-form).
3-ribosylxanthine-5'-phosphate (*syn*-form), 1-ribosylallopurinol-5'-phos-
phate (N in *ortho*). Almost inactive is 9-ribosylxanthine-5'-phosphate which
neither bears N in *ortho* nor is it locked in *syn* (from Ref. 288).

analogs are substrates and react (although in most cases at a slower rate
than their natural analogs).

**The chiral phosphorus is the crucial characteristic of phosphorothioate
nucleotides.** The phosphorus atom in a normal phosphate mono- or diester
is prochiral and therefore the stereochemistry of substrate and reaction
products does not reveal whether the reaction proceeds with inversion or
with retention of configuration at the phosphorus.

To answer such questions, the phosphorus atom must be converted
into a chiral center. This is most elegantly done by substituting sulfur
(698) or an oxygen isotope (697) for one of the two unesterified oxygen
atoms of a phosphate in a phosphodiester or pyrophosphate ester. If, as in
terminal phosphate groups, more than one substitution is required to
produce a chiral phosphorus, an isotope is used in combination with sulfur. Since the nucleoside bound to such a modified phosphate is chiral, the
two stereochemically distinct phosphorothioates are diastereoisomers,
not enantiomers.

In the course of the chemical synthesis, both isomers are obtained and

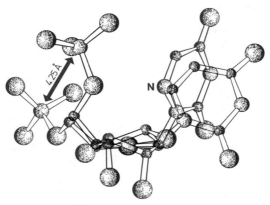

Figure 7-14. Comparison of molecular structures of uridine 5′-phosphate (black) and 6-azauridine 5′-phosphate (color). Both are superimposed such that ribose atoms $C_{1'}$–$O_{4'}$–$C_{4'}$ coincide. Uridine 5′-phosphate adopts standard conformation with sugar $C_{3'}$-*endo*, and torsion angles χ, *anti;* γ, (+*sc*). For the 6-azaanalog, sugar is $C_{2'}$-*endo*, χ is *high-anti* (−*sc*), and γ is in the unusual *ap* range never observed for other 5′-ribonucleotides in the crystalline state except when complexed with drugs, metals, or proteins. After (288).

can later be separated either by chromatographic techniques or by using partial digestion by suitable enzymes. As an illustration of the application of nucleoside phosphorothioates, the elucidation of the mechanism of pancreatic RNase is described in Box 7-1.

The main structural concerns when dealing with thiophosphate groups are some peculiarities associated with sulfur–oxygen substitution. The P–S bond, 1.8 Å, is about 0.3 Å longer than P–O and the van der Waals radius of S, 185 Å, exceeds that of O by 0.45 Å (699), (Table 3-1). Both features add, and the sulfur projects outward about 0.8 Å more than oxygen. As the electronegativities of oxygen (3.5) and sulfur (2.5) differ appreciably, in thiophosphate the negative charge is not equally distributed between these two atoms: the negative charge resides preferentially on oxygen, leaving sulfur almost neutral and, as a consequence, the P–S bond has mainly double-bond character. In addition, sulfur is more easily polarized and more hydrophobic than oxygen. Thus on the whole the thiophosphate group is quite different in character and in structure than phosphate. This is recognized by enzymes which accept nucleotides with sulfur-substituted phosphate as substrates but turn them over more slowly than normal nucleotides. The direction of attack at the active site, however, remains the same and justifies the use of nucleoside phosphorothioates as biochemical probes.

7.6 The Pyrophosphate Group in Nucleoside Di- and Triphosphates and in Nucleotide Coenzymes

The pyrophosphate moiety plays a role in the energy-rich nucleoside di- and triphosphates like ATP and ADP and in some nucleotide coenzymes like NAD^+, cytidyldiphosphocholine and coenzyme A. As bond distances and angles within pyrophosphate have already been discussed in Section 4.11, we concentrate here on conformational aspects. In Section 8.4 these will be extended to discussion of Λ and Δ stereoisomers found in metal chelates of polyphosphates.

In the nucleotides mentioned above, the pyrophosphate or triphosphate group is attached exclusively to the sugar 5'-hydroxyl and no nucleotide has been thus far found with pyrophosphate linked to the $O_{3'}$ or $O_{2'}$ hydroxyl except for the "magic spot" ppGpp. The conformation of the nucleotide moiety itself appears unaffected by this modification. The sugar puckering is preferentially $C_{2'}$- or $C_{3'}$-endo and the orientations about glycosyl and $C_{4'}$–$C_{5'}$ bonds conform to anti and +sc ranges, in agreement with the rigid nucleotide concept (Section 4.10). A direct intramolecular hydrogen bond between terminal phosphate and base has never been observed although quantum chemical calculations predicted such interaction for the in vacuo state (700). However, both in solution and in the crystalline state metal ion-mediated bridges were detected between terminal phosphate and N_7 of purine bases, as illustrated in Figures 7-15 and 7-16 and as discussed in more detail in Chapter 8 (701-703).

The conformational characteristics of pyrophosphate are determined by rotations of phosphate groups about P–O bridge bonds of the pyrophosphate link (59). Of the three conformational states shown in Figure 7-17, the one with cis-planar arrangement of the O–P–O–P–O string is improbable owing to steric interference between the terminal oxygen atoms. Rotation by 60° about one P–O$_{bridge}$ bond results in a favorable conformation with the two phosphate groups staggered looking along the P–P vector. Additional rotation by 60° about the other P–O$_{bridge}$ bond leads to a conformation with the two phosphate groups fully eclipsed. Both situations and the intermediate cases have been observed in crystal structure analyses of pyrophosphate groups. As oxygen atoms of the two phosphate groups are further apart in the staggered than in the eclipsed form, the latter is especially suited for bidentate metal complex formation (Section 8.4).

In general, the pyrophosphate group is flexible (59,704) and transitions between staggered and eclipsed forms can occur rather freely. Depending on the P–O$_{bridge}$ torsion angle ranges, the atomic chain running from the sugar $C_{4'}$ atom out into the terminal phosphate can adopt either a linear,

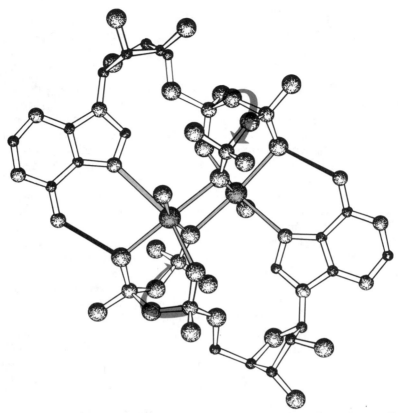

Figure 7-15. The structure of the $Na_2H_2ATP \cdot 3H_2O$ complex. Two sodium ions are coordinated to N_7 and terminal (γ) phosphate of two different ATP molecules and complete their octahedral coordination shell by binding, in addition, to water molecules. Note that the two ATPs display different conformations: $C_{2'}$-*endo* and $C_{3'}$-*endo*–$C_{2'}$-*exo* sugar puckers and triphosphate chains are folded into left- and right-handed helical forms, respectively; see colored arrows. After (721).

Figure 7-16 (opposite). (a) Two views of the molecular structure of the coenzyme NAD^+ as found in the crystalline complex with Li^+. For chemical structure of NAD^+, see Figure 2-2. Shown is a NAD^+ molecule coordinated to two (symmetry related) Li^+ cations which are indicated in color, together with coordinative bonds in tetrahedral arrangement and N, O atoms belonging to adjacent NAD^+ molecules. Both nucleotides of NAD^+ occur in standard conformation with sugar puckers $C_{2'}$-*endo* (adenosine) and $C_{3'}$-*endo* (nicotinamide–riboside), torsion angles χ in *anti* and γ in +*sc* ranges. Torsion angle β ($P-O_{5'}-C_{5'}-C_{4'}$) of adenosine-5'-phosphate is in +*ac* (120°) rather than in the usually observed *ap* (180°) range, probably because intramolecular Li^+-coordination to N_7 and pyrophosphate induces steric strain. (b) In aqueous solution, NAD^+ can adopt two conformations, one extended and the other folded into left- or right-handed helical structure. In the folded form, nicotinamide and adenine bases are stacked at 3.4 Å distance. After (704,709).   Note: Figure 7-16 appears on page 195.

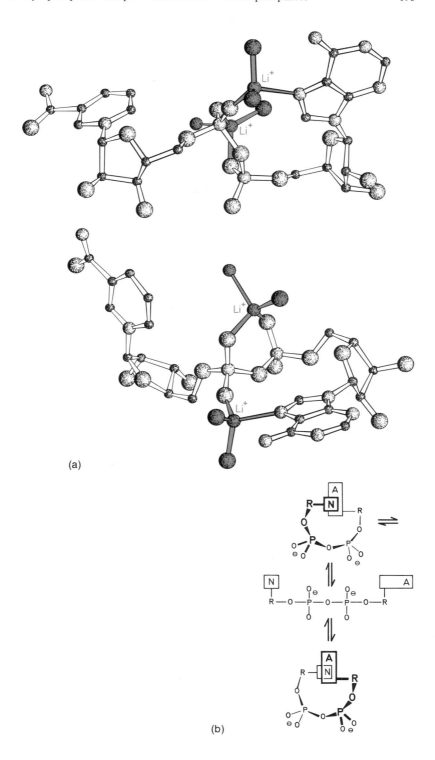

(a)

(b)

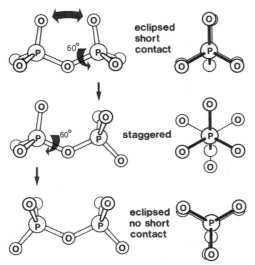

Figure 7-17. Three different arrangements of the pyrophosphate group. The *cis-planar* (eclipsed) orientation (top) is unfavorable because, as the colored double arrow indicates, steric strain between oxygen atoms occurs. Rotation of one or both phosphates about P–O$_{bridge}$ bonds results in staggered or all-*trans* eclipsed (middle and bottom) arrangements which are energetically allowed. The latter is especially suited for bidentate metal complexation: see the triphosphate chain in Figure 7-16a.

extended, or a left/right-handed screw structure. In crystallographic studies, thus far only the screw type has been observed, an excellent example being provided by the sodium complex of ATP or, more accurately, of Na$_2$H$_2$ATP · 3H$_3$O, where two ATP molecules occupy the asymmetric unit (Figure 7-15). These two molecules adopt *anti* and +*sc* conformations for glycosyl and C$_{4'}$–C$_{5'}$ bonds; yet they exhibit different sugar puckerings, C$_{2'}$-*endo* and C$_{3'}$-*endo*–C$_{2'}$-*exo*. The two triphosphate chains differ in their handedness, the C$_{2'}$-*endo* puckered nucleotide is left-handed, and the other is right-handed. To make the story even more complicated, in each ATP the three phosphates are coordinated to a sodium ion to form tridentate chelates of opposite chirality (Section 8.4). The sodium ion is further intramolecularly complexed to adenine N$_7$, to a water molecule, and to an oxygen of another ATP in order to complete its octahedral coordination shell. This complicated molecular structure might leave the reader unsatisfied, shaking his head and asking why this fundamentally important molecule displays such intricate stereochemistry. To comfort the reader, mention should be made of the fact that some metal ions like Mg^{2+} appear to bind only to β,γ-phosphate oxygens and not to adenine N$_7$, and that recent studies on kinetically stable cobalt and chromium complexes of ADP and of ATP shed some light (Section 8.4).

The intrinsic flexibility of the pyrophosphate group has been investigated in detail in the coenzyme NAD.$^+$ This molecule is an essential cofactor in some enzymes involved in oxidation/reduction processes. It can be looked upon as being composed of two 5'-nucleotides, AMP and nicotinamide ribose 5'-phosphate, which are fused head to head via a pyrophosphate bridge (Figures 2-2, 7-16). In solution, this molecule as well as the reduced species, NADH, adopts two main structural forms, one extended with adenine and nicotinamide bases about 12 Å apart and the other one folded with the two heterocycles stacked at 3.4 Å distance (Figure 7-16b). In addition, the folded form consists of two different conformational isomers, depending on whether adenine stacks on nicotinamide or vice versa, leading to right- and left-handed helical structures. All these conformationally distinct states are in rapid equilibrium. In aqueous solution, the extended form prevails; the amount of folded NAD$^+$ is sensitive to solvent conditions like pH, the presence of alcohols, and temperature.

Spectroscopic studies (175,705–707), potential energy calculations (708), and crystal structure analysis (704,709) revealed that the two nucleotide moieties of NAD$^+$ conform to the known standards in structural terms; i.e., the riboses are puckered $C_{2'}$- or $C_{3'}$-*endo,* the orientation about $C_{4'}-C_{5'}$ is $+sc$, and the heterocycles are in *anti* position. For nicotinamide, the *syn*⇌*anti* conversion is nearly unhindered because the *ortho* substituent known from pyrimidine nucleotides is lacking. This appears to be a prerequisite for A- and B-type dehydrogenases which bind NAD$^+$ in these two conformations and therefore accept hydride on different sides of the nicotinamide heterocycle (see also Chapter 18). Otherwise, the two nucleotides behave as structurally rigid entities in a broad sense, the conformational flexibility of NAD$^+$ residing mainly in the pyrophosphate link.

7.7 Nucleoside Antibiotics: Puromycin as Example

The arabinonucleosides, an important class of nucleoside antibiotics, have already been mentioned in Section 7.3. There are many other antibiotics, the chemical functions of which are treated in (587) and the structural aspects of which are discussed in Section 4.2. As a clear structure–function correlation is obvious for puromycin, this antibiotic is described here as representative of these pharmaceutically important compounds.

Puromycin, a broad-spectrum antibiotic isolated from *Streptomyces albo-niger*, is composed of nucleic acid and amino acid fragments, namely, 3'-deoxy-3'-amino-N_6-dimethyladenosine to which, at the 3'-amino group, *p*-methoxy-L-phenylalanine is attached via a peptide bond (Figure 7.18).

Puromycin is a structure analog of the 3' terminus of aminoacyl tRNA.

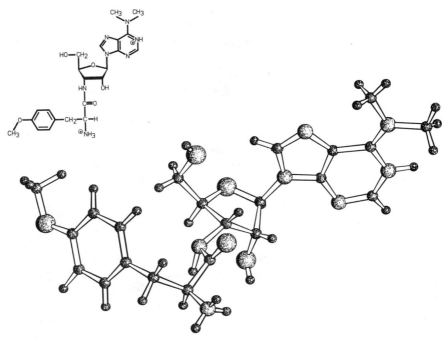

Figure 7-18. Chemical and molecular structures of the antibiotic puromycin (as dihydrochloride). Note that the adenosine portion is in standard conformation with $C_{3'}$-*endo* sugar, χ *anti*, and γ + *sc*. The *syn*-periplanar orientation of $H_{3'}-C_{3'}-N_{3'}-$ C and the extended form of $N_{3'}-C-C_\alpha-N$ chains appear to be a characteristic structural feature and are assumed to prevail also at the CCA terminus of tRNAs when charged with amino acids (with $N_{3'}$ replaced by $O_{3'}$). After (711).

The crystallographic analysis of puromycin was carried out for its dihydrochloride form, with protons attached at adenine N_1 and at the phenylalanyl amino group (710,711). The study revealed that the nucleoside moiety has the standard conformation with orientations about $C_{4'}-C_{5'}$ and glycosyl bonds in +*sc* and *anti* ranges and the furanose puckered in a $C_{3'}$-*endo* mode ($^3T_2$). The peptide group formed by $C_{3'}-N_{3'}-C-C_\alpha$ is *anti-periplanar* and projects straight from the five-membered sugar ring, with torsion angle $H_{3'}-C_{3'}-N_{3'}-C$ in *syn-periplanar* range. This particular atomic arrangement appears to be a characteristic feature of (planar) carboxyl or peptidyl groups attached to the $C_{3'}$ carbon of furanoses—in 3'-*O*-acetylnucleosides, a similar situation is found (195,712,713). The conformation of the *p*-methoxyphenylalanyl moiety is not unusual, with torsion angles $N-C_\alpha-C_\beta-C_\gamma$ and $N_{3'}-C-C_\alpha-C_\beta$ in the common extended (*ap*) and staggered (− *sc*) ranges.

Puromycin mimics the amino acid-charged CCA terminus of tRNAs. Because the structural features of the furanose–peptide or –ester link turn out to be rather constrained, we infer that an extended conformation similar to that of puromycin will be preferred, irrespective of the amino acid attached to the 3'-hydroxyl of a nucleoside ribose. If we extrapolate to tRNA charged at the 3'-terminal adenosine (the CCA end) with an amino acid, we find that puromycin will, on a structural basis, mimic that portion of tRNA. The CCA end is of prime importance in protein biosynthesis because it carries to the ribosome an amino acid to be incorporated into the growing polypeptide chain. In a second step, the polypeptide chain is transferred to this CCA end. If the chain is attached to puromycin instead of tRNA, a dead-end situation results, and protein synthesis leads to incomplete fragments.

The puromycin crystal structure provides a model for nucleic acid–amino acid interaction. As an extra bonus, the crystal structure of puromycin shows piles of alternating N_6-dimethyladenine and p-methoxyphenyl residues stacked parallel to each other at 3.4 Å distance or, in other words, p-methoxyphenyls intercalated between adenines. There are two kinds of stacking overlaps. Both involve the pyrimidine half of adenine, one showing interactions with methoxy oxygen, the other with the benzene ring of the amino acid. This type of stacking provides a good model for intercalation of aromatic amino acid side chains into nucleic acid helices, a general phenomenon discussed in greater detail in Chapter 18.

Summary

Chemically modified nucleosides and nucleotides are used extensively as probes to study enzyme mechanisms and as chemotherapeutic agents. Discussed here are only a few examples of structural interest. Thus, nucleosides with covalent bonds connecting base and sugar in different positions are conformationally rigid and serve as calipers for spectroscopic methods. In nature, cyclic nucleotides with 2',3' or 3',5' diester links play an important role, the former as intermediates in ribonuclease-catalyzed hydrolysis of RNA, the latter in cyclic AMP and GMP, the "second hormonal messengers." Rates of hydrolysis of these cyclic phosphates can be correlated with internal strain. Out of a number of nucleosides with substituted sugars only structural characteristics of halogeno-, arabino-, and α-nucleosides are discussed, the latter behaving as "enantiomers" of the β-nucleoside series. Nucleosides with modified bases play roles as cytokinins and as minor constituents of tRNA, with chemical implications of thioketo substitution, ring nitrogen alkylation, and saturation of the pyrimidine $C_5 {=\!=} C_6$ double bond rather well understood. Nucleosides with C_8–H in purine or C_6–H in pyrimidine bases replaced by aza-N display unusual *high-anti* conformation about glycosyl links, a property explaining in part their biochemical behavior. If oxygen in phosphate groups is substituted by sulfur, phosphodiester groups become chiral and can be used as probes to study reaction mechanisms of enzymes interacting with nucleotides. The exchange of P–O versus P–S, however, involves some drastic changes in stereochemistry, which are tolerated by enzymes. The pyrophosphate groups occurring in "energy-rich" di- and triphosphates and in several coenzymes confer considerable flexibility and, on a structural basis, do not indicate any energy-carrying strain (Chapter 4). Of the many nucleoside antibiotics, only puromycin is treated here. Structurally, it mimics the CCA end of tRNA and therefore inhibits protein biosynthesis at the ribosome.

Chapter 8

Metal Ion Binding to Nucleic Acids

Nucleic acids contain four different potential sites for binding of metal ions: the negatively charged phosphate oxygen atoms, the ribose hydroxyls, the base ring nitrogens, and the exocyclic base keto groups (Figure 8-1). Because metal ions like Mg(II), Ca(II), Na(I), and K(I) are present in the body in millimolar concentrations (Table 8-1), nucleic acids and nucleotides generally occur as complexes coordinated with metal ions. These complexes are of importance for the biological action of nucleic acids, nucleotides, coenzymes, and nucleoside di- and triphosphates and are the topic of this chapter.

8.1 Importance of Metal Ion Binding for Biological Properties of Nucleic Acids

The significance of metal binding to nucleic acids and to enzymes involved in processes related to DNA replication, transcription, messenger RNA translation has long been recognized (723–726). The action of metal ions in these enzyme-catalyzed reactions and also in DNA and RNA degradation by nucleases is only partially understood. In the stabilization of the tertiary structure of tRNA, it is clearly documented that Mg(II) is important because it cross-links phosphate groups (727,728). As shown in Figure 6-17, increasing alkali metal ion concentrations shift the thermally induced helix–coil transition of DNAs to higher temperatures T_m, probably because repulsive forces between negatively charged phosphate groups are compensated for by cations.

Metal ions, however, can also have a destabilizing effect on DNA double-helical structure if they interact with bases rather than with phosphate groups. In the series Mg(II), Co(II), Ni(II), Mn(II), Zn(II), Cd(II), Cu(II), the affinity for base complexation relative to phosphate binding increases from left to right. Addition of Mg(II) raises the T_m, indicating elevated stability of the double-helical structure. Cu(II), however, markedly decreases the T_m and promotes transition into the random coiled state. On the other hand, Cu(II) also facilitates renaturation (729). Behavior similar to that of Cu(II) is also found for Hg(II) and Ag(I) (724,730–734). In RNA, divalent metal ions like Cu(II), Zn(II) and Pb(II) can have an even more de-

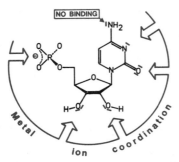

Figure 8-1. Schematic description of metal ion coordination sites on a nucleotide.

structive consequence on structure in that they catalyze nonenzymatic phosphodiester cleavage (735, 736, 736a).

   Research on metal ion–nucleic acid complexes was advanced when antitumor activities of *cis*-dichlorodiammineplatinum(II) were discovered (737–739). *In vitro* studies clearly demonstrated that this reagent attacks guanine at N_7 and also involves O_6 (740–742). The exact mechanism of the *in vivo* action is still a matter of debate (743); it is, however, well established that the template function of DNA is impaired and DNA synthesis inhibited. Owing to these properties, platinum(II) compounds are now being tested in the treatment of several forms of cancer (744,745).

   Active interest in metal ion–nucleic acid complexes has produced a number of review articles describing the stereochemistry of metal ion binding to bases, to nucleosides and, to nucleotides (725,746–750), and two publications concentrated on platinum binding to mono- and polynucleotides (751,752). Thermodynamic and kinetic properties of metal ion–

Table 8-1. Occurrence of Metal Ions in the Human Body, in Blood Plasma, and in Intracellular Fluid

Cation	Total concentration in body (g/70 kg)	Blood plasma (mmole/liter)	Intracellular (mmole/liter)
Na	100	142	10
K	140	4	160
Ca	1100	3	1
Mg	35	1	13
Fe	4	0.018	
Cu	0.15	0.016	
Zn	3	0.018	
Mn	0.02		
Co	0.001		
Mo	<0.001		

nucleotide complexes were mainly derived on the basis of spectroscopic data and are summarized in Ref. (753–759).

8.2 Modes of Metal Ion Binding to Nucleotides and Preferred Coordination Sites

In this chapter, only stereochemical aspects of metal ion–nucleotide complexes as obtained from crystallographic studies will be discussed. Data based on solution spectroscopic methods are sometimes more sensitive and can detect even relatively weak complexes. But they cannot, in most cases, give an unambiguous answer concerning coordination number, coordination sites, and arrangement of ligands around the metal ion.

Binding sites for metal ions offered by the four naturally occurring nucleotides are oxygen and nitrogen atoms, and both belong to class a, or "hard," ligands (760,761) (Figure 8-1). Therefore, only certain types of metal ions are able to bind, viz., alkali, alkaline earth and transition metal ions. These are listed in Table 8-2 and correlated with most preferred nucleotide binding sites derived from crystalline complexes.

Coordination sites, distances, and nucleoside and nucleotide conformations are described for a number of typical complexes with metal ions in Table 8-3. The essential, structurally most interesting characteristics of these compounds can be summarized as follows:

A. Phosphate groups. These can, in general, bind to all the metal ions listed in Table 8-2 to form salt-like complexes between positively charged

Table 8-2. Metal Ions Interacting and Cocrystallizing with Nucleic Acid Constituents and Preferred Binding Sites

Alkaline	Li, Na, K, Rb, Cs
Alkaline earth	Mg, Ca, Sr, Ba
Transition metals	Mn(II), Ru(III), Os(VI, as osmate), Co(III), Co(II), Ni(II), Pd(II), Pt(II), Cu(II), Ag(I), Au(III), Zn(II), Cd(II), Hg(II)

Preferred binding sites:

Base keto oxygen and ring nitrogen	Phosphate oxygen	Sugar hydroxyl

‿‾‾‾‾‾‿
Transition metals
‿‾‾‾‾‾‾‾‾‾‾‾‾‾‾‾‾‾‾‾‾‿
Alkaline and Alkaline Earth

Note. Os is exceptional because it reacts, as osmate(IV), with pyrimidine $C_5\!=\!C_6$ double bonds and with ribose *cis*-hydroxyls to form cyclic esters. For a classification of metal ions reacting with different sites in DNA, see (724).

Table 8-3. A Selection of Data on Metal–Nucleoside and Metal–Nucleotide Complexes Obtained by Crystallographic Studies

Complex[a]	Geometry about metal M[b]	Coordination site L	Distance M–L[c] (Å)	Sugar puckering	Conformational range		pH of crystallization
					γ	χ	
Purine nucleosides							
[Os(Ado-2H)(pyridine)$_2$O$_2$]	oct	O$_{2'}$	1.91	C$_{2'}$-endo	+sc	syn	>7
		O$_{3'}$	1.99				
[Co(dAdo)(acac)$_2$(NO$_2$)]	oct	N$_7$	1.99	C$_{2'}$-endo	+sc	anti	~7
				C$_{3'}$-endo	+sc	anti	
[Pt(Guo)$_2$(en)]$^{2+}$	sq pl	N$_7$	1.97	C$_{3'}$-endo	+sc	anti	<7
cis-[Pt(Guo)$_2$(NH$_3$)$_2$]$^{2+}$	sq pl	N$_7$	2.01	C$_{3'}$-endo	+sc	anti	~7
[Hg(Guo(μ-chloro)Cl)]$_n$	dist sq pl	N$_7$	2.16	C$_{2'}$-endo	–sc	anti	~3
Pyrimidine nucleosides							
[Cu(Cyd)(glygly)]	sq pyr	N$_3$	2.01	C$_{3'}$-endo	+sc	anti	<7
		O$_2$	2.76				
trans-[Pt(Cyd)(Me$_2$SO)Cl$_2$]	sq pl	N$_3$	2.03				~7
Purine nucleotides							
[Cd(5′-GMP)(H$_2$O)$_5$]	oct	N$_7$	2.37	C$_{3'}$-endo	+sc	anti	4.5
[Co(5′-IMP)(H$_2$O)$_5$]	oct	N$_7$	2.16	C$_{3'}$-endo	+sc	anti	4.5
[Ni(5′-IMP)(H$_2$O)$_5$]	oct	N$_7$	2.11	C$_{3'}$-endo	+sc	anti	4.5
[Ni(5′-AMP)(H$_2$O)$_5$]	oct	N$_7$	2.08				5.0
cis-[Pt(5′-IMP)$_2$(NH$_3$)$_2$]$^{2-}$	sq pl	N$_7$	2.02	C$_{2'}$-endo	+sc	anti	6.85
[Pt(5′-IMP)$_2$(en)]$^{2-}$	sq pl	N$_7$	2.07	C$_{2'}$-endo	+sc	anti	~7
[Cu(5′-IMP)$_2$(dien)]$^{2-}$	oct	N$_7$	1.92	C$_{2'}$-endo	+sc	anti	>7
[Cu(3′-GMP)(o-phen)(H$_2$O)]$_2$	sq pyr	O(phos)	1.93	C$_{2'}$-endo	–sc	anti	6.8
[Cu(5′-IMPH)(bipy)(H$_2$O)$_2$]$^+$	sq pyr	N$_7$	1.99	C$_{3'}$-endo	+sc	anti	6

[Zn(5'-IMP)]$_n$	tet	N$_7$	1.99				4
		O(phos)	1.95				
[Cu(5'-GMP)$_3$(H$_2$O)$_8$]$_n$	sq pyr	N$_7$	2.24	C$_{3'}$-endo	+sc	anti	~7
		O(phos)	1.95	C$_{3'}$-endo	+sc	anti	
				C$_{2'}$-endo	+sc	anti	
[Cd(5'-IMP)(5'-IMPH)$_2$	oct	N$_7$	2.36	C$_{2'}$-endo	+sc	anti	3.88
(H$_2$O)$_6$]$_n$		O(phos)	2.23	C$_{2'}$-endo	+sc	anti	
		O$_{2'}$	2.42				
		O$_{3'}$	2.32				
Pyrimidine nucleotides							
[Cu(5'-UMP)(dpa)(H$_2$O)]$_2$	sq pyr	O(phos)	1.93	O$_{4'}$-endo	−sc	anti	7
				C$_{2'}$-exo	−sc	syn	
[Co$_2$(5'-UMP)$_2$(H$_2$O)$_4$]$_n$	oct	O(phos)	2.12	—	—	—	6.8
[Cd(5'-CMP)(H$_2$O)]$_n$	dist sq pyr	N$_3$	2.33	C$_{3'}$-endo	+sc	anti	4.6
		O(phos)	2.25				
[Co(5'-CMP)(H$_2$O)]$_n$	tet	N$_3$	1.96	C$_{2'}$-endo	+sc	anti	5.4
		O(phos)	1.95				
[Zn(5'-CMP)(H$_2$O)]$_n$	tet	N$_3$	2.04	C$_{2'}$-endo	+sc	anti	3.5
		O(phos)	1.90				
[Pt(5'-CMP)(en)]$_2$	sq pl	N$_3$	2.06	C$_{2'}$-endo	−sc	anti	6–7
		O(phos)	1.97	C$_{2'}$-endo	+sc	anti	
[Mn(5'-CMP)(H$_2$O)]$_n$	dist oct	O$_2$	2.08	C$_{3'}$-endo	+sc	anti	5.2
		O(phos)	2.21				

Source. From Ref. (749). For references see therein.

[a] Abbreviations used: acac, acetylacetonate; bipy, 2,2'-bipyridyl; dien, diethylenetriamine; dpa, 2,2'-dipyridylamine; en, ethylenediamine; glygly, glycylglycinate; o-phen, ortho-phenanthroline.

[b] Abbreviations used: dist, distorted; oct, octahedral; pl, planar; pyr, pyramidal; sq, square; tet, tetrahedral.

[c] Distances corresponding to equivalent metal–ligand coordination sites have been averaged.

metal ions and negatively charged phosphate oxygens (Figures 7-15, 7-16, 8-2).

B. Sugar hydroxyls. These interact preferentially with alkaline and alkaline earth cations but not with transition metals. Exception thus far known is Cd(II), described in (762). In all these complexes, the hydroxyl group enters the metal ion coordination sphere with the lone electron pair of the oxygen, as observed in many other adducts between sugar hydroxyls and alkaline or alkaline earth cations (763,764) (Figure 8-2).

C. Heterocyclic nitrogens of bases. Since these atoms carry lone electron pairs, they are good ligands for alkali and transition metals and also for alkaline earth cations (764). As described in Table 8-4, coordination to purine bases with unsubstituted N_9 is preferentially first to N_9 (in bridge to N_3), then to N_7 and then to N_1, with binding to the imidazole ring in general favored over binding to the pyrimidine moiety. Coordination to N_1-unsubstituted pyrimidine bases is first to N_1 rather than to N_3 in thymine and uracil, but for cytosine the reverse trend is found. In both purine and pyrimidine bases, metal ion binding shifts the proton tautomeric equilibrium toward one side, to N_7–H for purines and to N_3–H for pyrimidines, thymine, uracil.

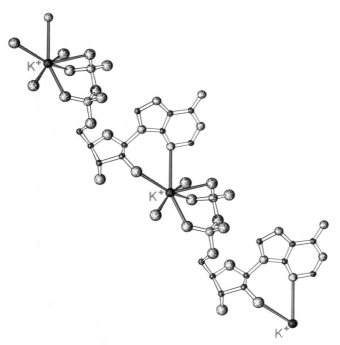

Figure 8-2. Simultaneous coordination of sugar ($O_{2'}$), base (N_3), unesterified phosphate oxygens, and water molecules by potassium ion in the adenosine diphosphate complex. The chelate formed by potassium ion and the pyrophosphate group has the Δ configuration; see p. 216. After (787).

Table 8-4. Preferred Binding Sites for Metal Ion Coordination in Nucleic Acid Bases

	Free base	Nucleoside
Purines N_9 or $N_9 \quad N_3$	$> N_7 \gg N_1, N_3$	$N_7 \gg N_1, N_3$
Thymine, Uracil $N_1 \gg N_3 > O_2$ or O_4 Cytosine $N_3 \gg N_1 > O_2$ or O_4	$N_3 > O_2$ or $N_3 \quad O_2$	Cytidine
	$O_2 > O_4$	Uracil, Thymine

Note. $\smile$ means metal ion bridges atoms N_9 and N_3 or N_3 and O_2, respectively.

When glycosyl links are formed, N_9 of purine and N_1 of pyrimidine bases are blocked. In purines, binding to N_7 is then preferred over that to N_1 and over N_3. In pyrimidine derivatives, only cytidine can offer N_3 for complexation: uridine and thymidine have no ring nitrogen available (for examples, see Figures 7-15, 7-16, 8-2 to 8-4).

D. Base keto substituents. These are able to contribute to complex formation with metal ions. In pyrimidine nucleotides, direct metal ion binding to O_2 (cytosine) or to O_2, O_4 (thymidine, uracil) takes place. In contrast, in the purine systems guanine and hypoxanthine, O_6 is generally not involved in direct metal binding, probably because N_7 is a better ligand

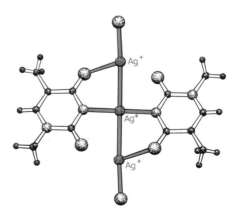

Figure 8-3. Coordination of Ag(I) in the complex with 1-methylthymine. The compound was formed under strongly alkaline conditions in order to remove the N_3–H proton. Ag(I) is bound to all three electronegative groups N_3, O_4 and (more weakly and not indicated here) O_2. After (788).

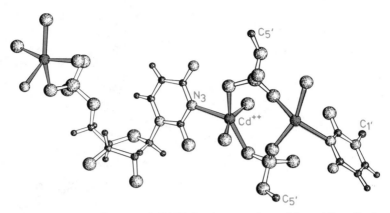

Figure 8-4. The ligands of Cd(II) in the complex with cytidine 5′-phosphate are N_3, unesterified phosphate oxygens, and water. After (789).

and simultaneous coordination of N_7 and O_6 would lead to unfavorable geometry (765,766). The keto oxygen O_6 can, on the other hand, accept hydrogen bonds from other ligands in the coordination sphere of the metal ion, and these interligand interactions indirectly contribute to complex formation (Figure 8-5; 767).

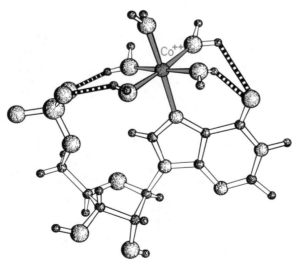

Figure 8-5. In the complex between cobalt(II), inosine-5′-phosphate, and water, the cation is directly coordinated with base nitrogen N_7 and indirectly via water-mediated hydrogen bonds to phosphate and guanine O_6 oxygens. In the actual crystal structure, hydrogen atoms of water ligands were not located and are drawn here to illustrate the coordination scheme. Also, only one hydrogen bond to guanine O_6 is observed; the situation depicted here would occur if one water molecule rotated slightly. After (767).

E. Base amino substituents. These are *never* involved in direct metal ion coordination because, as we have seen in Section 4.1, the lone electron pair is delocalized over the π-bonding system of the attached heterocycle and is not available for metal ion binding. As discussed under point D for keto oxygen, amino groups can, however, act as hydrogen bond donors to other ligands of the metal ion and thus indirectly facilitate complexation. Under very alkaline conditions, amino groups are deprotonated and become good ligands for transition metal ions (Figure 8-3).

F. Base thioketo substituents. In 6-mercaptopurine, in 2-thiocytosine, and in 2- and 4-thiouracil these are "soft" groups and therefore very good ligands for transition metal ions (760,761). They are far better candidates than purine or pyrimidine nitrogen and oxygen atoms as could be demonstrated by the exclusive and strong affinity of mercury and platinum reagents to the only 4-thiouridine in $tRNA^{Val}_{E.coli}$ (751).

G. Osmium(VI) binding coordinatively and as cyclic diester. When applied in the form of the $OsO_3(pyridine)_2$ complex to $tRNA^{Phe}_{yeast}$, osmium was found coordinated with N_7 of a guanine. Additionally, it was bound as cyclic osmate diester to the *cis*-hydroxyls of the terminal ribose (768), and to the C_5–C_6 double bond of a cytidine (769). The latter interaction appears to be typical for pyrimidine bases because it was also observed with thymine (770–772).

H. Correlation of coordination number and binding sites in metal ion complexes of pyrimidine and purine bases (748). In purines, N_7 is rather exposed and allows sixfold (octahedral) or even higher coordination of the metal ion. In contrast, binding to N_3 in cytosine is sterically restricted due to the two flanking keto and amino groups which limit the available spatial volume and consequently lead to preference for lower, four- or fivefold coordination number.

I. Metal ion coordination versus protonation. Protonation and metal ion coordination sites in purine and pyrimidine bases are similar for guanine and cytosine but different for adenine and uracil (thymine) (Table 8-5). Another dissimilarity in these two reactions is found in the increase of the endocyclic C–N–C angle with protonation of the vertex nitrogen atom, discussed in Section 4.1. Coordination to metal ions leaves the geometry about the respective liganded nitrogen practically unchanged (747).

Table 8-5. Comparison of Protonation and Metal Coordination Sites in Nucleosides

	Coordination	Protonation
Adenine	$N_7 > N_1$	$N_1 > N_7$
Thymine	$O_2 > O_4$	$O_4 > O_2$
Guanine	$N_7 > N_3$	$N_7 > N_3$
Cytosine	N_3	N_3

J. Selective metal ion coordination of bases. If topics discussed under points A to I are summarized, it becomes clear that metal ion coordination to bases is rather specific. Thioketo substituents (point F) as soft ligands are very good candidates for coordination to transition metal ions and override hard ligands nitrogen and oxygen. Binding to base ring nitrogen atoms manifests subtle selectivity because N_3 of cytosine is sterically shielded by keto and amino groups and is therefore a ligand for metal ions with coordination number lower than 6, whereas N_7 of purine bases can accommodate all kinds of alkaline, alkaline earth, and transition metal ions (point G). Finally, interligand hydrogen bonding between N_6 amino group in adenine or O_6 keto group in guanine and other ligands of the metal ion adds another degree of selectivity (points B,C) (748).

K. Effect of metal ion complexation on nucleotide conformation. A glance at Table 8-3 shows that, in general, sugar puckering modes obey the normal $C_{2'}$-*endo* and $C_{3'}$-*endo* preference, and orientations about the glycosyl bond (described by torsion angle χ) are largely *anti*, the *syn* structure of $[Cu(5'\text{-}UMP)(dpa)(H_2O)]_2$ being a remarkable exception (773). As described in Chapter 4, in nucleotides the conformation about the $C_{4'}$–$C_{5'}$ bond is such that $O_{5'}$ resides over the ribose and torsion angle γ is $+ sc$. In metal ion complexes of nucleotides, however, this strict rule (the rigid nucleotide) no longer applies. Although $+ sc$ is preferred, in some cases $- sc$ is found, in both the pyrimidine and purine series, with the metal ion in all cases bound to phosphate oxygen and to a large ternary ligand like ortho-phenanthroline or 2,2'-dipyridylamine. We therefore infer that binding of metal ions in connection with bulky groups to phosphate might distort nucleotide geometry; the influence of metal ions when coordinated only to base or to ribose is less obvious and cannot be discussed on the basis of the available data.

8.3 Platinum Coordination

Platinum binding to DNA has outstanding biochemical and pharmacological importance. For the compounds summarized in Table 8-6, two mechanisms for interaction with DNA are observed. First, the platinum complexes containing a planar terpyridine ligand surrounding a centrally located platinum(II) or palladium(II) do not bind coordinatively but rather intercalate between stacked DNA base-pairs (see Chapter 16). Second, platinum compounds of the *cis*- and *trans*-dichlorodiammine type coordinate to nucleic acid constituents, with N_7 (and simultaneously O_6) of guanine most preferred (741,752). They can also react with uracil and thymine to form "platinum blue," an oligomeric species which can further attack DNA. The structure of "platinum blue" is still unknown but is probably analogous to that derived for the crystalline *cis*-diammineplatinum–α–pyridone blue complex (751).

The possible mode of action of *cis*-dichlorodiammineplatinum(II) is illustrated in Table 8-7. The relatively low doses of platinum needed to inhibit tumor growth suggest a solvent-assisted pathway for the drug. Thus, in extracellular plasma the high concentration (~ 100 mM) of chloride ion will suppress the reactivity of the drug. However, after diffusion into the cell where the cytoplasmic chloride ion concentration is only 4 mM, it can transform into the hydrolysis product *cis*-$[(NH_3)_2Pt(H_2O)(OH)]^+$ (774) which coordinates to DNA more readily because water is a good leaving group (751). It could as well be that uracil or thymine platinum blue complexes first form and then react with DNA.

Why only *cis*- and not *trans*-dichlorodiammineplatinum(II) exhibits antitumor activity is not clear. Substitution of a chloride by base ring nitrogen in the *cis*- isomer might perhaps labilize and activate the *trans*-positioned ammine. This would give rise to further Pt–DNA coordination and lead ultimately to a Pt ion bound to four DNA sites. However, significant quantities of such species are not detected by NMR spectroscopy using *cis*-dichlorodiammineplatinum(II) and poly(I) and poly(I)·poly(C). Rather, the data indicate only single- and double substitution of Pt by inosine N_7 (775), the latter in line with crystallographic data on a complex between this Pt(II)-compound and a DNA-dodecamer (Figure 8-6).

8.4 Coordination of Metal Ions to Nucleoside Di- and Triphosphates: Nomenclature of Bidentate Λ/Δ and of Tridentate $\Lambda/\Delta/endo/exo$ Chelate Geometry

The concentration of free Mg(II) in living cells is around 20mmole/liter. Under these conditions the equilibria

$$ADP^{3-} + Mg^{2+} \rightleftharpoons [MgADP]^-, \qquad \log K_{ass} = 3.21,$$
$$ATP^{4-} + Mg^{2+} \rightleftharpoons [MgATP]^{2-}, \qquad \log K_{ass} = 4.05$$

are shifted toward complex formation. ADP, ATP, and their nucleotide analogs are therefore present mainly as Mg(II) chelates (776). In this form, they play a dominant role as substrates and as suppliers of energy in many enzymatic processes.

The molecular structure of the $[MgATP]^{2-}$ complex was early proposed by Szent-Györgyi as the metal ion coordinated to both N_7 of adenine and the oxygens of the polyphosphate chain (777). This attractive hypothesis gained support from immunochemical studies suggesting that, in the presence of Mg(II), ADP and ATP have conformations different from that of AMP (778). More recent spectroscopic data indicate, however, that Mg(II) and other divalent cations do not ligate simultaneously to base and phosphate but rather preferentially bind to phosphate (779–782).

Table 8-6. Chemical Formulas of Some Platinum Compounds Intercalating Between DNA Base-Pairs and Binding to Guanine-N_7

Intercalation into DNA double helix

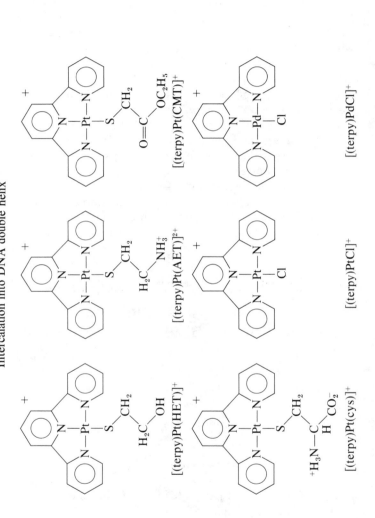

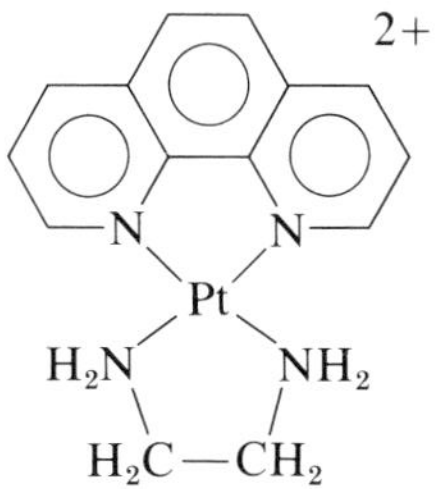

$[(o\text{-phen})Pt(en)]^{2+}$

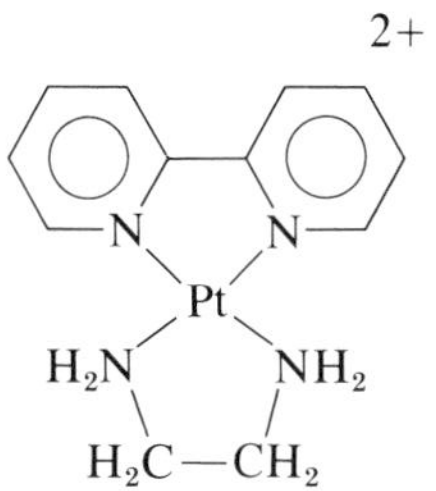

$[(bipy)Pt(en)]^{2+}$

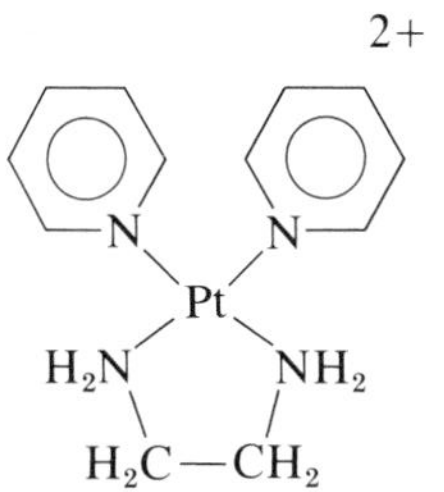

$[(py)_2Pt(en)]^{2+}$

Binding to guanine

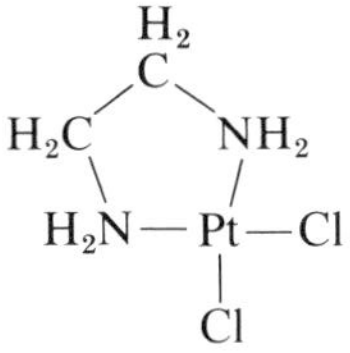

$[(en)PtCl_2]$

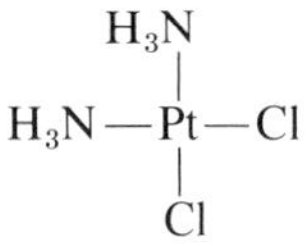

$cis\text{-}[(NH_3)_2PtCl_2]$

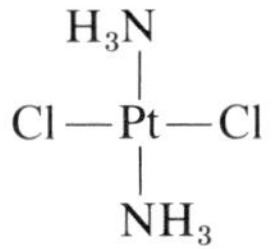

$trans\text{-}[(NH_3)_2PtCl_2]$

Note. Platinum complexes with large, planar organic ligands such as terpyridine (terpy), *ortho*-phenanthroline (*o*-phen), or bipyridyl (bipy) are able to intercalate between base-pairs in double-helical DNA. If smaller, exchangeable ligands such as Cl or NH_3 are involved, direct binding to N_7 of guanine takes place. For formulas of more complexes, see (738,739,752).
Source. From (751).

Table 8-7. Mode of Action of *cis*-Dichloroplatinum(II) Complexes

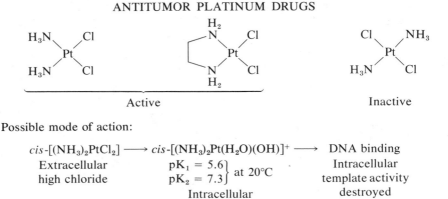

ANTITUMOR PLATINUM DRUGS

Active   Inactive

Possible mode of action:

cis-[(NH$_3$)$_2$PtCl$_2$] $\longrightarrow$ cis-[(NH$_3$)$_2$Pt(H$_2$O)(OH)]$^+$ $\longrightarrow$ DNA binding
Extracellular   pK$_1$ = 5.6 ⎫ at 20°C   Intracellular
high chloride   pK$_2$ = 7.3 ⎭   template activity
                        Intracellular   destroyed
                        low chloride

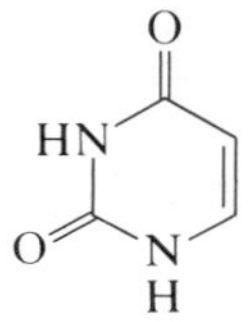

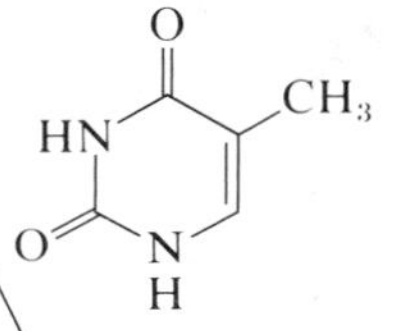

Uracil platinum blue   Active drugs with lower   Thymine platinum blue
"(NH$_3$)$_2$Pt(U)(OH)"   kidney toxicity   "(NH$_3$)$_2$Pt(T)(OH)"
oligomeric

Note. The extracellular high chloride concentration ($\sim$100 mM) suppresses the reactivity of the drug which in a cytoplasmic chloride concentration of $\sim$4 mM partially hydrolyzes into an active species. The latter either attacks DNA directly or forms an intermediary "platinum blue" complex with uracil, thymine which then reacts with DNA.
Source. From (751).

In solution, the Mg(II) complexes of nucleoside di- and triphosphates consist of a variety of different coordination isomers which rapidly transform from one to the other at a rate of 10^3 to 10^5 sec^{-1}. Nuclear magnetic resonance spectra therefore give only time-averaged signals. These cannot be resolved, and the structures of the individual chelate species remain hidden (783).

The situation changes if metal ions such as Cr(III) and Co(III) are utilized. They both yield structurally stable coordination complexes with nucleoside di- and triphosphates. In contrast to other metal ions which form salt-like ion pair bonds, coordinative bonds between phosphate oxygens and Cr(III) or Co(III) are largely covalent in character. This particular behavior is evidenced by a drastic reduction in secondary phosphate pK, around 7 under normal conditions (see Figure 5-2), to about 3 to 4 in pres-

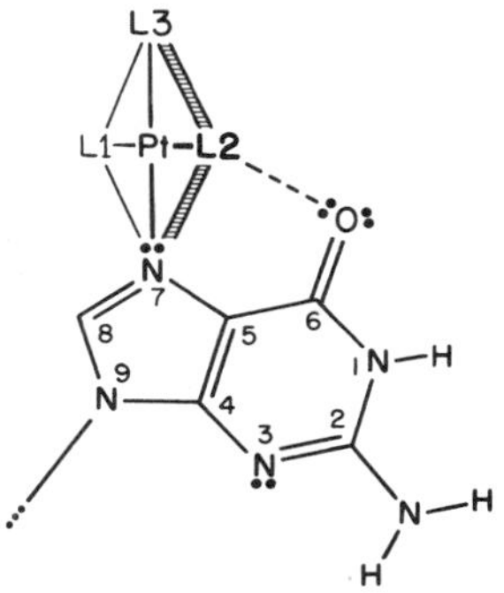

Figure 8.6. Coordination of *cis*-dichlorodiammineplatinum(II) to guanine in the DNA dodecamer CGCGAATTCGCG. The structure of the complex is similar to that described in Figure 8-5, with ligands L1, L2, and L3 either H_2O or NH_3. From (742).

ence of these metal ions (783). The Cr(III)– and Co(III)–polyphosphate complexes are especially stable if the coordination sphere is filled by ammine rather than by aquo ligands. In these circumstances, they can be isolated and the various configurational species separated using column chromatography (783).

Applying proton and ^{31}P nuclear magnetic resonance techniques, the structural isomers can be identified for cobalt–ammine complexes of the general formulas $[Co(NH_3)_nHATP]$ ($n = 2$, 3, or 4) and $[Co(NH_3)_nADP]$ ($n = 4$ or 5). The ATP complexes exist in seven stereochemically distinct variations: four tridentate complexes with α-, β-, and γ-phosphates each liganded to the same metal ion by means of a single oxygen; two bidentates with β- and γ-phosphates coordinated; and one monodentate with only the terminal γ-phosphate involved in metal binding. For ADP only three stereochemically distinct forms can be detected, namely, one oxygen atom each of α- and β-phosphate chelated to Co(III) in two bidentate complexes and, in a monodentate, only the β-phosphate coordinated (see Table 8-8) (784).

Table 8-8. Coordination Complexes of ADP and ATP with Cobalt(III)

Compound	Denticity	Coordination to phosphate groups	Number of configurational isomers
$[Co(NH_3)_4ADP]$	Bidentate	α and β	2
$[Co(NH_3)_5ADP]$	Monodentate	β	1
$[Co(NH_3)_2HATP]$	Tridentate	α, β, and γ	4
$[Co(NH_3)_3HATP]$	Bidentate	β and γ	2
$[Co(NH_3)_4HATP]$	Monodentate	γ	1

What do these coordination compounds of ADP and ATP look like? In the monodentate complexes, only the terminal phosphate group is involved in binding and therefore they are, stereochemically, of no special interest. In the bidentates, however, two configurational isomers were found which differ in the way the polyphosphate wraps around the metal ion. The stereochemistry of these complexes can, in principle, be clearly defined by the well-known R,S nomenclature. This is impracticable, however, because the atomic number for Mg(II), for instance, is lower than that of P but the numbers for Co(III), or similar metals are higher and therefore chelate rings of the same chirality would correspond to symbols R,S, indicating different handedness. In order to avoid this problem, a new concept was proposed. Symbols Λ and Δ were introduced to define the chirality of the respective isomers as follows (784,785): Place the (six-membered) chelate ring so that the bond connecting it to the rest of the ligand points toward the viewer, as shown in Figure 8-7. If the shortest path from the metal around the ring to this bond is clockwise, the isomer is called right-handed (Δ, $d$elta, for $d$exter = right). If the shortest path is counterclockwise, it is left-handed (Λ, $l$ambda, for $l$aevus = left).

If, in nucleoside triphosphates, the β,γ-bidentate chelate is augmented by additional coordination of a free, unesterified oxygen atom of the α-phosphate group, the chirality of the already existing chelate ring determines its position on the coordination sphere of the metal ion, giving rise to an additional six-membered chelate ring fused to the first one through the central β-phosphate and metal ion. The nucleoside moiety can be arranged in two ways: pointing toward the bicyclic chelate (*endo*) or away from it (*exo*). Because these two stereochemically different structures can occur with both the Λ and Δ bidentate chelates, four tridentate isomers are expected and have been experimentally identified for the [Co(NH$_3$)$_4$ATP] complex (Figure 8-7; Box 8-1).

The Λ/Δ*endo*/*exo* nomenclature can be applied to the known crystal structures of ADP, ATP, and the coenzyme NAD$^+$, which contains an ADP fragment (Figures 7-15, 7-16, 8-2). In the potassium salt of ADP (Figure 8-2) potassium is coordinated to oxygen atoms of the α- and β-phosphates and forms a six-membered chelate ring corresponding to the Δ diastereomer. For the lithium salt of NAD$^+$ (Figure 7-16), the same handedness (Δ) is observed for both the ADP and nicotinamide ribose diphosphate portions of the molecule. In disodium ATP (Figure 7-15), two sodium ions are coordinated to the same ATP molecule. One Na$^+$ is bound simultaneously to base N$_7$ and to the terminal γ-phosphate and the other Na$^+$ forms a tridentate coordination complex with α-, β-, and γ-phosphate. The six-membered chelate formed by β- and γ-phosphate has Λ handedness, the α-phosphate is additionally bound such that adenosine is placed *endo* with respect to the chelate; that is, the configuration is Λ, *endo* for the ATP shown in lower right in Figure 7-15. For the other ATP molecule in the complex, however, the configuration differs and is described as Δ, *endo* because, as has been discussed in Section 7.6, the con-

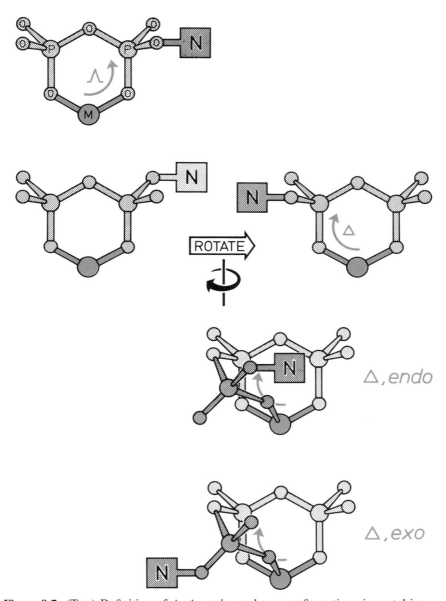

Figure 8-7. (Top) Definition of Λ, Δ, *endo,* and *exo* configurations in metal ion–pyrophosphate complexes. The six-membered pyrophosphate chelate ring is placed on the paper such that the nucleoside points toward the viewer. If the shortest path from the metal to the nucleoside P–O$_{5'}$ bond is clockwise, the complex is Δ; if it is counterclockwise, the complex is Λ. See text. (Bottom) In triphosphates, an additional pyrophosphate entails another configurational possibility, *endo* or *exo,* depending on whether the nucleotide O$_{5'}$ is located over (*endo*) or away (*exo*) from the chelate formed by β- and γ-phosphates. Colored sphere is metal ion, boxed N represents 5'-linked nucleoside.

Box 8-1. Isolation and Structure Elucidation of the Λ- Isomer of [Co(NH₃)₄ATP]

Because the Co(III) and Cr(III) complexes of nucleoside di- and tri-phosphates are so stable, they can be used as probes in the study of stereochemical requirements at the active centers of enzymes which use these substances as substrates or cofactors.

In the case of yeast hexokinase [and of other kinases and polymerases (783,786)], the two Δ and Λ bidentate diastereomers of [Co(NH₃)₄ATP] were each reacted in the presence of glucose. The enzyme specifically transfers the γ-phosphate of one [Co(NH₃)₄ATP] isomer to glucose to yield glucose 6-phosphate and the resulting [Co(NH₃)₄ADP] complex was separated from the inactive, unreacted isomer of [Co(NH₃)₄ATP]. The ADP complex was again subjected to reaction with hexokinase and the added glucose 6-phosphate transferred its phosphate group to produce an active [Co(NH₃)₄ATP]. This was isolated, the ATP was cleaved by treatment with periodate followed by aniline, the resulting [Co(NH₃)₄PPP] complex was crystallized, and its absolute configuration (as Λ isomer) was established by X-ray analysis (785).

$$[\text{Co(NH}_3)_4\text{ATP}]_{\Delta+\Lambda} + \text{glucose}$$

⇅ (yeast hexokinase)

$$[\text{Co(NH}_3)_4\text{ADP}]_\Lambda + \text{glucose 6-phosphate}$$

⇅ (yeast hexokinase)

$$[\text{Co(NH}_3)_4\text{ATP}]_\Lambda$$

↓ periodate, aniline

$$[\text{Co(NH}_3)_4\text{PPP}]_\lambda$$

In a similar scheme, other enzymes utilizing ADP and ATP metal complexes can be investigated for their stereospecificity.

formations of the triphosphate chains are enantiomorphic to one another (they describe right- and left-handed screws). In two more recently determined ATP complexes, Zn(II)–ATP–2,2′-bipyridyl, and Cu(II)-ATP-phenanthroline, the zinc and copper ions are involved in tridentate coordination with ATP and the configurations are Λ, *exo* in both cases (786a,786b).

Summary

Nucleotides offer four different binding sites for metal ions: phosphate oxygens, ribose hydroxyls, base ring nitrogens, and keto substituents. Base amino groups do not participate in metal coordination. While transition metals bind to phosphate oxygens and base nitrogens, alkaline earth cations prefer phosphate oxygens and sugar hydroxyls over base nitrogens. Alkaline metal ions complex equally well to all three kinds of ligands. The affinity of metal ions for nucleotide-binding sites is rather specific, complexation to purine N_7 accommodating sixfold coordinated metals whereas pyrimidine N_3 allows only lower coordination numbers. Moreover, interligand interaction through hydrogen bonding to purine base keto and amino groups contributes to specificity. As a rule of thumb, in purine bases, imidazole nitrogens are better ligands than pyrimidine nitrogens. Metal ion binding can distort nucleotide geometry (orientation about $C_{4'}-C_{5'}$ bond) if phosphate oxygens and large ternary ligands are involved; otherwise, the preferred nucleotide geometry appears unaffected and is not changed in any obvious manner.

Nucleoside di- and triphosphates coordinate to metal ions preferentially through their phosphate groups and not with simultaneous cooperation of the base. In triphosphates, metal ions can be liganded as mono-, di-, or tridentate complexes in one, two (Λ, Δ), or four (Λ, Δ, *endo, exo*) stereochemically distinct configurations. These have been established for the stable Cr(III) and Co(III) complexes and are probably similar to complexes with other metal ions, especially with Mg(II).

Chapter 9

Polymorphism of DNA versus Structural Conservatism of RNA: Classification of A-, B-, and Z-Type Double Helices

In Chapters 10 to 15, the molecular structures of oligo- and polynucleotides are discussed together because they are intimately related. In most of the known oligonucleotide crystal structures, self-complementary, double-helical arrangements are observed. While limited in length between 2 and 12 base-pairs, these reveal fine details of polynucleotide self-assembly, hydration, and metal ion coordination. In addition, oligonucleotides sometimes crystallize as nonhelical entities with extended or looped configurations; this indicates how flexible the nucleotide building block or the phosphodiester swivel link between the individual nucleotides may be. As described in Chapter 15, the crystal structure of tRNA offers a wealth of structural information both on double-helical and on nonhelical polynucleotide domains.

Keep in mind that, as outlined in Chapter 3, results of X-ray diffraction methods applied to structure analysis of polynucleotide fibers and of oligonucleotide single crystals differ qualitatively and quantitatively. In the following six chapters, solid structural information derived for oligonucleotides is intermingled with soft information gathered for polynucleotides. Although details of the latter might be subject to change within the next years, we shall see that, in general, a picture emerges that gives confidence to the structural assignments of helical, polymeric RNAs and DNAs and at the same time reveals the variety of existing double helices.

Reviews pertinent to this topic are given in (38,790–793). In addition, the readers' attention is drawn to a few recent publications selected out of a vast literature concerning theoretical investigations on polynucleotide conformation (794–808), the flexibility of the double helix (809–817), and spectroscopic investigations of polynucleotide structures (818–824). The biochemical structure–function correlations are treated in reviews for DNA (792) and for RNA (793).

9.1 Polymorphism of Polynucleotide Double Helices

The full palette of double-helical structures observed for naturally occurring and synthetically available RNA and DNA is displayed in Table 9-1. Specimens for X-ray fiber diffraction studies are prepared as lithium or sodium salts and the concentration of excess salt is controlled by varying the relative humidity of the environment (165). Under certain experimental conditions, different double-helical structures represented as A, A', B, α-B', β-B', C, C', C'', D, E, and Z were observed (824). The letters A to Z denote primarily structural polymorphism. Prefixes α and β refer to mere packing differences associated with crystal lattice symmetries, and additional primes indicate small variations within one type of structure. Thus, while naturally occurring DNA with random sequence adopts under certain conditions the C form with 9.33 nucleotides per turn (see below), the synthetic, repetitive DNA poly(dA–dG–dC)·poly(dG–dC–dT), having a trinucleotide as asymmetric unit, forms C''-DNA with 9 nucleotides per turn. Poly(dA–dG)·poly(dC–dT) repeats exactly only after two turns (9 × 2 nucleotides) and is referred to as C'-DNA.

The symmetries of the various double helices are represented in the form of two numbers N_m. In this nomenclature adapted from crystallographers (Box 9-1), m indicates the number of helical turns (pitch heights), and N the number of nucleotides after which exact repeat along the helix axis is achieved. In most cases, one helical turn is equivalent to one repeat ($m = 1$) but, as in C-DNA with symmetry 9.33_1, the motif repeats exactly only after three turns comprising $3 \times 9.33 = 28$ nucleotides, also expressed by the symbol 28_3. It should be emphasized that exact repeat in a strict sense is only possible for synthetic polynucleotides with defined sequence. For naturally occurring DNAs and RNAs with random sequence, the term *exact repeat* refers to the sugar phosphate backbone rather than to individual base-pairs which are taken as averaged, Watson–Crick-type isomorphous A–T(U) and G–C structures (Chapter 6).

Polymorphism of DNA versus structural conservatism of RNA. It is obvious from the entries in Table 9-1 that under a variety of experimental conditions, double-helical RNA, of natural or synthetic origin, is confined to the closely related A and A' forms. In contrast, DNAs display a rather wide range of structures (Tables 9-1, 9-2). While DNAs with random sequence are only found in A, B, and C forms as shown in the generalized scheme in Figure 9-1, material with strictly repetitive oligonucleotide sequences can adopt, in addition, D, E, and Z forms, the latter even describing a left-handed double-helical DNA.

Structures of synthetic DNAs with well-defined oligonucleotide building blocks are sensitive to sequence. Inspection of Table 9-1 suggests that DNAs of like composition but of unlike sequence exhibit different struc-

Table 9-1. Principal Crystalline Forms of DNA and RNA in Fibers: Dependence of Form and Helix Symmetry on Counterion and Relative Humidity (Equivalent to Salt Concentration)[a]

Polynucleotide	Counterion	Relative humidity (%)	Form	Helix symmetry	Reference
Native DNA	Na	75	A	11_1	(842)
	Na	92	B	10_1	(101)
	Li	57–66	C	$9.33_1(28_3)$	(94)
	Li	44	C	$9.33_1(28_3)$	(94)
	Li	66	B	10_1	(824)
Poly(dA)·poly(dT)	Na	70	β–B'	10_1	(843)
	Na	92	α–B'	10_1	(843)
Poly(dG)·poly(dC)	Na	75	A	11_1	(844)
	Na	92	B^c	10_1	(844)
Poly(dA–dT)·poly(dA–dT)	Na	75	D	8_1	(845)
	Na	Up to 98	A	11_1	(846)
	Li	66	B	10_1	(846)
Poly(dA–dC)·poly(dG–dT)	Na	66	A	11_1	(824)
	Na	66–92	B	10_1	(824)
	Na	66	Z	6_5	(304)

Poly(dA–dG)·poly(dC–dT)	Na	66	C″	9_1	(824)
	Na	95	B	10_1	(824)
Poly(dG–dC)·poly(dG–dC)	Na	43	Z	6_5	(304)
	Na	Up to 92	A	11_1	(824)
	Li	81	B	10_1	(824)
Poly(dA–dA–dT)·poly(dA–dT–dT)	Na	66	D	8_1	(847)
	Na	92	B	10_1	(847)
Poly(dA–dG–dT)·poly(dA–dC–dT)	Na	Up to 98	A	11_1	(824)
	Li	98	B	10_1	(824)
	Li	66	C	9_1	(824)
Poly(dA–dI–dT)·poly(dA–dC–dT)	Na	66	D	8_1	(824)
	Na	81	C	$9.33_1 (28_3)$	(824)
	Na	92	B	10_1	(824)
Poly(dI–dC)·poly(dI–dC)	Na	66	B	10_1	(388)
	Na	75	D	8_1	(388)
Native RNA (reovirus)	Na	Up to 92	A	11_1	(848)
Poly(A)·poly(U)	Na	Up to 92	A	11_1	(848)
Poly(I)·poly(C)	Na	Up to 92	A′	12_1	(848)
Hybrid poly(rI)·poly(dC)[b]	Na	75	A′	12_1	(849)
Hybrid DNA–RNA	Na	33–92	A	11_1	(850)
Phage T2 DNA	Na	60	T	8_1	(858)

[a] In the original literature (824), more data are entered, and in (304) the notation S-DNA instead of Z-DNA is used.

[b] Helical parameters not mentioned in (849) but in (851).

[c] As described in (844), poly(dG)·poly(dC) transforms only incompletely from A- to B-DNA.

Box 9-1. Helical Symmetries and Crystallographic Screw Axes.

The helical symmetries entered in Table 9-1 are designated by indexed numbers N_m, adapted from crystallographic symbols for screw axes operations. In contrast to simple rotation axes N, which imply that a motif repeats after rotation about an axis by $2\pi/N$, screw axes operations involve an additional translation, parallel to the rotation axis, by m/N of the repeat distance along the screw axis. In crystallography, there are 2-, 3-, 4-, and 6-fold screw operations with symbols

$$2_1 \quad 3_1 \quad 3_2 \quad 4_1 \quad 4_2 \quad 4_3 \quad 6_1 \quad 6_2 \quad 6_3 \quad 6_4 \quad 6_5.$$

The operation 3_1 rotates a motif by $2\pi/3 = 120°$ in a right-handed screw sense and elevates it by $1/3$ axial repeat etc. In 3_2, rotation is again $2\pi/3$ in right-handed sense but elevation is now $2/3$ axial repeat, yielding motifs 1,2,3,4, etc. (see Figure). The missing motif at $1/3$ axial repeat is produced by translation of the motif 3, at $2 \times 2/3$ elevation, by one unit ($3/3$); i.e., we obtain ($4/3 - 3/3 = 1/3$). As illustrated in the Figure, connection of the different motifs by a continuous line leads to right-handed screw sense for 3_1 but left-handed sense for 3_2. Analogously, 4_1, 6_1, and 6_2 are right-handed and 4_3, 6_4, and 6_5 are corresponding left-handed screw operations.

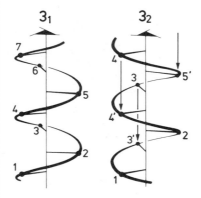

In helical symmetries of nucleic acids, the same symbols N_m are used with the same meaning as in crystallographic terms. In this situation, the motif is, in general, a nucleotide or a base-pair and helical symmetry as in A-DNA, 11_1, indicates that from one nucleotide to the next, rotation is through $2\pi/11$ (32.7°) and elevation is $1/11$ the repeat unit (=pitch). In other wording, A-DNA is 11-fold, with 11 nucleotides in one turn.

In C-DNA with symmetry 9.33_1, a nonintegral helix occurs. Exact repeat is only met after 3 turns with $3 \times 9.33 = 28$ nucleotides, often expressed as 28_3, a nomenclature not strictly in agreement with crystallographic usage but convenient to describe nonintegral helices.

In alternating, repeating sequences such as poly(dA–dT)·poly(dA–dT) or poly(dG–dC)·poly(dG–dC) the repeating motif is, in a proper description, a dinucleotide. Therefore, the 8_1 symmetry given in Table 9-1 for D-DNA in alternating copolymers should actually be designated 4_1 and Z-DNA with 12 base-pairs per turn has symmetry 6_5. Except for Z-DNA, the reduced symmetry symbols are not used in Table 9-1 in order to render it more general, with one nucleotide or one base-pair as repeating motif.

tural properties. Thus, while poly(dA–dC)·poly(dG–dT) adopts A, B, and Z forms, the compositional analog with all purines and all pyrimidines in separate strands, poly(dA–dG)·poly(dC–dT) occurs only in B and C″ forms and is unable to transform into an A-DNA-like structure.

An even more striking example is poly(dA)·poly(dT) and its compositional isomers: poly(dA–dT)·poly(dA–dT) and poly(dA–dA–dT)·poly(dA–dT–dT). The former exists exclusively in B form above 70% relative humidity and at lowered humidity transforms into a metastable A form which disproportionates into a triple-stranded poly(dA)·2 poly(dT) complex with A-type conformation (Section 11.4). In contrast, poly(dA–dA–dT)·poly(dA–dT–dT) adopts, as sodium salt, only B and D forms and poly(dA–dT)·poly(dA–dT) behaves similarly but in addition is able to exist in a metastable A form.

Base composition also plays a role in determining structures of synthetic DNAs. The pairs poly(dA–dT)·poly(dA–dT) and poly(dG–dC)·poly(dG–dC) (Table 9-1) display the same purine–pyrimidine alternating sequence but different composition. The sodium salt of poly(dA–dT)·poly(dA–dT) adopts metastable A and stable D forms, but poly(dG–dC)·poly(dG–dC) up to 92% relative humidity occurs exclusively in the A form. If, however, polynucleotides with inosine replacing guanosine are considered, it is striking to see that under no conditions was an A-DNA form obtained. An explanation for this behavior is given in Chapter 17 where hydration of DNAs is discussed. Moreover, the spectacular transformation of right-handed B-DNA to left-handed Z-DNA occurs with poly(dG–dC) but has not been observed thus far with poly (dA–dT).

It appears that sequence and compositional effects do not influence the average overall structures of natural DNAs. X-Ray scattering (825), X-ray

Table 9-2. Structure Types and Helical Parameters Observed by X-Ray Fiber Diffraction Methods for Natural DNAs with "Random" Sequence and for Synthetic DNAs with Homopolymer or Repetitive Nucleotide Sequence

| Structure type | Pitch (Å) | Helical symmetry[a] | Axial rise (h) and turn angle t per nucleotide residue | | Groove width[g] (Å) | | Groove depth[g] (Å) | |
			h (Å)	t (°)	Minor	Major	Minor	Major
Natural and synthetic DNAs								
A	28.2	11_1	2.56	32.7	11.0	2.7	2.8	13.5
B	33.8	10_1	3.38	36.0	5.7	11.7	7.5	8.5
C	31.0	$9.33_1(28_3)^b$	3.32	38.6	4.8	10.5	7.9	7.5
Synthetic DNAs with nonrandom sequence								
B'	32.9	10_1	3.29	36.0				
C'	29.5	9_1	3.28	40.0				
C"	29.1	$9_1(18_2)^c$	3.23	40.0				
D	24.3	8_1	3.04	45.0	1.3	8.9	6.7	5.8
E	24.35	$7.5_1(15_2)^d$	3.25	48.0				
S	43.4	6_5^e	3.63	-30.0^h				
Z^f	45	6_5^e	3.7	-30.0^h	2.7^i		9.0^j	

Source. From (824).
[a] Given are number of residues per one turn and in parentheses number of residues after which exact repeat is achieved. Differs in some entries from (824).
[b] Helix turn is 9.33 residues; repeats after three turns with $9.33 \times 3 = 28$ residues.
[c] Dinucleotide sequence, repeats exactly after two turns with 18 residues.
[d] Trinucleotide sequence, repeats after two turns with 15 residues.
[e] For dinucleotide as repeating unit.
[f] Obtained by extension of the d(G–C) hexanucleotide, Ref. (917).
[g] From Ref. (893). Groove width gives perpendicular separation of helix strands drawn through phosphate groups, diminished by 5.8Å to account for van der Waals radii of phosphate groups. Groove depth is also based on van der Waals radii.
[h] This gives the *average* turn angle for a base-pair in the dinucleotide repeat. The turn angle for this repeating motif is $-60°$; for the individual base-pairs it is not equal, but different for GpC and CpG sequences. See Chapter 12.
[i] From Ref. (303): there is no major groove. For minor groove width, 5.8 Å have been subtracted from the shortest P---P distance, 8.5 Å, to account for van der Waals radii.

fiber diffraction (826), infrared dichroism (827), and ultraviolet circular dichroism (828) of DNAs with varying A–T content and on highly repetitive satellite DNA (Box 9-2) appeared to indicate that, besides the traditional A, B, and C forms, further structural isomorphs are possible. These findings, however, are called into serious question by studies employing ultraviolet circular dichroism (829) as well as X-ray fiber diffraction techniques (824,830–832). In essence, for naturally occurring DNAs, repetitive satellite DNA or not, only A, B, and C forms were confirmed. Because these methods are limited to a picture of the *overall* structural

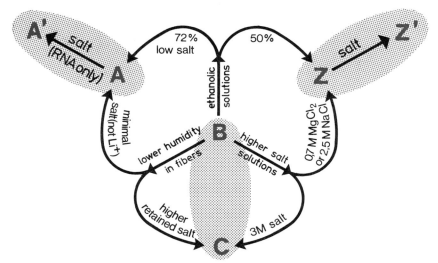

Figure 9-1. Generalized scheme showing families of naturally occurring DNAs and of RNA. Transitions between members of one family, within oval shading, are influenced by salt concentration. Transitions between families are induced by changing relative humidity and salt content in fibers or films, or changing ionic strength or solvent polarity in solution. Critical salt and ethanol concentrations from Refs. (914,915,839) give midpoints rather than endpoints of transition. Note that RNA is limited only to the A family and that the Z family with left-handed helical sense may be restricted only to certain alternating purine/pyrimidine sequences. Redrawn from (302).

Box 9-2. What are satellite DNAs?

If DNA of eukaryotic organisms is fragmented by shearing and subsequently banded in a CsCl gradient, a main peak is obtained which is flanked by "satellite" peaks, indicating DNA of different density and hence different base composition. Analysis of these satellite DNAs has shown that they contain serially repeated sequences the length and complexity of which are characteristic for the species investigated.

In crabs, a satellite DNA was found composed of 93% alternating A and T, or poly(dA–dT), with only 2.7 mole% (dG+dC) (860). From *Drosophila virilis,* three different satellite DNAs with heptanucleotide sequences such as poly(dA–dC–dA–dA–dA–dC–dT)·poly(dA–dG–dT–dT–dT–dG–dT) were isolated (861).

Satellite DNAs are located in the centromeric chromatin. They act probably as spacers between genomes but their actual function is unknown (862). Evidently, they do not code for proteins and they are not even transcribed (349).

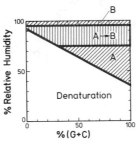

Figure 9-2. "Phase diagram" relating conformational changes of natural DNAs with different (dG+dC) content with varying relative humidity. At high relative humidity (95%), only the B form exists which, in the range 95 to 73%, transforms into A-DNA. Below 73%, only the A form is observed and, as an additional condition, the percentage (dG + dC) must be >30% in order to prevent denaturation. The diagram was obtained from DNA films prepared with 3–4% NaCl and investigated by infrared spectroscopy. Redrawn from (833).

features of a molecule but do not trace fine details, we cannot exclude that local, subtle changes do occur, as proposed, e.g., for the "alternating B-DNA" structure (Section 11.3).

This does not mean that DNAs with widely varying (dA+dT) content do not display different thermodynamic properties. In fact, it has been shown (833) that the B→A structural transition is greatly inhibited when the (dG+dC) content is below 30% and is increasingly favored when the (dG+dC) proportion is raised, as illustrated in the phase diagram (Figure 9-2).

DNA structure is sensitive to cation type and to temperature. There are many examples in Table 9-1 to show that sodium and lithium salts of natural and synthetic DNAs adopt different helical structures in the fiber state. This behavior is paralleled in solution where type and concentration of cations and changes in temperature are of influence. Different types of anions, however, appear not to have any measurable effect (834,835). By and large, all structural changes induced by environmental conditions are restricted to B-DNA forms; A-DNA is obviously more rigid in its molecular conformation, as is RNA which only occurs in A and A' forms and resists all kinds of sequence, compositional, and environmental influences.

9.2 The Variety of Polynucleotide Helices with Right-Handed Screw Classified into Two Generically Different Families: A and B

Turning our attention to Tables 9-3 and 9-4, and ignoring for the moment the exceptional left-handed Z- or S-DNA treated in Chapter 12, we find that polynucleotides of both the ribo and deoxyribo series have common structural features. In general, right-handed DNA and RNA helical structures, be they single, double, triple, or quadruple stranded, can be broadly classified into two generically different categories called A and B families.

The essential distinction between A- and B-type polynucleotide helices lies in the sugar puckering modes, $C_{3'}$-*endo* for A families and $C_{2'}$-*endo* (or the equivalent $C_{3'}$-*exo*) for B families. The different sugar puckerings imply variation in distance between adjacent phosphates of the same polynucleotide chain, ranging from 5.9 Å for $C_{3'}$-*endo* to 7.0 Å for $C_{2'}$-*endo* conformations (Figure 9-3). As a consequence, the helical arrangements are macroscopically different, as further discussed below.

The sense of the base-pair tilt θ_T is associated with sugar puckering. In double-helical polynucleotides, the normals to the base-pairs are not exactly parallel to the helix axis but inclined to it by up to 20°. Characteristically, the sense of this tilt θ_T defined in Figure 2-14 is *positive in A-type* but *negative in B-type* helices, and hence is correlated with sugar puckering. As we shall see later, this difference in tilt angle combined with the right-handed helical screw sense leads to alterations in stacking overlap between adjacent bases.

Base tilt angle γ_T is correlated with rise per nucleotide residue, h. If the bases in base-pairs were coplanar and the base-pairs exactly perpendicular to the helix axis, the axial rise per nucleotide, h, should correspond to the van der Waals distance, 3.4 Å. However, in all the entries in Tables 9-2 and 9-3 for double-helical Watson–Crick-type helices, the base-pairs are tilted and the axial rise per nucleotide is smaller than 3.4 Å. On purely geometrical grounds (8), there exists a linear relationship between h, the axial rise per nucleotide and tilt angle, γ_T, which describes the angle between the normal to an *individual* base and the helix axis (Figure 9-4). This value is, in general, slightly different from the base-pair tilt θ_T involving the best plane through both bases (Figure 2-14). It is curious that, by extrapolating the line in Figure 9-4 to 0° tilt angle, an axial rise per residue of 3.7 Å is obtained, a value significantly different from the expected 3.4 Å. Apparently the ever-present base-pair propeller twist which optimizes base–base interactions in helical polynucleotides (836) counteracts a coplanar arrangement of bases in base-pairs. Consequently, if one base is at $\gamma_T = 0$°, the partner is tilted $\gamma_T \neq 0$ and serves as a lever separating adjacent base-pairs beyond 3.4 Å.

Rise per residue, h, is more variable in A-DNA than in B-DNA whereas rotation per residue shows the reverse trend. By and large, the axial rise per nucleotide residue, h, is in the range 2.59 to 3.29 Å in A-type Watson–Crick double helices and is associated with a relative small variation in rotation from one residue to the next, 30.0° to 32.7° (see Tables 9-4, 9-5). In B-type double helices, however, h is rather constant and changes only from 3.03 to 3.37 Å from B- to D-DNA, whereas the rotation per residue varies in a broader range, from 36° to 45°. As a result, A-type double helices are rather uniform in overall shape whereas those of the B-type are

Table 9-3. Conformational Parameters for Polynucleotides Belonging to the A and B Families

Polynucleotide		Helix type	α (P–O$_{5'}$)	β (O$_5$–C$_{5'}$)	γ (C$_{5'}$–C$_{4'}$)	δ (C$_{4'}$–C$_{3'}$)	ϵ (C$_{3'}$–O$_{3'}$)	ζ (O$_{3'}$–P)	χ	Reference
				A Family						
Heteropolymer double helices										
α-A-RNA } reovirus		A	−88	−155	54	83	−179	−52	−155	(852)
β-A-RNA } reovirus		A	−92	−151	52	83	174	−46	−146	
Poly(A)·poly(U)		A	−62	180	48	83	−151	−74	−166	(848)
Poly(I)·poly(C)		A′	−65	−167	44	83	−168	−60	−161	
A-RNA^a		A	−68	178	54	82	−153	−71	−158	
A-DNA^a		A	−50	172	41	79	−146	−78	−154	
A-DNA		A	−90	−149	47	83	−175	−45	−154	(304)
A-DNA		A	−75	−179	59	79	−155	−67	−158	(843)
Poly(2-methylthio-A)·poly(U)										
with Hoogsteen } Poly(ms²A)		A	−102	153	100	79	−132	−72	4	(887)
base-pairs } Poly(U)		A	−93	−177	76	85	−149	−72	8	
Heteropolymer triple helices										
Poly(U)·poly(A)·poly(U)										
92% relative humidity	U	A′	−28	171	24	83	−156	−75	−154	
	A	A′	−66	−179	53	83	−163	−67	−150	
	U	A′	−40	167	37	83	−149	−83	−156	
72% relative humidity	U	A	−55	173	47	83	−152	−74	−158	(853)
	A	A	−76	176	64	83	−156	−69	−152	
	U	A	−63	174	53	83	−149	−76	−158	

DNA–RNA hybrids									
Poly(I)·poly(dC)	A'	−71	−176	56	80	−159	−64	−165	(849)
Poly(U)·poly(dA)·poly(U)									
U	A	−57	172	49	83	−152	−74	−158	(853)
dA	A	−71	178	58	83	−160	−67	−149	
U	A	−62	174	53	83	−149	−76	−157	
Homopolymer complexes									
Poly(AH⁺)·poly(AH⁺)	A	−75	168	69	82	−144	−67	−176	(458)
Poly(s²U)·poly(s²U)									
α-chain	A	−43	172	37	77	−146	−77	−160	(461)
β-chain	A	−43	163	41	81	−148	−80	−150	
Poly(X)·poly(X)									
pH 5.7	A	−75	171	63	87	−142	−80	−156	(854)
pH 8.0	A	−66	−179	51	81	−153	−71	−162	
[poly(I)]₄, [poly(G)]₄	A	−103	176	92	83	−156	−69	−169	(855)
Homopolymer single helices									
Poly(C)	A	−57	173	47	83	−129	−65	−159	(947)
Poly(O₂′-methyl-C)	A	−88	−175	66	83	−138	−58	−154	
Poly(A)									
Twofold	—	99	183	46	154	−88	39	−110	(963)
Ninefold	A	−64	154	53	83	−145	−70	−160	(957)

[a] New fiber data, private communication Drs. Arnott and Chandrasekaran (1982).

Table 9-3. (*Continued*)

Polynucleotide	Helix type	α (P–O$_{5'}$)	β (O$_5$–C$_{5'}$)	γ (C$_{5'}$–C$_{4'}$)	δ (C$_{4'}$–C$_{3'}$)	ϵ (C$_{3'}$–O$_{3'}$)	ζ (O$_{3'}$–P)	χ	Reference
				B Family					
B-DNA	B	−46	−147	36	157	155	−96	−98	(856)
B-DNA	B	−41	136	38	139	−133	−157	−102	(304)
C-DNA	C	−39	−160	37	157	161	−106	−97	(857)
Poly(dA–dT)·poly(dA–dT)	D	−62	−152	69	157	141	−101	−97	(845)
T2-DNA	D	−73	−137	89	146	136	−85	−110	(858)
Poly(dA)·poly(dT)	B′	−52	−136	39	157	145	−87	−96	(859)
"Alternating" B-DNA									
Fragment A–T		−60			110°(A)		−60		(900)
Fragment T–A		−50			155°(T)		−120		
Homopolymer single helix									
Poly(dT)		−72	179	43	151	−105	164	−148	(358)
DNA·RNA hybrid at high relative humidity									
Poly(A)·poly(dT)									
Poly(A)	A	−88	174	75	97	−146	−77	−175	(912)
Poly(dT)	B	180	96	166	152	−178	−36	−152	

more variable. The reason is that the different rotation angles, from 36° to 45°, imply a change in number of residues per turn from 10 to 8 which considerably alters the appearance.

Polynucleotides of the A family obey structural conservatism. Inspection of Tables 9-3 and 9-4 shows that A-family double helices with $C_{3'}$-*endo* puckering mode are only observed with the structurally closely related A- and A'-RNA and with A-DNA. Because this puckering associates with a short interphosphate distance of only 5.9 Å (Figure 9-3), the polynucleotide chain is underwound compared to B-type double helices and consists of 11 to 12 nucleotides (or base-pairs) per turn, corresponding to a 32.7° to 30° rotation per nucleotide. The base-pairs are inclined at 8° to 20° with respect to the helix axis, producing an axial rise per nucleotide between 3.29 and 2.56 Å.

Polynucleotides of the B family display structural variety. The helical structures belonging to the B family are formed exclusively by DNAs; none is observed for RNA. The $C_{2'}$-*endo* or equivalent $C_{3'}$-*exo* sugar puckering mode observed for B-family helices pushes adjacent phosphates of one polynucleotide chain about 7 Å apart (Figure 9-3), giving rise to overwound helices compared with those of the A family. As a consequence, the rotation per nucleotide now ranges from 36° to 45° and causes the axial rise per residue, 3.03 to 3.37 Å, to become generally larger than in A-type helices. Associated with these characteristics is a smaller, yet *negative* base-pair tilt between −16.4° and −5.9° (Table 9-5).

It must be stressed that, compared with the A-family double helices, the B families consisting of B-, C-, and D-DNAs show a larger structural variety. The main reason lies in the macroscopic arrangement of the A- and B-type double helices which, in the A form, smoothly allow underwinding from 11- to 12-fold helices, but in the B form, overwinding from 10- to 8-fold is more dramatic. This does not mean that strain energies are involved in these helical transitions but, as we will discuss in a moment, base-pair dislocations and grooves along the helical surfaces are a dominant structural feature.

Deep and shallow grooves and base-pair dislocation—Macroscopic features distinguishing A- and B-type helices. Besides sugar puckering modes and base-pair tilt sense, the most important parameter distinguishing between A- and B-type helices is the dislocation D, which describes the displacement of base-pairs away from the helix axis (Figure 9-5). In B-DNA, the base-pair is located astride the helix axis ($D \sim -0.2$ Å) while in D-DNA, the axis is moved toward the minor groove side of the base-pair ($D \sim -1.8$ Å). In A-DNA, however, the helix axis is pushed far out into the major groove side, with D amounting to 4.4 to 4.9 Å.

The consequence of the displacements is that, in A-type double helices, the polynucleotide chains wrap around the helix axis like a ribbon, leaving an open cylinder along the helix axis with a van der Waals diameter of ∼3.5 Å. The base-pairs are driven out toward the periphery of the double helix, giving rise to a very deep and narrow major groove but a

Table 9-4. Helical Parameters for Polynucleotides Displaying $C_{3'}$-endo Sugar Puckering and Therefore Belonging to the A Family

Polynucleotide	Helix type	Helix symmetry[a]	Pitch height (Å)	Axial rise per residue (Å)	Rotation per residue (°)	Dislocation[b] D (Å)	Tilt[c] (°)	Reference
Heteropolymer double helices								
α-A-RNA } reovirus	A	11_1	30.0	2.73	32.7	4.4	17.0 }	(852)
β-A-RNA }	A	11_1	30.0	2.73	32.7	4.4	19.2 }	
Poly(A)·poly(U)	A	11_1	30.9	2.81	32.7	4.4	16.0 }	(848)
Poly(I)·poly(C)	A'	12_1	36.0	3.00	30.0	4.4	10.0 }	(843)
A-DNA	A	11_1	28.2	2.56	32.7	4.7	20.0	
Poly(2-methylthio-A)·poly(U) with Hoogsteen Poly(ms²A) base–pairs } Poly(U)	A	10_1	31.6	3.16	36.0	—	13	(887)
	A						10	
Heteropolymer triple helices								
Poly(dT)·poly(dA)·poly(dT) Watson–Crick, Hoogsteen	A'	12_1	39.1	3.26	30.0	2.80 Watson–Crick	8.5	
Poly(U)·poly(A)·poly(U)	A'	12_1	36.5	3.04	30.0		12.0	(853)
92% relative humidity	A	11_1	33.4	3.04	32.7		12.0	
72% relative humidity Poly(I)·poly(A)·poly(I)	A	11_1	36.2	3.29	32.7	2.8	8.0	
DNA–RNA hybrids								
Poly(rI)·poly(dC)	A'	12_1	35.6	2.97	30	4.9	5.68	(849)
Poly(U)·poly(dA)·poly(U)	A	11_1	33.6	3.06	32.7		12.0	(853)
Poly(dC)·poly(dI)·poly(dC)	A	11_1	34.8	3.16	32.7		10.0	

Homopolymer complexes								
Poly(AH$^+$)·poly(AH$^+$)	A	8_1	30.4	3.8	45		10	(458)
Poly(s^{2}U)·poly(s^{2}U)	A	11_1	28.8	2.62	32.7		18	(461)
Poly(X)·poly(X)								
pH 5.7	A	10_1	30.1	3.01	36		21	(854)
pH 8.0	A	11_1	27.7	2.52	32.7		11	(855)
$[Poly(I)]_4^a$, $[Poly(G)]_4$	A	23_2	39.2	3.41	31.3		11	
Homopolymer single helices								
Poly(C)	A	6_1	18.6	3.1	60	—	21	(947)
Poly(O$_2$-methyl-C)	A	6_1	18.9	3.2	60	—	28	
Poly(A)								
Twofold	—	2_1	5.8	2.9	180	—	36	(963)
Ninefold	A	9_1	25.4	2.82	40	—	24	(957)

(Table 9-4 continues)

Table 9-4. (*Continued*) Helical Parameters for Polynucleotides Displaying $C_{3'}$-*exo* (or $C_{2'}$-*endo*) Sugar Puckering and Therefore Belonging to the B Family

Polynucleotide	Helix type	Helix symmetry[a]	Pitch height (Å)	Axial rise per residue (Å)	Rotation per residue (°)	Dislocation[b] D (Å)	Tilt[c] (°)	Reference
B-DNA	B	10_1	33.7	3.37	36	−0.14	−5.9/−2.1	(843,856)
C-DNA	C	28_3	30.9	3.31	38.6	−1.0	−8.0/1.0	(842,857)
Poly(dA-dT)·poly(dA-dT)	D	8_1	24.2	3.03	45	−1.8	−16.4	(845)
T2-DNA	D	8_1	27.2	3.40	45	−1.43	−6/4	(858)
Poly(dA)·poly(dT) "Alternating" B-DNA	B'	10_1	32.9	3.29	36	−0.02	−7.9/−1.0	(859)
Homopolymer single helix								
Poly(dT)	—	7.2_1	25.2	3.5	50		132	(358)
DNA·RNA hybrid at high relative humidity								
Poly(A)·poly(dT)								
Poly(A) ($C_{3'}$-*endo* sugar pucker)	B	9.7_1	33.7	3.46	37.0	—	—	(912)
Poly (dT)	B							

[a] For explanation, see Box 9-1.
[b] Defined in Figure 2-14.
[c] Gives the angle between normal to plane of base and helix axis. If two numbers separated by a slash are given, the first denotes base-pair tilt θ_T, the second propeller twist θ_P of bases relative to each other (see Figure 2-14).
[d] Original assignment as $C_{2'}$-*endo*-type sugar (993) later proved erroneous (855).

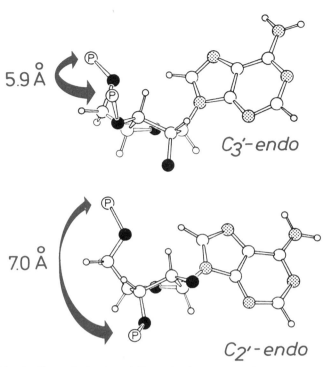

Figure 9-3. Nucleotides in C$_{3'}$-*endo* (top) and C$_{2'}$-*endo* (bottom) conformations as observed in A- and B-type polynucleotide helices. Phosphate···phosphate distances are indicated in angstroms. From (863).

reduced, shallow, wide minor groove (section 2.8, Figure 2-13, Table 9-2). In B- and C-DNAs the base-pairs are pierced by the helix axis and the grooves are less pronounced than with A-type double helices. The groove depths are nearly identical, yet the widths differ, being wider for the major than for the minor groove. In D-DNA where the helix axis is pushed into

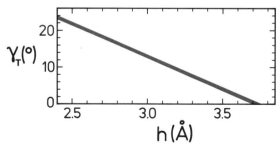

Figure 9-4. In polynucleotide double helices, the axial rise per nucleotide, h, is linearly related to γ_T, the angle between the normal to an *individual* base plane and the helix axis. After (791).

Table 9-5. Comparison of Structural Characteristics of A- and B-Type Polynucleotide Double Helices

Family type	A	B
Sugar pucker	$C_{3'}$-endo	$C_{2'}$-endo
Intrastrand phosphate · · · phosphate distance	5.9 Å	7.0 Å
Dislocation of base-pairs from helix axis	4.4 to 4.9 Å in major groove	−0.2 to −1.8 Å in minor groove
Rotation per nucleotide	30° to 32.7°	36° to 45°
Axial rise per nucleotide	2.56 to 3.29 Å	3.03 to 3.37 Å
Base-pair tilt	Positive, 10° to 20.2°	Negative, −5.9° to −16.4°

the minor groove side, the groove depths are about equal and, since the 8-fold helix is rather overwound with respect to B-DNA, the minor groove is very narrow. In contrast, the major groove is rather wide and open.

Different groove sizes are associated with special complexing properties Depending on the base-pair tilt and, consequently, on the axial rise per residue h, the A-type double helix displays different geometries of its major groove. If the axial rise per residue is small, around 2.6 Å for A-DNA and A-RNA, the major groove is deep and narrow and only accessible to water molecules and metal ions. On the other hand, if h expands to 3.29 Å, the major groove is still deep but opened up to accommodate another polynucleotide chain, as we will see in Section 10.2 for poly(U)·poly(A)·poly(U) and in Section 11.3 for poly(dT)·poly(dA)·poly(dT). In addition, a model has been advanced for binding of a polypeptide in antiparallel β-pleated sheet structure in the shallow minor groove of A-RNA, the complex being stabilized by hydrogen bonds between ribose $O_{2'}H$ hydroxyls and polypeptide carbonyl oxygens (837). Analogously, a model of

Figure 9-5. Illustration of helix axis dislocation relative to base-pairs in A-, B-, C-, and D-DNAs. Crosses (X) mark positions where helix axes pierce base-pair planes.

the histone H2A can be fitted into the narrow, minor groove of B-DNA and fixed to it via hydrogen bonds involving peptide NH and polynucleotide $O_{3'}$ oxygens (838) (Section 18.3).

The sodium ion complexation in the ApU crystal structure to be discussed in Section 10.4 sheds light on sequence specific cation$\cdots O_2$(uracil) coordination in the minor groove. In polymeric double-helical nucleic acids, such kind of interactions could lead to cation-dependent formation of DNA structures and explains why, in Table 9-1, some DNA modifications are only obtained with certain cations. On the other hand, cations will interact with the phosphate groups and evidence has been presented for some kind of specificity (839). Thus, in the alkali series, the radius of cations increases form Li^+ to Cs^+; yet the actual radius of the *hydrated* species decreases from 7.4 Å for Li^+ to 3.6 Å for Cs^+. In structural terms, this implies that Li^+ fits well into the wide minor groove of B-DNA while Cs^+, on the other side of the series, fits into the narrow minor groove of D-DNA, in agreement with the measured winding capability of these ions.

Electrostatic potential around nucleic acids confers binding specificity. In interactions between DNA or RNA and other molecules, not only steric but also electronic complementarity plays an essential role. When electrostatic potentials for a number of DNA oligo- and polymers are calculated with Na^+ counterions placed at the bisector of each PO_2^- group, a sequence-dependent pattern results (839a). The minor groove is lined with negative potential in case of A/T sequence because adenine N_3 and thymidine O_2 both contribute with partial negative charges. In contrast, for G/C sequence, a more neutral situation is encountered, the negative guanine N_3 and cytosine O_2 being counterbalanced by the positive guanine N_2 amino group. In the major groove, the potentials for G/C and A/T are again different. Looking into the groove, we see a pattern $(--+)$ for $(N_7O_6\cdots HN_4$ of G-C), a reversed pattern $(+--)$ for $(N_4H\cdots O_6N_7$ of C-G), yet for A-T or T-A the same pattern $(-+-)$ for $(N_7N_6H\cdots O_4$ or $O_4\cdots HN_6N_7)$ occurs. This means that G-C and C-G can be distinguished by their electrostatic potential but A-T and T-A look the same. The calculation of electrostatic potentials around the surface of macromolecules has just begun and will be a major breakthrough in understanding intermolecular interactions. We will touch upon electrostatic potentials later when discussing nucleic acid-protein complexes and hydration of nucleic acids (see also Refs. 839b,c).

In A- and B-type double helices, base stacking overlaps differ greatly (Figure 9-6). In essence, stacking is largely limited to interactions between bases in the same polynucleotide chain in B-type double helices (*intra*strand), but in A-type double helices stacking also involves bases belonging to two different chains (both *intra*- and *inter*strand stacking occur). The reasons for this phenomenon are twofold. First, the rotation per nucleotide in A-type helices, 30° to 32.7°, is less than the rotation of 36° to 45° in B-type helices; this favors both intra- and interstrand stacking

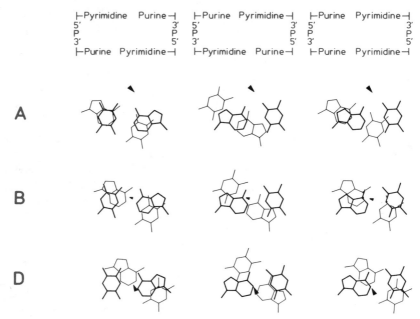

Figure 9-6. Base stacking overlaps in A- and B-type double helices, illustrated from top to bottom for A-, B-, and D-DNA. A–T base-pairs are stacked in sequences T–A on A–T, A–T on T–A, and A–T on A–T. The T–A on T–A stacking is comparable to A–T on A–T, with top and bottom reversed. Because views are perpendicular to base-pairs and not along the helix axes, the latter project as lines, indicated by arrows pointing from upper to lower base-pair. Note that in A- and D-DNA, arrows point in opposite directions because in the former the tilt is positive and negative in the latter. The arrow in B-DNA is smallest because the tilt is only slightly negative. For C-DNA the stacking patterns are between B- and D-DNA. Redrawn from (845).

in A-type helices. Second, the base-pair tilt which is characteristically positive in A-type, but negative in B-type helices, leads, in combination with right-handed helical screw sense, to increased overlap in A-type, but to reduced overlap in B-type helices. Both effects, produced by residue rotation, base-pair tilt, and right-handed screw sense, combine and give the different stacking overlaps displayed in Figure 9-6.

Native DNA adopts the B form. The exact 10-fold repeat of B-DNA is bound to the crystalline state and does *not* persist in solution. DNA in the B form in crystalline fibers has 10 base-pairs per pitch; i.e., the double helix is integral with symmetry 10_1 (Table 9-1). This molecular structure is correlated with the crystal lattice (840) and does not change substantially even if the fiber is extremely wet (841). However, as soon as the crystal lattice breaks down into solution and direct helix–helix contacts are absent, the structural limitations imposed by crystal packing forces no longer exist and the DNA molecule slightly underwinds, yielding 10.3 to 10.6 base-pairs per turn. Evidence for this number is provided by the crystal structure analysis of a dodecanucleotide (Section 11.2), which crystallizes with one complete turn of B-type DNA in the asymmetric unit (464), by wide-angle X-ray scattering studies (462), by enzymatic digestion (463), by circular dichroism measurements (835) applied to DNA in aqueous solution, and by theoretical calculations (836).

Because this B form of DNA is associated with high humidity in fibers or with aqueous solutions of DNA, it is generally believed to represent the native DNA occurring in living organisms. As discussed in Chapter 19, a similar molecular structure is observed in DNA wound around the histone core of nucleosomes.

Summary

This chapter illustrates the main structural features of double-stranded nucleic acid helices. It is shown that RNA duplexes are conformationally confined to two closely related forms A and A' (Chapter 10), whereas the DNA double helix undergoes a variety of structural changes, from the right-handed A, B, C, and D forms (Chapter 11) to the left-handed Z form (Chapter 12). The DNA structures depend greatly on environmental conditions (salt, alcohol), but (repetitive) nucleotide sequence and base composition also come into play. The right-handed forms are grouped into two families, A (with $C_{3'}$-endo sugar pucker) and B (with $C_{2'}$-endo sugars); the second family also consists of C- and D-DNA. Associated with the different sugar puckering modes are different overall geometries of double helices, expressed in relative disposition of base-pairs with respect to helix axis, tilt of base-pairs, and size of minor and major grooves.

Chapter 10

RNA Structure

Depending on their biological function, naturally occurring RNAs either display long, double-helical structures or they are globular, with short double-helical domains connected by single-stranded stretches. Double-helical domains can, in many cases, be predicted from primary nucleotide sequences and special computer algorithms have been developed for this purpose (864). Short double helices are found in tRNA (Chapter 15), in ribosomal RNAs (865–868), in the genes coding for the coat proteins of bacteriophages MS2 (869) and R17 (870), in globin mRNA (871), and probably in other mRNAs as well (872).

Pronounced, well-developed double helices occur in rice stunt virus RNA (873), reovirus RNA (874–877), wound tumor virus RNA (878), and biosynthetic polynucleotide complexes (852,853,878). Fibers of reovirus RNA display two crystal forms (α and β), which differ mainly in crystal packing but little in molecular structure (874–877). The synthetic homopolymer RNAs poly(A)·poly(U) and poly(I)·poly(C) crystallize isomorphously with β-reovirus RNA, i.e., their structures resemble those of the natural RNA species (852,853). As described in Section 10.2, poly(A)·poly(U) and the all-purine analog poly(A)·poly(I) disproportionate as the salt concentration is raised and form triple-stranded helices in which both Watson–Crick and Hoogsteen base-pairing are simultaneously realized (853,880,881). If, in these complexes, the 2 position in adenine is substituted by a bulky group, only Hoogsteen-type duplexes form. All of these examples, however, exhibit only polynucleotide chains of the A family and demonstrate the remarkable conformational conservatism of RNA.

10.1 A-RNA and A'-RNA Double Helices Are Similar

RNA double helices display two major, structurally similar conformations, depending on the salt concentration of the environment. At low ionic strength, the A-RNA double helix with 11 base-pairs per helix turn predominates. If the salt concentration is raised to in excess of 20%, A-RNA is transformed into A'-RNA with a 12-fold helix. A third conformation, A"-RNA with presumably a nonintegral helix containing 11.3 ± 0.5 base-pairs per turn has been proposed but not yet subjected to thorough analysis (879).

In the 1960s it was thought that A-RNA should be a 10-fold double helix (92,876). Subsequently, more elaborate investigations showed the A-RNA double helix to be truly 11-fold and recent X-ray fiber diffraction studies have confirmed this finding. Geometrical details of the structure of the A-RNA helix were derived on the basis of data from the biosynthetic materials poly(A)·poly(U) and poly(I)·poly(C), and the atomic parameters of the A'-RNA structure were obtained from the poly(I)·poly(C) complex at higher salt concentration (Table 9-3) (851,852).

Both A- and A'-RNA structures exhibit features typical of Watson–Crick base-pairs. The polynucleotide chains are arranged antiparallel and form a right-handed double helix. Because the base-pairs are displaced by 4.4 Å from the helix axis, a very deep major groove and a rather shallow minor groove are created (see Chapter 9). The main differences between A- and A'-RNA are in the pitch heights, about 30 Å for A-RNA but 36 Å for A'-RNA; these are correlated with the number of nucleotides per turn, 11 in A-RNA and 12 in A'-RNA. The axial rise per residue in A-RNA, 2.73 to 2.81 Å, is smaller than for A'-RNA, 3.0 Å, a difference that is reflected in the base-pair tilt angle, 16° to 19° in A-RNA and 10° in A'-RNA. Otherwise, the nucleotide conformations in both A- and A'-RNA are the same, as indicated by the torsion angles (Table 9-3) and by Figures 10-1 and 10-2.

10.2 RNA Triple Helices Simultaneously Display Watson–Crick and Hoogsteen Base-Pairing

Disproportionation of homopolymer hetero duplexes yields triple-stranded helices. Both poly(A)·poly(U) and poly(I)·poly(C) undergo the structural transformation A-RNA⇌A'-RNA as the salt concentration is raised (848). Poly(I)·poly(C) is stable in the A'-RNA configuration. Poly(A)·poly(U), however, suffers disproportionation into a poly(U)·poly(A)·poly(U) triple helix and a single poly(A) strand (880,881) because the adenine heterocycle is simultaneously able to engage in both Watson–Crick and Hoogsteen base-pairs (Figure 6-7):

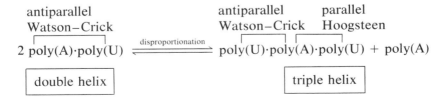

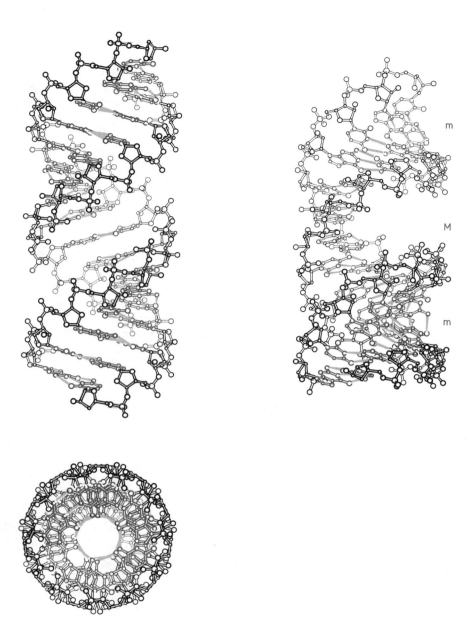

Figure 10-1. Illustration of the molecular structure of A-RNA, drawn with computer program SCHAKAL (1376) according to coordinates given in (848). Circles of increasing radius indicate C,N,O atoms; hydrogens are omitted for clarity. The ribose–phosphate backbone is drawn in black, the bases in color, and the base-pair hydrogen bonds in gray. Scale: 1 cm = 5 Å. Parts closer to the viewer are indicated in heavier lines. (left) Views are perpendicular and parallel to the helix axis. (right) The double helix is tilted by 32° to show the depth of the major groove (M) and the shallowness of the minor groove (m).

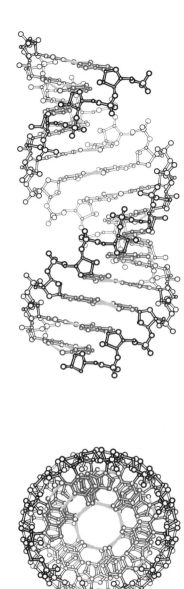

Figure 10-2. The structure of A'-RNA, drawn with atomic coordinates published in (848). For further details, see the legend to Figure 10-1.

The structure of the ternary complex consists of an antiparallel poly(A)·poly(U) double helix with Watson–Crick base-pairs which adopts, at 92% relative humidity, the A'-RNA form and changes near 75% into A-RNA. The extra poly(U) strand fits into the deep major groove of this double helix and is involved in a Hoogsteen-type base-pair with N_7 and N_6–H of the poly(A) chain. The two strands of the Hoogsteen poly(A)·poly(U) double helix run parallel to each other (Figure 11-8).

Hypoxanthine has the same functional carbonyl and imino groups as uracil. Therefore it is not surprising to find that the ternary complex poly(U)·poly(A)·poly(U) has an all-purine analog poly(I)·poly(A)·poly(I) (853), with Hoogsteen and wobble Watson–Crick base-pairs (Figure 6-7). The A:I Watson–Crick-type base-pair with interglycosyl distance $C_{1'}...C_{1'}$ of 13.0 Å is substantially longer than the A:I Hoogsteen base-pair with a distance between $C_{1'}$ atoms of 10.8 Å, a value comparable to that of a normal Watson–Crick base-pair.

All three strands have nearly the same conformation. In all triple-helix structures the base-plane tilts of either the Watson–Crick or Hoogsteen base-pairs are nearly identical and are significantly smaller (7 to 13°) than the double-helical A-RNA tilt (16 to 19°). As a consequence, the axial rise per nucleotide residue h is increased to more than 3 Å (see Figure 9-4) and all these changes lead to opening of the deep, major groove to enable accommodation of the extra polynucleotide chain. In addition, the dislocation of the helix axis is reduced from 4.4 Å in duplex A-RNA to 2.8 Å in the triple-stranded poly(U)·poly(A)·poly(U) helix. This change is due to the presence of the third chain which can be fitted with reasonable geometry to the Watson–Crick duplex only if the helix axis is approximately in the rational "center" of the base-triplet and renders the three $C_{1'}$–N glycosyl bonds nearly equivalent. This in turn leads to similar conformations for all nucleotides involved as the torsion angles indicate (Table 9-3).

Poly(I)·poly(C) is a potent interferon-inducing agent and has been used in the treatment of herpes simplex virus infections (882). The structurally isomorphous poly(A)·poly(U) is reported to be much less effective in inducing interferon (883), presumably because it occurs not as a double-helical structure under the ionic conditions of living organisms, but rather disproportionates.

10.3 A Double Helix with Parallel Chains and Hoogsteen Base-Pairs Formed by Poly(U) and 2-Substituted Poly(A)

If Watson–Crick base-pairing is sterically prevented, poly(A)·poly(U) may occur in Hoogsteen form. With triple helices of the poly(A)·2poly(U) type, one may ask whether they can release the poly(U) chain bound in the Watson–Crick mode and retain the resulting double helix with Hoogsteen-type base-pairs and polynucleotide chains in parallel orientation.

Such duplexes have actually been observed, not with normal poly(A) and poly(U) but with poly(A) substituted in the 2 position of adenine by methyl-, methylthio-, ethylthio-, and dimethylamino groups in order to provide a sterically bulky group which prohibits Watson–Crick-type base-pairing.

A Hoogsteen-type double helix with nonequivalent, parallel-oriented chains. Spectroscopic (884–886) as well as X-ray fiber diffraction methods (887) have been employed to derive the physical and structural characteristics of duplexes containing 2-substituted poly(A). These duplexes are thermodynamically less stable than their poly(A) analogs. X-Ray diffraction analysis of the blurred reflection pattern obtained from fibers has been interpreted in terms of a Hoogsteen-type double helix with 10 base-pairs per pitch of 31.6 Å in the case of the poly(2-methylthio-A)·poly(U) complex, with a base-pair tilt of nearly 10°. Interestingly, the two polynucleotide chains are conformationally nonequivalent, displaying $C_{3'}$-*endo* sugar puckering for poly(ms^{2}A) and $C_{3'}$-*endo*–$C_{2'}$-*exo* for poly(U); this classifies the duplex as a member of the A family (Tables 9-2, 9-3).

The helix structure is largely stabilized by base stacking (Figure 10-3). The 2-methylthio group of one base-pair is located over N_3 of an adenine "below." These S···N interactions appear to contribute significantly to the thermal stability of this complex because the T_m-value is higher than that of other comparable polynucleotides with no sulfur in 2-position. As outlined in Section 13·5, the enormous thermal stability of double helical poly(s^{2}U) with T_m about 57° raised with respect to that of unsubstituted poly(U) is due to a similar S···N overlap.

10.4 Mini-Double Helices Formed by ApU and GpC

The self-complementary dinucleoside monophosphates ApU and GpC each crystallize from aqueous solutions as a sodium salt (465,466). In addition, ApU can be obtained as an ammonium salt (888) and GpC as a calcium salt (889). Under all conditions, mini-double helices are formed with structural features reminiscent of A-RNA (Figure 10-4).

ApU and GpC duplexes display all essential features of double-helical A-RNA. Base–base hydrogen bonding interactions both in ApU and in GpC crystal structures are of the Watson–Crick type. This is expected for the latter from the structures encountered for monomeric, crystalline G–C pairs. But, as we have seen in Chapter 6, monomeric adenine and uracil derivatives cocrystallize only with Hoogsteen base-pairing (XXIII or XXIV in Figure 6-1). In ApU, however, limitations imposed by the sugar–phosphate backbone conformation and, in particular, stacking forces lead exclusively to the formation of Watson–Crick base-pairs.

The overall conformational features of the two dinucleoside monophosphates are comparable to those described in Table 9-3 for A- and A'-

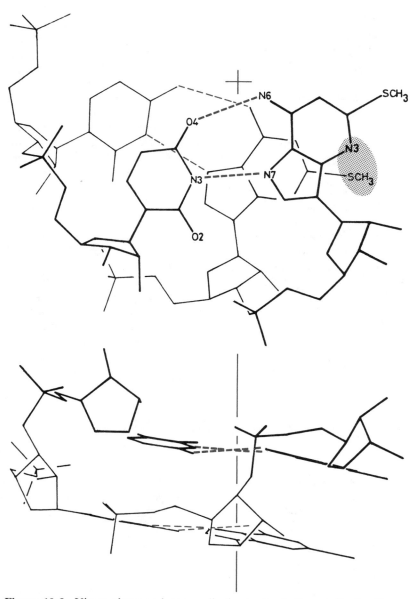

Figure 10-3. Views along and perpendicular to the helix axis in the Hoogsteen-type duplex poly(2-methylthio-A)·poly(U). Portions closer to the viewer are indicated by heavier lines. Note S···N₃ stacking interaction between adjacent adenine residues (hatched colored area) which probably adds to the stability of this complex. *From* (887).

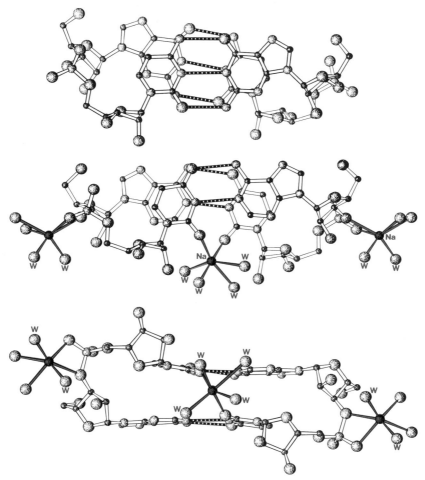

Figure 10-4. Molecular structures of dinucleoside phosphates obtained by single-crystal analysis of ApU and GpC. In GpC (top), a crystallographic dyad relates the two partners of the complex and in ApU (middle and bottom), a pseudo dyad is evident. Sodium ions and their coordination to the O_2 keto oxygen of uracil, phosphate oxygen, and water (W) are indicated in color. One of the two sodium ions in ApU is coordinated to both uracil O_2 oxygens, representing an interchain, sequence-specific nucleic acid–cation recognition. Drawn with coordinates from (465,466); hydrogen atoms and noncoordinated water molecules are omitted for clarity.

RNA. Since both ApU and GpC duplexes represent two steps of the RNA double helix, extension into "infinite" helical structures is done mathematically (890), resulting in the arrangements given in Table 10-1. In general, the synthetic ApU and GpC helices exhibit similar structural features characteristic of double helices belonging to the A-type rather than

Table 10-1. Comparison of Parameters of A- and B-Type Double Helices with Those Derived Synthetically for "Infinite" ApU and GpC Helices

	ApU	GpC	A-RNA	A'-RNA	A-DNA	B-DNA
Translation (Å)	2.36	2.59	2.81	3.0	2.59	3.38
Rotation (°)	30.2	34.7	32.7	30.0	32.7	36.0
Pitch (Å)	28.1	26.9	30.9	36.0	28.5	33.8
Residues/turn	11.9	10.4	11.0	12.0	11.0	10.0
Base tilt (°)	23	28	13	14	19	−6

Source. From (890).

the B-type double helices. This behavior suggests that the architecture of a polynucleotide double helix is intrinsically related to the microscopic conformational and self-associative properties of the individual mono- or dinucleoside phosphate rather than to the overall properties of the whole system.

ApU and GpC crystal structures all have in common heavy hydration; they resemble concentrated aqueous solutions. For this reason, one can expect that the observed structural features are not dictated by crystal packing forces but rather are to be found in aqueous solution as well. In this context it is of interest that in the sodium salt of GpC a crystallographic dyad axis passes through the dimer and relates one GpC molecule with the base-paired partner (Figure 10-4), just as indicated by fiber diffraction analyses for RNA and DNA double-helical structures. In the other crystal structures of GpC and ApU, such crystallographic dyads are not present but pseudodyads can be visualized which relate one molecule with its base-paired partner, the deviation from ideal symmetry being quite small.

Sodium ion coordination in ApU indicates specificity. In the sodium salt of ApU, one of the two sodium ions present occupies a position on the pseudodyad and is liganded by both uracil O_2 keto oxygens located in the minor groove. In addition, it is surrounded by four water molecules in order to complete its octahedral coordination shell (Figure 10-4). The other sodium ion has water and phosphate oxygens as ligands, a situation similar to the sodium and calcium ligands found in the GpC crystal structures.

The particular sodium–uracil coordination in ApU is sequence specific because the O_2 oxygens are only 4 Å apart, a distance ideal for twofold $Na^+\cdots O$ interactions. In a comparable, isomorphous UpA duplex the $O_2\cdots O_2$ separation would be about 8 Å. In G,C sequences, the guanine amino group would sterically interfere with cation binding to cytosine O_2 (465).

Sequence-specific base–cation interactions of the type found for sodium ApU are potentially important for the recognition of nucleic acids by proteins. Conceivably they modulate local structure in double helices and can have an influence on overall double-helical configuration, especially in the DNA series (Table 9-1).

10.5 Turns and Bends in UpAH⁺

UpA is, like ApU, a self-complementary dinucleoside phosphate and in principle could form a minihelix with Watson–Crick base-pairs. However, if crystallized from acidic solutions (10^{-3} M HCl), adenine becomes protonated at N_1 resulting in a zwitterion Up^-AH^+, which is unable to engage in such base–base interactions. Instead it exists in nonhelical configuration with two crystallographically independent and different molecules A and B in the unit cell (317,318,891).

Standard nucleoside geometry combined with nonhelical P–O torsions. For both UpAH⁺ molecules (Figure 10-5) the individual nucleoside moieties occur in standard conformation with $C_{3'}$-*endo* sugar puckering, base in *anti* orientation, and torsion angle γ about the $C_{4'}-C_{5'}$ bond in $+sc$ range, a geometry stabilized by short $C-H\cdots O_{5'}$-type interactions (Figure 4-20). The P–O torsion angles α and ζ, however, differ significantly for the two molecules. In molecule A, both are in the $+sc$ range, resulting in opposite orientations of the sugar residues (Figure 10-5). In molecule B, the α angle is in the $-sc$ range typical of a helical arrangement, but ζ is *ap*,

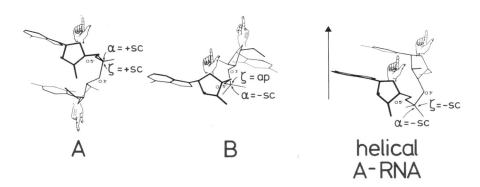

Figure 10-5. Relative orientations of nucleoside moieties in UpAH⁺ molecules A and B (left), and in A-RNA (right; helix axis indicated by arrow). In order to clarify ribose alignments, hands are attached to $O_{4'}$ oxygens. Note sharp turn with oppositely oriented hands in molecule A and parallel yet vertically rotated hands in molecule B. A right-handed rotaton of uridine by 110° will cast molecule B into helical form. From (891). For another illustration of UpAH⁺ molecule B, see Figure 4-20.

a conformation 110° away from being helical. As discussed in Section 15·4, the former structure represents a π_1 turn found in some tRNA loops and characteristic of sharp turns in polynucleotide alignment. The latter is better described as a nonhelical bend.

Intermolecular base–base and base–sugar contacts. Because the nonhelical P–O torsions in the UpAH$^+$ molecules do not allow bases to approach closely enough for *intra*molecular stacking interactions, adenine forms an *inter*molecular stack with uracil. Additionally, all bases are in stacking contact with O$_{4'}$ atoms of adjacent riboses, a feature previously found in several nucleoside crystal structures (Figure 6-8) and also realized in left-handed helical duplexes (Chapter 13). In addition to these vertical interactions, bases are associated horizontally in combinations AH$^+$·AH$^+$ and U·U, which are shown as base-pairs II and XVI in Figure 6-1.

Summary

The closely related A- and A′-RNA double helices of Watson–Crick type are found in synthetic, complementary RNA duplexes and in many naturally occurring RNAs such as tRNA, ribosomal RNAs, several viral RNAs, and messenger RNAs. Whereas A-RNA is an 11-fold double helix, the high-salt form A′-RNA is 12-fold. In both, sugar puckering is C$_{3'}$-*endo* and therefore they belong to the A-family double helices. Poly(A)·poly(U) can disproportionate to form a triple-helix poly(A)·2poly(U) displaying simultaneously Watson–Crick and Hoogsteen-type base-pairing. The latter is also observed in a complex between poly(U) and poly(2-methylthio-A). The self-complementary ApU and GpC form mini-double helices of the A-RNA type whereas UpAH$^+$ cannot form a Watson–Crick duplex and adopts an unusually bent structure with individual nucleotides in "standard" conformation but P–O torsion angles α, ζ in nonhelical orientations.

Chapter 11

DNA Structure

RNAs are found only in two related conformations A and A′, which both belong to the A-family double-helical structures (Chapter 10). In contrast, DNAs can adopt several other conformations depending on environmental conditions such as counterion and relative humidity and, in synthetic polynucleotides with defined, repetitive oligonucleotide building blocks, on sequence and on composition (Chapter 9). DNA double helices are classified either as A type with A-DNA the only representative or as B type encompassing B-, B′-, C-, C′-, C″, D-, E-, and T-DNA. Besides these right-handed double helices, a left-handed variety (Z-DNA) has been discovered (Chapter 12), adding considerably to the chameleon-like, adaptable character of DNAs.

A, B, C, and D are the traditional polymorphs of DNA. As illustrated in Tables 9-1 and 9-2 and outlined in Chapter 9, further polymorphs were identified by X-ray fiber diffraction methods but they represent only minor modifications of the four principal structures.

It is of interest that in DNA isolated from phage T2, cytosine is replaced by 5-hydroxymethylcytosine and glycosylated to 70%, with an additional 5% diglycosylated (Scheme 11-1, p. 272). This DNA occurs at high relative humidity in a B form. Under reduced humidity, it transforms directly into T-DNA (858), a D-like double helix with 8_1 symmetry, without passing through the A form. This behavior is reminiscent of synthetic DNAs in which guanosine is substituted by inosine (824). If, however, the glycosyl residues are removed, phage T2 DNA resembles ordinary DNA and, depending on environmental conditions, adopts the A, B, and C forms to be discussed below.

Double helical DNA structures display systematic sequence-dependent modulations: Calladine's rules. In oligo- and polynucleotide double helices, base-pair propeller twists are in general positive, i.e., looking down the long axis of a base-pair, the near base is rotated clockwise with respect to the far one (Figure 2-14 top). As suggested by theoretical considerations (836), propeller twist improves stacking interactions between bases along the double helix. However, as Calladine has pointed out (892a), propeller twist also causes steric clashes between adjacent purines in opposite polynucleotide strands. The clashes are due to interference between guanine O_6/adenine N_6 atoms in the major groove for purine–$(3′,5′)$–pyrimidine sequences, whereas for pyrimidine–$(3′,5′)$–purine sequences, clashes occur in the minor groove between guanine N_3, N_2/adenine N_3 atoms (Figure 11-1). The clashes are about twice as severe compared with those in the major groove. In general, clashes are mainly

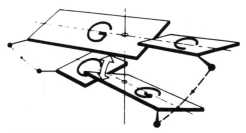

PYRIMIDINE - PURINE

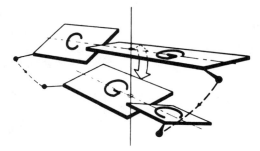

PURINE - PYRIMIDINE

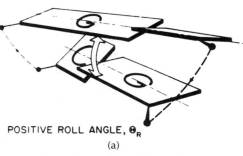

POSITIVE ROLL ANGLE, θ_R

(a)

Figure 11-1. Illustration of Calladine's rules (892a), taken from (892b). (a) Adjacent G-C base-pairs in a DNA double helix with positive propeller twist display sequence-dependent interactions between guanines on opposite polynucleotide chains. (Top) sequence pyrimidine–(3′,5′)–purine leads to clashes in minor groove (open arrow), whereas sequence purine–(3′,5′)–pyrimidine (middle) causes clashes in the major groove. Clashes can be avoided by reduction of propeller twist, opening up of roll angle θ_R (bottom), reduction of helical twist t shown in (c), or shift of base-pairs such that purine (guanine) is pulled out of helical stack, shown in (b). (b) Purine–purine clash displayed in (a) can be reduced by lateral shift of G-C base-pair so that G is pulled out of the helical stack (arrow). In order

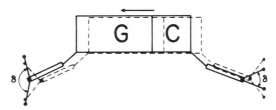

ANTICOMPLEMENTARITY OF TORSION ANGLES, δ

(b)

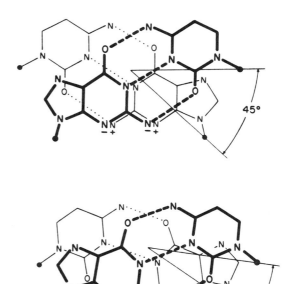

(c)

to maintain helical P ··· P distances at constant values, sugar puckers are changed, yielding larger δ angle for purine and smaller one for pyrimidine. Associated with these changes are χ rotations as required by the anticorrelation principle, section 11.2. (c) Increase of helical twist from 36° to 45° (top) shortens distance between N_3 atoms in minor groove (top), whereas reduction to 27° (bottom) widens this distance. If coupled with positive propeller twist (defined in Figure 2-14), N_3 atoms in top base-pairs (heavy lines) move down (indicated by minus signs) and those in bottom base-pairs (light lines) move up, leading to unfavorable N_3 ··· N_3 clashes. These are diminished by helical *un*-twisting to 27° (bottom Figure).

observed in mixed purine/pyrimidine sequences, whereas in homopurine or homopyrimidine sequences, they are negligible.

In general, all clashes can be avoided if a combination of four different maneuvers is carried out:

1. Reduce base-pair propeller twist locally
2. Open roll angles θ_R between base-pairs such that clash is diminished (Figure 11-1a)
3. Shift base-pair parallel to its long axis so that purine is pulled out of the helical stack (Figure 11-1b)
4. Decrease local twist angle between base-pairs to minimize stack (Figure 11-1c).

These rules were derived on the basis of the B-DNA dodecamer crystal structure described in Section 11.2. Rules 1, 2, and 4 lead to changes in helical and propeller twists (t and θ_T) and roll angle θ_R (see Figure 2-14). Strategy 3 is associated with alterations in sugar torsion angles δ and gives an interpretation of the anticorrelation principle, section 11.2.

In A-DNA, the situation is different compared with B-DNA because the helix axis is removed from the base-pairs and located in the major groove; moreover, base-pairs are tilted by about 20°. Therefore, purine–purine overlap between opposite polynucleotide chains is only appreciable in the minor groove side for the sequences pyrimidine–(3',5')–purine (Figure 9-6), whereas in the major groove no clashes are apparent.

11.1 A-DNA, The Only Member of the A Family: Three Crystalline A-Type Oligonucleotides d(CCGG), d(GGTATACC), and d(GGCCGGCC)

The conformation of A-DNA closely resembles that of A-RNA (Table 9-3) (892). In both molecular structures, 11 nucleotides complete one helical turn, but the axial rise per residue in A-DNA, 2.56 Å, is less than that of A-RNA, 2.81 Å. Base-pairs are tilted by 20° relative to the helix axis and the axis is dislocated by 4.7 Å into the major groove, yielding the open core around the axis and the deep major but shallow minor groove already described in Chapter 9 and in Section 10.1 (Figure 11-2). Most strikingly, the sugar puckering mode for A-DNA is $C_{3'}$-endo whereas for all other DNA modifications $C_{2'}$-endo-type pucker is reported. In addition, the orientation about the $C_{1'}$–N glycosyl bond is characteristically different, with the χ angle around $-160°$ ($-ap$) for A-type duplexes but around $-100°$ ($-ac$) for B-type duplexes.

Crystal structure analyses of three oligonucleotide duplexes belonging to the A family. The A form of DNA is also found in three crystallographic

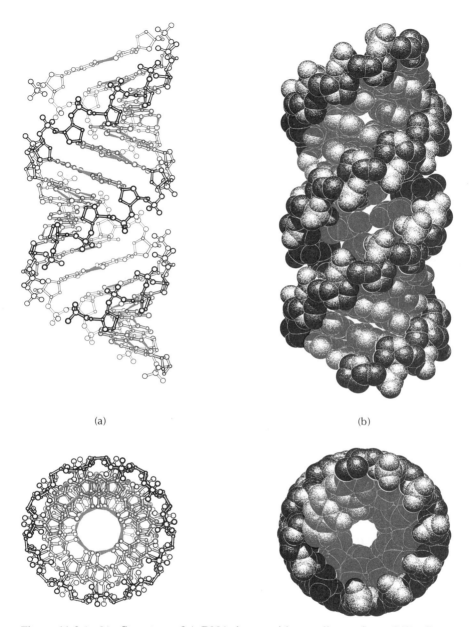

(a)   (b)

Figure 11-2 (a, b). Structure of A-DNA drawn with coordinates from (843). Computer plots are side and top views of A-DNA in (a) ball and stick, and (b) space filling representations. If the helix is tilted about 30° from the viewer to give an impression of the shapes of minor and major grooves, a picture similar to Figure 10-1 (right) is obtained. For graphical details and scale, see the legend to Figure 10-1. In the space filling plot (b), the same input as for the ball-and-stick drawing is used; atomic radii are C = 1.6 Å, N = 1.5 Å, and O = 1.4 Å. Hydrogen atoms are omitted. A light source shines on the molecule under an angle of 45° from the top. Note that the deep, narrow major groove appears dark because it is in shadow.

studies on self-complementary oligonucleotides d(CCGG), d(GGCC-GGCC) and d(GGTATACC) (893, 893a, 894, 894a). The oligomers were crystallized with the addition of alcohol, i.e., conditions which dehydrate and induce B → A transition (895). Since, however, other oligonucleotides in B-form were also crystallized from alcohol (section 11.2), it appears that the GG-sequence common to all three A-type oligomers contributes predominantly to the stabilization of the A-form (893a).

As summarized in Table 11-1, average helical parameters of the oligonucleotides adopt values similar to those derived from fiber diffraction work for DNA in the A-form. There are, however, local variations in these parameters as well as in base-pair propeller twist and roll angles which are sequence dependent and agree largely with Calladine's rules. Thus, in pyrimidine–(3'-5')–purine steps, purine atoms clash in the minor groove and the roll angle θ_R opens accordingly by 10–15°. In all other dinucleotide steps in these A-DNA type oligomers, no systematic trend is observed, in agreement with prediction and in contrast to the B-DNA type dodecamer, section 11.2.

Sugar pucker restricted to $C_{3'}$-endo and α, γ torsion angle correlation. In all of the three A-type oligonucleotides, average P ⋯ P distances are around 6 Å. They are characteristic of $C_{3'}$-endo sugar pucker (Figure 9-3), which is also indicated by torsion angles δ falling into the narrow range 80–91° (894a, Table 11-2).

The movement of base-pairs relative to each other in order to avoid purine–purine clashes according to Calladine's rules is not achieved by variations in sugar pucker as in B-DNA (section 11.2) but by rotations in torsion angles α, γ, and χ, with strong correlation observed between α and γ (894a; see also section 4.13). As indicated by Table 11-2, other torsion angles along the sugar–phosphate chain are rather constant. This behavior contrasts with B-DNA (Table 11-3), where torsion angles at the 3'-ends of the nucleotides (δ, ε, ζ) are more flexible than those at the 5'-end.

In contrast to B-DNA oligonucleotides, individual residues in A-DNA-type fragments display uniform structural features. The sugar units in d(CCGG) and d(GGTATACC) duplexes are all in standard $C_{3'}$-endo puckering mode. The sequence-dependent modulations of the helical structure, so obvious of B-DNA-type oligomers (see below), is not apparent except for d(GGCCGGCC), where some deviations occur with the central four nucleotides. Rather it seems that nucleotides in A-DNA double helices are characterized by more narrowly confined conformations shown in Table 9-3. The same holds for A-RNA because in the two large helical regions of tRNA (Chapter 15), no obvious and systematic sequence-dependent variation of conformation or of secondary structure is found (896).

DNA–RNA hybrids adopt the A form common to both. Because A-DNA and A-RNA double helices are conformationally isomorphous, it is not

Table 11-1. Average Helix Parameters in A-DNA Type Oligonucleotides and in A-DNA. Taken from (894a).

Oligonucleotide	Twist t (°) per base-pair	Rise (Å) per base-pair	Base tilt (°)	Displace-ment D (Å)	Groove width (Å) major[a]	minor[a]
d(GGTATACC)	32.2(1)	2.87(1)	13.5	4.0	10.2	6.3
d(I^CCGG)[b]	33.9(2)	2.85(3)	14.0	3.7	10.6	4.8
d(GGCCGGCC)	32.6(3)	3.03(4)	12.0	3.6	9.6	7.9
A-DNA[c]	32.7	2.56	20	4.5	10.9	3.7

[a] Estimated from distances between phosphates less 5.8Å to give van der Waals groove widths.

[b] The 5'-terminal C is iodinated at position 5.

[c] Data from Arnott and Chandrasekaran (1982, unpublished); these data differ from entries in Table 9-3.

Table 11-2. Average Torsion Angles (°) in A-DNA Type Oligonucleotides. Standard Deviations Given in Parentheses. Taken from (894a).

Oligonucleotide	α (P–O$_{5'}$)	β (O$_{5'}$–C$_{5'}$)	γ (C$_{5'}$–C$_{4'}$)	δ (C$_{4'}$–C$_{3'}$)	ϵ (C$_{3'}$–O$_{3'}$)	ζ (O$_{5'}$–P)	χ (C$_{1'}$–N)
d(GGTATACC)	−62(12)	173(8)	52(14)	88(3)	−152(8)	−78(7)	−160(8)
d(^{i}CCGG)[a]	−73(4)	180(8)	64(10)	80(6)	−161(7)	−67(4)	−161(7)
d(GGCCGGCC)[b]	−75(36)	185(13)	56(22)	91(18)	−166(19)	−75(19)	−149(10)
r(GCG)d(TATACGC)	−69(31)	175(14)	55(22)	82(9)	−151(15)	−75(16)	−162(10)
A-DNA[c]	−50	172	41	79	−146	−78	−154
A-RNA[c]	−68	178	54	82	−153	−71	−158

[a] The 5′-terminal C is iodinated at position 5.

[b] Crystal structure determined at −8°C δ values of disordered sugars G1 and G2 omitted for averaging.

[c] Data from Arnott and Chandrasekaran (1982, unpublished); these data differ from the entries given in Table 9-3.

surprising to find that polynucleotide chains of DNA–RNA hybrids exist in this form. B-Type DNA must transform into A-type *prior* to hybrid formation with RNA (during transcription) except in the case of special, synthetic polymers (Section 11.7).

11.2 B-DNA Structures Exhibited by Polymeric DNA and by the Dodecanucleotide d(CGCGAATTCGCG): Introduction to B-Family Duplexes

While A-DNA had some characteristic features in common with the A-RNA double-helix family, B-DNA and the associated C- and D-DNAs clearly belong to another genus of double helices, the B family. In these, the furanose rings are puckered $C_{2'}$-*endo* (or the related $C_{3'}$-*exo*), the distance between adjacent phosphate groups is increased to 7.0 Å (Figure 9-3), and the χ angle describing rotation about the glycosyl link $C_{1'}$–N is in the $-ac$ range, $-90°$ to $-120°$. In B-DNA, there are 10 base-pairs per pitch with an axial rise per nucleotide of 3.3 to 3.4 Å, resulting in a small negative tilt $(-6°)$ of bases. These characteristics lead to the macroscopic appearance of the B-DNA double helix described in detail in Chapter 9 and illustrated in Figure 11-3, with base-pairs located on the helix axis, major and minor grooves more nearly equal in size, and base stacking limited to intrastrand, without interstrand overlap interactions (Figure 9-6).

The crystal structure analysis of the dodecamer d(CGCGAATTCGCG) confirmed the structure of B-DNA proposed on the basis of X-ray fiber diffraction data. The dodecamer, which has a central restriction endonuclease recognition site, d(GAATTC), was crystallized at pH 7.5 from aqueous solution containing Mg^{2+}, spermine, and 20 to 30% 2-methyl-2,4-pentanediol, an alcohol frequently used to crystallize proteins. The self-complementary molecule crystallized as a duplex with Watson–Crick base-pairs and a B-DNA-like double helix consisting of a little more than one turn. In closer detail, however, there are some remarkable new findings:

(1) The overall helix is not straight but rather is bent by 19° corresponding to a radius of curvature of 112 Å (Figure 11-4). This distortion is probably induced by crystal packing forces rather than being an intrinsic structural property of the complex (464). Assuming that the DNA duplex is an elastically deformable rod, it can be estimated from the known persistence length of DNA in solution (815,816,897,898), that only 0.25 to 0.50 kcal/mole dodecamer would be sufficient to bring about the observed curvature. As we will see in Chapter 19, bending of DNA becomes of importance when DNA packaging in chromatin and DNA twisting in superhelices are considered.

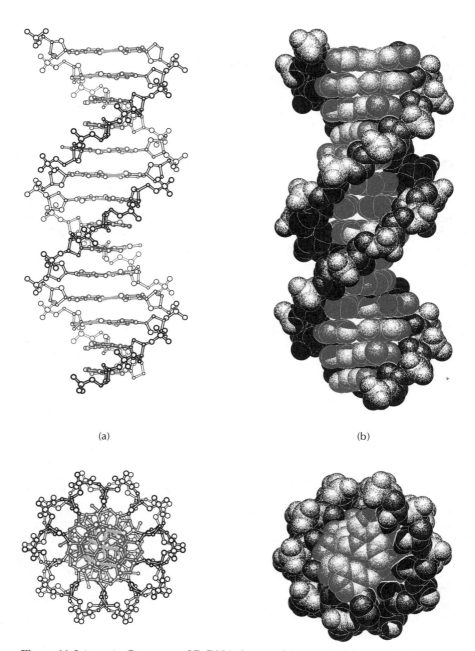

(a)   (b)

Figure 11-3 (a to c). Structure of B-DNA drawn with coordinates supplied by Drs. S. Arnott and R. Chandrasekaran (1981). For graphical details and scale, see legend to Figure 11-2. (a and b) Side and top view of B-DNA in ball-and-stick and space filling representation; (c) the helix tilted 32° from the viewer to show minor (m) and major (M) grooves.

Figure 11-3 (*Continued*)

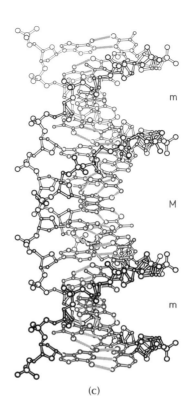

(c)

(2) The innermost part of the dodecamer comprising the *EcoRI* recognition site d(GAATTC) displays B-type DNA structural characteristics with 9.8 base-pairs per turn. However, the terminal nucleotide triplets attached on both sides of this hexamer show more complex features. The base-pairs adjacent to the central hexamer are displaced away from the local helix vector into the minor groove side as in A-DNA (Figure 11-5). Second-next neighbors show rotations about the local axis of about 45° relative to each other, reminiscent of D-DNA. In spite of this tendency to interrupt the regular B-DNA structure, the uniform P–P separation along the chain, 6.68 Å, is indicative of B-DNA (for ideal A-DNA, 5.62 Å would be expected) (895,899) (Table 11-3).

(3) The sugar puckering is not constant over the whole dodecanucleotide double helix but varies between $C_{3'}$-*exo* and $O_{4'}$-*endo* (Table 11-3). Expressed in torsion angle δ (Chapter 2), sugar conformations are distributed in Gaussian fashion about $C_{1'}$-*exo* with a mean $\delta = 123°$. As displayed in Figure 11-6, sugar conformations are strongly correlated with glycosyl torsion angles χ and further, pyrimidine nucleosides tend to favor low δ ($O_{4'}$-*endo*), whereas for purine nucleosides, high δ ($C_{2'}$-*endo*) is preferred. Strikingly, there is an *anticorrelation* defined as tendency of

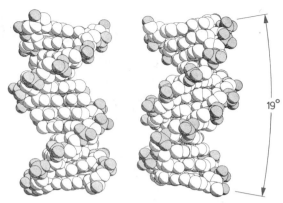

Figure 11-4. Two views of the molecular structure of the d(CGCGAATTCGCG) double helix as revealed by single-crystal X-ray diffraction. Views are at 90° to each other, showing (left) the major and (right) the minor groove sides. The curvature of the duplex is evident from the picture at right. Phosphate groups are shown stippled; hydrogen atoms are omitted for the sake of clarity. From (464).

the two nucleotides in a base-pair to adopt χ and δ values distributed at about equal distances to either side of the midpoint of Figure 11-5.

(4) **Calladine's rules and anticorrelation principle are equivalent.** As outlined at the beginning of this Chapter, the sequence-dependent modulation of double helical DNA structures can be explained on the basis of purely mechanical considerations. Opening of roll angles θ_R toward the minor groove occurs for the sequences pyrimidine–(3',5')–purine and toward the major groove for purine–(3',5')–pyrimidine, whereas homo-sequences do not influence roll angles as explained by rule 2 on page 256. Rule 3 directly correlates the lateral shift of base-pairs with torsion angle δ. Since this shift is such that the purine is pulled out of the double helical base stack, torsion angle δ widens to ~135° (Figure 11-6) for purine, whereas it decreases to ~110° for pyrimidine in order to maintain the phosphate ⋯ phosphate distance constant, around 6.8 Å. See also Figure 11-1b.

(5) **The base-pairs have a propeller twist,** 11° for dG–dC and 17° for dA–dT pairs, in contrast to the average 2.1° derived from X-ray fiber diagrams (843). As can be seen from the dodecamer crystal structure, the tendency for dG–dC base-pairs to be less propeller-twisted than dA–dT is not due to the extra hydrogen bond but rather appears to be caused by the guanine amino group clashing with adjacent bases "above" or "below" if larger twists are adopted.

(6) **In the sequence dT–dC–dG, dC slips away from dT** and stacks with more overlap on dG (see Figure 11-5). Note that this parallels the preferences outlined in Section 6 for base stack formation in aqueous solution.

In essence, the crystal structure of the B-DNA dodecamer has shown that *a DNA double helix should not be considered as the uniform, rod-like entity derived from X-ray fiber diffraction data. Rather, there are sys-*

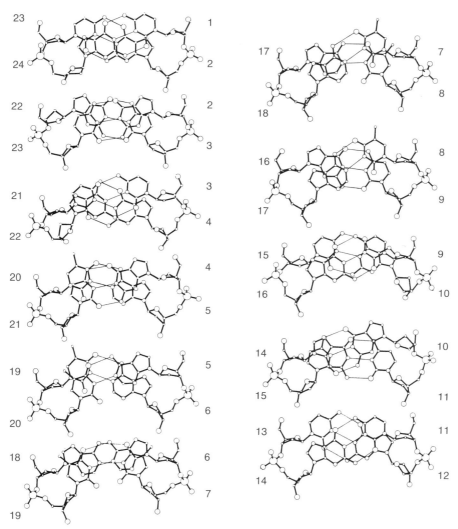

Figure 11-5. Views of the 11 base-pair overlaps in the d(CGCGAATTCGCG) do-decamer. Base sequences are indicated in color in one strand from 1 to 12, in the other from 13 to 24. Note the pattern of base overlap and sugar conformation which is rather regular in the central hexamer d(GAATTC) (steps 4 to 8) and more irregular in the outer base-pairs; compare steps 9 (A-DNA like) and 10 (D-DNA like). All base-pairs except the terminal ones appear twice and are shown in identical orientations toward the viewer. From (895).

tematic sequence-dependent structural modulations which could be features essential for protein–nucleic acid recognition. On the other hand, A-type DNA and RNA double helices appear to be more uniform in character and they do not display the polymorphism found for B-DNA which we will now elaborate further.

Table 11-3. Conformational (A), and Local Helical Parameters (B) for the B-DNA Dodecamer d(CGCGAATTCGCG). From (899,899a).

(A)

Residue	χ	α	β	γ	δ	ε	ζ	Sugar pucker	Adjacent P-atom separation (Å)
						Angles, degrees			
C1	−105	—	—	174	157	−141	−144	$C_{2'}$-endo	
G2	−111	−66	170	40	128	−186	−98	$C_{1'}$-exo	6.64
C3	−135	−63	172	59	98	−177	−88	$O_{4'}$-exo	6.47
G4	−93	−63	180	57	156	−155	−153	$C_{2'}$-endo	6.83
A5	−126	−43	143	52	120	−180	−92	$C_{1'}$-exo	6.88
A6	−122	−73	180	66	121	−186	−89	$C_{1'}$-exo	6.90
T7	−127	−57	181	52	99	−186	−86	$O_{4'}$-endo	6.29
T8	−126	−59	173	64	109	−189	−89	$C_{1'}$-exo	6.87
C9	−120	−58	180	60	129	−157	−94	$C_{1'}$-exo	6.70
G10	−90	−67	169	47	143	−103	−210	$C_{2'}$-endo	6.55
C11	−125	−74	139	56	136	−162	−90	$C_{2'}$-endo	7.05
G12	−112	−82	176	57	111	—	—	$C_{1'}$-exo	
C13	−128	—	—	56	137	−159	−125	$C_{2'}$-endo	
G14	−116	−51	164	49	122	−182	−93	$C_{1'}$-exo	6.62
C15	−134	−63	169	60	86	−185	−86	$O_{4'}$-endo	6.45
G16	−115	−69	171	73	136	−186	−98	$C_{2'}$-endo	7.12
A17	−106	−57	190	54	147	−183	−97	$C_{2'}$-endo	6.77
A18	−108	−57	186	48	130	−186	−101	$C_{2'}$-endo	6.71
T19	−131	−58	174	60	109	−181	−88	$C_{1'}$-exo	6.70
T20	−120	−59	179	55	122	−181	−94	$C_{1'}$-exo	6.70
C21	−114	−59	185	45	110	−177	−86	$C_{1'}$-exo	6.17
G22	−88	−67	179	50	150	−100	−188	$C_{2'}$-endo	6.60
C23	−125	−72	139	45	113	−174	−97	$C_{1'}$-exo	6.68
G24	−135	−65	171	47	79	—	—	$C_{3'}$-endo	
Mean	−117	−63	171	54^a	123	−169	−108		
±SD	14	8	14	8	21	25	34		
B DNA^b	−119	−61	180	57	122	−187	−91		
B_FDNA^c	−102	−41	136	38	139	−133	−157		
A_FDNA^c	−154	−90	−149	47	83	−175	−45		

11.3 "Alternating B-DNA" and the Tetranucleotide d(pATAT); d(TpA), Dinucleoside Phosphate Mimicking Double Helical Arrangement

The alternating oligodeoxynucleotide d(pATAT) has been crystallized as the ammonium salt from aqueous solution by alcohol addition (359). A priori one might expect this self-complementary tetramer to form an anti-

Table 11-3. (*Continued*) (B) Local Helical Parameters

Base-pairs	Propeller twist, $°\theta_p$	Helix twist angle, $°t^d$	Base-pairs per turn (n)	Rise per base pair, Å (h)
C1/G24	13.2 ± 2.0			
G2/C23	11.7 ± 2.1	38.3 ± 1.1	9.40 ± 0.27	3.36 ± 0.01
C3/G22	7.2 ± 2.1	39.6 ± 6.1	9.09 ± 1.40	3.38 ± 0.08
G4/C21	13.2 ± 1.9	33.5 ± 2.1	10.75 ± 0.67	3.26 ± 0.05
A5/T20	17.1 ± 2.1	37.4 ± 1.7	9.63 ± 0.44	3.30 ± 0.10
A6/T19	17.8 ± 2.1	37.5 ± 0.9	9.60 ± 0.23	3.27 ± 0.02
T7/A18	17.1 ± 1.9	32.2 ± 2.1	11.18 ± 0.73	3.31 ± 0.03
T8/A17	17.1 ± 2.0	36.0 ± 2.8	10.00 ± 0.78	3.29 ± 0.01
C9/G16	18.6 ± 1.9	41.4 ± 2.1	8.70 ± 0.42	3.14 ± 0.02
G10/C15	4.9 ± 1.9	32.3 ± 1.3	11.11 ± 0.45	3.56 ± 0.07
C11/G14	17.2 ± 1.9	44.7 ± 5.4	8.05 ± 0.97	3.21 ± 0.18
G12/C13	6.2 ± 2.3	37.0 ± 1.9	9.73 ± 0.50	3.54 ± 0.19
Mean	13.4 ± 4.9	37.3 ± 3.8	9.75 ± 0.98	3.33 ± 0.13
A DNA		32.7	11.0	2.56
B DNA		36.0	10.0	3.38
C DNA		38.6	9.33	3.31
D DNA		45.0	8.0	3.03

[a] C1 value omitted because it represents end effect.

[b] From the energy-refined B-DNA (836).

[c] From X-ray fiber diffraction data (304).

[d] Helical parameters obtained by using vectors between atoms $C_{1'}$ and attached N of one base and the equivalent atoms of the next nucleotide along the chain. Helical parameters are defined in Chapter 2.

parallel, Watson–Crick-type double-helical structure similar to those just discussed or to ApU or GpC (465,466). This assumption, however, is only partially fulfilled because each d(pATAT) tetramer is base-paired in Watson–Crick mode with *two* symmetry-related, frame-shifted partners. As a consequence, two short d(pApT) duplexes are formed and the central d(pTpA) fragment is in nonhelical conformation (Figure 11-7).

Conformations of phosphodiester links show variation. Torsion angles α and ζ describing orientations about $P–O_{5'}$ and $P–O_{3'}$ bonds are in the $-sc$ range for the two double-helical d(pApT) fragments of each tetrameric molecule, as required for helical geometry. The central phosphodiester link, however, displays torsion angles ap for ζ and $-sc$ for α, characteristic of structure P_2 in Figure 4-26 and giving rise to an extended backbone structure with the disrupted all-helical arrangement illustrated in Figure 11-7. If the ζ angle were to be changed from its actual value, 168°, to about $-100°$, a double-helical conformation results and is described in more detail below.

Furanoses associated with purines are $C_{3'}$-endo, with pyrimidines $C_{2'}$-endo. The most surprising structural feature of the d(pATAT) molecular

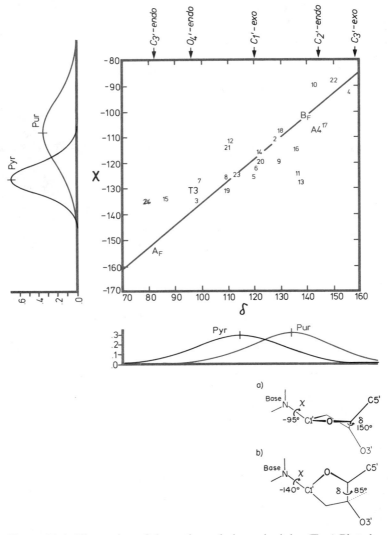

Figure 11-6. Illustration of the anticorrelation principle. (Top) Plot showing near-linear dependence of glycosyl torsion angle χ on sugar pucker defined by torsion δ ($C_{5'}-C_{4'}-C_{3'}-O_{3'}$). Data points for dodecamer nucleotides in sequential numbers, for the central step of the daunomycin complex with d(CGTACG) (Section 16.3) given as A4, T3, and for classical A-DNA (A_F) and B-DNA (B_F) as obtained from fiber diffraction. Color for purines, black for pyrimidines. Normalized Gaussian distributions (far left and bottom) indicate that purine nucleotides prefer higher χ and δ values than pyrimidine analogs. Data for each base-pair are almost symmetrically distributed about the midpoint at δ = 123°: compare base-pair partners 1/24, 2/23, 3/22, etc. Sugar pucker given on top of graph. (Bottom) Mechanical explanation for the χ/δ correlation. In order to maintain the $O_{3'}\cdots C_{5'}$ distance constant so that the average P$\cdots$P separation is about 6.68 Å over the whole length of the dodecamer, rotation of the sugar from (a) to (b) is associated with reduction of δ from *ac* to *sc* range (from 150° to 85°). As explained in Section 4.5, and obvious from the χ distribution above, purines tend to adopt less negative χ angles than pyrimidines. From (895).

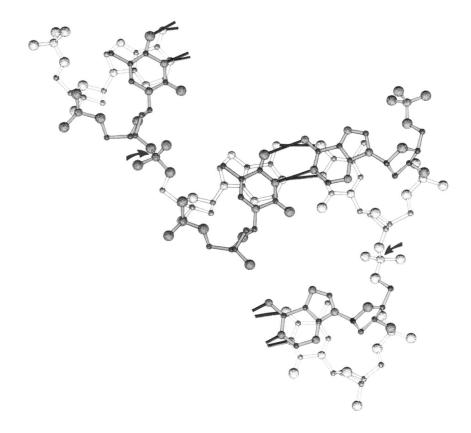

Figure 11-7. Crystal structure of the oligonucleotide d(pATAT). Nucleotides closer to the viewer are shaded, hydrogen bonds are in color. Arrows mark P-O$_{3'}$ torsion angles ζ, which, owing to their *ap* range, destroy the all-helical arrangement. After (359).

complex is that the sugars have alternating puckering modes. As a consequence, glycosyl torsion angles vary considerably, being around $-160°$ for adenosine and around $-110°$ for thymidine moieties.

Why is d(pATAT) not a four-base-pair double helix? Comparison with d(TpA). The reason for the nonhelical structure of the tetranucleotide probably lies in the poor base overlap in the case of a helical d(TpA) sequence which contrasts the favorable overlap in the two d(ApT) fragments; compare B-DNA overlaps for (pyrimidine–3′,5′–purine) and (purine–3′,5′–pyrimidine) in Figure 9-6. For a hypothetical d(pATAT) double helix with alternating sugar puckering, the (pyrimidine–3′,5′–purine) base overlap is even more reduced (900). Support for this interpretation is provided by nuclear magnetic resonance data on the two dinucleotides d(pTpA) and d(pApT) in aqueous solution. There is a clear indication that stacking in the latter molecule is a dominating structural feature whereas d(pTpA) is largely unstacked (901).

An unstacked molecule was also proposed in order to explain the "polymeric" structure of the dinucleoside phosphate, d(TpA) (902,902a). If an aqueous solution of the ammonium salt of d(TpA) is evaporated, films are obtained which produce X-ray diffraction patterns reminiscent of A-DNA fibers. The suggested double-helix formation with d(TpA) dimers, however, utilizes Hoogsteen rather than Watson–Crick base-pairs. Each d(TpA) molecule interacts via base-pairs with *two* partners, leading to a helix-like structure with d(TpA) units. The whole, polymeric complex is stabilized mainly by dA–dA stacking interactions which are more favorable than the dT–*on*–dA stacking required if a Watson–Crick structure were formed (see stacking preferences outlined in Section 6.6).

"Alternating B-DNA" has sequence-dependent conformation. Inspired by the particular alternating conformation in the tetranucleotide d(pATAT), a double-helical, B-DNA-like model was hypothesized (900,900a). In order to construct such a model on the basis of the tetranucleotide, the P-$O_{3'}$ torsion angle of the central d(TpA) fragment had to be rotated from *ap* to $-sc$ range. This being done, a B-type duplex was in fact generated which was further subjected to energy minimization. After this procedure, the resulting model displayed structural characteristics deviating significantly from those found in "classical" B-DNA. Thus, the sugar puckerings, as adopted from the tetranucleotide, are alternating $C_{3'}$-*endo* for adenosine and $C_{2'}$-*endo* for thymidine, and orientations about glycosyl bonds vary accordingly. While in B-DNA, P–O torsion angles α and ζ are $-60°$ and $-90°$ for all nucleotides, they are in the "alternating B-DNA" $\alpha = \zeta = -60°$ for d(ApT) and $\alpha = -50°$, $\zeta = -120°$ for d(TpA) fragments.

"Alternating B-DNA" is supported by physical properties and has interesting biological implications. There is some experimental evidence that "alternating B-DNA" does in fact occur. The findings include ^{31}P nuclear magnetic resonance studies of alternating poly- and oligodeoxynucleotides in fibers and in solution which display a doublet for phosphorus and indicate the existence of two phosphates in different chemical environments, corresponding to the alternating P–O torsion angles (903–905). In addition, two biochemical results lend support to the hypothesis of an alternating conformation in double-helical poly(dA–dT)·poly(dA–dT). First, enzymatic digestion with pancreatic DNase I produces almost exclusively fragments having dT at their 5' ends (906), although the enzyme cleaves natural double-helical DNA at random (907). Second, *lac* repressor protein binds to poly(dA–dT)·poly(dA–dT) about 1000-fold more tightly than to calf thymus DNA (908). This difference in binding properties could be taken as being due to specific nucleotide sequence recognition by the protein. However, *lac* repressor also complexes very well with a series of poly(dA–dT) analogs modified chemically at the thymine-5-position by substitution of halogen or hydrophobic residues for the methyl group. From these studies it was inferred that an alternating B-DNA structure is induced and stabilized by favorable stacking interactions between these substituents and adjacent adenines (900).

11.4 The Conformationally Stiff Unique Poly(dA)·Poly(dT) Double Helix and Its Transformation into Triple Helix

In aqueous solution, poly(dA)·poly(dT) occurs as a unique double helix with 10.0 base-pairs per turn. The homopolymers poly(dA) and poly(dT) can form a 1:1 Watson–Crick complex, which, at low salt concentration and high relative humidity, exhibits a typical B-DNA molecular structure. Other DNAs of natural origin or the analogous poly(dA–dT) with alternating sequence all show around 10.5 ± 0.1 base-pairs per turn in aqueous solution, but the poly(dA)·poly(dT) duplex displays 10.1 ± 0.1 base-pairs per turn (909,910,910a). This is the only case where B-DNA structural features in fibrous and solution state are fully consistent. That poly(dA)·poly(dT) is distinctly different from DNAs with alternating or random distributions of bases is also evident from its resistance to transform into other helical forms, its peculiar UV circular dichroism spectra, and its inability to form reconstituted nucleosomes when combined with histone octamers (910).

Disproportionation into triple-helix poly(dA)·2poly(dT). The B-DNA-like structure of poly(dA)·poly(dT) was called B′-DNA (843) because the helices are packed in the crystal lattice differently from ordinary B-DNA, and the helix pitch is reduced by 0.08 Å (Table 9-4). If the salt concentration of the environment is raised (or the relative humidity lowered), the normal B→A transformation does not take place. Instead a disproportionation of the double helix into a triple helix and a single polynucleotide chain is observed, similar to the poly(A)·poly(U) system (Section 10.2). In both cases, a Watson–Crick duplex of the A family (A-DNA) accommodates in its major, deep groove an extra poly(pyrimidine) strand which is hydrogen-bonded to the Hoogsteen base-pairing side of the poly(purine). These two latter strands run parallel whereas the polynucleotides in the Watson–Crick duplex are antiparallel. The overall conformational features of the triple helices in the ribo and deoxyribo series are similar (Figure 11-8; Table 9-4).

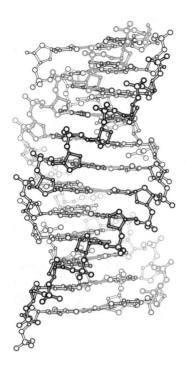

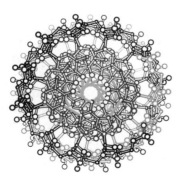

Figure 11-8. The DNA triple-helix poly(dT)·poly(dA)·poly(dT). An A-DNA double helix with Watson–Crick base-pairs and antiparallel polynucleotides accommodates in its major groove an extra poly(dT) chain drawn in light color. The latter is Hoogsteen base-paired with poly(dA) and the two chains are oriented parallel. For graphical details and scale, see the legend to Figure 10-1. Drawn with atomic coordinates from (859).

11.5 C-DNA Double Helix Formed by Natural and Synthetic DNA

X-Ray fiber diffraction studies revealed that a new form, C-DNA, is produced if the lithium salt of natural DNA is maintained at rather low relative humidity, 44 to 66% (Figure 11-9; Table 9-1). The same DNA structure can also be obtained with sodium as counterion at high salt content and at humidity conditions intermediate between those required to produce A- and B-DNA (857). Similar results were evidenced by infrared linear dichroism studies on oriented films of DNA isolated from various sources and differing largely in base composition. As an additional factor, it appears that low d(G+C) content facilitates B→C transition (911) (Figure 9-1).

Whereas C-DNA thus far has been considered as an intermediate between A and B forms trapped under special salt/humidity conditions, more recent data point out that C-DNA is as important as A and B (911a). If Na^+ concentration and humidity were reduced, it was possible to obtain C-DNA from natural DNA in the d(G+C) content range 31% to 72% and even from the glycosylated T2 DNA. If water is added slowly, a reversible C→A→B transformation takes place.

C-DNA structure resembles B-DNA structure. The nonintegral C-DNA double helix with 9.33 base-pairs per pitch of 30.9 Å repeats exactly after three turns with 28 base-pairs, hence the symmetry notation 28_3 (Table 9-1; Box 9-1). The deoxyribose puckering is $C_{3'}$-*exo* (or the almost equivalent $C_{2'}$-*endo*) as in B-DNA; the base-pairs are slightly tilted by $-8°$, allowing the axial rise per residue, 3.31 Å, to be a little smaller than in B-DNA. The helix axis is located near the minor groove of the double helix at $D = -1.0$ Å: i.e., in A-DNA the helix axis is located in the major groove, in B-DNA it is in the base-pairs; but in C-DNA it is displaced toward the minor groove (Figure 9-5). The grooves are similar in size as in B-DNA, the major groove being a little more shallow and the minor groove a little more deep. In general, C-DNA resembles B-DNA, with conformational parameters of the nucleotide building blocks changed only slightly (Table 9-3).

Besides natural DNAs, synthetic DNAs with defined sequences can also transform into C-type structures. In these, investigated by X-ray fiber diffraction, not only the nonintegral 9.33_1 double helix with symmetry 28_3 is adopted but integral 9_1 and another 18_2 duplex were found with trinucleotide and dinucleotide repeating units, respectively (824) (Table 9-1). In synthetic, repeating DNAs with inosine substituting for guanosine, C- and D-type conformations replace the A-DNA type which does not form even if the conditions are widely varied. The reason for this behavior appears to be connected with hydration of DNA as discussed in Chapter 17.

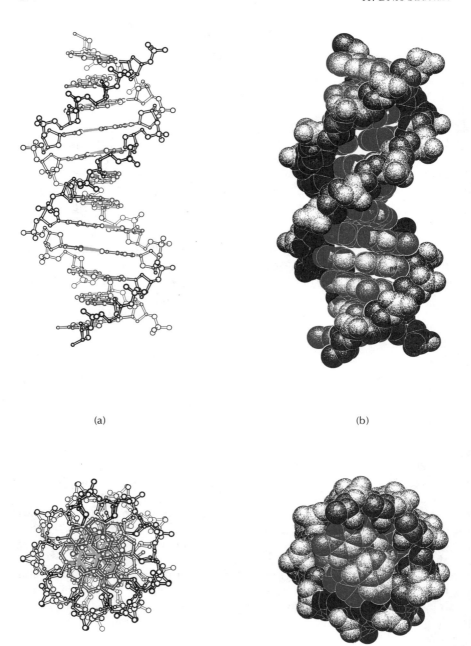

(a)

(b)

Figure 11-9 (a to c). Structure of C-DNA drawn with atomic coordinates given in (94). For graphical details and scale, see the legend to Figure 11-2. (a and b) Side and top view of C-DNA in ball-and-stick and space filling representation. (c) The helix tilted 32° from the viewer to show minor (m) and major (M) grooves.

Figure 11-9 (*Continued*)

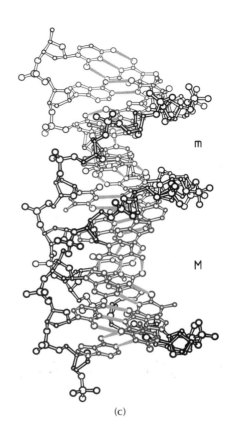

m

M

(c)

11.6 D-DNA Is Only Formed by DNA with Alternating A, T-Sequence and by Phage T2 DNA

Natural DNAs do not adopt D-type double helices. An exception are A, T-rich regions in DNA with alternating sequence and phage T2 DNA which is highly modified. Phage T2 DNA contains 5-hydroxymethylcyto-sine instead of cytosine, and is, in addition, glycosylated to over 70% (Scheme 11-1). The conformational parameters of this T-DNA are similar to those of D-DNA (Tables 9-3, 9-4).

In synthetic DNAs, D-DNA substitutes for A-DNA when guanosine is absent. In poly(dA–dT)·poly(dA–dT), the A-DNA conformation appears to represent only a metastable structural state (845, 845a) whereas in the poly(dG–dC) analog, it prevails up to 92% relative humidity (Table 9-1). In poly(dA–dA–dT)·poly(dA–dT–dT), only B- and D-type structures are observed and, as outlined above, in complementary polydeoxynucleotides with inosine replacing guanosine, the A form is not found at all (Section 11.5).

Compared with B-DNA, the D-DNA double helix is overwound and displays a very deep, narrow minor groove, a good cavity for trapping water and cations. As in other B-type double helices, the furanose puckering is $C_{3'}$-*exo* in D-DNA. The structure is considerably overwound with respect to B-DNA, with only eight base-pairs completing one turn of 24.3 Å pitch height; i.e., the rotation from one nucleotide to the adjacent one is 45° and the axial rise per residue is 3.03 Å.

The helix axis is positioned in the minor groove of the duplex at $D = -1.8$ Å (Figure 9-5), rendering the helix van der Waals diameter only 21 Å, and the base-pairs are now tilted $-16°$ toward the helix axis, contrasting the $+20°$ A-DNA (Figure 11-10; Table 9-4). Because the helix axis

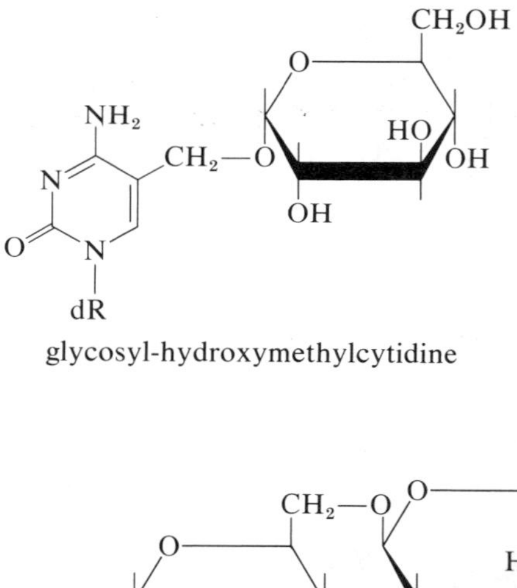

glycosyl-hydroxymethylcytidine

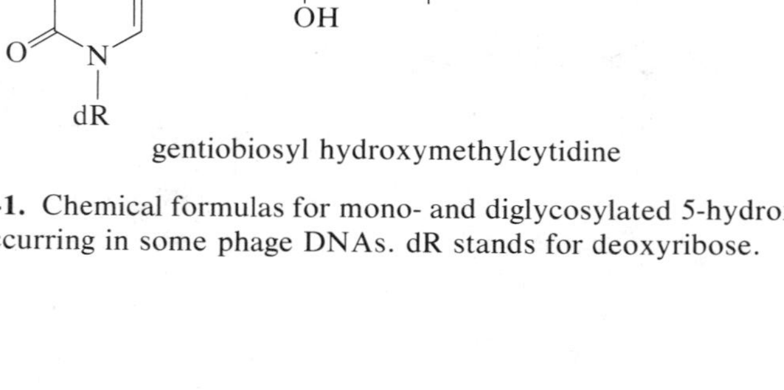

gentiobiosyl hydroxymethylcytidine

Scheme 11-1. Chemical formulas for mono- and diglycosylated 5-hydroxymethylcytidine occurring in some phage DNAs. dR stands for deoxyribose.

is located in the minor groove side, the major groove becomes more shallow with respect to B- and C-DNA and the minor groove is now very deep and, in addition, very narrow. This, in turn, causes the phosphate groups and O_2 keto groups of opposing polynucleotide strands to approach each other closely so that water molecules and cations could possibly cross-link and stabilize the D-DNA conformation. In very recent work on poly(dA–dT), occurrence of a left-handed helix was proposed (845a).

Phage T2 DNA is the only known natural DNA that can occur in D-DNA type conformation. In fibers we find that this phage DNA exists in a B form with 10 nucleotides per pitch at relative humidity above 95%. As the humidity is lowered, the number of base-pairs per pitch is *gradually* reduced, until at below 60% relative humidity, one finds a helix with 8 base-pairs per pitch. From the gradual course of the B→T helical transformation it can be inferred that there is no internal potential barrier to be overcome and that T2 DNA smoothly adjusts its dimensions to the relative humidity over a wide range. In contrast, the B→A transition of normal DNA occurs abruptly, indicating the energy barrier between $C_{3'}$-*exo* and $C_{3'}$-*endo* sugar puckering (858).

The 27.2-Å pitch height of T2 phage DNA is associated with eight nucleotides and thus allows for an axial rise per residue of 3.4 Å. It requires, therefore, only a gentle negative tilt of $-6°$ which contrasts the $-16°$ found in D-DNA (Figure 11-11). The sugar pucker is again of the $C_{2'}$-*endo* (or $C_{3'}$-*exo*) type, leading to a dislocated helix axis positioned at -1.43 Å from the base-pair in the minor groove side. This gives T2 DNA a deep, narrow minor and shallow, wide major groove.

11.7 DNA–RNA Hybrids Restricted to RNA-Type Double-Helices: A and A'. Polymers and r(GCG)d(TATACGC). The B-DNA Form of Poly(A)·Poly(dT).

Duplexes between DNA and RNA are of biological importance. They occur when DNA base sequences are transcribed into complementary messenger RNA. This process is catalyzed by RNA polymerase, an enzyme that interacts with double-helical DNA, separates the strands, and synthesizes, at only one DNA strand, new RNA. At least at the active site of polymerase, which is estimated to be 40 nucleotides long, therefore, a DNA–RNA hybrid is actually formed (1017; Section 14.4). Moreover, such hybrids occur if DNA is produced from a (phage) RNA by the action of reverse transcriptase, and if a short RNA sequence primes for DNA replication. X-Ray fiber diffraction studies have been carried out with a DNA·RNA complex (850) and also with three synthetic hybrids:

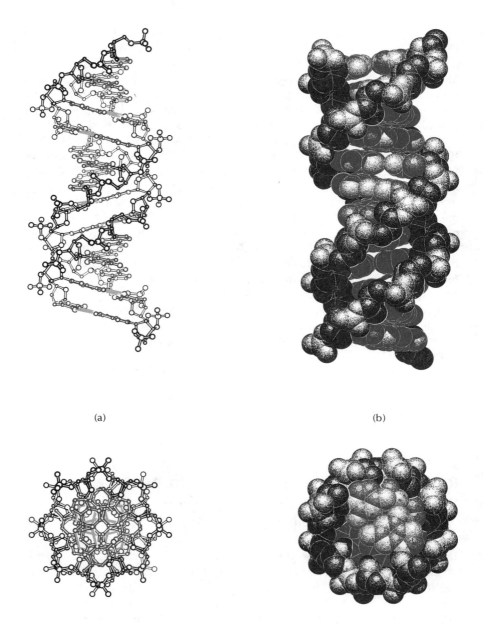

(a)

(b)

Figure 11-10 (a to c). Structure of D-DNA, drawn with atomic coordinates for poly(dA–dT) given in (845). For graphical details and scale, see the legend to Figure 11-2. (a and b) Side and top view of D-DNA as ball-and-stick and space filling plots; (c) the helix tilted 32° from the viewer to display geometry of minor (m) and major (M) grooves.

Figure 11-10 (*Continued*)

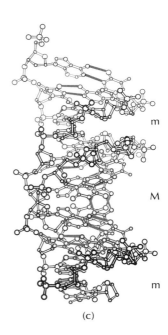

(c)

poly(A)·poly(dT) (912), poly(dI)·poly(C) (851), and poly(I)·poly(dC) (849). The former two adopt an 11-fold A-RNA-type structure whereas the latter two more closely resemble the 12-fold A'-RNA.

The poly(I)·poly(dC) complex has been subjected to detailed structural study (849). It forms an A'-RNA-type 12-fold double helix with pitch height 35.6 Å; the base-pairs are dislocated by $D = 4.9$ Å from the helix axis and are tilted in a positive sense by 5.7° (Table 9-4; Figure 11-12). Both RNA and DNA polynucleotide chains are composed of nucleotides in the $C_{3'}$-*endo* sugar puckering mode and have identical conformation angles.

DNA–RNA hybrids are in general unable to transform into structures of the B family. This means that structural conservatism of RNA is not a consequence of RNA duplex formation alone but is operational even if only one single RNA strand is present in a double helix.

In case of hybrids of homopolymer chains like poly(dA)·poly(U), changes of environmental conditions can, in fact, induce other structures, namely, triple helices in the forms discussed in Sections 10.2 and 11.4. Here again, however, the A-family conformational features persist (913).

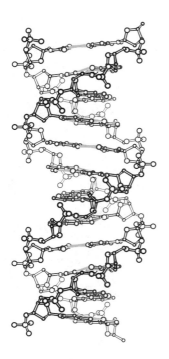

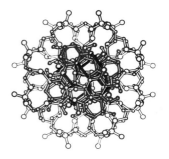

Figure 11-11. Structure of T-DNA, a geometrical D-DNA analog formed by phage T2 DNA with 5-glucosylated cytosine residues. Side and top views of eightfold double helix drawn with coordinates given in (858). Substituents at C_5 of cytosines are omitted. For graphical details and scale, see the legend to Figure 10-1.

r(GCG)d(TATACGC), a self-complementary decamer RNA/DNA hybrid was studied by crystallographic techniques (913a). It occurs as antiparallel, Watson–Crick base-paired double helical complex displaying structural characteristics of A-RNA. Base-pairs have an average helical rotation of 33°, a rise along the helix axis of 2.6 Å. There are 10.9 base-pairs per turn and the tilt is about 20°. The bases in the pairs describe a propeller twist of 14°, with the same handedness as observed in the dodecamer (Section 11.2). The A-RNA form is stabilized by intramolecular, water mediated hydrogen bonds $O_{2'}H \cdots$ water $\cdots O_2$ (cytosine) and there is no obvious discontinuity between ribo and deoxyribo sections in the duplex. Conformational parameters are entered in Table 11-2.

The hybrid poly(A)·poly(dT) is an exception because it can transform into a B form. An X-ray fiber diffraction study on the hybrid complex poly(A)·poly(dT) revealed that at high relative humidity, a transformation from an A′-RNA-type double helix to a form resembling B-DNA took place (912). This finding appears to contradict the notion that hybrids occur exclusively in A-DNA conformation. However, it could well be that a special situation is encountered with the two homopolymers

Figure 11-12. Structure of a DNA-RNA hybrid in the A'-form, poly(I)·poly(dC). Side and top view drawn with atomic coordinates given in (849), for graphical details and scale see the legend to Figure 10-1.

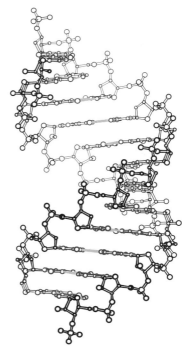

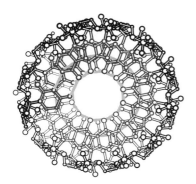

poly(A) and poly(dT). First, as outlined in Section 11.4, the all-deoxyribo analog poly(dA)·poly(dT) is unique in that it adopts a B-DNA form with 10.0 base-pairs per turn (in fibers and in solution) and resists transformation into other conformations. This behavior could explain in part why the hybrid poly(A)·poly(dT) consisting of two homopolymers transforms into the B-DNA form. Second, the exclusive occurrence of A/T bases favors a certain hydration scheme which, as outlined in Chapter 17, again pushes the conformation toward the B form. It appears therefore that in the case of poly(A)·poly(dT) the A→B transition is forced by the particular base composition and would not occur with DNA·RNA hybrids with G/C bases or with "random" sequence.

Summary

Right-handed double-stranded DNA helices in the A-, B-, C-, D- and T-forms and DNA-RNA hybrids are discussed in this chapter. For A-DNA, the only member of the A family (Chapter 9) besides polymeric nucleic acids of natural and synthetic origin, three self-complementary tetra- and octameric oligonucleotides were investigated. Similarly, for the B family (B-, C-, D-, T-DNA), a dodecamer was studied, its single-crystal study revealing a number of new details not accessible with fiber diffraction studies of polymeric samples of natural or synthetic origin. The oligo-deoxynucleotide crystal structures suggest that in A-type double helices, nucleotides adopt rather uniform conformations. This contrasts with B-type double helices, where sequence-dependent structural modulations are observed according to Calladine's rules. The crystal structure of the tetramer d(pATAT) was used to propose the "alternating B-DNA" with nonequivalent nucleotide conformations, and the dinucleoside phosphate d(TpA) self-assembles in helix-like arrangement without proper Watson–Crick base-pairing but well-defined base stacking. Poly(dA)· poly(dT) is conformationally "stiff," prefers B form, and transforms into a triple helix poly(dA)·2poly(dT) rather than adopting an A-type structure. C- and D-DNA forms both belong to the B family displaying the $C_{2'}$-*endo* sugar puckering mode and forming nine- and eight-fold double helices, respectively. If DNA-RNA hybrids are generated, they belong to the A family and resist transformation into B form. Poly(A)·poly(dT) appears to be exceptional but its behavior can be explained with its particular base composition, rendering it similar to the "stiff" poly(dA)·poly(dT).

Chapter 12

Left-Handed, Complementary Double Helices—A Heresy? The Z-DNA Family

Guided by X-ray fiber diagrams of DNA and RNA double helices, molecular models with right-handed screw sense were developed. Because this experimental evidence cannot, per se, distinguish between image/mirror image or right/left-handed screw molecules, in the past serious doubts as to the DNA double-helical structure have been raised (107–114). Moreover, theoretical considerations suggested that the nucleotide unit in its standard conformation can be fitted into a left-handed double-helical screw with Watson–Crick-type base-pairing (380, 383). Only in recent years has the right-handed screw structure been fully supported by means of single-crystal X-ray studies on oligonucleotides (Sections 10.4, 11.2) and tRNA (Chapter 15).

Are there any experimental indications for a left-handed double helix? Evidence for a dramatically different DNA structure is found for the alternating, synthetic poly(dG–dC). In aqueous, low salt solution, this DNA displays properties reminiscent of other right-handed double-helical DNAs as shown by circular dichroism spectra and a single peak in ^{31}P nuclear magnetic resonance experiments (905,914).

However, with the addition of 0.7 M MgCl$_2$ or 2.5 M NaCl (914) or alcohol (915), a cooperative transition into another structure is observed and can be followed by inversion of the circular dichroism spectrum (Figure 12-1) or by splitting of the ^{31}P resonance into a doublet (905,914,915). The NMR data suggest a double helix with two nonequivalent phosphate groups and the circular dichroism spectrum was not interpretable with confidence because well-defined reference structures were lacking. It should be stressed that only nucleic acids like poly(dG–dC), poly(dG–dm⁵C), poly(dA–dC), poly(dT–dG), poly(dA–ds⁴T) and the halogenated poly(dI–dBr⁵U) exhibit this kind of conformational transition. The compositional isomer poly(dG)·poly(dC) and the ribo analog preserve a "normal" behavior as do other alternating copolymers with guanine replaced by other purines such as poly(dA–dT), poly(dA–dBr⁵U), poly(dA–dI⁵U), and poly(dI–dC).

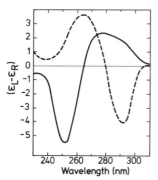

Figure 12-1. Circular dichroism spectra of poly(dG–dC) at 0.2 M NaCl, pH 7, 25° C (——) and after addition of solid NaCl (- - - - -). From (914).

12.1 Crystal Structures of Oligo(dC–dG) Display Left-Handed Double Helix

The provocative issue of the existence of a left-handed double helix was finally settled when two single-crystal X-ray diffraction studies of oligonucleotides with the same alternating sequence as poly(dG–dC), namely, the tetramer d(CpGpCpG) or d(CGCG) and the hexamer d(CpGpCpGpCpG) or d(CGCGCG), were published (302,303). They gave unambiguous proof for the existence of a left-handed double-helical DNA (916) and displayed near-identical molecular conformations although the crystallization conditions differed greatly: low salt for the hexamer (15 mM MgCl$_2$, 10 mM spermine tetrachloride, and 5% isopropanol) and high salt for the tetramer (0.2 M MgCl$_2$ and 10% 2-methyl-2,4-pentanediol) (917,818). The tetramer also crystallizes into a left-handed duplex at low salt (917,919). Because these conditions are far below the transition point of poly(dG–dC) in solution, 0.7 M MgCl$_2$ (914), extra stabilization of the left-handed structure due to crystal packing forces has been invoked (919).

syn-**Purine and** *anti*-**pyrimidine in left-handed DNA.** In both crystal structures of the tetramer and hexamer oligo(dC–dG), the self-complementary oligonucleotides are arranged into antiparallel, left-handed helical duplexes such that 12 Watson–Crick-type G–C base-pairs would complete a turn of 44.6 to 45.7 Å (Figure 12-2). As illustrated in Figure 12-3 and summarized in Table 12-1, the geometries of the individual nucleotides dG and dC in the left-handed duplex differ greatly from each other, in contrast to right-handed DNAs, where—except for alternating B-DNA—only one conformation for all nucleotides is found. Thus, while deoxycytidine adopts a standard conformation with C$_{2'}$-*endo* sugar pucker, base *anti* and orientation γ about the C$_{4'}$–C$_{5'}$ bond in +*sc* range,

Table 12-1. Structural Parameters for Z-DNAs

	Helix		
	Z_I	Z_{II}	Z_F
Bases per turn	12	12	12
Pitch height (Å)	44.6	44.6	43.5
Rotation per dinucleotide repeat (°)	−60	−60	−60
Rise per dinucleotide repeat (Å)	7.43	7.43	7.25
Base tilt (°)	−7	−7	−5
Radius (Å) of phosphate			
d(CpG)	6.3	6.1	
d(GpC)	7.3	8.0	

Torsion angle (°)

	Z_I		Z_{II}		Z_F		Z'		Z	
	pG	pC	pG	pC	pG	pC	pG	pC	pG	pC
α	47	−137	92	146	52	−110	75	177	70	−175
β	179	−139	−167	164	−153	−168	175	−166	−174	−168
γ	−165	56	157	66	178	54	−178	51	176	61
ϑ	99	138	94	147	76	147	122	141	97	143
ϵ	−104	−94	−179	−100	−72	−103	−150	−85	−142	−97
ζ	−69	80	55	74	102	−91	−17	71	−7	77
χ	68	−159	62	−148	89	−159	70	−160	65	−154

Note. Data for Z_I and Z_{II} were derived on the basis of the d(CpGpCpGpCpG) crystal structure; those for Z- and Z'-DNA are from the analogous tetramer (302,303,917,918), and data for Z_F are from a fiber diffraction study (304).

the deoxyguanosine sugar occurs in $C_{3'}$-*endo* form and, most strikingly, the base is *syn* and the rotation γ about $C_{4'}$–$C_{5'}$ is *ap*.

In Z-DNA, torsion angles δ and χ defining sugar pucker and torsion about glycosyl bonds are not correlated (920). In contrast to B-DNA where a δ, χ plot for the dodecamer crystal structure showed linear dependence of δ and χ (Figure 11-5), for Z-DNA, such behavior is not observed (Figure 12-4). If individual δ, χ values are plotted for the known Z-DNA-type crystal structures, they fall into two well-separated domains for purine and pyrimidine nucleotides; the former have χ *syn* ($\sim 70°$), the latter *anti* ($\sim -160°$). Within the domains, the distribution of χ and δ values is again characteristically different: for cytidine, the sugar pucker appears to be locked in $C_{2'}$-*endo* (δ around 140°), whereas for guanosine, a distribution in the range between $C_{2'}$-*endo* and $C_{2'}$-*exo* is found (δ from 100° to 150°); for torsion angles χ the trend is reversed, guanosines adopting a well-defined *syn* form with χ between 62° and 78°, whereas cytidines occur in a broader *anti* range, with χ between − 145° and − 180°. A rationale for this behavior can be given on purely steric grounds. Most surprising, the

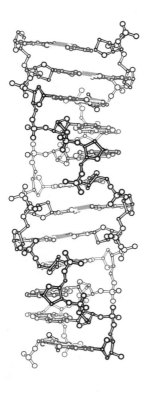

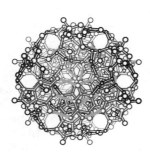

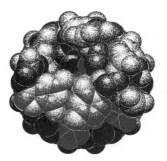

Figure 12-2. Molecular structure of the left-handed double-helical poly(DG–dC)· poly(dG–dC). Drawn with coordinates for Z_1-DNA derived on the basis of the hexanucleotide d(CGCGCG) in side and top view. For graphical details and scale, see legend to Figure 11-2.

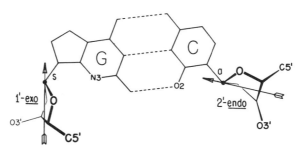

Figure 12-3. Structure of a G–C base-pair in Z-DNA. Arrows indicate different orientations of sugar moieties, which are due to *syn*-deoxyguanosine and *anti*-deoxycytidine and destroy twofold symmetry observed in Watson–Crick base-pairs. From (895).

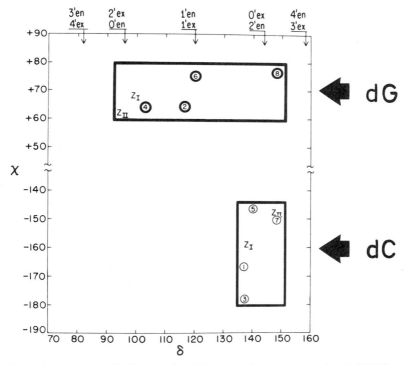

Figure 12-4. χ/δ torsion angle diagram for Z-DNA. Points 1 to 8 are for d(CGCG), whereas Z_I and Z_{II} indicate data derived for the respective Z-DNA helices. Compared with Figure 11-6, the distribution of data does not follow a linear χ/δ correlation because dG is *syn* and dC is *anti*. Sugar puckering modes (3'-en = $C_{3'}$-endo, etc.) given on top of diagram. From (895).

correlation between χ and δ so clearly demonstrated for B-DNA in Figure 11-5 is not evident in Z-DNA because δ is only variable if χ is constant and vice versa.

The alternating *syn–anti* nucleotide conformations have consequences for the overall geometry of the left-handed double helix. In right-handed double helices with Watson–Crick-type base-pairs and all nucleotides in *anti* conformation, dyad axes are located in and between base-pairs and relate one polynucleotide chain with the antiparallel partner. In left-handed geometry, the same Watson–Crick base-pair is present and the two chains are antiparallel. However, the nucleotides are in *syn* and *anti* conformations, hence destroying the dyad within the base-pairs (Figure 12-3) while that between the base-pairs is still operative. Further, as a consequence of the alternating *syn–anti* conformations, the sugar units in one polynucleotide chain are oriented alternately up and down, contrasting greatly with the right-handed screw geometry where all sugars point in the same direction.

The base-stacking pattern in the left-handed double helix shows novel features (Figure 12-5). The spatial relation between adjacent bases is of the usual screw-rotation type only for the d(GpC) sequence which also exhibits "normal" intrastrand stacks between G and C. However, for d(CpG), a shearing of base-pairs is evident. In that case, cytosines form an interstrand stack and guanines are not stacked at all with bases but rather with furanose $O_{4'}$ oxygen of adjacent deoxycytidines, an interaction observed previously in several nucleoside and nucleotide crystal structures (Figure 6-8) but not in polynucleotide helices. These differences in base-pair stacking patterns are reflected in local twist angles, about $-45°$ for d(GpC) and only $-15°$ for the sheared d(CpG) sequence. In total, they add up to the $-60°$ listed in Table 12-1 as rotation per dinucleotide repeat.

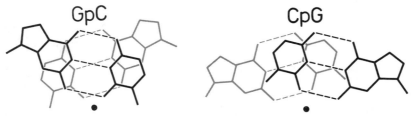

Figure 12-5. Stacking of adjacent bases in left-handed poly(dG–dC) poly (dG–dC) for sequences d(GpC), left, and d(CpG), right. Note that in d(GpC), intrastrand stacking between guanine and cytosine occurs and base-pairs are related by screw rotation. In contrast, in d(CpG) only interstrand stacking between cytosines is indicated and guanines interact with $O_{4'}$ of adjacent deoxycytidine. Dots in minor groove represent location of helix axis. Drawn with Z_1-DNA coordinates given in (917).

Phosphates in sequences d(CpG) and d(GpC) are nonequivalent. In all right-handed, complementary DNA and RNA duplexes (except in "alternating B-DNA"), phosphates along the polynucleotide chain are equivalent; i.e., they are located at the same helical radius and experience identical chemical surroundings. In contrast, the left-handed screw in alternating poly(dG–dC) sequences associates a smaller radius with d(CpG) phosphate, ~ 6.2 Å, than for d(GpC), ~ 7.6 Å (Table 12-1). Therefore, the two phosphates are chemically nonequivalent, in agreement with ^{31}P nuclear magnetic resonance results mentioned earlier (905).

Why is left-handed DNA called Z-DNA? Because phosphates are located on two different, sequentially alternating radii and neighboring sugar units point in opposite directions, a line connecting phosphorus atoms does not have the smooth appearance of double helices with all equivalent nucleotides. Rather, it follows a zig-zag course, hence Z-DNA (303) (Figure 12-6).

The left-handed double helix has only the minor groove; the major groove is filled with cytosine C_5 and guanine N_7, C_8 atoms. As depicted in Figure 12-2, the G–C base-pairs are not symmetrically located near the helix axis. Rather, they are shifted toward the periphery, exposing cytosine C_5 and guanine N_7, C_8 atoms at the major groove and rendering it convex instead of, as usual, concave. At the minor groove side, which

Figure 12-6. Zig-zag course of line connecting phosphate groups in Z-DNA. Vertical line represents helix axis; nearly horizontal lines indicate base-pairs and sugars. Path is vertical past guanines and horizontal when passing cytosines. From (895).

now accommodates the helix axis, the situation is reminiscent of C- and D-DNA: the groove is deep, narrow, and lined with phosphate groups.

Does Z-DNA adopt the same structure in solution as observed in the crystals? There is often the tendency to believe that crystal structure data are bound to the solid state and the situation could be different in solution. For the Z-DNA form observed with poly(dG–dC) in aqueous solution, the initial circular dichroism and ^{31}P nuclear magnetic resonance spectra already indicated a new, fundamentally different DNA structure. In a more detailed proton magnetic resonance study, crystallographically derived atomic coordinates were employed to calculate a spectrum taking into account ring current contributions and effects from dia- and paramagnetic components of atomic anisotropy. Since the observed spectrum obtained with poly(dG–dC) at 4 M NaCl is consistent with prediction, Z-DNA probably exists in the same or at least very similar form both in the solid state and in solution. Conversely, the observed spectrum of B-DNA was different from that of Z-DNA but agreed with that calculated on the basis of published B-DNA coordinates (921).

The preferential cutting sites of micrococcal nuclease on poly(dG–dC) and on poly(dG–dm⁵C) absorbed on calcium phosphate surface was determined as ~13.5 base-pairs, a number significantly different from the 12.0 observed in fibers (922). This alteration of overall length of the Z-DNA pitch is not enough to produce a noticeable change in the ^{31}P nuclear magnetic resonance spectrum. By and large, it appears that both B-DNA and Z-DNA underwind if in solution, a phenomenon probably due to flexibility and motion which should loosen (and increase) the pitch.

12.2 Extrapolation from Oligo- to Polynucleotides. The Z-DNA Family: Z-, Z_I-, Z_{II}-, and Z'-DNA

The crystal structures of d(CGCG) and of the hexanucleotide were used to mathematically construct polynucleotide analogs. This led to molecular models which bear a likeness to each other but still display significant differences, similar to the relationship between A- and A'-RNA or between B- and C-DNA (303,917).

Common to all Z-type double helices are the characteristics described above and summarized in Table 12-1, like 12-fold double helix with pitch height of 44.6 to 45.7 Å, axial rise per dinucleotide repeat unit of about 7.4 Å, associated with −60° rotation. The bases display a gentle negative tilt of −7° and the left-handed helix is slimmer than the right-handed counterparts, with an 18.1 to 18.4-Å van der Waals diameter contrasting 23 Å (A-RNA) and 19.3 Å (B-DNA).

Differences in molecular structure among Z, Z', Z_I, and Z_{II}-DNA. In crystal structures of the hexanucleotide with different cations, the molecules exhibit nearly identical d(CpG) residues. For d(GpC), however, two main forms Z_I and Z_{II} are observed, depending essentially on coordination of the phosphate group to water or to hydrated magnesium ion. The differences reside in rotation and hence displacement of the d(GpC) phosphate by about 1 Å, giving rise to differences in P–O torsion angles α and ζ indicated in Table 12-1. These two phosphate orientations are stabilized by water-mediated intranucleotide hydrogen bonds linking the guanine N_2 amino group with oxygen at the deoxyguanosine 3'-phosphate in the step d(GpC). In Z_I, one such water suffices to provide this interaction while in Z_{II}, with phosphate rotated "outward" by 1 Å, two water molecules are needed.

The situation is quite different for Z- and Z'-DNA. These varieties of the Z family were obtained based on the tetramer d(CGCG) crystallized under both low and high salt conditions mentioned earlier. In the Z'-DNA double helix, the deoxyguanosine sugar is puckered $C_{1'}$-exo, a variant of $C_{2'}$-endo and contrasting with the $C_{3'}$-endo found in Z-, Z_I-, and Z_{II}-DNA. An explanation for this variation in puckering mode can be given on the basis of relatively short intramolecular $C_{2'} \cdots N_3$ contacts in the syn-deoxyguanosine of Z'-DNA which, in $C_{2'}$-endo form, would be even shorter and therefore prohibited by van der Waals criteria. The system avoids this by a slight twisting into $C_{1'}$-exo and in the hexanucleotide (Z_I and Z_{II} forms) it even adopts $C_{3'}$-endo puckering.

Specific solvent interactions. This difference in sugar puckering for Z- and Z'-DNA can also be explained by variation in solvent interactions around the guanine N_2 amino group. As shown in Figure 12-7, a water molecule bridges N_2 amino and 3'-phosphate groups of deoxyguanosine intramolecularly via hydrogen bonding. In the high salt d(CGCG), however, the water is replaced by chloride ion in hydrogen-bonding contact with this amino group, but the Cl^- repels the phosphate. This brings about a rotation of δ from 82° to 122° and, hence, a change in sugar puckering mode from $C_{3'}$-endo (Z-DNA) to $C_{1'}$-exo (Z'-DNA). We witness here a specific solvent–nucleic acid interaction which stabilizes a certain helix conformation.

Transition into a Z-DNA like form is also evidenced by circular dichroism spectra for poly(dI–dBr⁵C), albeit at considerably higher salt concentration, 3.3 M NaCl, compared to 2.5 M NaCl at the midpoint of transition for the (dG–dC)₈ oligomer. It appears that the 5-bromo substituent contributes to the stabilization of the Z-DNA double helix. In other systems not containing the guanine N_2 amino group—like poly(dI–dC), poly(dA–dT), poly(dA–dBr⁵U), poly(dA–dI⁵U), and, in addition, the ribo analog poly(G–C)—no transition into a left-handed DNA is observed; in this line, poly(dA–ds⁴T) with 4-thio substituted thymidine is an

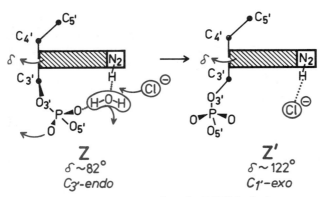

Figure 12-7. Solvent interactions in Z-DNA. Left: a water molecule mediates an intranucleotide hydrogen bond between N_2 and the phosphate group at the 3' end of *syn*-deoxyguanosine. Right: the water molecule is replaced by chloride which repels the phosphate. As a consequence, torsion angle δ changes from 82° ($C_{3'}$-*endo* sugar puckering) to 122° ($C_{1'}$-*exo*). Redrawn from (302).

exception because it is able to adopt a Z-DNA like conformation. In general, the N_2 amino group of guanosine and alternating purine–pyrimidine sequences appear to be essential prerequisites for Z-DNA. The former provides for favorable solvent interaction and the latter assists Z-DNA formation because purines adopt the *syn* conformation easier than pyrimidines (923).

The Z-DNA forms differ in backbone structure rather than in base-pair overlap. Comparing base-pair stacking between adjacent pairs in the four Z-DNAs, it is apparent that there is no systematic variation except as described above for d(GpC) and d(CpG) steps (920). However, the sugar–phosphate backbones display some differences, be it in sugar pucker for deoxyguanosines in Z-DNA ($C_{3'}$-*endo*) and Z'-DNA ($C_{1'}exo$) or in phosphate orientation for d(GpC) steps; see the respective δ, ϵ, and ζ values in Table 12-1.

This behavior contrasts findings for right-handed DNAs described by B, C, and D forms (Chapter 11). In these, the sugar–phosphate backbones are essentially the same; yet base-pair overlaps are markedly different because the orientations of the base-pairs relative to the helix axis and base-pair tilts vary from B- to D-DNA.

Transformation from B-DNA into Z-DNA does not require strand separation. Conversion of right-handed B-DNA into left-handed Z-DNA probably proceeds intrahelically and does not depend on strand separation. After an initial separation of base-pairs, guanine flips over into *syn* conformation and the entire deoxycytidine rotates, this movement retaining the *anti* orientation of the base. The bases then rejoin in Watson–Crick-type hydrogen bonding (303). This means that complete strand separation is not required and that B→Z transformation can travel along the helix as a bubble.

If, in B-DNA, a transformation into Z-DNA takes place, it has been envisaged that sequence plays a role. Since the phosphate–phosphate distance across the helix, ~ 17.5 Å in B-DNA, is closer to the separation found for d(GpC), ~ 15 Å, than to that for d(CpG), ~ 12.5, a B→Z transition is likely to occur at sequence d(GpC). In any case, there will be a discontinuity in base-pair stacking at the junction of B- and Z-DNA and the left–right interface will exhibit at least one flipped-out base-pair. This proposed model is experimentally supported by the large activation energy of 21 kcal/mole measured for the right–left conversion (914).

Thermodynamic investigations have shown that the B⇌Z transition is cooperative (914). This implies that once a Z-form nucleus is formed in a B-DNA double helix, it induces structural transformations in adjacent nucleotides and thus propagates along the polynucleotide chain. The transition is practically independent of temperature, $\Delta H \sim 0$ kcal/mole, as is also the B⇌A transition, while B⇌C, with $\Delta H = -10$ kcal/mole, favors the C form with decreasing temperature.

12.3 Left-Handed Z-DNA Visualized in Fibers of Three Alternating Polydeoxynucleotides

After the existence of the provocative left-handed Z-DNA was verified by the oligo(dC–dG) crystal structures, hitherto unexplainable X-ray fiber diffraction patterns of poly(dG–dC) obtained at 44% relative humidity and 3–6% retained salt could be interpreted (304). The pitch height, 43.5 Å, is somewhat smaller than in the double helices derived from oligo(dC–dG) but, otherwise, the conformational parameters appear similar.

A Z-DNA-like X-ray fiber diffraction pattern was also observed for the alternating poly(dG–dT) · poly(dA–dC). This shows that, while poly-(dA–dT) · poly(dA–dT) is unable to transform into Z-DNA and to adopt the D-DNA structure, replacement of one A by G suffices to bring about a change in structural properties, with Z-DNA now replacing D-DNA. The rather featureless fiber diagram of poly(dA–ds⁴T) · poly(dA–ds⁴T), first interpreted in terms of a right-handed double helix with dinucleotide repeat unit and repeating after two turns of seven nucleotide units each (924), can now also be explained in terms of a left-handed structure (304). It must be stressed, however, that in the latter case additional information should be supplied because a poor fiber diagram can be interpreted in different ways.

12.4 Factors Stabilizing Z-DNA

Spermine at 2 μM stabilizes Z-DNA. Is Z-DNA only stable at high salt and alcohol concentrations as outlined in Section 12-1 or is it possible to induce B→Z transition under near-physiological conditions? This question is of particular interest if biological experiments with Z-DNA are designed. In fact, the high salt required for B→Z transformation can be significantly reduced if alcohol or divalent ions or spermine at concentrations as low as 2 μM are added (Table 12-2).

Z-DNA can also be produced by chemical modification. If guanine residues of poly(dG-dC) are substituted in the 8 position by N-2-acetylaminofluorene or brominated, the sterical implications discussed in Section 4.7 arise and, as illustrated in Figure 2-10a, the *syn* orientation of the base and hence the Z-DNA form is preferred (925,926). Z-DNA is also stabilized if cytosine is methylated in the 5 position (926). The transition point is then reached at salt concentrations much lower than those for the unmethylated polymer (Table 12-2). In both modifications, conditions for B→Z transition are near or at physiological conditions and these polymers can be used in *in vivo* experiments to test for Z-DNA.

Table 12-2. Concentrations of Cations or Ethanol at Midpoints of the B $\rightleftharpoons$ Z Transition in Poly(dG-dC)·poly(dG-dC) and the Analog Containing Cytosine Methylated at C_5 (m^5C)

Ion	Poly(dG-dC)· poly(dG-dC)		Poly(dG-dm^{5}C)· poly(dG-dm^{5}C)	
[a]Na$^+$	2500	mM	700	mM
Mg^{2+}	700	mM	0.6	mM
Ca^{2+}	100	mM	0.6	mM
Ba^{2+}	40	mM	0.6	mM
Co(NH$_3$)$_6^{3+}$	0.02	mM	0.005	mM
Spermidine^{3+}	Aggregates		0.05	mM
Spermine^{4+}	Aggregates		0.002	mM
Ethanol	60%(v/v)		20% (v/v)	
[b]Mg^{2+} + 20% Ethanol	0.4	mM		
Mg^{2+} + 10% Ethanol	4	mM		

[a] From (927).
[b] From (924a).

12.5 Does Z-DNA Have a Biological Significance?

Is left-handed Z-DNA an *in vitro* artifact or does it also occur *in vivo*? If so, what are the implications for biological function? Although the issue is too recent to answer these questions in detail, some results are available which suggest that Z-DNA is of biological importance.

(dm⁵C–dG) sequences are prone to undergo B→Z transition. In eukaryotic DNA, the sequence (dm^5C-dG) occurs quite frequently, is the major result of DNA methylation, and is involved in regulation of gene transcription (928). As illustrated in Table 12-2, poly$(dG-dm^5C)$· poly$(dG-dm^5C)$ transforms into Z-DNA under physiological conditions and therefore it is possible that longer $(dG-dm^5C)$ sequences adopt the Z form. What, however, are the implications of Z-DNA in the gene?

Z-DNA in plasmid DNA influences topology. Circularly closed, double-stranded DNA called plasmid DNA was cut, a poly$(dG-dC)$ sequence was inserted, and the ring was closed again (929, 929a). As we shall see in Chapter 19, circularly closed DNA can form supercoils (i.e., the DNA double helix itself is again helically twisted). Under the influence of high supercoil density, the B→Z conversion in the $(dG-dC)_n$ blocks of such plasmid is induced even at physiological salt concentrations (200mM NaCl). Moreover, in agreement with the above statement, dm⁵C substituting for dC in the $(dG-dC)_n$ segments also facilitates the transition and, most surprisingly, even less supercoiling is required for producing Z-DNA if the NaCl concentration is reduced. As pointed out in (929), the B→Z transition is incomplete. It appears that a B/Z junction site of about 11 base-pairs remains in a state somewhere between left- and right-handed DNA.

Most noteworthy, it was found that d(G–C) segments of the order of only 1.3% of the total plasmid DNA are sufficient to dramatically influence the supercoiling properties of the plasmid. As outlined in Chapter 19, a B→Z interchange alters the linking number which in turn is correlated with writhing and twisting, parameters determining the overall topology of circularly closed DNA. Thus, B→Z transition modulates overall DNA structure and could be of importance in regulating DNA expression. The last question to be answered now is: Does Z-DNA occur *in vivo*?

Z-DNA visualized in polytene chromosomes. By injecting rabbits with chemically modified poly$(dG-dC)$ in the Z-DNA form under physiological conditions (section 12.4) anti-Z-DNA antibodies can be produced. These were isolated, labeled with fluorescent dye, and incubated with polytene chromosomes from *Drosophila melanogaster* (925). This special form of chromosomes consists of a thousand or more individual chromatids aligned in precise register and therefore provides for an amplification of signal not found with normal chromosomes. As a result, bright bands of stain could be seen which clearly indicated that stretches of Z-DNA are located at interband regions of these chromosomes.

This experiment proves that Z-DNA actually occurs in chromosomes. Although still an *in vitro* investigation, one may nevertheless assume that chromosomes *in vivo* also contain Z-DNA. In this connection it is of interest, that poly(dG–dm⁵C), when at low salt in the B-DNA form, binds to histone octamers and forms nucleosomes (925a). In high salt, however, if in the left-handed Z-DNA conformation, this polymer still binds to histones, yet is unable to produce the characteristic nucleosome particles (see Chapter 19). This finding suggests that only B-DNA is able to form the nucleosome structure, and that transformation into Z-DNA disrupts nucleosomes and with them the normal chromatin assembly.

The role of Z-DNA is still not entirely clear but, as outlined above, a regulatory function is feasible, especially because the B→Z transformation is reversible. Regulation of expression may involve supercoiling (925), the binding of proteins specific for Z-DNA, binding of certain cations like spermidine (Table 12-2), or modification by methylation of (dC).

Left-handed Z-DNA is transcribed and interacts with drugs. It was shown that $MgCl_2$ and ethanol act synergistically and transform poly(dG–dC) at very low concentrations, (Table 12-2, 924a). The obtained Z-DNA sediments faster than Z-DNA produced under high salt conditions and was called Z*. This Z*-DNA serves as template for *E. coli* RNA polymerase, with about half the rate measured for the right-handed form of poly(dG–dC). It is not clear yet whether RNA polymerase really transcribes Z*-DNA or whether the left-handed structure is transiently changed to right-handed and then transcribed. Moreover, Z*-DNA binds to the intercalating drugs ethidium and actinomycin D, and also forms a complex with the non-intercalating mithramycin.

In short, Z-DNA is another milestone in the history of DNA structure. Research on its biological implications has just begun and we can hope that it will shed some light on the only partially understood mechanism of gene expression.

Summary

If poly(dG–dC) in aqueous solution is treated with high concentrations of $MgCl_2$, NaCl, or alcohol, it transforms into the left-handed double-helical Z-DNA. It still exhibits Watson–Crick-type base-pairing; yet guanosine occurs in $C_{3'}$-*endo* sugar pucker, and orientation about χ is *syn* and about γ *ap*, whereas these data for cytidine are $C_{2'}$-*endo*, *anti*, and +*sc*. There is not only one Z-DNA but rather a family, Z-, Z_I-, Z_{II}-, and Z'-DNA, depending on environmental conditions. Z-DNA is stabilized by spermine or divalent cations and by chemical modification and appears to be of biological importance. It has been visualized in *Drosophila* chromosomes and probably is involved in regulation of DNA supercoiling (Chapter 19).

Chapter 13

Synthetic, Homopolymer Nucleic Acids Structures

Looking back at Figure 6-1 reveals that base–base recognition through hydrogen bonding is not restricted to interactions between complementary bases. Like bases can also associate with themselves in a great variety of arrangements which, although thermodynamically less stable than their complementary counterparts (Section 6.4), have nevertheless been confirmed in many cases by single-crystal studies (192).

Antiparallel and parallel arrangement of strands in homopolymer duplexes. In view of the possible pairing schemes involving like bases, it is not surprising to find that complexes between like nucleic acid homopolymers are formed. Double-stranded structures can display asymmetric base-pairs like that observed for poly(s^2U)·poly(s^2U), with both glycosyl $C_{1'}$–N linkages only approximately related by a dyad located *within* the plane of the base-pair (Figure 13-1). This arrangement forces the polynucleotide chains into antiparallel orientation, as observed for Watson–Crick-type double helices. The situation is drastically different in duplexes with symmetrical base-pairing as, for instance, poly(AH^+)·poly(AH^+), a double helix formed at acidic pH. Here the dyad axis relating the glycosyl links is *perpendicular* to the plane of the base-pair and coincides with the helix axis (Figure 13-1). As a consequence, sugar–phosphate backbones are 180° apart and in parallel orientation; only one type groove at the periphery of the double helix results.

Parallel strands in quadruple helices. Another striking property is met with poly(G) and its 2-deamino analog poly(I) both of which form quadruple helices. In these, a fourfold axis coincides with the helix axis, producing a cord consisting of four identical, parallel polynucleotide chains, and again only one type of groove is evident.

Single-stranded helices. The story is even more complicated because homopolynucleotides can also occur as single-stranded helices. In solution, they exhibit at least a partially ordered structure which would not be formed if the nucleotides failed to retain their standard conformations (818, 930–932). This is also true if solutions are heated to break base-stacking interactions. Even apurinic acid, a polydeoxyribonucleotide with about half the bases excised, displays essentially the same behavior (933). This suggests that it is the rotational restriction of bonds in the sugar–phosphate backbone which confers ordered structure, rather than base–base interactions.

Homopolymer nucleic acids are not just synthetic model systems; they are also found in nature as stretches of poly(A) about 200 nucleotides

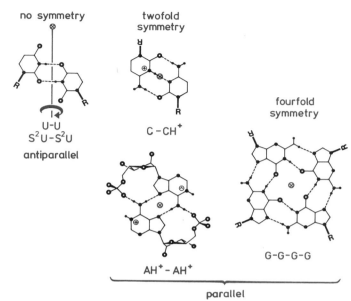

Figure 13-1. Base-pairing schemes found in homopolymer nucleic acid self-complexes. The unsymmetrical U–U and s^{2}U–s^{2}U base-pairs lead to *antiparallel* orientation of polynucleotide chains similar to the A-RNA double helix. In the symmetrical AH$^+$–AH$^+$ and hemiprotonated C–CH$^+$ base-pairs, the dyad axis coincides with the helix axis and gives rise to *parallel* polynucleotide chains. The base quadruple G–G–G–G is formed by guanosine or guanylic acid gels and is observed for [poly(G)]$_4$. A fourfold rotation axis coincides with the helix axis, resulting in parallel orientation of substituents at N$_9$, i.e., sugar moieties in gels and sugar–phosphate chains in polynucleotides.

long. These are linked covalently to the 3' end of polydisperse nuclear RNA, messenger RNA of animal cells, and many animal viruses (934,935).

As we shall see, structures of nucleic acid homopolymers display much more variety than Watson–Crick RNA and DNA. This suggests that complementary base-pair formation imposes stereochemical constraints, which ultimately limit structural variety of double helices.

13.1 Right-Handed, Base-Stacked Single Helix Revealed for Poly(C) and the O$_{2'}$-Methylated Analog

The homopolymer poly(C) has long been known to occur in helical configuration in aqueous solution (930,936–942). It adopts two different, highly ordered forms, one at pH 7.0 and the other at around the pK of cytidine, 4.5. For the low pH structure, a double helix with parallel chains and hemiprotonated base-pair CH^+-C displayed in Figure 13-1 has been proposed. Infrared spectroscopic data confirmed the existence of this base-pair both in solution and in the solid state (943), and this is additionally corroborated by single-crystal studies on cytosineacetic acid (944) and 1-methylcytosine hydroiodide (945). An early X-ray diffraction study on poly(C) fibers drawn at pH ~5 favored such a complex (103) but it was later interpreted in terms of a single-stranded helix (104). It must be emphasized, however, that these more recent X-ray results have not ruled out the possible existence of a double-helical $poly(CH^+) \cdot poly(C)$ complex. In fact, the latter might be found in the X-ray fiber pattern of the ammonium salt of poly(C), which is quite different from those of single-helical poly(C) but still too diffuse to allow unambiguous interpretation (104).

The poly(C) single helix belongs to the A family. Six nucleotides ascend $h = 3.11$ Å per step and give a right-handed turn as illustrated in Figure 13-2. The nucleotide is in the standard $C_{3'}$-*endo* conformation with torsion angles represented in Table 9-3. In general, these data agree quite well with those for A-RNA except for the $C_{3'}-O_{3'}$ torsion angle ϵ, $-129°$ in poly(C). This value is near that observed in mononucleotides rather than A-type double helices, indicating that single helices retain standard nucleotide conformation even better than the stereochemically more demanding double helices. The base tilt angle in poly(C), $\gamma_T = 21°$, is quite large (Table 9-4) and does not fit the $h-\gamma_T$ diagram in Figure 9-4, suggesting that the $h-\gamma_T$ relationship is only valid for duplexes with complementary base-pairs.

In the poly(C) single helix, there are no other obviously stabilizing elements besides base-stacking and the sterical restrictions imposed on the sugar–phosphate backbone by standard nucleotide geometry. The $O_{2'}-H \cdots O_{4'}$ hydrogen bond between adjacent nucleotides has been invoked in many discussions. Because the $O_{2'} \cdots O_{4'}$ distance is far greater than 3 Å, it cannot be formed directly but could be mediated by a water molecule as discussed in Section 13.5.

This conclusion is supported by physical and structural properties of poly(2'-*O*-methyl-C), a poly(C) analog with all $O_{2'}$ hydroxyls methylated. The thermodynamic and spectroscopic behavior of the two polymers is nearly identical (239,946). An X-ray fiber diffraction study suggested that the structure of poly(2'-*O*-methyl-C) greatly duplicates that of poly(C), with methyl groups protruding from the periphery of the single-stranded helix (Figure 13-2) (947). The sugar puckering is $C_{3'}$-*endo* as indicated by torsion angle δ in Table 9-3 and consistent with predictions from classical potential energy calculations on the dinucleoside phosphate GpC (948). For the monomeric 2'-*O*-methylcytidine, however, both theory (949) and crystallographic study (950) revealed $C_{2'}$-*endo* ribose, showing that phosphorylation at $O_{3'}$ and/or polymer formation can change sugar puckering preferences.

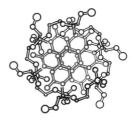

Figure 13-2. The structure of the poly(2'-O-methyl-C) single helix, drawn with co-ordinates in (947). Methyl groups are indicated by large spheres. The structure of unsubstituted poly(C) is almost identical. For graphical details and scale, see the legend to Figure 10-1.

The story about the helical structure of single stranded poly (C) cannot be ended without mention of a very recent NMR study (950a). It was shown that poly(C) in neutral aqueous solution probably occurs as a left-handed single helix with eight nucleotides per turn. The axial rise per residue is given as 2.9 Å, resulting in a pitch height of 23.2 Å. These geometrical data are so different from those derived on the basis of the fiber diffraction of poly(C) that a novel kind of helical structure is likely to exist. This is also evident from the overall shape of the left-handed helix: the ribose–phosphate backbone is at its inside and the bases are at the outer periphery, similar as proposed for the right-handed helix of poly(dT) (Section 13.4). The cytosines are therefore not stacked but adjacent cytosines interact through hydrogen bonding $N_4H \cdots O_2$. The proposed (hypothetical) helical structure is further stabilized by another hydrogen bond between $O_{2'}H$ and the phosphate attached to $O_{3'}$ of the same nucleotide. Concerning the conformation angles, sugar pucker is $C_{3'}$-endo, bases are in *anti* orientation, rotation about the $C_{4'}$–$C_{5'}$ bond (γ) is an unusual *sp* and both P–O torsions α, ζ are in *-sc*.

13.2 Bases Turned "in" and "out" in Nine- and Twofold Single-Stranded Helices of Poly(A)

Depending on pH, poly(A) exists in single- or double-stranded helical form. Like poly(C), poly(A) also adopts two kinds of ordered structures in aqueous solution: at neutral or alkaline pH, a single-helical poly(A) species prevails which transforms into a double helix if the adenines are protonated (at N_1) at below pH ~4. These two modifications of poly(A) have been the target of a number of thermodynamic studies using spectroscopic methods (331,930–932,951,952) and theoretical approaches (955,956). The single helix⇌double helix transformation depends on temperature, pH, and salt concentration (954). The transition from double-helix into random-coil structure is cooperative but for the single helix, only noncooperative behavior was established (952).

Structural details of ninefold poly(A) single helix. Solution spectroscopic data suggest that the poly(A) single helix is right handed, with nucleotides in $C_{3'}$-endo conformation, classifying this helix as a member of the A family. Bases are in *anti* orientation and stacked parallel to each other. Consistent with these characteristics, the structural model illustrated in Figure 13-3 was derived on the basis of the single-crystal study of the trinucleoside diphosphate ApAH$^+$pAH$^+$. As will be detailed in Section 13.3, this oligomeric zwitterion exhibits a helical part, ApAH$^+$, which was used to construct the helix by mathematical extension of this two-step fragment (957).

The poly(A) single helix thus obtained contains nine nucleotides per

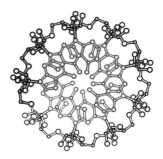

Figure 13-3. Model for the poly(A) single helix proposed on the basis of the ApAH⁺pAH⁺ crystal structure. Drawn with coordinates in (957). For graphical details and scale, see the legend to Figure 10-1.

pitch of 25.4 Å. The corresponding 2.82 Å rise per nucleotide and the base-tilt of 24° with respect to the helix axis are again inconsistent with the linear $h - \gamma_T$ plot in Figure 9-4. The phosphates are arranged at a radius of 8.07 Å, a little less than the 8.89 Å in double-helical A-RNA. As observed in poly(C), torsion angle ϵ, $-145°$, is closer to the value found for mononucleotides rather than the $\sim 180°$ for A-RNA-type double helices (Table 9-3).

Different sugar puckering in the deoxyribo analog poly(dA). The $C_{3'}$-*endo* conformation and base-stacking of helical poly(A) has also been verified for oligo(A) in aqueous solution (331). In the case of the deoxyanalog, nuclear magnetic resonance data suggest preference for $C_{2'}$-*endo* puckering and pronounced stacking (958). With growing oligomer length, the $C_{2'}$-*endo* population increases, consistent with higher conformational purity mentioned in Table 4-3. The structural difference between poly(A) and poly(dA) can also be demonstrated by immunological assays (959).

A twofold single-helical poly(A) in denaturing solvents. The situation changes dramatically if poly(A) is dissolved not in water but instead in denaturing solvents such as formamide, dimethylsulfoxide, or ethanol. These solvents are known to force bases to unstack and therefore the thermodynamic and spectroscopic properties of nucleic acids are different (960,961) [for 80% ethanol, see (962)]. Surprisingly enough, even under such conditions, poly(A) can still be drawn into ordered fibers which produce near-identical X-ray diagrams for all these solvents, indicating that the polymer adopts a defined, ordered secondary structure (963). What does such a "denatured" structure with no base-stacking look like?

The X-ray diffraction pattern of poly(A) fibers prepared from formamide solution is very different from that of a Watson-Crick-type double helix (Figures 3-9c and 13-4). It nevertheless suggests that the molecular structure is helical, with a pitch of 5.8 Å corresponding to two nucleotides. Packing considerations combined with birefringence data point to significantly tilted ($\sim 36°$) bases, with probably no base–base interaction whatsoever and heavy solvation brought about by 3.5 formamide molecules per nucleotide (Figure 13-5).

A succession of π turns. Torsion angles pertinent to the twofold poly(A) helix are given in Table 9-3. Notably, the sugar pucker is $C_{3'}$-*exo* and torsion angles about P–O ester bonds are both in the $+sc$ range, differing drastically from the $-sc$ range normally observed for helical polynucleotides.

This particular conformation about P–O bonds had previously been described for nonhelical oligonucleotides UpAH$^+$ and ApAH$^+$pAH$^+$ and also in tRNA. Because it implies reversal of chain direction as in π turns of polynucleotides, it was analogously termed a π turn (964) (see Section 15.4). Here, it should suffice to say that the whole poly(A) single helix formed under denaturing conditions can be regarded as a series of consecutive π turns.

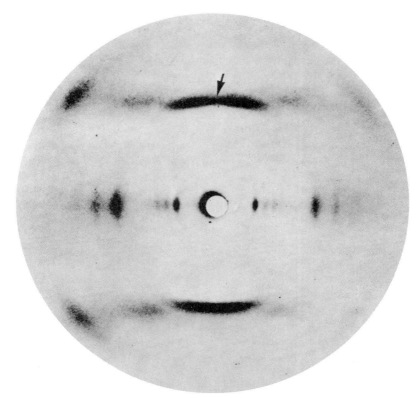

Figure 13-4. X-Ray fiber diffraction pattern for poly(A) from formamide solution (963). Fiber tilted 8° from the perpendicular to the X-ray beam. The axial rise per residue is 2.9 Å and indicated by a meridional reflection (not shown here); the layer-line separation is 5.8 Å and gives the pitch height. The arrow points to a meridional minimum on the first layer line, showing that this is in fact not identical with the axial rise per residue.

13.3 A Double Helix with Parallel Strands for Poly(AH⁺)·Poly(AH⁺) Forms under Acidic Conditions. Helix, Loop, and Base-Pair Stacks in ApAH⁺pAH⁺ Dimer

At pH < 4, poly(A) is protonated at adenine N_1. The polymer aggregates to form a double helix which is preserved in aqueous solution (965), can be precipitated in the cold (966), and which can be drawn into fibers (458,967).

A polymeric zwitterion displaying a base-pair with rotational symmetry and base–phosphate interaction. The X-ray fiber diffraction pattern of

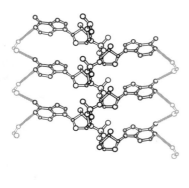

Figure 13-5. Structure of the poly(A) twofold single helix obtained under denaturing conditions and derived from the X-ray pattern in Figure 13-4. Views are with helix axis in the paper plane (top) and almost vertical (bottom). Solvent molecules (formamide) are hydrogen bonded to adenines and indicated by shading. Drawn with coordinates in (963). For graphical details and scale, see the legend to Figure 10-1.

poly(AH$^+$)·poly(AH$^+$) indicates that the helix is eightfold, implying a 45° turn angle per nucleotide, and that the pitch height is 30.4 Å, yielding a rise per nucleotide of 3.8 Å (458) (Figure 13-6). A dyad axis coincides with the helix axis, resulting in the rotational symmetric base-pair illustrated in Figure 13-1. The dyad further arranges the two intertwined polynucleotide chains parallel to each other, as discussed at the outset of this chapter. Of particular interest is the interaction N$_6$–H $\cdots$ O$^-$–P between the N$_6$ amino group and a free phosphate oxygen (Figure 13-6). This hydrogen bond is probably stabilized by protonation at adenine N$_1$ which, in addition, renders the poly(AH$^+$)·poly(AH$^+$) duplex a polymeric zwitterion. Solution studies employing ^{1}H and ^{31}P NMR methods are also in agreement with this model (967a).

ApAH$^+$pAH$^+$, an oligomeric zwitterion duplex that contains helical, looped, and base-stacked structures. The poly(A) fragment ApApA crys-

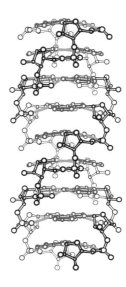

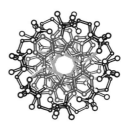

Figure 13-6. Structure of the poly(AH$^+$)·poly(AH$^+$) double helix formed under acidic conditions. A dyad axis coincides with the helix axis and arranges the polynucleotide chains in parallel orientation. Note that there is only one type of groove because the base-pair has rotational symmetry (Figure 13-1). Hydrogen bonds indicated in gray are formed between bases and between N$_1$H · · · O$^-$–P. For graphical details and scale, see the legend to Figure 10-1. Drawn with coordinates given in (458).

tallized under slightly acidic conditions in the form of the zwitterionic duplex $Ap^-AH^+p^-AH^+ \cdot Ap^-AH^+p^-AH^+$, with the two negative charges on the phosphate groups counterbalanced by protonation at N_1 of the adenines in the second and third places in the oligonucleotide chain (986); for another interpretation of the protonation pattern, see (191). A priori, one could expect the dimer to adopt a double-helical form like that of $poly(AH^+) \cdot poly(AH^+)$ (Figure 13-7). However, the individual molecule actually exhibits a rather complicated configuration, consisting of elements of both helical and folded structure. This finding is in agreement with circular dichroism studies of oligomers of type $Ap(Ap)_nA$, suggesting that only above a critical chain length of $n = 6$ is a double helix formed under acidic conditions (952). Below $n = 6$, nonhelical duplexes of yet undefined structure occur, suggesting that for initiation of a $poly(AH^+) \cdot poly(AH^+)$ double helix with parallel strands, the chain length required for nucleation followed by propagation is longer than the three base-pairs derived for Watson–Crick-type helices (Figure 6-12).

In the $ApAH^+pAH^+$ dimer, the two molecules are arranged *antiparallel* (Figure 13-7C). They are, however, associated through AH^+–AH^+ base-pairs which, as described in Figure 13-1, require *parallel* nucleotide

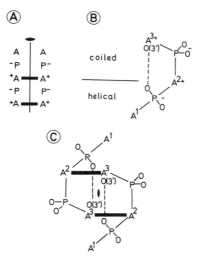

Figure 13-7. Expected (A) and actually observed (B,C) molecular configurations in $ApAH^+pAH^+$ (968). In (A), the $ApAH^+pAH^+$ dimer is arranged in a short double helix, with base-pair as in Figure 13-1, structure as in Figure 13-6, and parallel oriented chains. However, in the crystalline duplex the molecules with helical and coiled pairs (B) are associated in antiparallel form (C), yet stabilized by the $AH^+ \cdots AH^+$ base-pair (thick bars) which requires parallel arrangement. This conflict leads to the looped structure of the oligonucleotide displayed in Figure 13-8.

orientation. This obvious conflict leads to folding of the AH⁺pAH⁺ part of
the molecule; the remaining ApAH⁺ fragment displays a helical configuration.

In spite of this complicated overall molecular structure, the individual
nucleotides are in standard conformation with $C_{3'}$-*endo* ribose pucker,
bases in *anti* and torsion angles γ in $+sc$, positioning $O_{5'}$ "over" the ribose. The orientations about the P–O ester bonds in the two dinucleoside
phosphate portions, however, differ strikingly, being characteristic of hel-

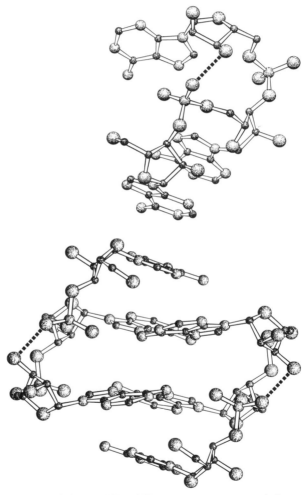

Figure 13-8. Molecular structure of the ApAH⁺pAH⁺ monomer (top) and dimer
(bottom) (968). Hydrogen bonds between $O_3H \cdots O^- - P$ are indicated by dashed
lines. Hydrogen atoms are omitted for clarity.

ical structures, $-sc$, in the ApAH$^+$ part but display the π-turn structure, $+sc$, in the folded AH$^+$pAH$^+$ fragment (964) (Section 15.4).

What forces are responsible for stabilization of the ApAH$^+$pAH$^+$ dimer? Figure 13-8 shows that the bases are thoroughly stacked, giving the duplex the appearance of a globular, almost helical structure. In addition, the AH$^+$pAH$^+$ folding is supported by a hydrogen bond between the terminal O$_{3'}$–H and the penultimate phosphate.

13.4 The Deoxydinucleotide d(pTpT) Suggests Single-Stranded Poly (dT) Helix with Nonstacked Bases Turned "out"

While for poly(U), double-helical and probably antiparallel secondary structure could be established (at least at low temperatures), judging from spectroscopic data the deoxyribo analog poly(dT) and its oligomeric fragments do not exhibit any ordered configurations (146,147,969,970).

The dinucleotide d(pTpT) was crystallized as a sodium salt from ethanolic aqueous solution (358). The salient features of the molecular structure of d(pTpT) are that the two nucleotides have virtually identical conformation with C$_{2'}$-*endo* sugar puckering, base in *anti* and C$_{4'}$–C$_{5'}$ torsion angle γ in the $+sc$ range. These characteristics are maintained in aqueous solution (970); this is not too surprising in view of the high hydration of 26 water molecules per dinucleotide in the crystal structure which gives the crystal more the appearance of a concentrated solution rather than a solid.

An extended molecular structure. The torsion angle about the phosphodiester link α is in the $-sc$ range as required for helical configurations whereas ζ is ap, resulting in an extended molecular structure (Figure 13-9). The bases are too far apart for intramolecular stacking, yet intermolecular stacking with adjacent molecules is very pronounced.

Sodium cations coordinated to thymine O$_2$. Phosphates and bases are heavily hydrated and in both thymines, O$_2$ is coordinated to a Na$^+$ whereas O$_4$ is only involved in hydrogen bonding to water molecules. If we recall that in the ApU dimer structure a similar situation was met, we may surmise that keto O$_2$ · · · Na$^+$ interactions appear to be rather specific.

A single helix with bases turned "out." Because both nucleotides adopt comparable conformations, the d(pTpT) molecular structure allows extension into a polynucleotide, with the sugar–phosphate backbone screwed in right-handed sense and the bases protruding at the periphery and tilted with respect to the helix axis by 48° (Figure 13-10). The helix is nonintegral with 7.2 residues per pitch of 25.2 Å, yielding a rotation and rise per residue of 50° and 3.5 Å.

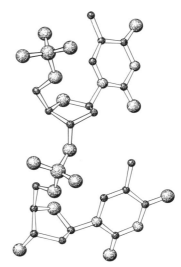

Figure 13-9. Molecular structure of d(pTpT). Within the crystal, molecules are heavily hydrated and both O_2 keto oxygens are coordinated to Na^+, which in turn is surrounded octahedrally by water and phosphate oxygens. The two nucleotides have essentially the same conformational parameters and therefore extension of the dinucleotide structure into a single helix was possible (see Figure 13-10). Drawn with coordinates in (358); H atoms are omitted for clarity.

13.5 The Antiparallel, A-RNA-Type Double Helices of Poly(U), Poly(s²U) and poly(X)

Folding back into hairpin-type duplex structures. In contrast to poly(dT) the riboanalog poly(U) exhibits a secondary structure at low temperature, ~5°, which can be monitored by UV hypochromic effects and by nuclear magnetic resonance, circular dichroism, light scattering, infrared spectroscopy, viscosity and kinetic parameters employing the temperature jump method (237,459,460,971,972). The structure is best defined as hairpin loop formed by folding of the poly(U) molecule back on itself, resulting in a double-helical, antiparallel arrangement with unsymmetrical base-pair displayed in Figure 13-1 (237,460). The relatively short poly(U) double helices melt cooperatively and they are stabilized by high counterion and polyamine concentrations (972,973).

Because the latter can raise the melting temperature of poly(U) to 28°, fibers with spermine as counterions can be drawn and subjected to X-ray diffraction analysis (974). The resulting, rather diffuse pattern gives a pitch of 24 Å but the number of nucleotides per turn, between 9 and 11, is unclear. It is probable, however, that the poly(U) structure is similar to that derived for the thioketo analog poly(s²U) (461).

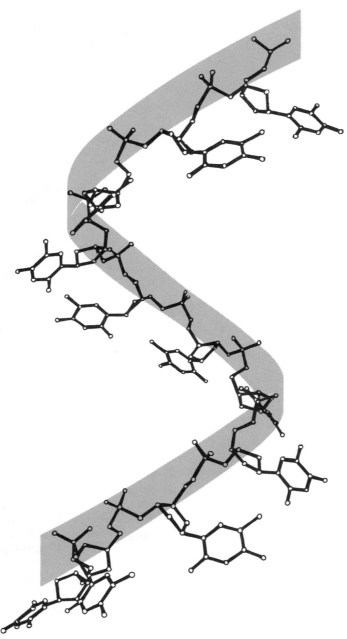

Figure 13-10. The single-helical structure proposed for poly(dT) on the basis of the d(pTpT) crystal structure (358). Helix is right-handed, 7.2 residues per pitch of 25.2 Å, and bases are turned ''out'' and therefore not stacked.

2-Thioketo substitution in nucleic acids raises melting temperature. In general, if a 2-thioketo group is introduced into pyrimidine bases of nucleic acids, the melting temperature rises. In poly(s^2U), the effect is especially dramatic, raising the $T_{m'}$ 5.8° for poly(U), to 68.5° for poly(s^2U) under similar ionic conditions (975). Substitution of the 4-keto group by sulfur has only slightly stabilizing or even destabilizing influence for reasons given below.

In contrast to poly(U), the X-ray diffraction pattern of poly(s^2U) is of good quality and, in its characteristic features, similar to that of A-DNA (461). This points to an antiparallel double helix with 11 base-pairs per pitch of 28.8 Å, base-pairs tilted in a negative sense with respect to the helix axis, and $C_{3'}$-*endo* sugar puckering typical of A family structures (Figure 13-11). Conformational parameters are entered in Table 9-3.

A double helix with no dyad symmetry. As a consequence of the nonequivalence of the two glycosyl links of the s^2U–s^2U base-pair with hydrogen bonds N_3–H $\cdots$ S_2 and N_3–H $\cdots$ O_4, a dyad relating the two polynucleotide chains with each other is lacking and therefore the not necessarily identical chains are referred to as α and β chains (461). In a more recent study, least-squares refinement methods have shown that the difference between the chains can be ironed out and mainly absorbed in rotations about the glycosyl links (976). On the other hand, one could also claim that one of the s^2U bases in a pair occurs as enol tautomer and thus enables a symmetrical base-pair to form. This possibility is improbable, however, for chemical reasons outlined in (461).

Why is poly(s^2U) so stable compared to poly(U)? If we look at the stacking pattern of poly(s^2U) illustrated in Figure 13-12, we find that S_2 stacks over N_1 of the adjacent nucleotide, an interaction that was already frequently observed as a recurring structural motif in crystal structures of nucleosides and bases containing thioketo groups (453). We therefore infer that this kind of stacking, assisted by the large polarizability and hydrophobic character of sulfur, causes the superior stability of all polymers containing 2-thioketopyrimidine moieties (975). In polymers with 4-thioketo substituents, the direct S $\cdots$ N interaction is not possible, hence no stabilizing effect.

Poly(dT), poly(rT), poly(dU), and poly(U) display different helical stabilities suggesting that $O_{2'}$ plays a key role. Why is the secondary structure of poly(dT) less stable than that of poly(U)? Nuclear magnetic resonance has shown that oligo(dT) adopts preferentially the $C_{2'}$-*endo* sugar pucker whereas oligo(U) is more in the $C_{3'}$-*endo* form (146,147,237,970). This difference in sugar conformation, also found in the adenosine analogs oligo(dA) and oligo(A) (331,958), can be interpreted with the electronegativity concept described in Section 4.3 and might bear on stacking geometry which certainly is different for $C_{2'}$-*endo* polynucleotides (B family) compared with $C_{3'}$-*endo* polynucleotides (A family).

Another factor contributing to the relative stability of the poly(U) dou-

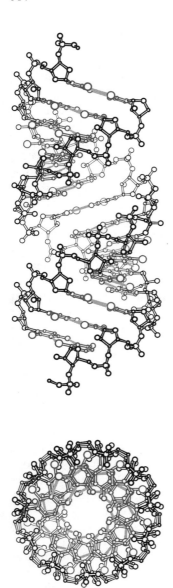

Figure 13-11. The structure of poly(s²U)·poly(s²U), exhibiting the s²U–s²U base-pair with comparable geometry as displayed for U–U in Figure 13-1. The poly(s²U) double helix repeats after 28.8 Å comprising 11 nucleotides, and has a structure reminiscent of A-DNA or A-RNA. Sulfur atoms are indicated by larger spheres; drawn with coordinates in (461). For graphical details and scale, see the legend for Figure 10-1.

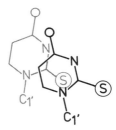

Figure 13-12. Base overlap in poly(s^{2}U)·poly(s^{2}U), showing S_2···N_1 interactions which contribute to the unusual stability of the double-helical structure ($T_m = 68.5°$).

ble helix could be hydrogen bonding between $O_{2'}$–H of one ribose and $O_{4'}$ of the adjacent one, either via direct contact or mediated by a water molecule (237). On the basis of the presently available data, we cannot decide which effect is more significant, but it is clear that the $O_{2'}$–H group is the crucial point. This is particularly evident if we contrast poly(dU) which has no secondary structure with poly(U) and, analogously, poly(dT) with poly(rT) where the melting temperature indicates an ordered structure even more stable than that of poly(U) (402).

An all-purine double helix formed by polyxanthylic acid. The purine base xanthine displays the same functional groups as uracil. Therefore the base-pairing situation as discussed for uracil or 2-thiouracil (s^{2}U) is met, with either the unsymmetrical pair (No. XVI in Figure 6-1) or a symmetrical pair with one of the two partners in enol tautomeric form. Whatever the base-pair might look like, X-ray fiber diffraction patterns show that, depending on the pH of the medium, two structurally related double helices are observed. One, at pH 5.7, has an axial rise per residue $h = 3.01$ Å, a 10-fold helix, and a diffraction pattern reminiscent of A-RNA. For the second at pH 8.0, h is 2.52 Å, there are 11 base-pairs per turn, and the pattern more nearly resembles that of A-DNA. In both double helices, the $C_{1'} \cdots C_{1'}$ separation in a base-pair is 2 Å greater than that of the Watson–Crick pair because only purines are involved (854).

13.6 Sticky Guanosine—Gel Structure of Guanosine and Guanylic Acid: Quadruple Helix Formed by Poly(G) and Poly(I)

Gel formation requires free hydrogen-bonding sites and high salt concentration. In contrast to all other nucleic acid constituents, guanosine, guanylic acid, and their derivatives (but not the base guanine alone) form gels from aqueous solutions. Inspection of over 60 guanosine derivatives led to the conclusion that (i) hydrogen-bonding sites N_1–H, N_2–H, N_7, and O_6 must be intact, (ii) the $O_{5'}H$ hydroxyl group must be either present or replaced by phosphate, sulfate, or chlorine (but not by other substituents), and (iii) a certain concentration ($\sim0.1\ M$) of excess alkali salt is essential (978).

With nucleotides, gel formation takes place only if we are dealing with the monoanion existing above the first pK of the phosphate group. At neutral pH, guanosine 5′-phosphate still forms ordered structures which, however, do not aggregate into a gel (979).

Addition of salt is crucial for gel formation. If no excess salt is present, the monosodium salt of guanosine 5′-phosphate and guanosine itself yield well-defined crystals (980,981) but no gel.

Piles of tetrameric, planar aggregates in gels of guanosine derivatives and of guanosine 3′-phosphate. The structural characteristics of gels formed by these guanosine-based molecules were elucidated by X-ray fiber diffraction techniques (982,983). In all cases, guanine bases are hydrogen bonded in the planar, tetrameric arrangement shown in Figure 13-1 and further piled on top of one another, with sugars protruding at the periphery. Interactions between adjacent layers are probably not only through stacking forces. Rather, the tolerated substitution of $O_{5'}H$ by phosphate, sulfate, and chlorine suggests that these groups are also involved and probably form hydrogen bonds with guanine residues in adjacent stacks (978,984).

The individual tetrameric layers do not stack in register but rather are rotated with respect to each other, the extent of rotation depending on the sugar moiety. In guanosine gels, successive layers are rotated by 45° and for 8-bromoguanosine, rotation by 54° is evidenced, probably caused by the *syn* orientation of the base. In deoxyguanosine, a rotation angle of $-54°$ is adopted, suggesting that hydrogen bonding to $O_{2'}H$ hydroxyl in guanosine is a structure-determining factor (978,984).

In these cases, the layers are arranged face to back, with a repeat distance of ~3.4 Å as determined from X-ray diagrams (984). For 2′,3′-O-diacetylguanosine and for guanosine 3′-phosphate, however, a face-to-face juxtaposition leads to reflections at twice that distance, ~6.8 Å.

Guanosine 5′-phosphate forms helical aggregates. In gels of guanosine 5′-phosphate, the tetrameric base association is nonplanar; it is broken at one side like a lock-washer such that a continuous, probably left-handed helix with 15 nucleotides in 4 turns is generated (982,983). This change in overall structure from stacks of planar tetramers to helical screw requires only small distortions of hydrogen bonds because each adjustment is amplified fourfold. The helix is probably stabilized by a hydrogen-bonding interaction between guanine N_2–H and 5′-phosphate of the adjacent nucleotide level, an interaction not possible in guanosine 3′-phosphate. If, however, the sodium salt of guanosine 5′-phosphate is kept at neutral pH in the presence of 1.3 M NaCl, fibrous specimens crystallize which, if oriented in an X-ray beam, produce a diffraction pattern reminiscent of right-handed, quadruple helical poly(G) discussed below (985).

Spectroscopic studies support X-ray interpretations. The structural assignments on gels formed by guanosine derivatives and by guanosine 3'- and 5'-phosphate are in general agreement with spectroscopic studies. UV absorption and vapor pressure osmometry show that gel formation is highly cooperative and that the mechanism follows the scheme monomer⇌tetramer⇌octamer⇌polymer (986).

Nuclear magnetic resonance spectra monitoring ^{1}H, ^{23}Na, and ^{31}P signals of sodium guanosine 5'-phosphate in neutral aqueous solution indicated that below 0.2 M nucleotide concentration, normal base stacking occurs. Above 0.3 M nucleotide this turns into hydrogen-bonded associations favoring tetramer, octamer, and hexadecamer aggregates. The self-assembly is enthalpy driven with $\Delta H = -17 \pm 2$ kcal/mole and $\Delta S = -51 \pm 6$ cal/mole·K^{-1}, rather than determined by positive entropy changes as would have been expected for normal stacking processes (987).

Why does gel formation of guanosine derivatives require excess salt? Nuclear magnetic resonance has provided evidence for the cation being an integral part of the tetrameric guanine association (Figure 13-13) (987,988). The cation is located in the central cavity lined by the four O_6 keto oxygens which, separated from the rotation axis by 2.28 Å in the planar arrangement (985), leave a void of 0.88 Å van der Waals radius. Considering the ionic radii of Li$^+$ 0.68 Å, Na$^+$ 0.97 Å, K$^+$ 1.33 Å, Rb$^+$ 1.47 Å, and Cs$^+$ 1.67 Å (196), it follows that Li$^+$ is too small to snugly fill the cavity, Na$^+$, K$^+$, and Rb$^+$ have just the right size either to penetrate or to position themselves slightly above and between adjacent layers, coordinating now to all eight surrounding oxygen atoms. Cs$^+$ appears too large to satisfy the spatial conditions and does not support tetrameric guanosine aggregation. In general, this rather specific cation complex formation is paralleled by a whole group of naturally occurring ionophores and by synthetically available crown ethers which provide ether or keto oxygens as ligands for specific cation binding, the selection being largely due to spatial effects (989).

Geometrical characteristics of gel structure are maintained in quadruple helices of [poly(G)]$_4$ and [poly(I)]$_4$. A variety of physical data have suggested that tetrameric molecular arrangements found in gels of guanosine derivatives are reproduced in quadruple helical structures obtained for poly(G) and for its 2-deamino analog poly(I) (990,991). Two independent X-ray fiber diffraction analyses on these two polymers arrived at the same conclusion and provided evidence for right-handed helical sense. The pitch height of 39.2 Å corresponds to 11.5 nucleotides with an axial rise per residue of 3.4 Å, or phrased in other words, exact repeat is after 23 nucleotides in two turns, helix symmetry 23$_2$ (990,992). In this model, conformational angles are similar and characteristic of A-family polynucleotide helices (Table 9-3). These studies could rule out an earlier four-fold helical structure with C$_{2'}$-*endo* sugar puckering (993) as well as a

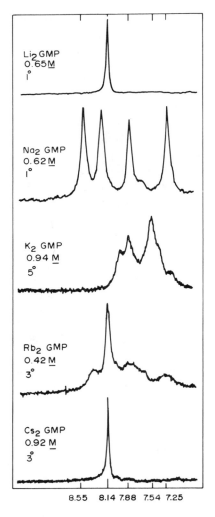

δ, ppm

Figure 13-13. 220 MHz nuclear magnetic resonance spectra showing H_8 signal of guanosine-5'-phosphate in the presence of different alkali metal ions. Concentrations are near saturation limits (0.4 to 0.9 mole/liter), temperature 1° to 5°, in D_2O. Resonance lines are unaffected by Li^+ and Cs^+ but display splitting due to formation of tetrameric complex with Na^+, K^+, Rb^+ located in a cavity provided by four guanosine O_6 oxygens. Taken from (988).

threefold variety proposed on the basis of X-ray fiber diffraction data (994) and, in addition, a left-handed turn deduced for these polymeric complexes from spectroscopic data (990). The structural similarity of poly(G) and poly(I) quadruple helices is also established by immunologi-

cal studies which moreover provide evidence for single-stranded entities
(995).

**Besides quadruple complexes, poly(I) and poly(G) also exist as ordered
single- and double-helical structures.** Supplementary to the above-mentioned
immunological data, circular dichroism (990) and infrared spectroscopic studies (996) indicated the presence at low salt concentrations of
an ordered single-stranded, stacked, and probably helical structure for
both kinds of polymers. This form is metastable for poly(G) with metal
ions but stable with the larger and therefore less screening tetramethylammonium ion, suggesting that electrostatic repulsion between adjacent
negatively charged phosphates within the same or between different polynucleotide chains contributes significantly to the helix stability (996).

Finally, a double-helical model with eight nucleotides per turn was advanced on the basis of X-ray data for poly(G) fibers drawn in the presence

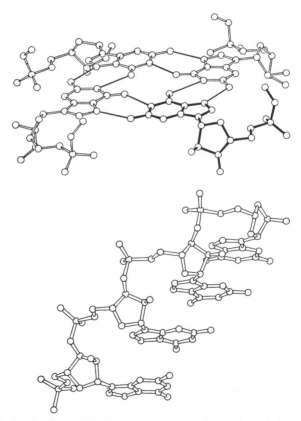

Figure 13-14. The quadruple helix of poly(G). Sugar puckering is $C_{3'}$-*endo* and the
helix repeats after two turns of 39.2 Å, comprising 11.5 nucleotides each, i.e.,
helix symmetry in 23_2. Shown is the quadruple base-pair (top) and one strand of
the fourfold helix (bottom). From (992).

of 0.03 M HCl (997). Under these conditions, N_7 is protonated and a GH^+ $-GH^+$ base-pair can be envisaged with $N_7-H \cdots O_6$ hydrogen bonding geometry, similar to the poly(AH^+)·poly(AH^+) duplex where $N_7 \cdots H-N_6$ hydrogen bonds are engaged (Figure 13-1). Although the X-ray diffraction diagram is too poor to allow detailed statements, the analogy between duplexes poly(AH^+)·poly(AH^+) and poly(GH^+)·poly(GH^+), both stable only if bases are protonated, becomes evident.

For the deoxyribose analogs, gels are formed as observed for monomeric guanosine and its derivatives (978). In the case of polymers, no detailed structure information is available except an electron microscope study suggesting that poly(dG) can adopt at least single-, double-, and triple-stranded forms which could, however, represent denatured states of a quadruple helix (998).

Summary

In contrast to natural RNA, *synthetic* homopolymer RNA can adopt several forms, ranging from single to quadruple right-handed helical. Thus poly(C) and the $O_{2'}$-methylated analog occur as sixfold single helices, and poly(A) under denaturing conditions (formamide) produces a twofold single helix in zig-zag form which under physiological conditions changes to ninefold, with right-handed twist. When ApApA was crystallized under acidic conditions, it assumed both helical and looped conformations, with $AH^+ \cdots AH^+$ base-pairing as found for double-helical poly(AH^+) under acidic conditions, the latter displaying parallel polynucleotide chains. The more standard *anti*parallel arrangement was found in poly(s^2U) as well as in poly(U) and poly(X) which adopt an A-RNA type form whereas the single helix of poly(dT) was derived on the basis of the crystal structure of d(pTpT). A fascinating story is found for guanosine which forms gels with stacked nucleosides in fourfold arrangement, a structure motif recurring in the helix of tetrameric $[poly(G)]_4$ and $[poly(I)]_4$.

Chapter 14

Hypotheses and Speculations: Side-by-Side Model, Kinky DNA, and "Vertical" Double Helix

To finish our discussion on ordered RNA and DNA helical structures, mention should be made of a few suggestions concerning their structure and dynamics. Of greatest interest among these are an alternative side-by-side model of duplex DNA which avoids twisting of one polynucleotide chain around the other and the "vertical" DNA suggested for double helices with bases in *high-anti(-sc)* position. Kinking of DNA was first introduced to visualize winding of the then assumed "stiff" DNA helix around the histone core in chromatin and was later extended to explain dynamic "breathing" of the DNA helix.

14.1 Side-by-Side Model—An Alternative?

A topological problem solved like the Gordian knot: By cutting. In the Watson–Crick-type double-helical DNA, the two polynucleotide strands are wound around the same helix axis and therefore intertwined. Because in DNA replication the two polynucleotide chains are separated and then associated with the newly formed, complementary daughter DNA (999), the parental DNA must unwind. Recalling that DNA is a very long, threadlike molecule (Chapter 1), we see immediately that unwinding, being accompanied by one rotation per turn of the helix, leads to immense movements, even ignoring the tight packaging of DNA in chromatin where the whole process takes place. Imagining this rotating, gyrating, and unwinding causes some headaches (1000), and the Gordian knot of coiled DNA was finally cut when a series of DNA processing enzymes called swivelases, gyrases, and nicking-closing enzymes were discovered to do just this job. These enzymes cut DNA to allow for local unwinding and, this being done, rejoin the ends to regain the original structure (1001).

Another solution to the problem: Side-by-side model of DNA. Is it possible to imagine a DNA complex that displays Watson–Crick-type base-pairing, has no strand intertwining, and gives the X-ray fiber diffraction pattern observed for DNA? The answer to this question came from two independent groups (112,113,1002–1004). They constructed DNA duplex models which mimick helical geometry and, instead of winding around in one "endless" screw sense, complete a right-hand screw only half-way, turn into a half left-hand screw, resume the right-hand screw, etc. (Figures 14-1 and 14-2).

How to accomplish the turning points stereochemically? In the regular, right- or left-handed screw sections, no particular stereochemical problems seem to arise. The turns from one into the other screw, however, present some difficulties which were solved by introducing nonstandard, yet stereochemically acceptable conformations involving rotations about $C_{4'}-C_{5'}$, glycosyl and phosphodiester bonds as well as changes in sugar puckering modes from $C_{2'}$-*endo* to $C_{3'}$-*endo* and vice versa. In a more recent model, even the orientations of adjacent sugars varied, pointing pairwise alternately up and down along the chain (113), reminiscent of the situation in left-handed Z-DNA (Chapter 12). Because rotation into right-handed screw is easier and more pronounced relative to the opposite sense, models of the side-by-side DNA duplex exhibit a long-range, overall right-handed superhelical twist (112).

The side-by-side model has the advantage that no DNA untwisting during replication is required; only hydrogen bonds have to be broken for strand separation (1004a). The side-by-side duplex displays considerable flexibility and it is probably easier to wind such a DNA into the narrow turns required in chromatin (Chapter 19) than the double-helical form. In addition, the side-by-side model exhibits both large and small grooves

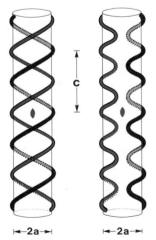

$\leftarrow$2a$\rightarrow$ $\leftarrow$2a$\rightarrow$

Figure 14-1. Idealized drawings of a double helix (left) and a side-by-side (SBS) structure (right). Note that overall dimensions (c = helical repeat and $2a$ = diameter) and dyad symmetries indicated by the symbol ● are preserved in both cases. Redrawn from (112).

Figure 14-2. Two schematic views of the side-by-side model. One of the chains is drawn in color to show that it follows a zig-zag course rather than a helical turn. From (112).

known from the double helix and, because the sugar–phosphate backbone winds along a "planar" surface, intermeshing between adjacent molecules is easily accomplished, again favoring close packing in chromatin.

Support for the side-by-side model has been advanced by X-ray crystallographic calculations (1005) and by biochemical studies showing that single-stranded, circular DNA molecules combine to give a duplex called V-DNA (1006). In this particular form, equal amounts of right- and left-handed double-stranded DNA were proposed, which can be envisaged as either the side-by-side form or right-helical stretches compensated for by equivalent left-helical lengths.

Evidence against the side-by-side model was provided by X-ray crystallographers (122). It points at the inadequacy of X-ray fiber diffraction analysis to cope conclusively with this problem, simply because the data are too poor. However, better evidence against a side-by-side model comes from gel electrophoresis experiments showing that supercoiled, circularly closed DNAs of the same size, yet different number of turns, migrate in a series of discrete bands (1007). This quantized property can be understood if we assume that all DNAs in one band have the same linking number (Chapter 19), a feature much easier to explain with the double-helical model rather than the side-by-side model.

From a structural point of view, finally, it is hard to see why, after five base-pairs of right-handed helix the screw sense should reverse instead of continue in its right-handed traveling mode. This holds especially if homopolynucleotides like poly(dA) · poly(dT) are considered—what tells the homopolynucleotide chain when to incorporate a turning point? And, further, how should we explain the occurrence of the different A-, B-, C-, and D-DNA forms if not by helical modifications?

The situation, of course, could again change if proteins associated with DNA come into play. In essence, the very existence of side-by-side DNA can, at present, not be denied with certainty but it is more likely to present a special rather than a general topological variant of duplex DNA.

14.2 Does DNA Fold by Kinking?

In chromatin, double-helical DNA in the B form is wound around a protein core and describes a superhelix with a radius of about 50 Å (Chapter 19). Because it appeared improbable that the DNA helix of diameter 20 Å winds smoothly into such compact form it was initially assumed that it kinks every 20 base-pairs (1008,1009). This hypothesis appeared to be supported by partial digestion of chromosomal DNA–protein complexes by DNase I. The enzyme produces DNA fragments with chain lengths in multiples of 10 base-pairs which was interpreted as preferential cleavage at kink sites (1008). These suggestions later proved erroneous (907) but the kink in the DNA helix is still an attractive structural possibility.

Geometrical description of a DNA kink. Model building studies led to the conclusion that the DNA double helix is most easily kinked if it is bent toward the minor groove side (Figure 14-3). Kink angles α as large as 100° can be achieved by rotating only torsion angle γ ($C_{4'}$–$C_{5'}$ bond) from the usually preferred $+sc$ range into ap. Energy calculations have shown that this conformational change associated with base-pair destacking should occur preferentially between A–T base-pairs. The energy required for this process is estimated as about 5 kcal/mole or less, depending on the theoretical approach (1010,1011).

Some macroscopic features of the kink are that (i) the axes of the two helical legs of the kink do not necessarily intersect but may be displaced by about 1 Å to either side (d in Figure 14-3), (ii) the two helical legs are related by a pseudodyad running along their bisector, (iii) the distance D defining the separation between helix intersection and the next base-pair can be up to 7–8 Å, and (iv) by and large, the kink causes the whole molecular structure to adopt a slight negative twist on the order of 15–20°.

If several kinks repeat at constant intervals, the DNA is folded in characteristic patterns. Thus, kinks every 5 base-pairs lead to a zig-zag form of DNA while kinks every $n \times 10$ base-pairs result in a left-handed superhelical arrangement, just as required for DNA folding in chromatin. As we

---

Figure 14-3. Kinked B-DNA, viewed from two angles 90° apart. Tilt angle α and separations d and D are explained. A pseudodyad (vertical in top and horizontal in bottom) relates the sugar–phosphate backbones in the two legs of the kink. From (1012). *(Figure appears on following page.)*

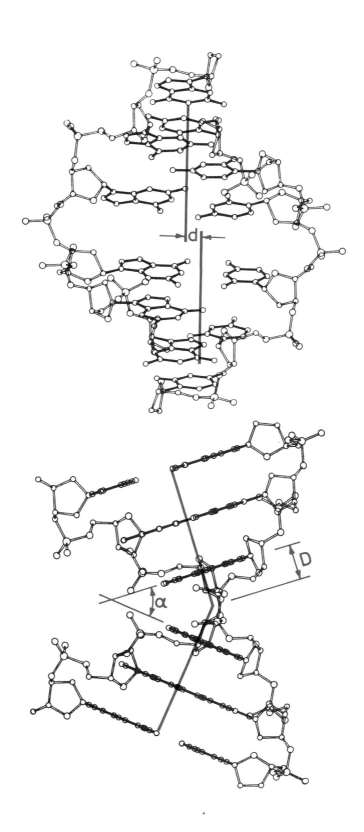

will see in Chapter 19, however, smooth bending of DNA appears more probable on the basis of data now available. Nevertheless, kinking as such has been invoked to describe dynamical "breathing" of the DNA double helix.

14.3 κ- and β-Kinked DNA: Breathing with the Speed of Sound

Because the DNA kinking hypothesis appeared to offer a reasonable solution to the packaging problem in chromatin, it was further extended on the basis of dinucleoside phosphate–drug intercalation complexes which are used as models for DNA–drug interaction (Chapter 16). In the model complexes which are predominantly of the ribose series, self-complementary dinucleoside monophosphate molecules form a mini-helix and base-pairs are separated by a gap of 3.4 Å to accommodate the drug. This process is in general accompanied by negative untwisting of the miniature double helix and, at least in the majority of the complexes, by a change in sugar puckering from $C_{3'}$-$endo$ to $C_{2'}$-$endo$. Although the situation might be different for deoxyribose oligo- or polynucleotides, it was suggested that in B-DNA, a kink of about 40° is produced by change of sugar pucker from $C_{2'}$-$endo$ to $C_{3'}$-$endo$ and unwinding of adjacent base-pairs at the kink site by $-10°$ (1012–1014). These kinks, as proposed, should occur every 10 base-pairs in chromatin ("κ-kinked B-DNA"). A kink every second base-pair should give rise to a more loose "β-kinked B-DNA" structure and exist prior to thermal DNA melting or transiently in A–T-rich regions, thus facilitating DNA–protein and DNA–drug interactions via intercalation of planar aromatic amino acid side chains or drug molecules.

β-Kinked B-DNA stretches are proposed to occur in DNA as a result of energy picked up by Brownian collision of solvent molecules. These events should give rise, in most cases, to anharmonic motions in the DNA structure but occasionally, if a solvent molecule hits DNA with the right momentum, a harmonic wave can be built up, gives rise to normal mode oscillations (breathing), and travels along the helix with the speed of sound (acoustic wave). This motion is associated with structural transitions from B-DNA into β-kinked B-DNA with every second sugar snapped into $C_{3'}$-$endo$ form (1014).

Here is not the place to decide whether this hypothesis is correct. It is clear, however, that the DNA double helix opens transiently because the hydrogen-deuterium exchange at 0°C is only about five minutes for DNA kept under conditions where the T_m is greater than 80°. Proposals have been brought forward that bubbles of about 10 open base-pairs travel along the DNA chain as solitary waves, excited by solitons. Whether the structure of DNA in these bubbles is reminiscent of β- or other kinked DNA, however, is yet unknown (1014a). The take-home lesson for us is that DNA is certainly not a rigid, stiff rod but should rather be looked at as a molecule with limited macroscopic flexibility and microscopic local motions.

14.4 Bends in DNA at Junctions of A- and B-Type Helices

Assuming that in double-helical DNA a similar tendency for exhaustive stacking prevails as in tRNA (Chapter 15), another proposal becomes of interest which does not involve kinking (= unstacking) in order to conduct the DNA helix around a corner but rather utilizes only changes in sugar puckering mode (381,1015).

Mixed sugar pucker, base stacks, and 26° bend at A–B junction. In an attempt to join 10-fold B-DNA and 11-fold A-DNA by model building, it was found that the junction is easily achieved if the base-pair at the interface exhibits a mixed sugar pucker. The best combination is met if, when going from A-DNA to B-DNA, the sugar in the strand with $5'\rightarrow3'$ polarity is $C_{2'}$-endo, and the sugar in the complementary nucleotide is $C_{3'}$-endo. In spite of this conformational change, base-stacking is fully preserved, all torsion angles are in the usual range, yet the helix bends by 26° (Figure 14-4). The B→A transformation leads to a rotation of $-3.3°$ for every transformed base-pair, resulting in an overall untwisting of DNA.

These proposals are supported by some experimental data. That the fully stacked, base-paired B–A junction really exists in polynucleotides could be demonstrated by nuclear magnetic resonance and circular dichroism spectroscopy studying the hybrid complex $[poly(dG)\cdot(rC)_{11}-(dC)_{16}]$ which in all-DNA section adopts the B form and exists in A form in the adjacent $(rC)_{11}$ segment (1015,1016). The untwisting associated with the B→A transition was actually observed for RNA polymerase holoenzyme binding to DNA (1017). Length estimates based on the untwisting angle of 140° suggested about 40 nucleotides per promoter binding site, a number greatly in agreement with analysis of promoter recognition sites (1018).

It is clear that a succession of such bends at A–B junctions can introduce a full turn in the DNA helix without having to invoke kinks and, in general, resembling more or less a smooth curvature. In addition, since A- and B-DNA are energetically similar (1019), it might well be that dynamics in DNA are brought about by A-DNA clusters traveling along the B-type helix rather than β-kinked clusters which would involve more drastic structural changes.

In this context, it should be noted that longer poly(dA) · poly(dT)

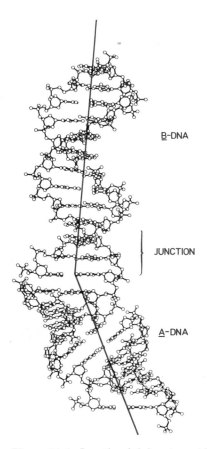

B-DNA

JUNCTION

A-DNA

Figure 14-4. Junction joining A- and B-DNA double helices. At the interface base-pair, one sugar is $C_{2'}$-*endo*, the other $C_{3'}$-*endo;* otherwise all torsion angles are as usual and bases are fully stacked, in contrast to kinks (Figure 14-3). Helix axes drawn in color describe a bend of 26°. From (1015).

sequences, which have actually been observed in the DNA of several eu-karyotes, are found experimentally to be terminators of DNA transcrip-tion (1016a). This has been interpreted in terms of DNA structure be-cause, as indicated in Section 11.4, poly(dA) · poly(dT) (and *not* the alter-nating poly(dA–dT) is conformationally stiff and occurs only in B′-DNA form. Hence it resists the B→A transition and the binding of RNA polymerase could thus be prevented (1016b).

Figure 14-5. Two views, perpendicular to and along the helix axis in the "verti-cal" double helix constructed for the poly(A) · poly(U) duplex. The poly(U) chain is drawn heavier than the poly(A) chain; hydrogen bonds are indicated by broken lines. Note near-vertical arrangement of bases in *high-anti(-sc)* orientation and large radius of helix. From (1022).   (*Figure appears on following page.*)

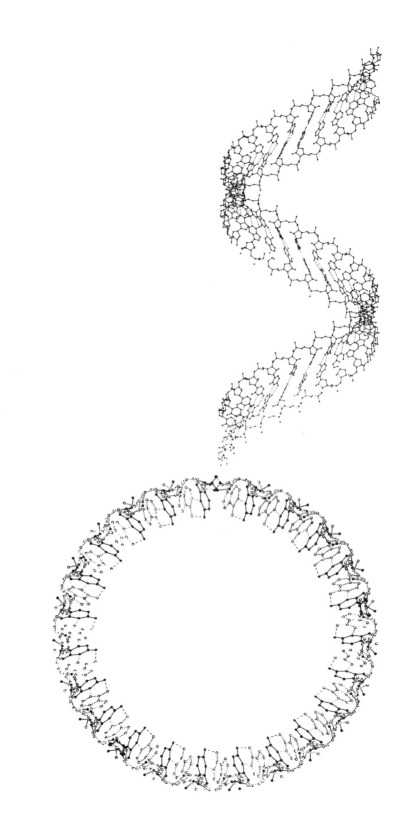

14.5 "Vertical" Double Helix for Polynucleotides in *high-anti* Conformation

If the bases in a polynucleotide are in *high-anti*(*-sc*) conformation, for instance by locking them chemically via cyclization of C_6 in pyrimidines and C_8 in purines with the sugar $C_{2'}$, single-stranded helices can nevertheless be constructed. These are of two kinds, either with left-handed screw and bases perpendicular to the helix axis (1020,1021) or, alternatively, with right-handed screw and bases nearly parallel to the helix axis.

The latter structure can further be extended into a "vertical" double helix with Watson–Crick-type base-pairs and structural characteristics differing markedly from the canonical DNA and RNA double helices (1022,1023) (Figure 14-5). The base-pairs are almost parallel to the helix axis and the sugar–phosphate backbone, with *high-anti*(*-sc*) torsion angle about the glycosyl links, winds around the helix axis on a wide radius of 8 to 18 Å, depending on which of several models is considered. In general, all torsion angles are in accord with those observed in double-helical DNAs and RNAs, and models are feasible both with $C_{3'}$-*endo* and with $C_{2'}$-*endo* sugar puckering modes. The proposal of the "vertical" double helix was based entirely on computerized model building studies but has found support from spectroscopic studies on homopolynucleotides fixed in the *high-anti*(*-sc*) conformation which are able to form single-, double-, and triple-helical structures reminiscent of the poly(A) · poly(U) system (1024–1026).

All of this goes to show that it is relatively easy to build stereochemically reasonable models, difficult to prove such models wrong, and even more difficult to prove them right.

Summary

This chapter describes some alternative or complementary suggestions for DNA/RNA structures and properties which were presented mainly on the basis of theoretical calculations. In order to avoid coiling of one DNA strand about the other during replication, a side-by-side model for the DNA duplex was proposed with five left-handed double-helical turns followed by five right-handed turns, etc. DNA winding around nucleosome cores was explained by kinking rather than smooth bending of double helices, kinking being produced by changes in sugar puckering and/or changes in torsion angle γ. Kinks in the DNA double helix were envisaged to explain premelting behavior and drug intercalation processes. Bends (or kinks) are also proposed to occur if A- and B-type DNA structures meet in a helix. A "vertical" DNA double helix was suggested if nucleotides occur in *high-anti*(*-sc*) conformation, this helix winding in a rather large radius around a wide and open cylindrical core.

Chapter 15

tRNA—A Treasury of Stereochemical Information

tRNA is one of the best and most thoroughly studied biological macro-molecules. Its structure–function relationships have been summarized in many books and articles (164,1027–1038), its spectroscopic behavior has been described with special reference to dynamical aspects (1027), and the thermodynamics of conformational transitions have been extensively investigated (449,554,1027,1028,1039,1040). Because discussing all of the structure–function aspects of tRNA would lead too far astray, the reader is referred to the cited literature and we focus on the questions of immediate structural interest: the primary, secondary, and tertiary structure of tRNA, the stabilization of the helical and looped domains by base stacking and base-pairing, the conformations of the nucleotides, and tRNA–metal ion interactions.

15.1 Primary and Secondary Structure of tRNA: The Cloverleaf

General conclusions based on 200 tRNA nucleotide sequences. Typically, tRNAs are polynucleotide chains 75 to 90 units long. They do contain not only the four standard nucleotides but also approximately 10% "rare" or "minor variant" nucleotides, some of which have been alluded to in Section 7.4. The primary sequences of the nearly 200 tRNAs known have been collected (1037,1041,1042) and all can be arranged in the cloverleaf secondary structure depicted in Figure 15-1.

The four stem regions of the cloverleaf each contain four to seven Watson–Crick base-pairs organized into double helices. These are called acceptor, anticodon, D, and T stems. The acceptor stem is the locus of both the 3' and 5' termini of the polynucleotide chain and carries, at the $C-C-A_{O_{3'}H}$ hydroxyl of the 3' end, the specific (cognate) amino acid corresponding to the sequence of the anticodon base-triplet in the anticodon loop (genetic code, Section 6.9). The D and T stems are so named because the neighboring loops invariably contain dihydrouridine (D) and ribothymidine (T). The number of nucleotides in stems and loops is generally

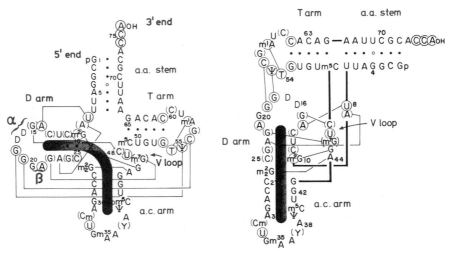

Figure 15-1. Nucleotide sequence of yeast tRNA^Phe folded (left) into secondary "cloverleaf" and (right) into tertiary "L-shape" arrangements. Invariant bases in class I tRNA are indicated by circles; those which are semi-invariant are enclosed in parentheses. Watson–Crick-type hydrogen bonds in stem regions are marked by dots, the G–U "wobble" pair (G_4–U_{69}) by a small circle. Number of nucleotides in variable loop V and in α and β regions can vary, depending on the tRNA species. Bases forming tertiary base–base hydrogen bonds are connected by thin lines. Note that most of the invariant and semi-invariant bases are positioned in the middle and hinge regions of the "L" and involved in tertiary hydrogen bonds which, together with base stacking illustrated in Figure 15-3, stabilize the tRNA structure. Shading in color and gray illustrates which parts of the "cloverleaf" join together to form the two legs of the "L". From (1031).

constant except for the variable loop which consists of 4 to 5 nucleotides in most tRNAs (class I) but can in some cases contain from 13 to 21 nucleotides (class II). The α and β regions in the D loop can vary from 1 to 3 nucleotides in different tRNA species.

Invariant and semi-invariant nucleotides. All tRNA sequences can be aligned in register so that, in some positions, the same nucleotide always occurs; these are called invariant nucleotides. In other places, only purines (R), pyrimidines (Y), or hypermodified purines (H) are found— these are the semi-invariants. Figure 15-1 indicates that most of the invariant and semi-invariant nucleotides are concentrated in the loop regions. A structural interpretation of this is based on the tertiary folding of tRNA molecules.

15.2 Folding of the Cloverleaf into Tertiary Structure: The L Shape

How are the four arms of the cloverleaf spatially arranged? X-ray studies reveal the tertiary structure. Even before the first X-ray crystal structure analysis of tRNA in 1973 (1043), chemical and physical evidence was used to design tRNA models and predict the tertiary folding (235,1044). Of these models, one proposed by Levitt was particularly similar to the now-established structure: even most of the tertiary base–base interactions were correctly placed (1045).

Crystals of tRNA have been obtained in several laboratories (1044), and successful X-ray crystal structure analyses have been carried out by three independent groups working with orthorhombic (396,1046) and monoclinic crystal forms of yeast tRNA[Phe] (85,87,1047). More recently, the structures of yeast tRNA[Asp] (1048), of initiator tRNAs[fMet] both from yeast (1049) and from *E. coli* (1050), and yeast tRNA[Gly] (1051) have been determined.

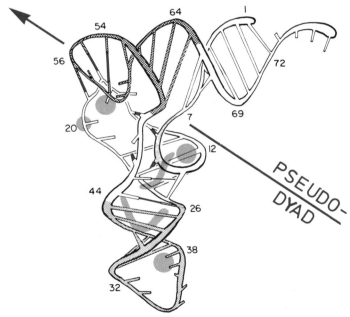

Figure 15-2. Schematic representation of the yeast tRNA[Phe] molecular structure, showing the ribose–phosphate backbone as a continuous tube. Base-pairs are indicated by long bars and single bases by short bars. Tertiary base–base hydrogen bonds, marked by rods (color), cluster in the hinge region and at the outer side of the corner of the "L". The four arms are shaded differently for clarity. Positions of four tightly bound Mg^{2+} cations are marked by gray spheres and those of two spermines by gray swivels. From (396).

The L shape is composed of two nearly perpendicular A-RNA helices. All the crystal structure analyses have revealed that tRNAs have more or less the same L-shaped configuration with each arm about 70 Å long and 20 Å thick; the latter dimension is the diameter of an A-RNA double helix. The molecular structures of the independently investigated yeast tRNA[Phe] molecules are practically identical although the packing of the molecules in monoclinic and orthorhombic crystals is different. Comparison of tRNA[Phe] with tRNA[Asp] and initiator tRNA[Met] shows small, yet significant, variations which are not considered here.

The two ends of the L are formed by the acceptor CCA end and by the anticodon loop which are about 80 Å apart (Figures 15-1, 15-2). The outer rim of the L corner is occupied by the T loop. Acceptor and T stems are stacked to form a continuous, 11-base-pair double helix of the A-RNA type. The anticodon and D stems are arranged in a similar way but the overall helix is not straight; there is a 26° kink between the two stem axes (396).

tRNA, a molecule with pseudodyad symmetry. In a broad sense, the two arms of the L are related by a pseudodyad bisecting the L angle and relating the anticodon/D helix to the acceptor/T helix (Figure 15-2). Because aminoacyl-tRNA synthetases, enzymes which charge tRNAs with their cognate amino acids, occur as monomers, dimers and higher subunit structures, this twofold symmetry of tRNA could, perhaps, be associated with the dimer (twofold) or higher aggregation state of the enzyme, facilitating enzyme–substrate recognition and giving the enzyme–tRNA complex again a twofold symmetrical structure (1052). This attractive hypothesis is still not proved since there is no crystal structure available of a tRNA-synthetase complex.

15.3 Stabilization of tRNA Secondary and Tertiary Structure by Horizontal and Vertical Base–Base Interactions

In addition to Watson–Crick-type base-pairs accounting for most of the horizontal base–base interactions (especially in the stem regions), there are a number of nonstandard base-pairs and base-triplets. The nonstandard base-pairing occurs predominantly at the outer corner and hinge region of the L where the two helical domains join. The nonstandard base-pairing stabilizes the folding of the cloverleaf secondary structure into the tertiary L structure. Extensive vertical base stacking, which is one of the most dominant characteristics of tRNA architecture (Figures 15-2, 15-3), reinforces the structure.

The non-Watson–Crick base interactions found in yeast tRNA[Phe] are displayed in Figure 15-4. One of them, the G_4–U_{49} "wobble" base-pair (Section 6.9), is integrated in the acceptor stem. Owing to its overall structural similarity to a Watson–Crick base-pair, it does not disrupt the double

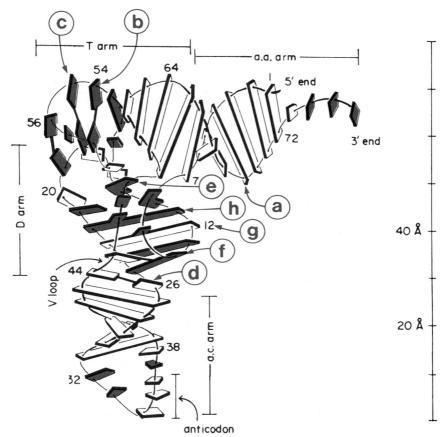

Figure 15-3 Illustration of base-pairing and stacking interactions in yeast tRNA[Phe]. Base-pairs stabilizing secondary and tertiary structure are drawn as bent or fused slabs or by connecting two slabs with rods. Invariant bases in full color and semi-invariant bases in shaded color cluster in anticodon loop, CCA end of acceptor stem, and at outer corner and hinge regions of the "L" (compare tertiary interactions marked in color in Figure 15-2). Note stacking of bases 34 to 38 in the anticodon loop, consisting of the anticodon Gm_{34}–A_{35}–A_{36}. Letters **a** to **g** indicate where base-pairs and triplets displayed in Figure 15-4 are located. From (1030,1046).

helix and causes only a slight bulge in the sugar–phosphate backbone. The G_{15}–C_{48} base-pair is of the reversed Watson–Crick type with both chains running parallel. Antiparallel orientation is observed in the reversed Hoogsteen base-pair m^1A_{58}–T_{54}. In this, the Watson–Crick site of m^1A_{58} is blocked by methylation. Therefore, Hoogsteen pairing is the only possible mode of hydrogen bonding with other bases, and it is probably also favored by the positive charge which results from methylation at N_1. The purine–purine base-pair $m_2^2G_{26}$–A_{44} is longer than a normal Watson–Crick pair and is also significantly nonplanar due to the steric repulsion

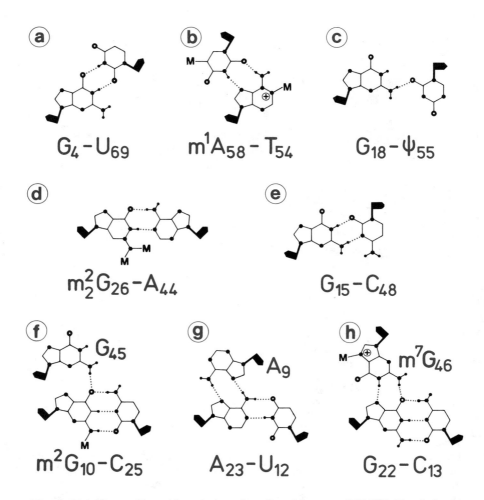

Figure 15-4 Unusual base–base interactions found in yeast tRNA[Phe]. For their positions in the tertiary structure, see Figure 15-3. All other base-pairs are of the Watson–Crick type. Flags at glycosyl links indicate backbone orientations. They are parallel in **c, e, f**(m²G₁₀, G₄₅), **g**(A₉, A₂₃), **h**(C₁₃, m⁷G₄₆) and antiparallel otherwise. All of the dimeric base–base interactions are already contained in Figure 6-1.

between the dimethylamino group of G_{26} and adenine C_2–H. Because this base-pair is at the junction of anticodon and D stems and is stacked with both, it is responsible for the 26° kink between these two helices.

There are also several base-triplets consisting of a Watson–Crick base-pair to which, in the major groove side, another base is associated via one hydrogen bond in m^2G_{10}–C_{25}–G_{45} or via two bonds in G_{22}–C_{13}–m^7G_{46} and in A_{23}–U_{12}–A_9, the latter displaying the same A–A base-pair found in the poly(AH⁺)·poly(AH⁺) duplex (Section 13.3). In all these base-triplets, polynucleotide chains are of necessity simultaneously parallel and anti-parallel.

Most of the tertiary hydrogen bonds are formed between constant bases. Looking at Figures 15-2 and 15-3, we find that most of the hydrogen bonds stabilizing the tertiary structure are between invariant or semi-invariant bases. Since all tRNAs can be arranged into the cloverleaf secondary structure, this finding suggests that the tertiary structure of all tRNAs is more or less the same L shape (1053,1054).

Base stacking as a dominant structural feature. In yeast tRNA^{Phe}, only 42 out of the 76 bases are involved in A-RNA-type helical structure. Yet 71 bases are engaged in stacking interactions. The non-stacked bases are G_{20}, U_{47}, the terminal A_{74}, and the dihydrouracils D_{16}, D_{17}. The latter, being non-planar and non-aromatic, inherently oppose stacking (Section 7.4).

The base-stacking pattern in the double-helical domains is nearly the same as that observed in A-RNA. However, the overlap is somewhat increased; a tightening of the helices equivalent to a helix of slightly less than 11 base-pairs per turn results (1052). Although bases U_7 and m^2C_{49} are not linked by a ribose–phosphate backbone, the helical structure at this junction between acceptor and T stems is preserved by the regular stacking of these two bases with adjacent base-pairs.

Interesting stacking patterns are found in the Watson–Crick pair m^2G_{10} $-C_{25}$ positioned over $m_2^2G_{26}$ of the "long" purine–purine pair mentioned above, and leading to the 26° kink between anticodon and D stems. Fur-

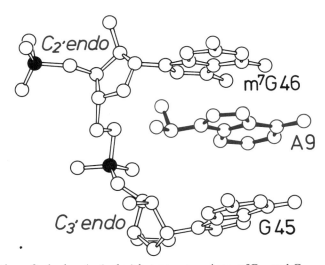

Figure 15-5. Intercalation of adenine A_9 (color) between guanines m^7G_{46} and G_{45}. This special kind of stacking is associated with separation of the two guanine bases by 3.4 Å to allow insertion of adenine, the separation being accomplished by a change in ribose puckering of m^7G_{46} from the preferred $C_{3'}$-endo to $C_{2'}$-endo. Phosphorus atoms are indicated by shading. Redrawn from (1054).

ther, in the T loop the reversed Hoogsteen pair between constant bases m^1A_{58}–T_{54} and the constant G_{18}–Ψ_{55} pair with only one hydrogen bond are stacked on the Watson–Crick pair G_{53}–C_{61}, the latter being also invariant in all tRNAs (1055). The position of the invariant $T_{54}\Psi_{55}C_{56}G_{57}$ sequence at the corner of the L-shaped tRNA offered an explanation for the apparent binding to the ribosomal 5S RNA (1056). More recent data, however, have disclosed that such interaction is probably not involved in protein biosynthesis (1056a).

Base intercalation: A special kind of base stacking. To optimize base–base stacking, in some instances individual bases of one strand are tucked between two bases of an adjacent strand. These interactions occur at positions where three strands meet: at the inside of the L corner and at the T-loop region. In addition to the case of A_9 between G_{45} and m^7G_{46} (Figure 15-5), three other base intercalations are observed: C_{13} between U_8 and A_9; G_{18} between m^1A_{58} and G_{57}; and G_{57} between G_{18} and G_{19} (1057). Because base insertion is associated with a base–base separation of 3.4 Å, the ribose–phosphate backbone must readjust accordingly. How this is done will now be described.

15.4 Change in Sugar Pucker, π Turn, and Loop with Phosphate–Base Stacking: Structural Features of General Importance

In order to enable the tRNA polynucleotide chain to fold into its complicated pattern of helices and loops, the nucleotide conformations deviate from the standard form in critical regions.

A $C_{3'}$-endo→$C_{2'}$-endo puckering change stretches the ribose–phosphate backbone, and widens the P···P distance from 5.9 to 7 Å (Figure 9-3). Such changes in sugar conformation are found at intercalation sites involving residues A_9, G_{19}, m^7G_{46}, and m^1A_{58} where adjacent bases are separated by 3.4 Å to permit intercalation (see Figure 15-5). $C_{2'}$-endo sugar puckers are also found at positions where chain foldings switch abruptly from helical to looped; at U_7, C_{48}, and C_{60}; and where chains are stretched as at D_{17} and G_{18} (1057,1058).

The π turn: P–O torsion angles change with a turn in the backbone folding. A characteristic deviation from helical P–O torsion angle conformation has been found in anticodon and T loops of yeast tRNA^{Phe}, in the $ApAH^+pAH^+$ duplex, in $UpAH^+$, and in the $ApU \cdot$9-aminoacridine complex (Sections 10.5,13.3,16.2). In all these molecular structures, two riboses connected by a phosphodiester link are not oriented in the same (helical) direction but rather point in opposite, antiparallel directions. This sharp turn, called a π turn by analogy with turns in polypeptide chains, is caused primarily by rotations about P–$O_{3'}$ or P–$O_{5'}$ bonds (Figure 15-6). In general, three classes of π turns called π_1, π_2, and π_3 are feasible, with ζ values centered around 85°, 170°, and 285° (Figure 15-7).

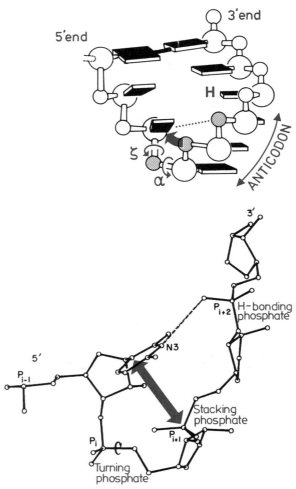

Figure 15-6. (Top) Schematic representation of the π_3-turn situation in anticodon and T loops of yeast tRNA[Phe]. Sugars and phosphate groups are simplified as large and small spheres, respectively; bases as slabs; base-pair interaction by a bar; H represents hypermodified base. Semi-invariant and invariant bases are stacked with the base-pair at the top end of the loop, and the base at the bottom of the anticodon loop is again stacked with the bases to the right. Phosphates shaded in color are actively engaged in the turn and form hydrogen bonds (dashed line) and stacking interaction (thick arrow) and adopt π_3-turn torsion angles α and ζ in *ap* and *-sc* ranges. Redrawn from (964). (Bottom) Detailed picture of loop showing "hydrogen bonding," "stacking," and "turning" phosphates shaded in color. Stacking interactions between phosphate and uracil base are indicated by double arrow; the hydrogen bond involving the N_3H or uracil and phosphate oxygen is drawn as dashed line. From (1060).

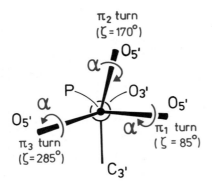

Figure 15-7. The three π turns are distinguished by their ζ angles, viewed along the P–$O_{3'}$ bond. The sharpness of the turn depends on how much α deviates from the helical conformation described by ~290°. Redrawn from (964).

The sharpness of the turn is dictated by angle α, which, in helical structures, is around 290°; i.e., a π_3 turn with α at 290° is the classical helical conformation. This turn is, of course, the most common in tRNA. Nonhelical π_3 turns with P–$O_{5'}$ torsion angle α in the $+sc$ or ap range are found at the 3' ends of nucleotides 9, 17, 33, 46, and 55, with those at positions 33 and 55 located in the anticodon and T loops and characteristically stabilized by base–phosphate interactions. In yeast tRNA^{Phe}, a π_1 turn is found at the 3' end of U_{47} and a π_2 turn is observed at G_{15}.

π_3 **Turns in T and anticodon loops are stabilized by base–phosphate stacking interactions.** In these two loops, a characteristic, virtually identical folding of the polynucleotide chain is observed (Figure 15-6) (1055). In both cases, the π_3 turn occurs at the 3'-phosphates of an invariant nucleotide, U_{33} or Ψ_{55}. These bases are stacked with the bases immediately preceding in the helically arranged sequence, the semi-invariant C_{m32} and the invariant T_{54}. The following nucleotides, G_{m34} and C_{56}, however, are displaced by the π_3 turn with α in ap and ζ in $-sc$. This distortion brings the phosphates attached to their $O_{3'}$ positions into stacking interaction with the preceding bases U_{33} and Ψ_{55} (1059). In addition, a hydrogen bond between N_3H of U_{33} and oxygen of phosphate attached to $O_{3'}$ of the second-next nucleotide down the chain, A_{35}, is formed and contributes to the stability of this arrangement. Because this kind of folding is associated with semi-invariant and invariant bases, it is expected to be a structural invariant in all class I tRNAs.

15.5 Some Stereochemical Correlations Involving Torsion Angles χ, δ and α, ζ

Stereochemistry of tRNA nucleotides has been extensively treated in several articles (87,396,1032,1046,1047,1058–1060). Here, only some of the most characteristic features associated with nucleotide conformation in tRNA are discussed.

Overall structural features represented as a conformational wheel. If conformation angles α to ζ and χ are plotted as a function of nucleotide sequence in a circular form, Figure 15-8 is obtained (1061,0162).

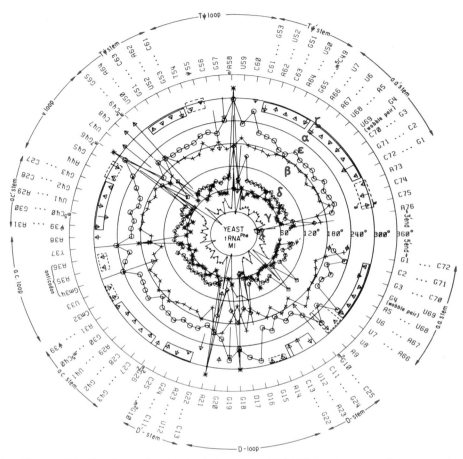

Figure 15-8. Conformational wheel for yeast tRNA^Phe in the monoclinic crystal form, drawn on the basis of coordinates published in (1058). On the periphery of the wheel, the nucleotide sequence is written with the Watson–Crick-type base-pairs in the helical stem regions indicated by dots. The wheel is subdivided into concentric circles drawn every 60°. Torsion angles are, reading from the periphery to the center of the wheel: α and ζ connected by arrows with α at the tip and ζ at the base; ϵ, ○ (circle); β, + (plus); δ, ◇ (diamond); γ, * (asterisk) and χ, . (dot), χ is based on torsion angle $O_{4'}-C_{1'}-N-(C_8$ or $C_6)$ and 180° should be added to conform with our nomenclature using $O_{4'}-C_{1'}-N-(C_4$ or $C_2)$. Helical phosphodiester conformations with both α and ζ in the $-sc$ range between 270° and 300° are outlined by solid arches in double and by dashed arches in single-stranded regions. Spikes indicate conformational irregularities and occur in D, V, and T loops. From (1062).

This summary representation suggests that in double- or single-stranded helical regions all torsion angles conform with standard nucleotide geometry as presented in Chapter 4. In the loop regions, however, large deviations occur, which correspond to the radial spikes in the circular plot. Looking more closely, we find that in most cases spikes are at positions where δ, indicating sugar puckering, increases from the preferred 90° range for $C_{3'}$-endo to the 140° range typical for $C_{2'}$-endo. This change is associated with concomitant variations in glycosyl torsion angle χ, now adopting in some cases a *high-anti* ($-sc$) conformation, and also with α, ζ, and γ, as shown in Figure 15-9.

Correlation of sugar pucker, orientation γ about $C_{4'}$–$C_{5'}$ bond, and P–O torsion angles α and ζ. The entries in Figure 15-9 were derived from geometries of diribose triphosphate fragments occurring in yeast tRNA[Phe] and cast into generalized form with the aid of theoretical calulations (1063). If puckering modes of one or both of the riboses in these fragments are $C_{2'}$-endo (that is, ^{2}E–^{3}E, ^{2}E–^{2}E, or ^{3}E–^{2}E), P–$O_{3'}$ torsion angles ζ of the central phosphate preferentially adopt the 120°, 180°, and 240° ranges, respectively. In helical regions with sugar puckerings confined to the ^{3}E–^{3}E mode, ζ is pushed into the typical 270° range. In case of the P–$O_{5'}$ torsion angle α, the orientation is more nearly correlated with the $C_{4'}$–$C_{5'}$ rotation γ, a $+sc$ conformation restricting α to the 270° range, whereas ap or $-sc$

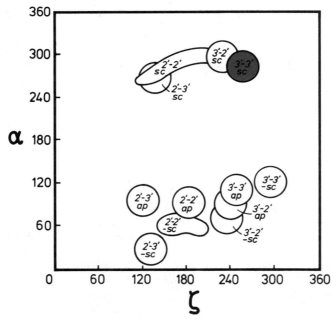

Figure 15-9. Correlation of P–O torsion angles α and ζ with sugar pucker ^{2}E, ^{3}E, and orientation about the exocyclic $C_{4'}$–$C_{5'}$ bond, γ, in dinucleoside phosphate fragments in tRNA. Puckers of the two pendant riboses are denoted as 2 for $C_{2'}$-endo and 3 for $C_{3'}$-endo, the combination 2–3 designating a ribose(^{2}E)–phosphate–ribose(^{3}E) fragment. Ranges of torsion angle γ are given as $+sc$, $-sc$, and ap. The standard helical configuration is marked by darker shading. Redrawn from (1063).

orientations of γ force α into a $+sc$ range. In contrast to ζ, torsion angle α seemingly is not dependent on sugar puckering but rather on exocyclic conformation γ.

Rules put forward in Chapter 4 for generalized nucleotide conformation also apply for tRNA. Looking at Figure 4-23 for torsion angles describing the polynucleotide backbone and glyosyl torsion χ, we find that on the whole distributions are consistent with the principles derived for mono- and oligonucleotides: the nucleotide stereochemistry is preserved even if the polymer is wound into a structure as complicated as tRNA. It is particularly evident that of the two P–O rotations, α is less restricted than ζ. Of the two C–O torsion angles, β is centered around ap whereas ϵ is pushed to $-ac$ and centered around 235°. Of the two C–C torsion angles within the backbone, γ displays an overwhelming preference for $+sc$, is sometimes in $+ac$, and is in $-sc$ and $-ac$ ranges in only a few cases. In contrast, δ is necessarily restricted to $+sc$ for $C_{3'}$-endo and $+ac$ for $C_{2'}$-endo puckerings, the predominance of the former evident from the pseudorotation angles P. Finally, the glycosyl torsion χ is largely restricted to anti with a few nucleosides in high-anti ($-sc$) form.

It ought to be stressed that the entries in Figures 4-23 and 15-8 are not very precise; they vary slightly from one structure refinement to the next. But the overall distribution remains unaffected even when data for yeast tRNA^Phe in different crystal lattices are compared.

15.6 Metal and Polyamine Cation Binding to tRNA

Since tRNA is a polyanion, it is not surprising to find that it binds metal ions and biological polyamines such as putrescine, spermidine, and spermine, Scheme 15-1, which under physiological conditions occur as polycations (1064,1065).

Binding of divalent metal ions is cooperative. What is surprising, however, is that the presence of certain cations is a necessary requirement for the stabilization of the native structure of tRNA (1027,1039,1066). Moreover, for divalent cations, strong and weak binding sites exist. This can be

---

Scheme 15-1. Chemical Formulas of Some Biological Polyamines at Physiological Conditions.

Putrescine $H_3N^+-(CH_2)_4-N^+H_3$

Spermidine $H_3N^+-(CH_2)_3-N^+H_2-(CH_2)_4-N^+H_3$

Spermine $H_3N^+-(CH_2)_3-N^+H_2-(CH_2)_4-N^+H_2-(CH_2)_3-N^+H_3$

shown by a variety of different methods: equilibrium dialysis, nuclear magnetic resonance, fluorescence quenching, and calorimetry (1067–1075).

The strong binding with association constants around 10^5 liters/mole is in general cooperative (1068) and, in the case of Mg^{2+}, is so slow that time constants are in the range of seconds. This is due to the high activation energy, around 30 kcal/mole, involved in displacing tightly bound Mg^{2+} (1074). These data suggest that when Mg^{2+} binds, tRNA folds from a denatured into a native conformation. Monovalent ions can substitute for Mg^{2+}, and, at least for sodium, the process proceeds stepwise in the order random coil→secondary→tertiary structure (Figure 15-10) (1076). The number of strong binding sites is not the same in all tRNAs and varies from 1 in *E. coli* tRNA[fMet] (1067) to 17 in yeast tRNA[Phe], of which only 5 are cooperative (1069). The Mg^{2+} ion can also be effectively replaced by Mn^{2+} (1069,1070) and by other transition metals in various coordination complexes (1071). The weak binding sites with association constants only approximately 10^3 liters/mole (1068) do not display cooperative behavior and are less specific than the strong ones.

Where do Mg^{2+} cations bind? Crystal structure analyses of yeast tRNA[Phe] in monoclinic and orthorhombic modifications show that Mg^{2+} binds as an octahedrally coordinated species $[Mg(H_2O)_6]^{2+}$ to nearly the

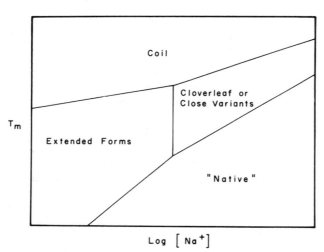

Figure 15-10. Phase diagram for *E. coli* tRNA[Phe] identifying the conformational states corresponding to different combinations of log[Na$^+$] and melting temperature T_m. Below a certain Na$^+$ concentration, tertiary folding of tRNA is abolished. At higher Na$^+$ concentration, the transition tertiary→secondary→coiled structure is observed as the temperature is gradually raised. From Cole, P. E., Yang, S. K., and Crothers, D. M., *Biochemistry 11*:43–58 (1972). Reproduced by permission of the publisher.

same sites in both crystal forms (85,1032,1077–1079). Interactions with tRNA^{Phe} are either via hydrogen bonding to water molecules within the hydration shell of Mg^{2+} or with one to two of these waters replaced by direct coordination to phosphate oxygens (Table 15-1; Figure 15-2). The four hydrated magnesium cations are all located in nonhelical regions and apparently serve to stabilize loops of the tertiary structure, in overall agreement with nuclear magnetic resonance studies (1079,1080).

Binding of transition metals. A number of cations other than Mg^{2+} are able to bind to tRNA. Sodium ion locations cannot be assigned unambiguously from crystal structure analyses. However, the binding sites of several transition metals have been established, these metals being of crystallographic rather than biochemical interest as they are used to solve the phase problem (Chapter 3). The binding of such cations is, in most cases, via coordination to bases as described in Chapter 8. Thus, *trans*-dichlorodiammine platinum(II) coordinates to N_7 and O_6 of G_{34}. Binding to N_7 of guanine is observed in several cases: osmium–bipyridine–G_{30}, Mn^{2+}–G_{20}, and Co^{2+}–G_{15}. In these cases water molecules form additional hydrogen bonds to O_6 keto groups of the guanine and to bases and phosphates of adjacent nucleotides (Figure 8-5) (1077). Binding of $AuCl_2^-$ is simultaneously to N_7 and N_6 of A_{31} and to N_7 and O_6 of G_{m34} (1047). The lanthanides, as outlined in detail for Sm^{3+}, are preferentially attracted by neighboring phosphate groups if these are close enough to form suitable cages (85,1077).

Binding of spermine. The polycation spermine (Scheme 15-1) is necessary for growing large, well-ordered crystals suitable for X-ray diffraction work. Therefore finding spermine associated with yeast tRNA^{Phe} in the crystal lattice was expected and several binding sites are found. The best defined site is located in the major groove near the junction of the anticodon and D stems, with the charged amino group of the spermine spanning the distance from phosphates A_{44}, G_{43}, and G_{42} on one strand to phosphates A_{23} and G_{24} of the complementary strand (Figure 15-2). The spermine molecule is not just inserted into the groove; it has a noticeable ef-

Table 15-1. Binding of Magnesium(II) Cations to Yeast tRNA^{Phe}

Species bound	Location	Direct coordination to	Hydrogen bonded to
$Mg(H_2O)_5^{2+}$	D loop	Phosphate oxygen of G_{19}	Phosphate oxygen of G_{19} and bases G_{20}, U_{59}, and C_{60}
$Mg(H_2O)_4^{2+}$	D loop	Phosphate oxygens of G_{20} and A_{21}	Bulk water
$Mg(H_2O)_6^{2+}$	Turn formed by U_8 to U_{12}	—	Phosphate oxygens U_8, A_9, C_{11}, and U_{12}
$Mg(H_2O)_5^{2+}$	Anticodon loop	Phosphate oxygen of Y_{37}	Bases C_{32}, Y_{37}, A_{38}, and Ψ_{39}

Note. The four sites given here represent strong binding (1079).

fect on its geometry. This is evident from the distance between phosphates on opposite sides of the major groove, 8.6 Å, compared to about 12 Å for the helical groove in the amino acid acceptor/T stem, where no spermine is present to contract the groove. In addition, it appears that this spermine stabilizes the 26° kink between anticodon and D stems (1079).

Another, less well defined spermine is found in the upper part of the D stem, near the sharp bend formed by residues 8 to 12. In this region there is a string of negative charges, with phosphate 10 only about 7 Å from phosphate 47. Spermine is needed for charge neutralization, being close to phosphates 9, 10, and 11 on one strand and to phosphates 45, 46, and 47 of the other (1079). These two spermines were observed in the orthorhombic yeast tRNA^{Phe} crystal form. For the monoclinic variety, additional locations associated with the acceptor stem have been mentioned (85).

For the interaction of an even longer polycation, the naturally occurring protamine AI, see Section 18.3.

15.7 Anticodon Preformed to Allow Rapid Recognition of Codon via Minihelix

In protein biosynthesis, charged tRNA bound to the ribosome must recognize its complementary codon on the messenger RNA through Watson–Crick and wobble base-pairing. Not knowing in detail what the ribosome looks like, can we nevertheless deduce a stereochemical model for this interaction?

Single-helical structure of stacked anticodon triplet. In a hypothesis put forward long before the three-dimensional structure of tRNA was available, it was proposed that the three anticodon bases stack in a single-helical array on the anticodon stem, with invariant U$_{34}$ and semi-invariant (pyrimidine)$_{33}$ looped out and stacked only weakly (579), see Figure 15-11. The structures actually observed for anticodon loops in yeast tRNA^{Phe} (85,86,1032), tRNA^{Asp} (1042), and the tRNAs^{fMet} isolated from yeast (1049) and from *E. coli* (1050), resembles this model remarkably. The basic element of the hypothesis concerning the anticodon structure is, essentially, correct. This particular shape of the anticodon loop appears to be generally preferred; it is preserved even under different crystallization conditions. Additional support comes from studies of binding between tRNAs and complementary nucleotide triplets which show that the anticodon adopts a rather "rigid" structure to facilitate codon–anticodon interaction (1081,1082).

A minihelix formed between codon and anticodon triplets. Continuing the hypothesis, it was proposed that the codon of the messenger RNA forms Watson–Crick and wobble base-pairs with the anticodon. A mini-

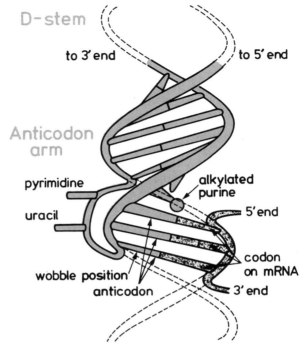

Figure 15-11. Proposed structure of the anticodon–mRNA interaction. Parts of the anticodon are in color; those of the mRNA are in gray. Note good agreement between this hypothetical anticodon structure and the one observed in tRNA crystals (Figures 15-2, 15-3). Redrawn from (579).

helix stacked on the anticodon stem (Figure 15-11) is thus created. Because the wobble base-pair is at the far end of the stack, the anticodon can adjust to achieve optimum base-pairing. In other words, the codon is not forced to adopt unusual conformations to pair with the wobble base, as has also been demonstrated by theoretical considerations (1083). This is in harmony with the expectation that codons should all be in the same structural configuration at the ribosome to allow for geometrical equivalence and rapid readout. The hypermodified, semi-invariant base at position 37 ("alkylated purine" in Figure 15-11), on the other hand, serves to locate the codon in proper register and to prevent misreading due to frame shifting.

Present knowledge indicates that the kink of 26° observed between anticodon and D stems in yeast tRNA[Phe] is not a constant, but rather is a variable feature depending on the tRNA under consideration. This suggests that the anticodon/D junction is used as a hinge point to allow for functional flexibility and proper adjustment of the anticodon during protein biosynthesis (1049).

15.8 The "Molten Architecture" of Yeast tRNA^{Gly}

The three-dimensional structure of yeast tRNA^{Gly} has the same relationship to yeast tRNA^{Phe} as does the molten architecture of Gaudi's famous cathedral in Barcelona to the other cathedrals in Europe: the overall shape as seen from a distance is the same but architectural details differ greatly.

Crystals of yeast tRNA^{Gly} were obtained under rather ususual conditions at 32°C in the presence of about 50% dioxane and after waiting for an average of about 6 months. The presence of Mg^{2+} and polyamines obviously helped to maintain the overall tertiary structure of this tRNA but could not prevent disintegration of the secondary structure, in agreement with spectroscopic studies showing that high concentrations of alcohols and of dioxane denature tRNA (1084).

Because the crystal structure of tRNA^{Gly} is not available at the same high resolution as for other tRNAs, only the structural outlines as illustrated in Figure 15-12 can be given (1051). In a broad sense, the familar L shape is retained in this tRNA and the overall tertiary folding is similar to that of other tRNAs. Residues 18 and 19 in the D loop are in close contact with invariant ψ_{55} and C_{56} in the T loop and form the outer corner

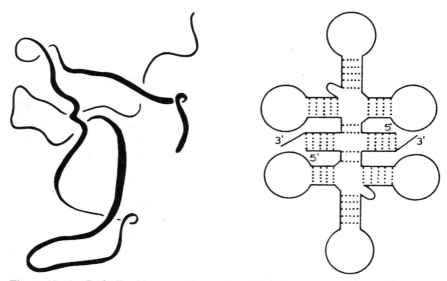

Figure 15-12. (Left) Backbone conformation of the "molten architecture" of yeast tRNA^{Gly}, displayed in the same orientation as the tRNA in Figures 15-2, and 15-3. Note the overall agreement in L-shape and tertiary interactions, contrasting the large differences in secondary structure. (Right) In the crystal lattice, two adjacent, symmetry-related molecules form a dimer with base-pairing between the acceptor stems. Redrawn from (1051).

of the L. The anticodon region also resembles other tRNAs. But the acceptor stem is strikingly different, with the 3' and 5' termini not in helical arrangement. They are unwound and diverge, giving rise to base-paired, double-helical contacts with an adjacent molecule in the crystal lattice, drawn schematically in Figure 15-12. Whether this dimer formation is of the proposed biological significance remains to be proven (1051, 1085).

Summary

Crystal structure analysis of tRNA has provided details of polynucleotide stereochemistry not accessible in other systems. Its primary sequence of 75 to 90 nucleotides contains about 10% minor constituents and can be arranged in the famous cloverleaf form with four double-helical stems and three loop regions. Some of the bases are conserved (invariant) in all tRNA sequences and they are mainly involved in tertiary interactions when tRNA folds into the native L-shaped structure with anticodon and dihydrouracil stems stacked to form one branch of the L, whereas thymidine and amino acid stems produce the other. Among these interactions are stacking, base intercalation, and unusual base-pairing involving even base-triplets. In double-helical stems nucleotides adopt conformations as in A-RNA but in loop regions, sugars are frequently in $C_{2'}$-*endo* puckering; torsion angle γ is sometimes in *ap* or $-sc$ range to allow for stretching of the chain. Base–phosphate interactions stabilize sharp turns called π_3 providing for chain reversal in thymine and anticodon loops. Binding of spermine is in major grooves of stems, with charge–charge interactions to phosphate groups. Divalent cations (Mg^{2+}) bind as $[Mg(H_2O)_6]^{2+}$ in cages formed by phosphate, sugar , and base atoms. The anticodon architecture appears rather rigid and allows rapid recognition with messenger RNA via minihelix. In contrast to tRNA[Phe], tRNA[Gly] crystallized from 50% dioxane displays some unusual tertiary foldings yet the main characteristics of the L shape are unaffected.

Chapter 16

Intercalation

DNA, the genetic material in living cells, can interact with certain classes of drugs, carcinogens, mutagens, and dyes all of which are characterized by extended (hetero)-cyclic aromatic chromophores. Owing to DNA's central role in biological replication and protein biosynthesis, modification by such interaction greatly alters cell metabolism, diminishing and in some cases terminating cell growth (1086–1090). These properties have generated great interest in such molecules during the past three decades. Applications in medicine have been found, and these compounds are extensively used in laboratory studies of DNA structure and function.

For one category of these DNA-active substances, the interaction with DNA is by means of chemical modification, guanine being the prime target (1091). Another category binds to the DNA double helix. This occurs either at the periphery or, in cases of drugs shown in Figure 16-1, by means of intercalation between adjacent base-pairs. This is accomplished without disrupting Watson–Crick hydrogen bonding. Since the stereochemistry of intercalation is rather well understood and of greater structural interest, we will focus on its characteristics (1089, 1090).

16.1 General Phenomena of Intercalation into DNA and RNA Double Helices

Intercalation changes the physical properties of a double helix. The first suggestion of the intercalation of planar aromatic molecules between and parallel to adjacent base-pairs was made on the basis of hydrodynamic and X-ray fiber diffraction studies of DNA in the presence of acridine dyes (1092). If an acridine dye is added to DNA and fibers are drawn for X-ray diffraction studies, patterns are obtained in which reflections due to regular helical structure are blurred or lost except for (i) equatorial reflections indicative of lateral molecular packing and (ii) the strong 3.4 Å meridional reflection caused by base-pair stacking (1092–1094). In essence, the aromatic 3.4 Å-thick drug molecules slide between base-pairs without disturbing the overall stacking pattern. However, because base-pairs must separate (unstack) vertically to allow for intercalation, the sugar–phosphate backbone is distorted and the regular helical structure is destroyed (Figure 16-2). According to such a model, the DNA must lengthen with increasing amounts of added drug. This is indeed observed as can be monitored by enhanced viscosity and the diminution of the sedimentation coefficient, effects which also suggest overall stiffening of the DNA duplex (1092).

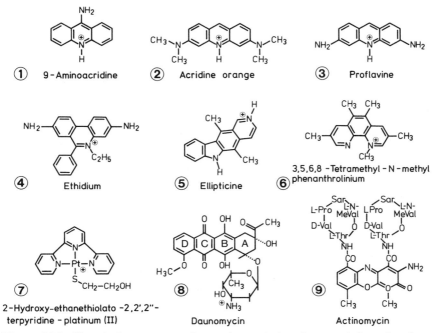

① 9-Aminoacridine   ② Acridine orange   ③ Proflavine

④ Ethidium   ⑤ Ellipticine   ⑥ 3,5,6,8 -Tetramethyl - N -methyl phenanthrolinium

⑦ 2-Hydroxy-ethanethiolato -2,2',2"- terpyridine - platinum (II)   ⑧ Daunomycin   ⑨ Actinomycin

Figure 16-1. Chemical structures of some intercalating drugs and dyes that form crystalline complexes with oligonucleotides and have been subjected to X-ray analysis.

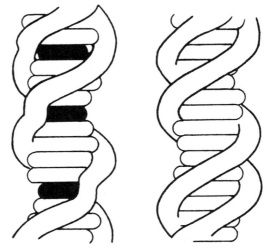

Figure 16-2. Intercalation of planar drug molecules (dark areas) into DNA double helix. Note that regular sugar–phosphate backbone (right) becomes distorted at intercalation sites (left). From (1089).

A two-stage binding process. Because of the chromophoric systems characteristic of intercalating agents, their interaction with DNA and RNA can be easily followed by spectroscopic methods (1095), which yield both the thermodynamics and kinetics of binding (1096,1097). A two-stage, anticooperative process was established from nonlinear Scatchard plots and from the kinetics. The kinetics show that intercalation in the millisecond time range is preceded by fast, diffusion-controlled binding at the outside of the double helix. These two binding modes are also distinguished by association constants which indicate tighter binding for intercalation than for outside aggregation (1095–1097).

The nearest-neighbor exclusion principle. Measurement of the lengthening of phage T2 DNA in the presence of proflavin shows that only 44% of potential intercalation sites between base-pairs are occupied by drug molecules (1098). This implies that every second (next neighbor) intercalation site along the DNA double helix remains empty, probably because nucleotides flanking the intercalator are geometrically distorted. Support for nearest-neighbor exclusion was also provided by X-ray fiber diffraction of a complex between DNA and the platinum-containing intercalative drug shown in Figure 16-1 (1099,1100). Doubts as to the general validity of the nearest-neighbor exclusion principle have been raised based on studies of bifunctional intercalators (1101); these, on the other hand, might well have structural peculiarities not relevant to monofunctional analogs (1102).

Intercalation causes DNA unwinding. The separation of base-pairs to

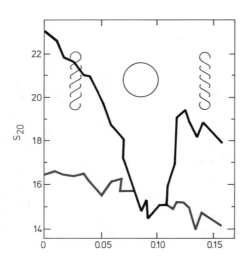

Daunomycin molecules bound/nucleotide

Figure 16-3. Effect of daunomycin on the sedimentation coefficient s_{20} of circular ϕX174RF DNA. Colored curve: nicked, noncircular DNA. Black curve: circular DNA with right-handed (left), no (center), and left-handed (right) superhelical twist. From (1103).

make room for the intercalator can be visualized as a combination of pulling along the B-DNA double-helix axis and left-handed unwinding in order not to break the sugar–phosphate backbone (1093). Experimental evidence for unwinding comes from intercalation of a series of drugs into circularly closed plasmid DNA (Chapter 19), a ring-shaped double-helical DNA further wound into right-handed superhelix. If a drug intercalates, the unwinding produces a 10° to 20° left-handed twist per intercalated molecule which counteracts the right-handed superhelix until balance is reached with a purely circular DNA. With the further addition of intercalator, unwinding continues and leads to a left-handed superhelix. The process is easily followed by sedimentation measurements (1103) (Figure 16-3).

Crystallographic analyses visualize the geometry of intercalation. The experiments described above suggested that planar, aromatic molecules are inserted between DNA and RNA base-pairs. Direct evidence of this complex formation and evaluation of the steric factors involved come from crystallographic studies. These have been summarized in several articles which also comment on the biological implications (1089,1090, 1104–1107,1012).

16.2 Stereochemistry of Intercalation into DNA- and RNA-Type Dinucleoside Phosphates

Information on intercalation geometry relevant to DNA and RNA double helices can be obtained from drug molecules inserted between at least two Watson–Crick-type base-pairs. Only such systems will be considered here—other systems with non-intercalative drug-base stacking and non-Watson–Crick base-pairs are not treated (1108–1110). In all cases (except daunomycin intercalated into a hexadeoxynucleotide (Section 16.3), the DNA or RNA component is a dinucleoside monophosphate. Therefore end effects which might lead to unjustified generalizations must be considered. Also, the majority of intercalator complexes are of the RNA type and extrapolation to DNA is questionable, especially if we recall that, structurally, the DNA double helix is quite flexible whereas RNA is fixed in the A-family geometry.

The specific pyrimidine–3′,5′-purine sequence (1111). In all known crystalline complexes of intercalators inserted between Watson–Crick pairs formed by self-complementary dinucleoside monophosphates for both the ribose and deoxyribose series, the nucleoside at the 3′ side of the phosphodiester link is always a pyrimidine and the one on the 5′ side a purine (Figure 16-4; Table 16-1). If this sequence is reversed, either no crystals are formed or the complex adopts a nonhelical structure (1110). The same pyrimidine–3′,5′-purine sequence was found even in the non-

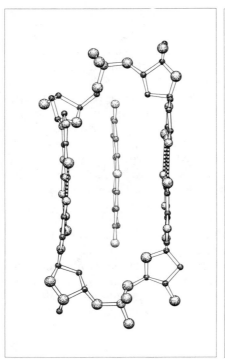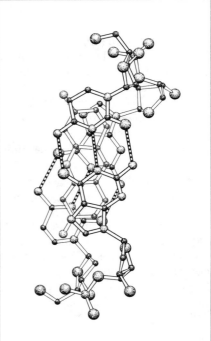

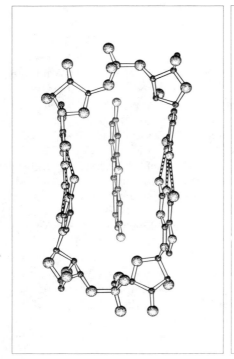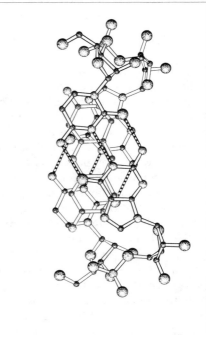

complementary CpA·proflavin complex formed by a parallel double helix with drug intercalated between A–A and C–C base-pairs (1109).

This particular sequence dependence for successful intercalation is also observed in aqueous solution using nuclear magnetic resonance spectroscopy (1112, 1113) and was further interpreted by means of theoretical considerations (1114–1118,1118a). These showed that, in DNA as well as in RNA, intercalation into the pyrimidine–3′,5′-purine sequence is 7 to 13 kcal/mole more favorable than purine–3′,5′-pyrimidine intercalation because base–intercalator overlap is more pronounced in the pyrimidine–(3′-5′)–purine sequence than in the reversed one (1119). This intermolecular interaction is supported by *intra*molecular electrostatic forces rendering pyrimidine–3′,5′-purine sequences more prone to intercalation (1115). In addition, stereochemically unfavorable interactions may play a role if the reversed base sequence is employed (1115,1117).

Characteristic sugar puckering distribution associated with intercalation. Of the 11 complexes listed in Table 16-1, *all* have the purine sugar at the 5′ position in $C_{3'}$-*endo* pucker. In most of them, the pyrimidine sugar at the 3′ end is $C_{2'}$*endo* with only a few cases of $C_{3'}$-*endo* found. The exceptions indicate that suggestions concerning a general $C_{3'}$-*endo*–3′,5′-$C_{2'}$-*endo* mixed puckering scheme (1011) are untenable and that the sugar puckering, except at the 3′ side, is a parameter not necessarily responsible for the opening of the base-pairs. This finding is further supported by characteristic changes in torsion angles χ and β (1120).

In Table 16-2, average torsion angles in DNAs, in RNA, and in minihelical dinucleoside monophosphates are compared with those observed in intercalator complexes. It is evident that, by and large, torsion angles are similar except for χ at the 3′ end and β. Upon intercalation, both χ and β increase by over 50°, χ being pushed into *high-anti* ($-sc$) range (1107,1120). Using computerized model building it can be shown that, starting from a Watson–Crick-type minihelix, these changes are sufficient to bring about the 3.4 Å-base-pair separation required for drug intercalation (Figure 16-5).

Unwinding and base-pair turns. Intercalation into circularly closed, superhelical DNA shows that this process is coupled with DNA unwinding, the unwinding angle being measured as the change in angle between $C_{1'} \cdots C_{1'}$ lines projected onto a mean plane. In dinucleoside monophosphate–intercalator complexes, no clear insight into characteristic unwinding angles is apparent (Table 16-1), probably because end effects obscure the picture. The daunomycin–hexadeoxynucleotide complex further supports this conclusion (Section 16.3). Model building studies

Figure 16-4 (opposite). Results of crystal structure analyses of complexes between proflavin dCpG (top), and CpG (bottom) shown in side and top views. These two pictures are representative of most dinucleoside phosphate–drug complexes entered in Table 16-1. Drug molecules are drawn in color; hydrogen atoms are omitted. For graphical details, see the legend to Figure 10-1. References are in Table 16-1.

Table 16-1. Conformational and Unwinding Angles for Dinucleoside Phosphate Intercalation Complexes and for L-DNA

Complex	α P–O$_{5'}$	β O$_{5'}$–C$_{5'}$	γ C$_{5'}$–C$_{4'}$	δ[a] C$_{4'}$–C$_{3'}$	ϵ C$_{3'}$–O$_{3'}$	ζ O$_{3'}$–P	$\chi(5')$[b] C$_{1'}$–N	$\chi(3')$[b] C$_{1'}$–N	Sugar pucker 5' end	Sugar pucker 3' end	Unwinding angles (°)	References
(a) Deoxyribonucleoside series												
Proflavin·dCpG	290	219	46		210	290	−164	−100	^3E	^3E	17	(1107)
	287	218	73		203	300	−170	−67	^3E	^2E		
Terpyridine–Pt·dCpG[c]	282	226	57		201	287	−148	−66	^3E	^2E	23	(1107, 1141)
	308	217	84		194	292	−146	−63	^3E	^2E		
DNA·[(bipy)Pt(en)]$^{2+}$, L-DNA	82	198	180	76;147	276	98	26	167	^3E	^2E	36	(1100)
	300	227	221	76;147	171	125	26	167	^3E	^2E		
(b) Ribonucleoside series												
Proflavin·CpG	287	234	53	75	204	292	−162	−93	^3E	^3E	0[d]	(1120)
	288	237	50	79	211	301	−171	−75	^3E	^2E		
Acridine orange·CpG	297	226	40		225	298	−172	−65	^3E	^2E	23	(1142)
Ethidium·5-iodoCpG	286	210	72	87;131	226	281	−151	−79	^3E	^2E	26	(1143)
	291	224	55	84;134	225	291	−156	−71	^3E	^2E		
Ethidium·5-iodoUpA	291	236	52	98;133	207	286	−154	−81	^3E	^2E	26	(1119)
	276	230	70	95;118	218	302	−166	−80	^3E	^2E		
9-Aminoacridine·5-iodo-CpG	280	220	67	72;110	236	311	−163	−61	^3E	^3E	26	(1144)
	300	229	38	85;130	216	296	−142	−81	^3E	^2E		
	295	208	58	99;106	216	310	−163	−97	^3E	^3E	23	
	294	222	45	105;156	209	287	−160	−78	^3E	^2E		

Proflavin·5-iodo-CpG	305	273	3	102;122	224	294	−176	−77	^3E	^3E	
	323	206	53	79;119	273	273	−161	−95	^3E	^3E	−3^e
Acridine orange·	291	235	58	84;145	217	294	−161	−85	^3E	^2E	23 (1132)
5-iodo-CpG	299	228	60	81;139	226	284	−165	−90	^3E	^2E	
Ellipticine·5-iodo-CpG	281	214	65	91;145	199	285	−155	−100	^3E	^2E	22
	315	194	47	104;141	234	258	−166	−106	^3E	^2E	
TMP·5-iodo-CpG^f	283	219	71	93;141	203	278	−158	−94	^3E	^2E	22 (1145)
	303	206	55	95;143	212	285	−167	−111	^3E	^2E	

Note. For formulas of intercalators, see Figure 16-1; L-DNA is (DNA·[(bipy)Pt(en)]$^{2+}$).

^{a} First value for pyrimidine at 3'-end, second for purine at 5' end.

^{b} If defined as $O_{4'}-C_{1'}-N-C_6/C_8$ in original literature, present values were obtained by subtracting 180°.

^{c} Terpyridine–platinum is 2,2';2',2''-terpyridine–platinum(II).

^{d} Instead of unwinding, base-pairs are disposed laterally by about 1.8 Å.

^{e} Overwinding instead of unwinding, base-pairs are disposed laterally by about 1.3 Å.

^{f} 3,5,6,8-Tetramethyl-N-methylphenanthrolinium.

Table 16-2. Comparison for Average Conformation Angles Observed in Some Dinucleoside Monophosphate–Intercalator Complexes, in Dinucleoside Phosphates, and in A-RNA, A-DNA, and B-DNA

	α	β	γ	δ	ϵ	ζ	$\chi(5')^a$	$\chi(3')^a$	Sugar pucker	
									5' end	3' end
Averaged data for intercalator complexes	290(8)	225(9)	59(14)		211(11)	293(7)	−162(10)	−76(12)	$C_{3'}$-endo	$C_{3'}$-endo $C_{2'}$-endo
Dinucleoside monophosphates	290(6)	174(7)	57(7)		217(5)	289(4)				
A-RNA	300	175	49	83	213	281	−166	−166	$C_{3'}$-endo	
A-DNA	270	211	47	83	175	315	−153	−153	$C_{3'}$-endo	
B-DNA	321	209	31	157	159	261	−95	−95	$C_{2'}$-endo	

Note. For individual data in intercalator complexes, see Table 16-1. Standard deviations are given in parentheses.
Source. From (1107).
^{a} Defined as $O_{4'}$–$C_{1'}$–N–C_6/C_8 in original literature. Present values were obtained by subtracting 180°.

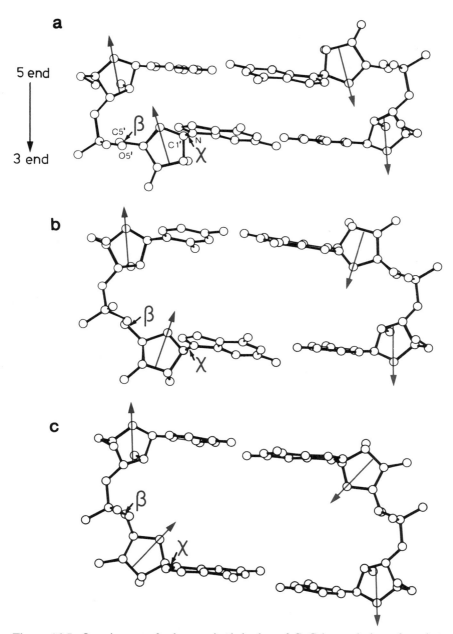

Figure 16-5. Opening up of a base-paired duplex of CpG by variation of torsion angles χ and β. **(a)** A-RNA geometry with $\chi = -166°$ and $\beta = 175°$; **(b)** conformation with $\chi = -140°$ and $\beta = 200°$; **(c)** opened duplex with $\chi = -100°$ and $\beta = 225°$. Colored arrows passing through $O_{4'}$ and bisecting line $C_{2'}–C_{3'}$ indicate orientations of sugar moieties. They are pairwise parallel in (a) due to helical configuration but differ largely in (c), lending support to the nearest-neighbor exclusion principle which requires geometrical distortions in base-pair stacks adjacent to intercalation site. From (1120).

suggested that the unwinding angle is dependent on a combination of small variations in backbone torsion angles and base-pair geometry expressed as bend and twist, and *not* just on sugar puckering (1120,1121). It is clear, however, that a correlation exists between unwinding angle and shape of intercalator agent (1122).

16.3 Improving the Model: The Daunomycin–d(CpGpTpApCpG) Complex

To avoid end effects, longer fragments of nucleic acids must be used. In the case of the crystalline daunomycin · d(CpGpTpApCpG) complex, this goal was achieved only in part. In the duplex formed by the self-complementary hexadeoxynucleotide, daunomycin (for formula, see Figure 16-1) is intercalated between the terminal CpG sequences. Since, however, adjacent molecules are stacked in continuous helical arrangement end effects could well be negligible (1123).

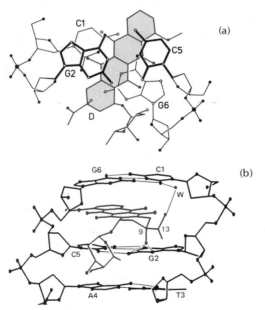

Figure 16-6. Two views of the daunomycin–d(CpGpTpApCpG) complex, looking **(a)** perpendicular to the base-pair planes and **(b)** approximately along base-pairs, daunomycin in color. In (b), only half the complex is drawn because the other half is related by a dyad located below the A_4–T_3 base-pair. Note hydrogen-bonding interactions between hydroxyl at C_9 of daunomycin and G_2 and between acetyl keto group C_{13} and O_2 of cytosine C_1, mediated by water molecule W. In (a), base-pair G_2–C_5 closer to the viewer is drawn heavier than C_1–G_6, further away. From (1123).

Overall geometry of daunomycin intercalation differs from that of other drugs. The daunomycin–hexadeoxynucleotide geometry (Figure 16-6) shows that the drug chromophore is inserted head-on between adjacent base-pairs, in contrast to all other intercalator complexes where edge-on alignment is observed. Ring D of daunomycin protrudes into the major groove, and ring A with the attached sugar moiety occupies the minor groove, an arrangement suggested earlier on the basis of spectroscopic data (1124–1126). In addition to stacking energies, the complex is stabilized by hydrogen-bonding interactions involving hydroxyl and carbonyl groups at C_9 of the daunomycin chromophore, the hydroxyl group being essential for activity (1127,1128). The daunomycin sugar moiety nearly covers the minor groove and this space filling suggests why not more than one daunomycin per three base-pairs is intercalated. The positively charged amino group at sugar $C_{3'}$ of daunomycin is not involved in binding to the oligonucleotide, in contrast to suggestions based on spectroscopic data (1125).

Evidence for long-range distortions. By and large, the hexadeoxynucleotide in this complex adopts a B-DNA-like structure. However, only the central two A–T base-pairs are precisely arranged as in B-DNA. Those next and second next to the intercalation site are displaced, the former by translation to the major groove side and shift of the helix axis by about 0.4 Å, the latter by 8° rotation in an unwinding mode. This unwinding, which is 0° at the immediate site of intercalation, is coupled primarily with changes of $C_{3'}-O_{3'}$ and $O_{3'}-P$ torsion angles ϵ and ζ. The unwinding angle of 8° is in overall agreement with the 11° estimated from experiments employing circular DNA (1103,1129).

The conformation of the backbone does not follow simple rules. Broadly, the overall sugar puckering of the hexadeoxynucleotide falls into the $C_{2'}$-*endo* range characteristic of B-DNA, with individual conformations (counting from deoxyribose 1 to 6) $C_{2'}$-*endo*, $C_{1'}$-*exo*, $O_{4'}$-*endo*, $C_{2'}$-*endo*, $C_{2'}$-*endo*, $C_{3'}$-*exo*. Associated with these variations are χ angles in a wide range: $-153°$, $-93°$, $-131°$, $-107°$, $-82°$, $-85°$ (the rather low value for the first residue is probably due to end effects).

Is daunomycin sequence specific? The hydrogen-bonding scheme between daunomycin and the CpG unit, (Figure 16-6) suggests that the drug could just as well be inserted into other sequences provided that N_3 of guanine is replaced by another hydrogen bond acceptor (N_3 of purine or O_2 of pyrimidine). The water-mediated interaction between the keto group and the O_2 of cytosine can also occur with other bases. This insensitivity to base sequence is coupled with a specificity for right-handed helices brought about by the asymmetric attachment of substituents at C_7 and C_9 of daunomycin. Therefore, it appears that daunomycin does not necessarily recognize specific sequences but needs only the right-handed double-helix geometry for intercalation.

16.4 Model Building Studies Extended to A- and B-DNA

Because of the inherent uncertainties due to end effects in dinucleoside monophosphate–intercalator complexes, models with polymer DNA in both A and B form were developed with the aid of computers, taking as basic assumptions the known stereochemistry of DNA and of the interaction between intercalators and base-pairs (1130,1131).

A-DNA and A-RNA: Base-pair separation without unwinding. In these two polynucleotides of the A family, all sugars can retain their $C_{3'}$-*endo* preference if the double helix is interrupted by an intercalator. Base-pair separation is accomplished mainly by changes in backbone conformation angles ζ and γ into the *ap* range, a process occurring without unwinding and probably modeled by the proflavin · CpG and proflavin · 5-iodo-CpG complexes (Table 16-1) (1120,1132). Because all sugars retain the same pucker and essentially the same orientation as without intercalation (for contrast, see Figure 16-4), it was suggested that intercalation into A-DNA and A-RNA does not obey the nearest-neighbor exclusion principle.

B-DNA is typical of intercalation phenomena. The situation is different in B-DNA, in which again intercalation forces two torsion angles, α and γ, into the *ap* range. In the most favored model (1131), the sugar puckering at the 5' side of intercalation is changed from $C_{2'}$-*endo* to $C_{3'}$-*endo,* in agreement with findings from dinucleoside monophosphate–intercalator complexes. The conformational changes associated with intercalation into B-DNA are not restricted to just the two base-pairs above and below the intercalation site. Rather, changes extend to neighbors farther out. Unwinding by 18° per drug involves over three base-pairs above and three base-pairs below the site of intercalation.

This long-range influence of intercalation is certainly observed in the daunomycin–hexadeoxynucleotide complex and is in harmony with conclusions from the nearest-neighbor exclusion principle. Similar long-range effects prevailing in solution are indicated by ^{31}P nuclear magnetic resonance spectroscopy of actinomycin–hexanucleotide intercalation (1133). Additionally, electric dichroism experiments show that the intercalator plane is not at right angles to the helix axis but rather is tilted by about 20°, thereby also distorting next and second-next base-pairs in a cooperative mode (1134,1135).

16.5 DNA Saturated with Platinum Drug Unwinds into a Ladder to Produce L-DNA

When calf thymus DNA is treated with excess ethylenediamine-(2,2'-bipyridine)platinum(II), a drug belonging to the family of platinum compounds described in Table 8-6 [(bipy)Pt(en)]$^{2+}$ (see Figure 16-1), a com-

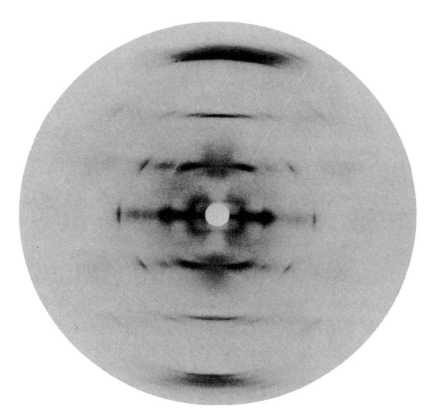

Figure 16-7. X-ray diffraction pattern obtained for calf thymus DNA saturated with intercalating $[(bipy)_2Pt(en)]^{2+}$. Layer-line spacings indicate 10.2 Å repeat along fiber axis, DNA is unwound into ladder-type structure (L-DNA). From (1100).

plex is formed which can be drawn into fibers suitable for X-ray diffraction studies. The pattern obtained is reproduced in Figure 16-7 and displays a unique distribution of X-ray intensities which bears no resemblance whatsoever to the patterns characteristic of double-helical DNAs (Figure 3-9c). The X-ray diagram shows a simple motif with layer-line separation indicating a 10.2 Å repeat unit along the fiber axis.

A ladder-like DNA with conformational features reminiscent of Z-DNA. Analysis of the X-ray pattern shows that with saturation level intercalation, the DNA double helix is completely unwound into a linear, duplex arrangement similar to a ladder and therefore named L-DNA (1100) (Figure 16-8). Bases are still hydrogen bonded in Watson–Crick fashion and are stacked at 3.4 Å distances, yet an intercalator molecule is inserted into every second, alternate site. In order to provide for the required base–base separation, nucleotides adopt quite unusual conformations

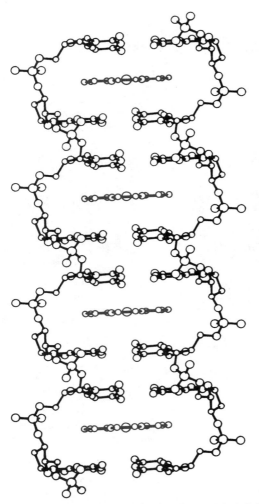

Figure 16-8. Structure of calf thymus DNA fully saturated with $[(bipy)_2Pt(en)]^{2+}$; see Table 8-6 for formula of drug. The DNA double helix is unwound and degenerated into a ladder-type duplex. In agreement with the nearest-neighbor exclusion principle, only every second potential binding site is filled with intercalator (shown in color). From (1100).

reminiscent of CpG steps in Z-DNA (Chapter 12) rather than A- or B-DNA (Table 16-1). Thus, in each Watson–Crick base-pair of L-DNA, one base is in *syn,* the partner in *anti* orientation. The *syn* base is associated with $C_{3'}$-*endo* sugar pucker, the *anti* with $C_{2'}$-*endo*. In Z-DNA a strict purine-*syn*, pyrimidine-*anti* correlation exists which also appears probable for L-DNA as far as stereochemistry is concerned. As there is no net winding in the ladder-type L-DNA, some torsion angles, involving especially those around $O_{3'}(\epsilon,\zeta)$ and around the glycosyl bond, χ, adopt unusual values differing from those observed in the CpG unit of Z-DNA (Table 16-1).

16.6 Actinomycin D: An Intercalator Specific for the GpC Sequence

Actinomycin D consists of a chromophoric phenoxazone ring system linked with two cyclic pentapeptides containing the unusual amino acids sarcosine, L-methylvaline, and D-valine (Figure 16-1). The antibiotic has been shown by biochemical studies to intercalate into DNA at sites containing specifically GpC (and not CpG) sequences. The 2-amino group of G is required for binding (1136,1137). This GpC specificity clearly contrasts the general pyrimidine–(3',5')-purine sequences required for intercalation of other drugs and dyes collected in Table 16-1. Why is actinomycin different?

The crystalline deoxyguanosine–actinomycin D complex shows specific peptide–base hydrogen bonds. The overall molecular structure of the antibiotic shows dyad symmetry with cyclic peptides in near-eqivalent conformation and held in register by hydrogen bonds between D-valine residues (1136,1138) (Figure 16-9). The dyad symmetry is even maintained in the complex because one deoxyguanosine is stacked "above" and one "below" the phenoxazone ring system. Specific hydrogen bonds between actinomycin and the peptides involve N_3 and amino N_2 of guanine and the peptide group of L-threonine. Obviously, this interaction successfully overrides the pyrimidine–(3',5')-purine sequence rule observed for more simple intercalators.

By adding deoxycytidines to both deoxyguanines in Watson–Crick-type hydrogen bonding and by simultaneous stacking with the phenoxazone ring system, it is possible to computer generate an actinomycin D–GpC complex. This is easily extended into an actinomycin–DNA intercalation, with the peptide rings fitting into the minor groove of the B-DNA double helix. Even dyad symmetry is maintained (1136,1139). This construction of a symmetrical DNA–peptide complex has led to several hypotheses concerning more elaborate protein systems interacting with DNA which in many cases are able to satisfy the dyad symmetry of a DNA double helix (1140).

Summary

Interaction of DNA with drugs is of particular pharmacological importance. Considered here are only intercalative complexes with planar drugs. Binding proceeds in two steps, the first one outside the helix and

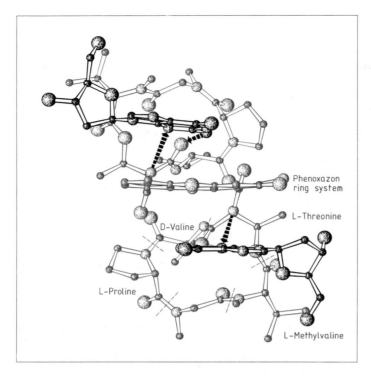

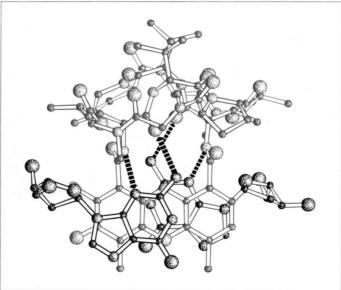

Figure 16-9. Structure of the actinomycin D–deoxyguanosine complex, with nucleosides drawn in black. Note specific hydrogen bonds between guanine and L-threonine residues. Drawn with coordinates published in (1138); for graphical details see the legend to Figure 10-1.

the second accounting for intercalation, with concomitant helix unwinding, lengthening, and general stiffening. According to the nearest-neighbor exclusion principle only about half of all potential sites are filled by drug molecules. Crystalline dinucleoside monophosphate–drug complexes all have the pyrimidine–3′,5′-purine sequence. Separation of bases to receive the drug is accomplished by a change in sugar puckering or by rotation of χ and β torsion angles. In the complex between daunomycin and a hexanucleotide long-range distortions are found, with some drug–DNA hydrogen bonds conveying specificity of binding. Planar drugs containing Pt chelated by bipyridyl deform DNA into a ladder (L-DNA) with probably every Watson–Crick base-pair displaying one nucleotide in *syn* and the other in *anti* form, reminiscent of left-handed Z-DNA. High specificity is observed for actinomycin D which binds exclusively at d(GpC) sequences because then favorable drug–DNA hydrogen bonds can be formed.

Chapter 17

Water and Nucleic Acids

Throughout this text, the importance of water surrounding nucleic acids has been emphasized. Water is not just a medium to keep the solutes dissolved. It interacts and, in the case of macromolecules, is mainly responsible for the stabilization of secondary and tertiary structure (525,1146–1148). This holds for proteins and even more so for nucleic acids because phosphate $\cdots$ phosphate electrostatic repulsion is diminished by the high dielectric constant of water and hydrated counterions. Moreover, the bases self-assemble into ordered structures, and this is partly due to hydrophobic forces which again involve the active participation of water molecules. The degree of hydration of DNA plays a key role in its conformation; high relative humidity favors the B form and reduced humidity or increased ionic strength leads to a transition from the B form to C-, A-, and, if sequence permits, D- and Z-DNA.

17.1 Experimental Evidence for Primary and Secondary Hydration Shells around DNA Double Helices

Γ is a measure of hydration. The solvation or net hydration of a macromolecule is given by Γ, which in DNA is a parameter indicating the number of moles of water per mole of nucleotide. The secondary structure of DNA is intimately related to Γ (1149,1150) and the latter is directly correlated with water activity a_w which is lowered if the salt concentration is raised (1151). Although the effect of cations on water activity a_w is given by ionic strength and largely independent of the nature of the cation, different cations have a pronounced effect on DNA secondary structure, indicating a somewhat specific DNA–cation interaction (1150).

Two hydration shells around DNA double helices. Experiments involving sedimentation equilibrium studies (1151–1155), isopiestic measurements (1156), gravimetric (1157) and infrared spectroscopic investigations (1158–1160), and X-ray fiber diffraction (Chapter 9) lead to the conclusion that DNA is heavily hydrated. The hydration is not homogeneous around the DNA polyelectrolyte molecule and can be described in terms of two discrete layers representing primary and secondary hydration shells.

The primary hydration shell is not ice-like and is impermeable to cations. According to the scheme described in Figure 17-1, primary hydration in double-helical DNA consists of at least 11 to 12 water molecules per nucleotide. These water molecules are grouped into three classes with de-

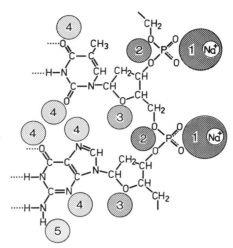

Figure 17-1. Preferred hydration sites in B-DNA. Numbers 1 to 5 indicate strength of binding, in decreasing order. Around phosphate groups, about five water molecules are found. Adapted from (1158).

creasing binding affinity for phosphate, phosphodiester plus sugar, and base. Infrared spectroscopic data for DNA at less than 65% relative humidity have been interpreted as hydration at phosphate oxygens, with 5 to 6 water molecules adsorbed per nucleotide. Below 60% relative humidity, phosphodiester and furanose $O_{4'}$ oxygens are also partly hydrated. Hydration of functional amino, imino, and keto groups of bases occurs above 65% relative humidity with the addition of 8 to 9 more water molecules. At about 80% relative humidity primary hydration of the DNA double helix is complete with about 20 water molecules per nucleotide. Further increase in hydration is accompanied by swelling of the sample, in agreement with X-ray fiber diffraction data (1161) (Figure 17-2).

The primary hydration shell of about 20 water molecules per nucleotide is different from bulk water, as indicated by infrared spectroscopic characteristics. The 20 water molecules are not all in direct contact with DNA, but 8 to 9 of them are bonded to 11 or 12 other water molecules in the inner, primary hydration shell. The latter is impermeable to cations (1155) and upon cooling to temperatures well below 0°C, this water does not freeze into an ice-like state (1160) (Figure 17-3). Instead of the regular structure of ice I (characterized by fused six-membered rings) other, more or less ordered structures of hydration shells are probably present (Section 17.5).

The second hydration shell is indistinguishable from bulk water as far as permeability to ions and crystallization into ice I is concerned. Its structure could, however, be slightly different from the bulk water located further away from the polyelectrolyte DNA because equilibria of the Donnan type could come into play (1151).

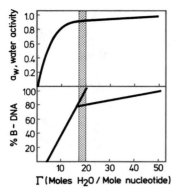

Figure 17-2. Two similar plots concerning net hydration, Γ, of DNA. (Top) Water activity is changed by different Cs^+-containing salts, and the Γ dependence suggests that with about 20 moles H_2O/mole nucleotide, hydration is essentially complete for the Cs^+ salt of DNA. Redrawn from (1153). (Bottom) Dependence of B-DNA content on net hydration, Γ, the latter adjusted with different alkali-chloride concentrations. Γ values for $B = 0\%$ and 100% indicate that minimum hydration of double-helical DNA (other than B type) is 3.6 water molecules per nucleotide and formation of B-DNA is complete if about 20 water molecules per nucleotide are present. If more water is added, the sample swells but is conformationally stable. Simplified after (1151).

17.2 Different Hydration States Associated with A-, B-, and C-DNA

In essence, the B form of DNA exists at high activity of water where cations do not influence and perturb the hydration and where the primary hydration shell consisting of about 20 water molecules per nucleotide is

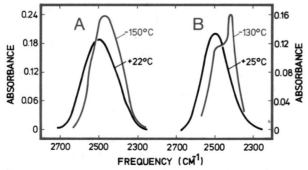

Figure 17-3. Two samples of DNA retained at 76% (A) and 86.5% (B) relative humidity of HDO show different OD stretching bands in infrared spectra. (A) About 9 water molecules are bound per nucleotide, constituting the innermost water layer of the first hydration shell, nonpermeable for cations and resisting ice formation. (B) About 14 water molecules are adsorbed per nucleotide of which about 5 are contained in the outer water layer, more loosely bound and subject to crystallization into ice I, indicated by a typical band at 2410 cm^{-1}. From (1160).

still intact. If the relative humidity in DNA fibers or films is reduced, or if the salt concentration of DNA solutions is raised, the hydration of DNA, Γ, is lowered. As a consequence, at a certain threshold corresponding to 20 water molecules per nucleotide, a sharp transformation from B-DNA into C and/or A forms is observed, depending in part on the nature of the cations involved (Figure 17-2). The B→C transition is continuous (1151), in agreement with the structural similarity of the two states. One should, however, expect that B→A or C→A transitions occur stepwise and co-operatively owing to the change in sugar pucker type $C_{2'}$-*endo* → $C_{3'}$-*endo*.

Structural transitions in high salt and in alcohol solutions. Transitions from one form of DNA to another can be easily monitored by changes in circular dichroism spectra (Figure 17-4). With the addition of salt, *continuous* structural transitions occur *within* a DNA family, e.g., B→C, and if sufficient salt is added, sharp cooperative transition *between* DNA families is observed e.g., C→A or B→A. If the polarity of the medium is altered by increasing amounts of ethanol, isopropanol, or dioxane up to around 80%, *cooperative* transitions *between* DNA families takes place, viz., B→A or C→A (839).

Specific cation–DNA interactions in 80% methanol. In the B-DNA family DNAs, cations interact nonspecifically in aqueous solution. But in 80% methanol, specificity is observed. The rotation angle between adjacent base-pairs varies from 45° for Cs^+ (D-DNA) to about 33° for Li^+ (B'-DNA), and can be interpreted as spatial fitting of hydrated cations into the narrow groove of the DNA double helix (839,1162).

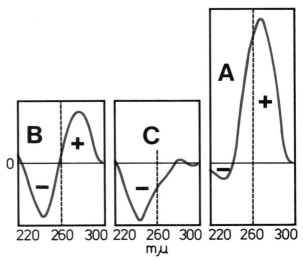

Figure 17-4. Schematic representation of circular dichroism spectra for A-, B-, and C-DNA. Dotted lines indicate positions of ultraviolet absorption maxima. From (839).

B-DNA in aqueous solution is slightly underwound. These circular dichroism studies and an X-ray scattering investigation (1163) indicate that DNA in solution is not the B-DNA with exactly 10 base-pairs per turn derived from X-ray fiber diffraction work (Chapter 10). Rather, an underwound B-DNA form with a smaller rotation angle between successive base-pairs exists, in agreement with theoretical, X-ray crystallographic, and enzymatic digestion studies mentioned in Section 9.2.

Stabilization of B-DNA structure by polyamines. The B-DNA-type double helix can be stabilized in aqueous solution if DNA is complexed with nucleoprotamine or with homopolymer polypeptides like polylysine or polyarginine (1164). These basic polymers are probably arranged in the minor groove and, as discussed for the polyamines spermine and spermidine (Section 15.6), they cross-link phosphates of two oppositely directed polynucleotide chains (1165–1167). In these two polyamines and in the related putrescine and cadaverine (Scheme 15-1), the $N^+ \cdots N^+$ distances vary and therefore different DNA structures are stabilized. Thus when measured in water/ethanol solutions (1165), the A form is preferred in the case of spermine and spermidine whereas the B form is found in the presence of putrescine and cadaverine.

These results from experiments carried out in the presence of ethanol are in conflict with details obtained on a B-DNA spermine complex crystallized by the addition of a higher alcohol (2,4-methylpentanediol (1168). In the dodecamer double helix formed by d(CGCGAATTCGCG) (Section 11.2), a spermine molecule straddles the *major* groove and is oriented nearly vertical to the helix axis, in contrast to previous suggestions (1165–1167). Besides forming salt bridges between terminal amino groups and phosphates on both polynucleotide chains, one of the spermine's internal ammonium nitrogens is in direct hydrogen-bonding contact with guanine O_6 jutting out into the major groove. One of the two salt bridges is rather loose due to contacts with a neighboring molecule. Thus, this particular spermine position might be influenced by crystal packing arrangement and therefore extrapolation to solution state is doubtful. On the other hand, interpretations of solution experiments are based on the assumption that spermine and its analogs are located in the minor groove and should be reconsidered in the light of these new findings.

17.3 Solvent Accessibilities in A- and B-DNA

The folding of polypeptides and of polynucleotides into ordered secondary and tertiary structures is associated with burial of atomic groups which would be exposed to solvent in a random polymer. To estimate the driving force of solvent effects in chain folding for proteins (525, 1146–1148), the concept of a solvent-accessible surface was introduced to quantitatively determine the proportion of buried and exposed atomic groups (1169,1170).

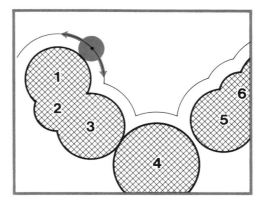

Figure 17-5. Schematic illustration of accessible surface area around a molecule of which six atoms (hatched) are indicated. A spherical probe of radius r_W (color) is moved along the surface of the molecule, its center describing the perimeter A of the accessible surface. The same procedure is repeated for slices above and below this plane, integration yielding the accessible surface area (Å^2). If r_W is increased, atoms lying in a concavity become more hidden. After (1172).

A spherical probe rolling around a molecule. Once the three-dimensional structure of a macromolecule is known, space filling can be visualized by drawing van der Waals spheres around atomic positions. If, now, a spherical (solvent) probe of radius r_W is rolled along the envelope of the van der Waals surface area at conveniently sectioned planes, the accessible surface of a solute molecule with respect to a solvent molecule or atom with radius r_W can be mapped out (Figure 17-5).

DNA becomes more polar if folded into a double helix. In two independent studies, the surface accessibility of tRNA[Phe] (1171,1172) and A- and B-DNA (1172) was evaluated. If the surface accessibility of fully extended DNA is calculated and compared with that of DNA in an A- or B-type double helix, characteristically the phosphate oxygens retain near-maximal exposure while the bases become 80% buried. Phrased differently, in DNA double helices, phosphates account for 45% of the total surface, bases for about 20%, and the remaining 35% is attributed to sugars (Figure 17-6). In general, therefore, the polar character of DNA *increases* if it is folded into double-helical structure. This is analogous to the three-dimensional shape of proteins which also have polar groups exposed at the periphery and non-polar, hydrophobic groups buried in the interior.

Solvent accessibilities explain solution properties of DNA. If r_W is assumed to be 1.4 Å corresponding to the radius of water, solvent accessibilities for A- and B-DNA are nearly equivalent. If, however, r_W is increased in order to simulate hydrated cations or amino acid side chains, phosphate solvent accessibility in B-DNA increases more rapidly than for A-DNA. This is because in B-DNA phosphate oxygens are more exposed. Correspondingly, carbon atoms in B-DNA become more rapidly buried relative to A-DNA (see Figure 17-7).

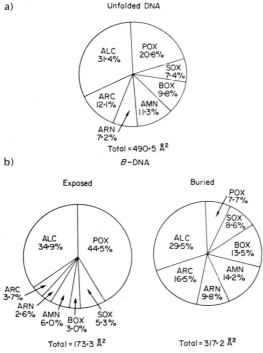

Figure 17-6. Group-type exposures in unfolded, extended-chain (**a**) and double-helical B-DNA (**b**) of random sequence. (b) Both exposed and buried groups in helical B-DNA are shown, adding up to yield (a). Folding of the polymer leaves phosphate oxygens exposed but bases become buried. Group types are ALC = aliphatic carbon, AMN = amino nitrogen, ARC = aromatic carbon, ARN = aromatic ring nitrogen, BOX = base carbonyl oxygen, POX = phosphate oxygen, SOX = sugar oxygen. From (1172).

These findings can be interpreted as follows (1172): The intersection of curves in Figure 17-7 suggests that total surface accessibility of A- and B-DNA double helices are virtually identical for water and that primary hydration layers are quantitatively equivalent in both forms. If r_W increases, B-DNA is more amenable to the formation of solvent clusters around phosphate groups because phosphate oxygens point away from the helix axis. In A-DNA, however, they are directed more toward the minor groove and are therefore less accessible. This general scheme is also supported if the number of water molecules which can be accommodated within minor and major grooves around one nucleotide is calculated: 10.5 for A-DNA and 19.3 for B-DNA, in good agreement with the 20 water molecules determined for primary hydration shell in B-DNA.

Another aspect of the surface accessibility studies explains the dependence of DNA conformation on A/T and G/C content (1172). Density gradient centrifugation studies (1155) and theoretical calculations (1173) have shown that an A–T pair can bind one or two water molecules more than a

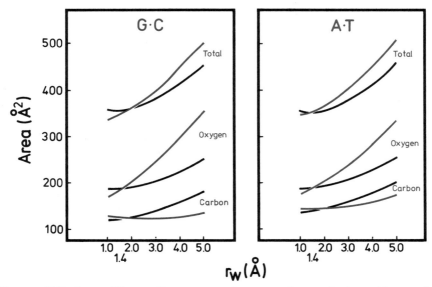

Figure 17-7. Accessible surface areas for two base-paired residues of poly(dG)·poly(dC) (left), and poly(dA)·poly(dT) (right) in A-DNA (black) and B-DNA (color) helical forms, calculated for varying probe radius r_W. For greater r_W, B-DNA shows relatively greater exposure of phosphate oxygens than A-DNA; carbon atoms show the reverse trend. From (1172).

G–C pair. Comparing these findings with surface accessibilities, it follows that for A–T pairs, the hydrophobic methyl group is greatly exposed and would, therefore, promote water–water aggregations, resulting in a wide major groove, in agreement with the known reluctance of A/T-rich DNA to transform into an A-DNA structure where the major groove is more narrow (Figure 9-2). On the other hand, in G–C pairs, hydrophilic amino groups line both grooves, encouraging base–water interactions and exhibiting no pronounced tendency toward one or the other DNA form; i.e., structural transformations occur more readily for G/C-rich than A/T-rich DNA (see also Section 17.6).

Consequences for DNA–protein interactions. Surface accessibilities calculated with a probe of radius $r_W = 3$ Å or greater show that dominant base exposure occurs in the major groove of B-DNA (cytosine N_4 and thymine methyl) and in the minor groove of A-DNA (guanine N_2). These findings suggest that for protein–nucleic acid interactions with intact DNA double helices, specific contacts between base-pairs and amino acid side chains occur preferentially in the major groove of B-DNA and in the minor groove of A-DNA. This is in agreement with data on B-DNA complexes with *E. coli* polymerase–lac promoter, with lac repressor–lac operator (1172), with lambda and *cro* repressors and with DNA–histone interaction in nucleosomes (Sections 18 and 19.3).

17.4 Theoretical Considerations

"Static" and "dynamic" pictures of DNA–water interactions. Two principal theoretical methods have been applied to describe the hydration of DNA, RNA, and their monomeric units (1173–1178). One, the supermolecule approach, uses self-consistent field calculations and empirical potential energy functions to obtain detailed static hydration around a molecule (1173–1176). In the other kind of study, Monte Carlo simulations are employed, using electrostatic potential contour maps of a molecule as a basis to compute the statistical distribution of water molecules. The water is moved around as a probe, yielding a dynamic view of solvent–solute interactions (1177–1180).

General agreement with experimental findings. Results of the Monte Carlo method showed that conclusions can be drawn concerning the position and orientation of a water molecule interacting with nucleic acid and with other nearby water molecules. For the DNA double helix, the first hydration shell shows a characteristic pattern, with phosphate–phosphate distances as the repeating motif. Between successive phosphates of one chain, water bridges formed by one single water molecule are highly probable. Water molecules are also found at PO_2^- groups, interacting preferentially with both oxygens simultaneously. In total, each phosphate attracts six water molecules in its first hydration shell, and this is in agreement with infrared spectroscopic data (Section 17.1). The distribution of water molecules in the first hydration shell is sensitive to the conformation of DNA (A or B type). In the typical pattern for B-DNA, water "filaments" bridging phosphates of opposing strands across the major groove are predicted (1180). The stabilization energy of about 15 kcal/mole water is so effective that a second and even a third water layer are attracted, increasing the total radius of hydrated DNA by 6 Å (for second layer) to 9 Å (for third layer) as compared to naked DNA. Water accommodated in major and minor grooves between the sugar–phosphate chains is more weakly bound, with stabilization energy of approximately 10 kcal/mole water (1177, 1179).

The supermolecule approach suggests that five water molecules are bound to phosphate groups of B-DNA, one of them about 4 kcal/mole more tightly bound than the other four (1175). In a G–C base-pair, three water molecules are bound in the major groove side to keto oxygen and amino nitrogen of guanine and cytosine and three more water molecules are located in the minor groove and hydrogen bonded to sugar $O_{4'}$, guanine N_3, and cytosine O_2. Thus the total number of water molecules in the first, innermost hydration shell is eight per nucleotide, in overall agreement with data mentioned above (Section 17.1).

How does hydration influence stabilization energies within the B-DNA double helix? The supermolecule calculation of B-DNA hydration can be used to analyze the forces responsible for the stabilization of the double helix (1175). Most prominent (15 kcal/mole base–pair) is the decrease of repulsion between adjacent, negatively charged phosphates. In addition, hydration increases the attraction between base and phosphate residues (12 kcal/mole) and has a marked influence on the stacking of neighboring bases (7–10 kcal/mole).

What forces are involved? Reduction of phosphate · · · phosphate repulsion is primarily due to the decrease of electrostatic interactions and only to a small extent due to an increase in polarization energy. The phosphate–base attraction depends on a strong increase of the electrostatic component and to a smaller concomitant increase of both polarization and dispersion energy. In base-stacking interactions, hydration contributes mainly to stabilization through electrostatic and dispersion effects. Polarization, as calculations revealed, is of only minor importance and a repulsion component of about 5 kcal/mole must be overcome prior to stack formation, resulting in overall stabilization of 7–10 kcal/mole for a G–C stack.

In A-RNA, the $O_{2'}$ hydroxyl takes part in structure stabilization. In RNA of the A form, the supermolecule approach has shown that the $O_{2'}-H$ hydroxyl can lead to three different kinds of interactions, depending on the orientation of the hydrogen atom (1176). First, a (weak) hydrogen bond $O_{2'}-H$ · · · $O_{4'}$ between adjacent riboses can be formed. Second, if the $O_{2'}-H$ group is *cis-planar* with respect to the $C_{2'}-C_{3'}$ bond, a water molecule can bridge intramolecularly $O_{2'}$ and O_2 of pyrimidine (or N_3 of purine), $O_{2'}$ · · · $H-O-H$ · · · $O_2(N_3)$. Third, a water molecule bridging $O_{2'}$ and $O_{3'}$ of the same ribose is very strongly bound. The 3' oxygen is rather electronegative because it is part of the phosphodiester linkage (see also discussion in Section 4.3).

Iceberg-like hydration shells are improbable. In neither DNA nor RNA double helices was the calculated arrangement of water molecules around nucleotide units iceberg-like as proposed for hydration shells around macromolecules (1181). An ordered structure as found in ice clathrates (1183) is probably too restrictive in its geometry to be formed around a molecule with a shape as complicated as that of a nucleotide or a nucleic acid. This does not preclude less ordered, yet systematic hydration structures as described in Section 17.6.

17.5 Hydration Schemes in Crystal Structures of A-DNA Tetramer and B-DNA Dodecamer Suggest Rationale for A→B Transition

Hydration monolayer and spine of water molecules in major and minor grooves of B-DNA. The crystal structure analysis of the B-DNA dodecamer d(CGCGAATTCGCG) not only provided information on geometrical features of the B-DNA double helix but also gave details concerning hydration. There are 72 more or less well ordered water molecules per dodecamer duplex in the crystal structure studied under ambient conditions (1168). A total of 48 more water molecules show up in defined positions if the concentration of the alcohol used for crystallisation (2-methyl-2,4-pentanediol) is increased to 60% or if the data are collected at 16 K (1183a).

In the ambient conditions study (1168), the phosphate groups are not systematically hydrated except for water molecules trapped between the thymine methyl group and 5'-phosphate of the same nucleotide. The high-alcohol and 16 K studies (1183a) showed in addition more immobilised water molecules bound to phosphate oxygens. There are one to five solvent molecules (average 3 ± 1.3) per phosphate oxygen, of which about 40% form bridges between two or more phosphate oxygens. This hydration occurs in the major groove which shows, in addition, monodentate binding of water molecules to keto oxygens and amino groups of purine and pyrimidine heterocycles and, in a few cases and to a lesser extent, to N_7 of purines. A second shell connected to this first one is too disordered to be clearly visible in the room temperature experiment (1168), yet some more insight into hydration is gained from the high-alcohol and 16 K data (1183a). Besides these monodentate interactions, there are also a number of the bidentate type, where the two ligands of the water molecules can be either hydrogen bond donors, both acceptors or of the mixed donor/acceptor type. Bidentate bridging is predominantly between N and O atoms on adjacent base-pair steps. Together with the phosphate-bound water molecules, these base-bound waters arrange in hydrogen-bonded strings connecting the opposite phosphate backbones across the major groove in a manner predicted earlier on the basis of theoretical considerations (1178,1179, "filaments," Section 17.4).

In the minor groove, there are well ordered water molecules along the central AATT sequence bridging thymine O_2 and adenine N_3. These two kinds of atoms belong not to the same but rather to adjacent base-pairs such that in a sequence X_pY, the two Y's are connected (Figure 17-8). This first layer of water molecules forms a continuous spine and is augmented by a second layer which serves to complete the tetrahedral environment. Water shells of the third and fourth order are less well defined and seem to engage in hydrogen bonding to phosphate groups lining the edge of the minor groove. This regular water structure is not found in the CGCG portions flanking the AATT sequence, probably because the guanine N_2-amino groups are too bulky and, acting as hydrogen bond donors, interfere with formation of the regular spine.

Water-mediated interstrand phosphate $\cdots$ phosphate contacts in the major groove of A-DNA. The hydration of A-DNA, as observed in the crystal structure of the iodinated tetramer d(^{I}CGCG) is considerably different from that of B-DNA (893). There is no spine of water molecules running along either groove, nor is there an obvious first hydration shell

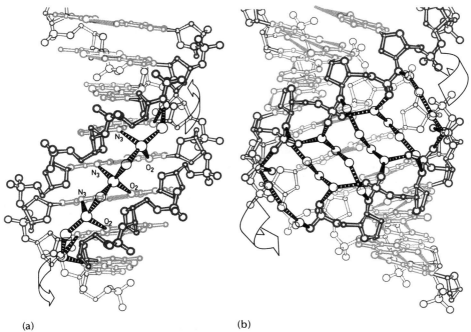

(a)   (b)

Figure 17-8. (a) Drawing shows schematically how water molecules are accommodated in the *minor* groove of sequence AATT of the B-DNA-type dodecamer (1168). The first hydration layer is hydrogen bonded with pyrimidine O_2 and purine N_3 and the second layer completes the tetrahedral coordination shell around water molecules. **(b)** In A-DNA, filaments or water molecules are formed across the *major* groove and stitch phosphate groups together (893).

around base or sugar atoms. However, phosphate groups are hydrated and, in contrast to B-DNA, display a well-ordered hydration scheme.

There is no known case of one water molecule simultaneously hydrogen bonded to both free oxygen atoms of a phosphate, as suggested by theoretical calculations (1177–1179). Yet, along a chain, adjacent phosphate groups are bridged by one to three water molecules. Additionally, as depicted schematically in Figure 17-8, there are chains of four water molecules crossing the major groove and linking $O_{3'}$ of one polynucleotide with O_{P2} of the other, the dyad symmetry generating double chains linking each phosphate group. (For a definition of O_{P2}, the unesterified phosphate oxygen atom pointing into the major groove, see Figure 2-3.) One can envisage an A-DNA (or A-RNA) double helix where such short chains of water molecules systematically link opposite phosphate groups, stitching the major groove together and thus stabilizing the A-form double helix. Such chains and the intrachain phosphate $\cdots$ phosphate interaction mediated by water molecules were predicted earlier on the basis of theoretical calculations (1180).

A ↔ B transformations explained in the light of hydration. Backed by experimentally located hydration water molecules, it is now possible to explain A↔B transformation in terms of water activity. Under conditions where water activity is high (low salt), all potential functional groups of phosphates, bases, and sugars are hydrated in a monolayer (and higher-order layers), favoring B-DNA. In the case of long A/T sequences a spine of water molecules runs along the minor groove and stabilizes this particular conformation. For this reason, poly(dA)·poly(dT) does not undergo B→A transformation (Section 11.4), and poly(dA−dT)·poly(dA−dT), adopts the A-DNA form only as a metastable state. In this context, it is important that polynucleotides with guanosine replaced by inosine (lacking the N_2 amino group) behave as A/T polymers and occur only in B-family double helices (Table 9-1).

If water activity is reduced hydration of base and sugar atoms breaks down and only the more polar phosphate oxygen atoms are hydrated. This change in overall hydration induces the B→A transition, the A-DNA form now being stabilized by water chains running between opposite phosphate oxygens and stitching the major groove together. If water activity is raised again, the DNA double helix becomes more hydrated and A→B transformation takes place.

17.6 Water Pentagons in Crystalline Dinucleoside Phosphate Intercalation Complex: The Generalized Concept of Circular Hydrogen Bonds and of Flip-Flop Dynamics

A well-ordered hydration shell in the d(CpG)·proflavin complex. In the crystal structure of the heavily hydrated intercalative complex between d(CpG) and proflavin, 26 water molecules are present in the asymmetric unit (1184). Of these, 15 are distributed in a rather regular scheme displaying four edge-linked, hydrogen-bonded five-membered circles. Such clusters of pentagons are linked from one unit cell to the adjacent one through one common water molecule. An additional pentagon consists of only four water molecules, the fifth corner being provided by a phosphate oxygen in hydrogen-bonding contact to two water molecules (Figure 17-9).

The pentagons are located near the major groove side and arranged to optimize hydrogen bonding at 3.4 Å intervals, the well-known base−base or base−drug stacking distance. This raster allows water−nucleotide hydrogen bonds to form primarily with base atoms in the major groove as well as with heteroatoms of the intercalated proflavin molecule. These hydrogen bonds are all formed within the planes defined by bases and drug, as predicted by theoretical considerations (1174).

Circular hydrogen bonds: A general concept. Such pentagonal, hydro-

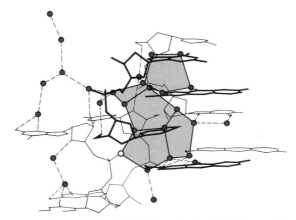

Figure 17-9. Schematic representation of the d(CpG)·proflavin·26H$_2$O crystal structure, with d(CpG) and proflavin indicated by line drawing and water molecules by colored circles. Of the latter, 15 are arranged, together with one phosphate oxygen (black, open circle), in five pentagons emphasized by colored shading. In these, all water molecules are in hydrogen-bonding contact of distance 2.8 to 3.1 Å and are further hydrogen bonded to d(CpG) and proflavin hetero atoms. Hydrogen atoms could not be located in this crystal structure owing to technical difficulties. After (1184).

gen-bonded motifs in irregular structures have been observed previously in another connection, namely, in hydrated α-cyclodextrin, an oligosaccharide consisting of six α(1→4)-linked glucoses with 6 primary and 12 secondary hydroxyl groups. Neutron and X-ray diffraction studies of cyclodextrin hexahydrate revealed that hydroxyl groups and water molecules form a complicated, yet systematic network comprising four-, five-, and six-membered circles as well as infinite chains composed exclusively of OH groups (1185, 1186). Because in this crystal structure hydrogen atoms were unambiguous, hydrogen-bonding directions are clearly defined and show that predominantly two forms of circular structures are formed (Figure 17-10). First, those with O–H · · · O–H · · · O–H hydrogen bonds running in the same direction (*homodromic*) and, second, those with two chains of O–H · · · O–H · · · O–H hydrogen bonds emanating from one water molecule, running in opposite directions and colliding at one oxygen atom (*antidromic*) (Figure 17-11). The *heterodromic* case with randomly oriented O–H · · · O hydrogen bonds is only rarely realized. As quantum chemical calculations have revealed, the *homodromic* variety is energetically more favorable than the *antidromic* one because the cooperative effect comes into play and stabilizes the "endless" chains of *homodromic*, linear O–H · · · O–H · · · O–H interactions (1187).

In contrast to the regular ice clathrates, circular hydrogen bonds are structurally rather irregular, with greatly distorted coordination geometry around water molecules and hydroxyl groups. It is, therefore, probable

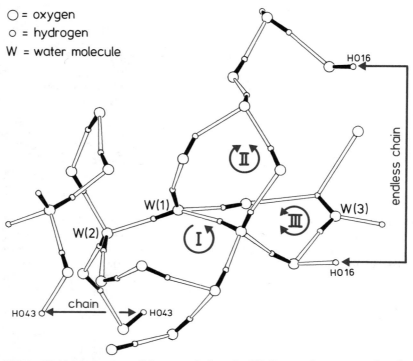

○ = oxygen
o = hydrogen
W = water molecule

Figure 17-10. A section of the α-cyclodextrin·6H$_2$O crystal structure showing hydrogen-bonding scheme. Small and large circles represent H, and O atoms; water molecules are indicated by W. Note that Circle I is *homodromic* as is the chain-like structure, whereas circles II and III are *antidromic;* for definition see Figure 17-11. From (1185).

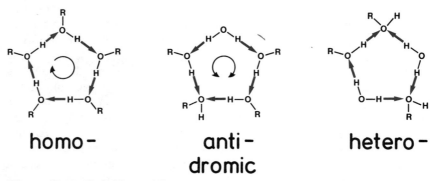

homo –   anti –   hetero –
                          dromic

Figure 17-11. Definition of *homo-, anti-,* and *heterodromic* circular hydrogen bonds displaying $\cdots$ O–H $\cdots$ O–H $\cdots$ O–H $\cdots$ hydrogen-bonding interactions in unidirectional, in counterrunning, and in random orientations. From (1186).

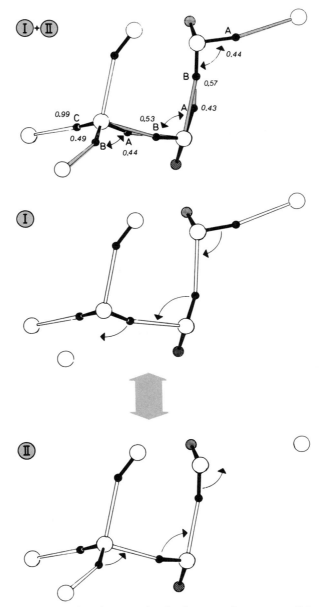

Figure 17-12. Flip-flop hydrogen bonds occurring in the crystal structure of β-cyclodextrin·12 H_2O. Carbon atoms shown stippled, hydrogens as small black spheres, oxygens as large spheres, and water oxygens marked W. (Top) Hydrogen positions A and B are only half occupied because they are hydrogens involved in flip-flop systems O–H · · · H–O. These can be deconvoluted into O–H · · · O and O · · · H–O hydrogen bonds displayed in (center) and (bottom) pictures. If a hydroxyl group rotates as indicated by double arrows in picture at top, all other hydroxyl groups in that flip-flop chain have to move in concerted motion, leading to a dynamical picture. Taken from (1188).

that similar structures coat macromolecules with a first and second, network-like hydration shell which is rather adaptable geometrically and certainly not well defined as required by the iceberg concept.

A glimpse at hydrogen bond dynamics: The flip-flop O–H $\cdots$ H–O systems. Another neutron diffraction study was carried out with the larger β-cyclodextrin·12H$_2$O containing several disordered sugar and water hydroxyl groups (1188). Besides normal O–H $\cdots$ O hydrogen bonds, the crystal structure shows 19 bonds of type O–H $\cdots$ H–O with O $\cdots$ O distances in the normally accepted range of 2.8 to 3.0 Å (Figure 17-12). However, the H $\cdots$ H contacts are so short, around 1 Å, that these atoms are mutually exclusive as is also indicated by their occupation parameters, around 0.5. What is observed in fact is a statistical average O–H $\cdots$ O $\rightleftharpoons$ O $\cdots$ H–O and several such hydrogen bonds are coupled to form larger systems. If, now, a change from O–H $\cdots$ O to O $\cdots$ H–O or vice versa occurs by rotation of an O–H group as indicated in Figure 17-12, the interconnected O–H $\cdots$ H–O bonds have to change H-bonding directions cooperatively; i.e., chains $\cdots$ O–H $\cdots$ O–H $\cdots$ O–H and $\cdots$ H–O $\cdots$ H–O $\cdots$ H–O $\cdots$ alternate with time, hence the name flip-flop. For entropic reasons, such systems should be more favored than isolated O–H $\cdots$ O bonds and, in connection with circular arrangements, a dynamical system of flip-flop circles, interconnected to form an oscillating hydration shell, is envisaged.

Summary

Hydration of nucleic acids plays an essential role in determining their structure and is responsible for the A $\rightleftharpoons$ B transition of DNA. There are two hydration shells around DNA, the primary one being not ice-like and impermeable to cations (20 water molecules per nucleotide) and the second shell is indistinguishable from bulk water. Solvent accessibilities can be derived theoretically on the basis of DNA or RNA structures. They show that DNA becomes more polar if folded into a double helix and explain some of the solution properties. Single-crystal studies carried out at ambient conditions suggest that in A/T-rich regions of B-DNA, the minor groove is filled with a spine of hydrogen-bonded water molecules whereas in A-DNA, filaments of water connect and stitch together phosphate groups lining the major groove. These observations explain B→A transformation if the water content of DNA is reduced. As a general scheme of hydration of macromolecules, four- to six-membered circles containing at their corners oxygen atoms or O–H groups have been envisaged in the d(CpG)·proflavin complex and in several cyclodextrin crystal structures. The hydrogen bonds in hydration shells are probably of the flip-flop type with the dynamical equilibrium O–H $\cdots$ O $\rightleftharpoons$ O $\cdots$ H–O favored entropically.

Chapter 18

Protein–Nucleic Acid Interactions

Our knowledge of interactions between proteins and nucleic acids is still rather limited. These interactions occur at all levels of DNA replication and expression and in numerous regulatory processes and are therefore of the very greatest importance in life. We do not understand how restriction endonucleases attach to DNA and cut it at specific sequences. Nor do we have a clear idea about the geometry of repressor–operator recognition. The selectivity of amino acid–tRNA synthetases for their cognate tRNAs is still obscure. The main problem is that these systems are all rather complex and require the simultaneous observation of two associated macromolecules. Spectroscopic methods are, except for a few favorable cases, inadequate, and crystallization of protein–nucleic acid aggregates is difficult. Recently, however, two cases have been reported where specific protein–DNA (1189,1190) and protein–tRNA (1191) complexes were obtained in a form suitable for X-ray diffraction studies. We can hope, therefore, to attend the unveiling of this mystery within the next few years.

In order to circumvent the difficulties inherent in these systems, model compounds have been investigated using both theoretical and experimental approaches. One major concern has been learning about the specificity of recognition between the four nucleic acid bases and the 20 amino acid side chains, using both monomeric constituents of both partners as well as polymers, in some cases polymers of synthetically available peptides or nucleic acids. Nucleotides bound as inhibitors or as coenzymes to active sites of enzymes have also been studied. More recently, proteins which bind specifically to DNA have been crystallized and their three-dimensional structures analyzed. That is the best we can do without having crystals of cognate protein–DNA complexes available. Such structures provide insight into the requirements of protein–nucleic acid interactions; conclusive evidence, however, means eventually investigating real complexes. Before discussing the known facts in this chapter, let us embark in more general thinking about protein–nucleic acid recognition and discuss some hypotheses which have been put forward.

18.1 General Considerations about Protein–Nucleic Acid Interactions

Looking at the amino acid side chains and at the polypeptide backbone (Figure 18-1), we find that there are essentially four kinds of potential interactions between proteins and nucleic acids (1192–1195) (Figure 18-2).

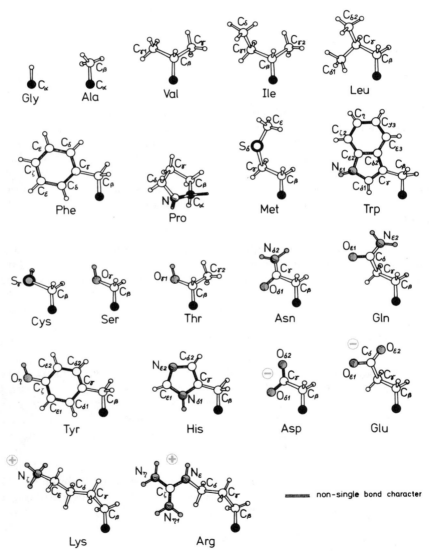

Figure 18-1. Summary of structures of amino acid side chains. Functional groups indicated in color and C_α carbon atoms are filled black. From (1289).

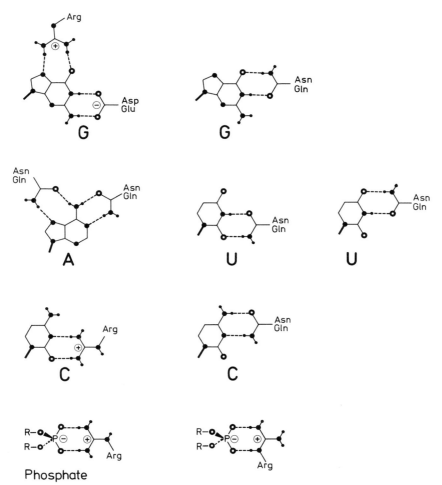

Phosphate

Figure 18-2. Some possible interactions among amino acid side chains, nucleic acid bases, and phosphate groups. The pattern becomes much more complex if peptide bonds and simultaneous binding of two amino acids to one nucleotide are considered.

(a) Salt briges formed between phosphate and positively charged amino acid side chains (N_ζ of Lys, guanidinium of Arg, and protonated His).

(b) Hydrogen bonding among phosphate, sugar, bases in nucleic acids, and peptide bonds or hydrophilic amino acid side chains in proteins.

(c) Stacking interactions involving aromatic amino acid side chains (Trp, Tyr, Phe, His) and bases.

(d) Hydrophobic interactions associating nucleic acid bases with nonpolar amino acid side chains.

Table 18-1. Electronic charge distribution in the peptide bond and in some amino acid side chains of interest for interaction with nucleic acids. For more data, see Ref. (1298), from where this compilation was taken.

Peptide Bond

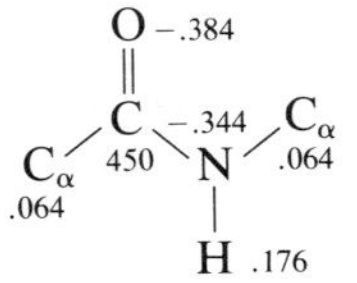

Cysteine $-CH_2-S-H$ (.015 .010)

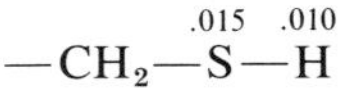

Serine $-CH_2-O-H$ (−.310 .170)

Lysine$^{\oplus}$ $-CH_2-CH_2-CH_2-CH_2-N-H$ ...

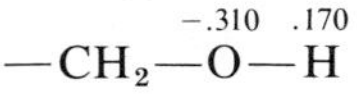

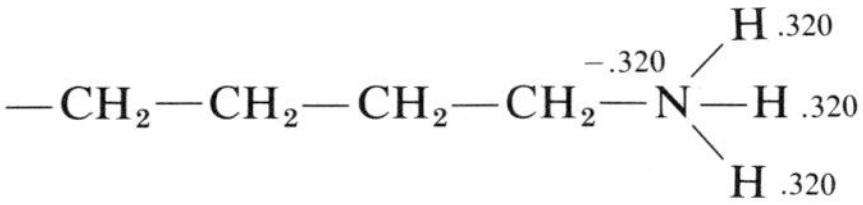

Arginine$^{\oplus}$

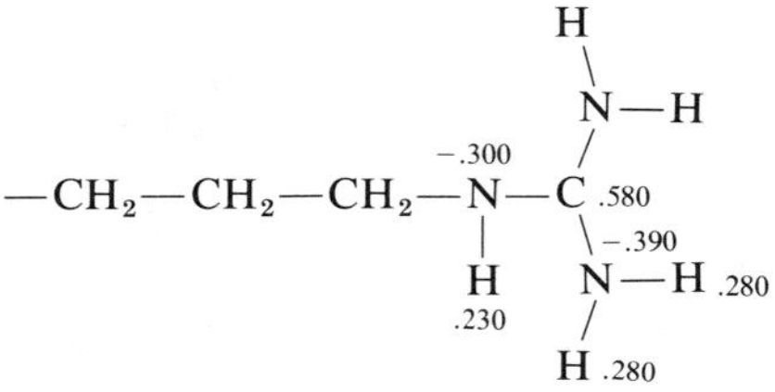

Aspartic and Glutamic Acid$^{\ominus}$

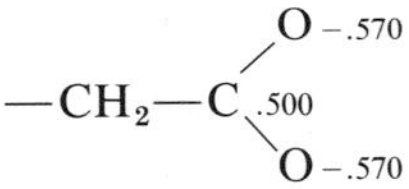

Asparagine and Glutamine

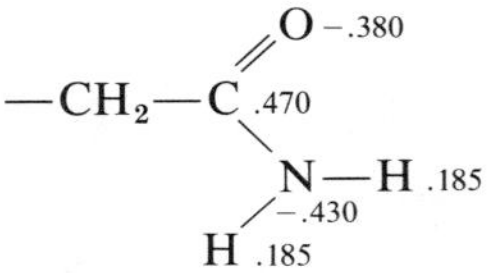

Histidine

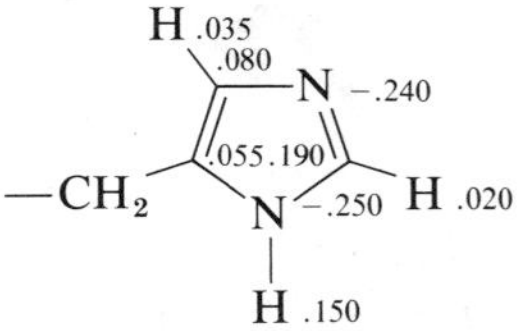

Histidine$^{\oplus}$

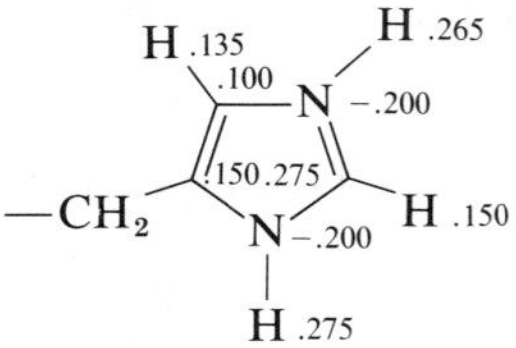

Tyrosine

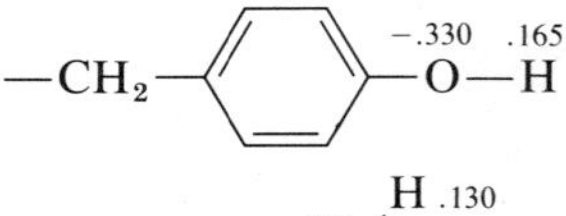

Tryptophan

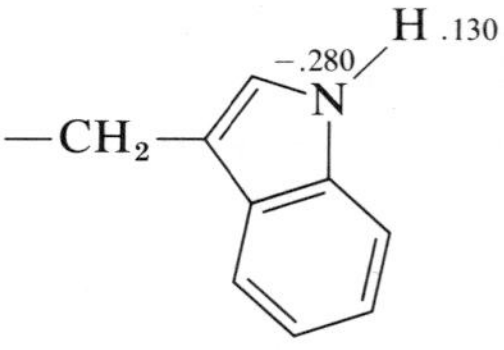

In a broad sense, the individual strength of these four kinds of interactions decreases from (a) to (d). Because charge–charge attractions play the predominant role, charge distributions of amino acid side chains and peptide groups are assembled in Table 18-1. They should be compared with data given in Table 5-1 for nucleic acid constituents. On this basis, selective interactions can be predicted. In nature, however, the situation is more complicated because multiple contacts are observed which can override charge–charge specificity as we will see in Section 18-6.

Interactions between individual units. If amino acids and nucleotides are considered as individual units, charge–charge and hydrogen-bonding forces rather than hydrophobic and stacking interactions seem to dominate structure and be best suited for selective recognition. One can write down hypothetical models of how amino acids and nucleic acid units could interact with each other (Figure 18-2). Quantitative measures of individual nucleic acid base–amino acid side chain association are now available (1196–1198) (Table 18-2), and these are supported by quantum chemical studies [at least for guanine and cytosine forming hydrogen-bonded complexes with arginine, glutamine, and lysine (1202)]. It appears

Table 18-2. Association Constants for Complexes Between Nucleic Acid Bases and Amino Acids or Suitable Analogues

Nucleoside or base[a]	Association constants K in [1/mole] for				
	L-arginine	L-lysine	L-glutamine	L-glutamate	L-glycine
Thymine	0.6 ± 0.1	——	0.31 ± 0.24	——	——
Cytosine	1.5 ± 0.2	0.63 ± 0.20	0.46 ± 0.21	——	0.27 ± 0.12
Inosine	1.26 ± 0.09	0.04 ± 0.15	1.02 ± 0.18	0.26 ± 0.15	0.12 ± 0.09
Adenosine	1.14 ± 0.08	0.06 ± 0.10	0.78 ± 0.14	——	——
Guanosine	1.91 ± 0.10	0.85 ± 0.14	1.05 ± 0.15	0.71 ± 0.12	0.51 ± 0.07

Base[b]	Amino acid or analogue	Conditions/ methods	K [1/mole]	Reference
9-Ethylguanine	Na-Acetate	DMSO/H_2O,303K/ NMR	110 ± 30	(1197)
9-Ethyladenine	Butyric acid	$CHCl_3$,308K/IR	130 ± 30	(1200)
Cyclohexyluracil	Butyric acid	$CHCl_3$,308K/IR	60 ± 20	(1200)
Cyclohexyluracil	p-Cresol ($\equiv$ Tyrosine)	$CDCl_3$,219K/NMR	69 ± 4	(1200)
Dimethyluracil	p-Cresol ($\equiv$ Tyrosine)	$CDCl_3$,300K/NMR	6.3 ± 0.2	(1200)
9-Ethyladenine	p-Cresol ($\equiv$ Tyrosine)	$CDCl_3$,303K/NMR	6.32 ± 0.15	(1201)

[a] Determined from solubility experiments in aqueous solution at neutral pH, 25°C (1198).
[b] Determined by spectroscopic methods in non-aqueous solvents.

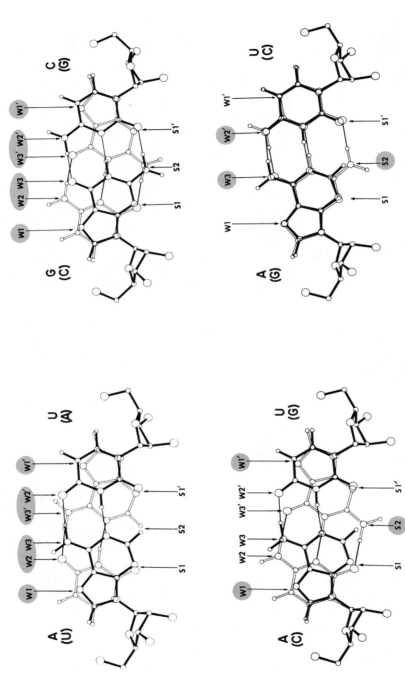

Figure 18-3. Superimposed A–U and G–C base-pairs in different combinations indicate where discrimination by hydrogen bonding can occur. The upper letter at the side refers to the pair drawn with solid lines, the lower letter (in parentheses) to the one in open lines. Sites W in the major (wide) and S in the minor (small) groove are related to W′ and S′ by the dyad symmetry inherent in DNA and RNA double helices. They are labeled in color if hydrogen bonding donor and acceptor characteristics change with base sequence and allow discrimination by amino acid side chains. From (1204).

that these interactions are especially favored if at least two hydrogen bonds are simultaneously formed, and this can be achieved in different ways (1203). This feature reminds us of base-pairing hydrogen bonds (Chapter 6), where one single interaction is too weak and unspecific for proper stabilization.

Interactions between Watson–Crick base-pairs and amino acid side chains. If DNA and RNA double helices rather than individual nucleic acid bases are considered, the situation changes because several hydrogen-bonding recognition sites on the bases are now hidden by Watson–Crick-type base-pairing (Figure 18-3). Model building (1204,1205) and theoretical studies (1206) indicate that, as mentioned above, specific association between base-pairs embedded in double helices and amino acid side chains (or peptide groups) requires again the simultaneous formation of at least two hydrogen bonds. The minor (small) groove has the potential for distinguishing only between guanine and other bases (due to the unique guanine N_2 amino group). The major (wide) groove offers the potential for more discrimination and should be the primary target if different base-pair combinations are to be distinguished from one another. On the other hand, the minor groove could well be a preferential site for metal ion and water binding, as outlined in Chapters 8 and 17.

Interactions between nucleic acid double helices and polypeptide antiparallel β-pleated sheets. Survey of a number of crystal structures of globular proteins has demonstrated that the antiparallel β-pleated sheet, (Figure 18-4) can naturally mold into a right-handed, twisted, helical structure (1207), with curvature corresponding very closely to that observed for the surface of double-stranded RNA (837) and DNA (1208). Combination of these two macromolecular components in model building studies immediately suggested that for A-RNA, a β-pleated sheet can fit into the minor groove. This arrangement is stabilized by direct hydrogen-bonding contacts between ribose hydroxyls $O_{2'}H$ and carbonyl oxygens of the peptide bonds, reinforced by additional, water-mediated hydrogen bonds between peptide N–H groups and ribose $O_{4'}$ and $O_{2'}$ oxygens, the latter belonging to the preceding ribose in the chain. The chain polarities, $5' \rightarrow 3'$ for RNA and $NH_2 \rightarrow COO^-$ for the polypeptide, run parallel for the two hydrogen bond-associated strands and, due to the intrinsic antiparallel character of the β-pleated ribbon, the overall dyad symmetry of RNA is also preserved in the complex.

Because the minor groove in B-DNA is shallower than that in A-RNA, it was at first thought improbable that the β-pleated sheet secondary structure could also be involved in protein–DNA recognition. However, assuming that proline and tryptophan are not pointing to the minor groove, two models can be built which satisfy all stereochemical requirements (1208). In one of these models, the polarities of DNA and the polypeptide strand are parallel (as in the complex with RNA described above) and in the other they are antiparallel. Yet in both models the hydrogen-

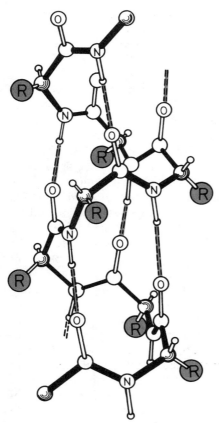

Figure 18-4. The basic protein structure elements. (Top) The α helix contains 3.6 residues per pitch of 5.4 Å and is stabilized by hydrogen bonds running nearly parallel to the helix axis. Side chains (colored circles) protrude approximately radially from the helix. (Opposite) In the β-pleated sheet (or ribbon) structure, polypeptide chains run either parallel or antiparallel to each other and are connected by hydrogen bonds. Side chains point alternately up and down. Arrows in thin lines indicate potential "outside" hydrogen-bonding sites (not possible in α helix), whereas arrows drawn as thick lines show that strands run antiparallel in this picture.

bonding interactions are similar, involving exclusively (peptide)NH $\cdots$ $O_{3'}$(nucleotide) contacts (Figure 18-5).

Complex formation between the β-pleated sheet structure of a protein and the B-DNA double helix probably occurs in *cro* repressor binding to DNA (Section 18.6) and could play a role in *lac* repressor–operator recognition. The amino acid sequence of the first 60 residues of the N terminus of *lac* repressor (the headpiece) suggests predominant β-pleated

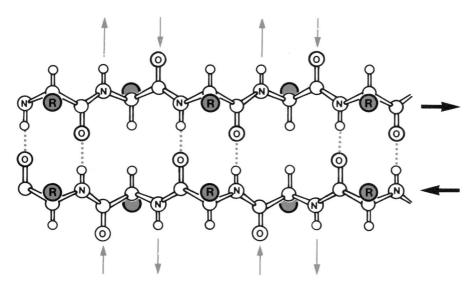

sheet secondary structure (1209) [for another interpretation, see (1209a)]. However, a β-pleated sheet bound to the B-DNA minor groove would not per se be expected to display the high specificity which is observed. Based on structural and binding data of two other peptide-like molecules, netropsin and distamycin, to A–T-rich regions in B-DNA, a proposal was put forward suggesting that the β-pleated ribbon adopts an asymmetric form when involved in the recognition of specific B-DNA sequences. The two strands of the DNA are thought to be in slightly different conformations which would allow the discrimination of A–T and G–C base-pairs provided that the amino acid side chains are in certain positions to recognize DNA sequence (1210).

Interactions between nucleic acid double helix and polypeptide α helix. Examples for aggregation and specific recognition of RNA and DNA double helices by protein α helices are given in recent literature and will be discussed in Section 18.6. Details of the interactions have not yet been worked out; peptide amide groups, however, will not be involved because they are all engaged in the stabilization of the α helix via hydrogen bonding (Figure 18-4). In this case, it is more probable that amino acid side chains protuding from the α helix are used for stabilization of the α helix–double helix complex.

In many cases nucleic acid–protein complexes display dyad symmetry. The antiparallel β-pleated ribbon interacting with the minor grooves of B-DNA and A-RNA possesses intrinsic dyad symmetry coincident with that of the nucleic acid double helix. The protein α helix has no such symmetry. However, *a pair* of antiparallel α helices can fulfill the requirements of dyad symmetry. As we will find in Section 18.6, three cases are

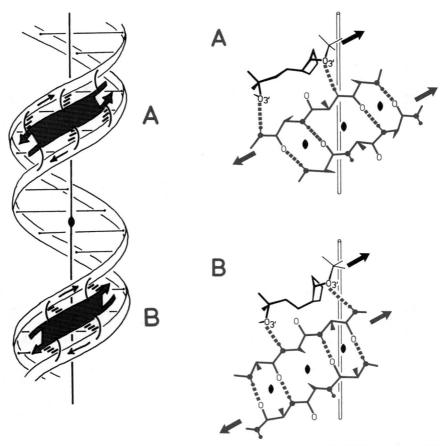

Figure 18-5. Schematic representation of complex between B-DNA double helix and β-pleated ribbon. The latter is located in the minor groove and its dyad symmetry corresponds to that of B-DNA. In A, polarities of nucleic acid (5′→3′) and of polypeptide (N→COO⁻) run antiparallel, whereas they are parallel in B. Only one chain of B-DNA is drawn, hydrogen bonds indicated by dashed lines. From (1208).

known where DNA binding proteins crystallize as dimers, with the α helices thought to interact with DNA major grooves in antiparallel orientation and nearly 30 Å apart [see also Ref. (1209a)]. If a specific nucleic acid sequence is recognized by the α helix, then the dimeric form of the protein requires a symmetrical (palindromic, Box 18-1) nucleic acid sequence. This is indeed observed for a number of DNA stretches known to bind specifically to proteins such as promoter and operator sites, as well as recognition sites for restriction and modification enzymes, summarized in (1211). In these cases, protein dimers (or tetramers) recognize double-stranded DNA with inherent dyad symmetry not only in the overall geometry of the double helix but also in nucleotide sequence. The exact sequential symmetry and deviations from it present a great tool for fine tuning of the interactions.

18.2 Model Systems Involving Nucleic Acid and Protein Constituents

In model systems single-crystal diffraction studies should give insight into the geometry of specific nucleotide–amino acid interactions. Spectroscopic methods show that such interactions are rather weak, (Table 18-2). Self-associations of the individual components usually predominate, and successful attempts at cocrystallization are rare. The few examples of the crystallization of the two partners or of covalently linked nucleoprotein constituents can be grouped into six classes.

Interactions between bases and carboxylic acids. These involve both dissociated and undissociated carboxyl groups, depending on the base. In cytosine, N_3 becomes protonated and the negatively charged carboxylate accepts two hydrogen bonds from N_4–H and N_3^+–H (1212–1214). In two examples with 5-bromocytosine, the bases are dimerized through two N_4–H $\cdots$ N_3 hydrogen bonds and N_3 is not available for protonation. Therefore the carboxyl group of the added glutamic acid derivative is not dissociated and hydrogen bonds (carboxyl)O–H $\cdots$ O_2 (cytosine) (1215, 1216). Undissociated carboxyl groups are also observed in thymidine 5′-carboxylic acid where thymine O_4 and N_3–H groups are hydrogen bonded with the carboxyl moiety (1217), and in a nucleoside dipeptide with the terminal carboxyl interacting with adenine N_6H and N_7 (1218).

Hydrogen-bonding interactions with indole NH. The indole ring of tryptophan bears one free NH group which, in crystalline complexes with 9-substituted adenine derivatives, hydrogen bonds to N_3 (1219,1220) and to N_7 (1221). In these crystal structures, stacking was observed only between bases, not between bases and indole. This is surprising because ultraviolet difference absorption spectroscopy in aqueous solution indicated charge-transfer-type stacking interaction between tryptophan and nucleoside 5′-phosphates and especially with AMP and ATP (1222).

Stacking between base and aromatic amino acid side chain. Continuing with indole as model for the tryptophan side chain, and with tyramine as model for tyrosine, stacking with bases can be achieved if the bases are modified to promote charge-transfer-type interactions. This was done primarily with 1,9-dimethyladenine (bearing a positive charge), nicotinamide, and isoalloxazine (1223–1228) in order to establish the geometrical relationship in complexes which have been shown by spectroscopic methods to be essential in apoenzyme–coenzyme associations (1229). In the crystal structures of these model systems, the heterocyclic partners

Box 18-1. Palindromic Sequences

Inspired by the clover-leaf secondary structure of tRNA where self-complementary base sequences allow folding into hairpin arrangement, Gierer proposed that in double helical DNA similar features should exist (1289a). For these to occur, sequences of inverted or "palindromic"* repeats are required which, owing to their inherent twofold symmetry, can form hairpin structures of simple cruciform-like or more complicated arrangement, called "Gierer trees."

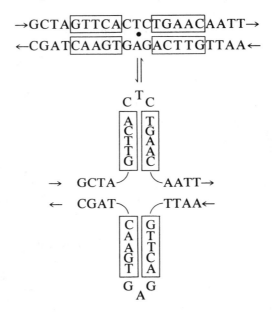

• = dyad axis for palindromic sequences indicated by boxes.

Later, palindromic sequences were found frequently, in some cases amounting to 3% of total DNA content and being from a few to several thousand nucleotides in length. As already suggested by Gierer, they aid in specific protein–DNA interactions and are involved in regulatory functions. Owing to the twofold symmetry of the palindromes, the respective proteins are at least dimeric in structure (1292). The question, however, remains whether palindromic DNA sequences are integrated into double-helical structure or whether Gierer trees really exist. This issue is of particular interest in connection with supercoiled DNA (Chapter 19) because supercoil twisting can be reduced and thus modified by formation of Gierer protrusions (1293 to 1295).

The two (or more) branches of a palindromic Gierer tree are again symmetrically related and can in principle be folded together to generate fourstranded helical arrangements. These can be stabilized by hydrogen bonds connecting functional amino and keto groups in the major grooves of Watson–Crick base-pairs and lead to base-quartets A–T:T–A and G–C:C–G (1296). The four-stranded arrangements have, thus far, been discussed mainly in theoretical terms but experimental proof for their occurrence and biological significance is still lacking (1297).

---

* A palindrome is a sentence reading the same in both directions, e.g.,: A man, a plan, a canal—Panama.

are usually stacked parallel to each other at or below 3.4-Å distance. The only case where two "natural" constituents, uracil and phenylalanine, are stacked is in the crystal structure of the nucleoside peptide 5-[N-(L-phenylalanyl)amino]uridine (1230), which in addition displays contacts between base and peptide group.

Base–peptide interactions have been found in the above mentioned nucleoside peptide (1230) and in another synthetic compound, 3-(adenin-9-yl)propionamide (1230a). In the former, N_3–H is hydrogen bonded to the peptide carbonyl oxygen and the terminal phenylalanine amino group – NH_2 interacts with uracil O_4. In the latter, adenine hydrogen bonds to adenine in base-pair V of Figure 6-1 and only one hydrogen bond, amide NH · · · N_3(adenine) is formed between base and "peptide." In strict terms, both are not peptide–base adducts of the predicted form. However, the geometry of the complex sheds light on the type of structure to be expected in base–protein recognition.

No interactions between nucleic acid and protein constituents. There are also a few crystal structures of covalently linked nucleic acid–protein moieties where the two components of the anticipated complex do not interact with each other (1231,1232). Although there is ample hydrogen bonding either to partner molecules or to water of hydration, nucleic acid and amino acid constituents do not display any obvious direct contact.

Salt bridges among phosphate, primary ammonium, and guanidinium groups (Figure 18-6). In this category, studies have been carried out with phosphate or diethyl phosphate cocrystallized with arginine, propylguanidine, and putrescine (1,4-diamino-n-butane) (1233–1235). The latter is an aliphatic diamine found in prokaryotic and eukaryotic cells as a precursor for spermine and spermidine (Box 15-1) and used as an analog of the terminal ammonium group of the lysine side chain.

In the crystal structures of putrescine cocrystallized with phosphate (1234) and with diethyl phosphate (1233), the terminal ammonium groups form three near-tetrahedrally oriented hydrogen bonds with free phos-

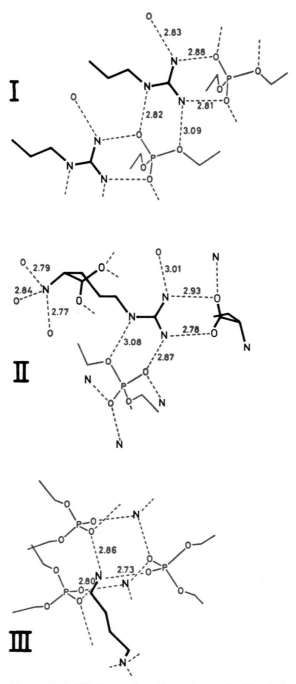

Figure 18-6. Hydrogen-bonding schemes in dietyl phosphate complexed with propylguanidine (I), arginine (II), and putrescine (III). Hydrogen bonds are indicated
by broken lines; N · · · O distances are in angstroms. From (1233).

phate oxygen atoms (Figure 18-6). Extended to more realistic systems such as polynucleotide–lysine interactions, the terminal lysine ammonium group could bind analogously to a free phosphate oxygen, with a second interaction to an ester oxygen of an adjacent phosphate along the chain. The third N–H would be free to attract a free phosphate oxygen of another polynucleotide chain (1233). Because polylysine binds preferentially to A–T-rich B-DNA (1236), it was also suggested that the terminal ammonium group of the lysine side chain binds not only to phosphate but also to $O_{4'}$, N_3, and O_2 of ApT or TpA sequences in B-DNA conformation. The three hydrogen bond acceptors are in the proper constellation for this interaction to occur (1234).

In case of the terminal guanidinium group of the arginine side chain, the situation is more complex. With phosphate, individual NH $\cdots$ O interactions were observed (1235), but with diethyl phosphate, a bidentate arrangement engaging one guanidinium group with two phosphates was found (Figure 18-6). These strong interactions are also expected to prevail in arginine–nucleic acid interactions and it was suggested that one guanidinium group can cross-link two DNA helices in fibers of DNA–polyarginine and DNA–protamine complexes (1233).

In general all of these studies on nucleic acid–protein model systems indicate that salt bridges between terminal lysine ammonium or arginine guanidinium groups interact strongly with phosphate oxygens. Preferential association is also observed if charge-transfer-type complexes can form. However, if hydrogen bonding or hydrophobic or stacking interactions are operative, multiple cooperative binding between nucleotides and protein surfaces or between nucleic acid double helix and protein β-sheet or α-helix is required.

18.3 Model Systems Combining Nucleic Acids and Synthetic Polypeptides or Protamines

Studies of this type rest mainly on spectroscopic and X-ray fiber diffraction methods although in one case, the tRNA–protamine complex, single crystals were investigated. Thus far, attempts to crystallize oligomeric nucleic acid–peptide systems have failed and therefore detailed information regarding the geometry of specific interactions between the two partners is still lacking. (For interaction between double-helical regions in tRNA and polyamines, see Section 15.6.)

Polyarginine and polylysine interact differently with polynucleotide phosphate groups. Because the synthetic homopolymers polyarginine and polylysine closely resemble protamines (Box 18-2) and histones (Box 19-1) in their predominantly basic character, they have been used exten-

Box 18-2. Oligoarginine Patterns in Protamine Sequences

Protamines are small proteins of molecular weight about 4200. Owing to the high content of arginine, they are highly basic in character and associate tightly with DNA and RNA. The arginines are clustered in sequences of four to six, and prolines occur in near-regular intervals, suggesting that the protamines adopt secondary structure composed of about four α helices.

In the table below, known sequences of protamines are arranged such that arginines (indicated as +) fall into register. Abbreviations used: A, alanine; E, glutamic acid; G, glycine; I, isoleucine; P, proline; Q, glutamine; S, serine; T, threonine; V, valine; Y, tyrosine. The protamines are assumed to adopt the secondary structure shown at the bottom, with four α helices joined by flexible hinges. Arginines are shown as black dots, and prolines and glycines as P and G. [From (1252)].

SALMINE AI	P + + + + S S S +	P V + + + + +	P + V S + + + + + +	G G + + + +
IRIDINE I(A)	P + + + + S S S +	P V + + + + +	P + + V S + + + + + +	G G + + + +
IRIDINE I(B)	P + + + + + + S S S +	P I + + + + +	P + + V S + + + + +	G G + + + +
IRIDINE II	P + + + + S S S +	P V + + + +	A + + V S + + + + + +	G G + + + +
THYNNIN Y1	P + + + + E A S +	P V + + + + +	Y + + S T A A + + + + +	V V + + + +
THYNNIN Y2	P + + + + Q A S +	P V + + + + +	Y + + S T A A + + + + +	V V + + + +
THYNNIN Z1	P + + + + + S S +	P V + + + + +	Y + + S T V A + + + + +	V V + + + +
THYNNIN Z2	P + + + + + S S +	P V + + + + +	Y + + S T A A + + + + +	V V + + + +
CLUPINE Z	P + + + + S + + A S +	P V + + + +	P + + V S + + + +	A + + + +
CLUPINE YI	A + + + + S S S +	P I + + + +	P + + + T T + + + +	A G + + + +
CLUPINE YII	P + + + T + + A S +	P V + + + +	P + + V S + + + +	A + + + +

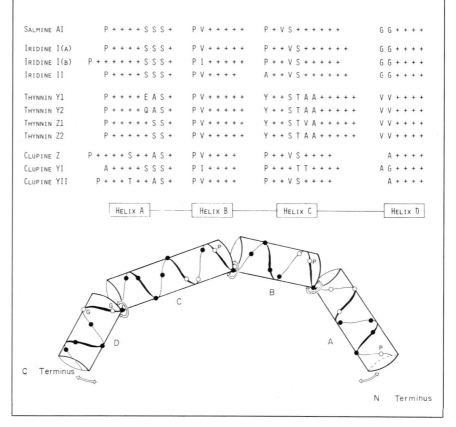

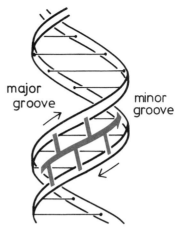

major
groove

minor
groove

Figure 18-7. Structure of B-DNA·protamine complex proposed from fiber diffraction studies, with the polypeptide arranged in the minor groove such that positive charges on adjacent amino acid side chains point alternatively up and down to neutralize negative charges on both DNA chains. From (1243).

sively as model compounds to study interactions between positively charged amino acid side chains and polynucleotide phosphate groups (1164,1192,1195,1237–1241).

Complex formation between polylysine or polyarginine and DNA is both cooperative and irreversible (1192,1239). The polypeptide side chains entirely neutralize the phosphate charges and, if concentrations are appropriate, lead to coacervation of the complex formed. In the case of DNA·polylysine, a condensation into rod-like and toroidal structures discussed in Section 19.1 is observed (1242). In all such complexes, even with addition of short oligopeptides, the melting temperatures T_m are elevated relative to free RNA or DNA (1239), indicating increased stability of the double helix due to shielding of negative charges.

As to the structures of the complexes, it has been suggested from X-ray fiber data that the polypeptide is located in the DNA minor groove in an extended form, with alternating side chains pointing "up" and "down" to neutralize phosphate groups on both polynucleotide strands (1239,1243, 1244) (Figure 18-7).

In DNA·polylysine a molar nucleotide:lysine ratio of 1:1 is reported (1192,1239). In complexes with RNA and with synthetic material such as poly(A)·poly(U) and poly(A)·2poly(U), nucleotide:lysine ratios of 3:2 and in the case of poly(I)·poly(C) of only 2:1 are found (1239). These different ratios were interpreted as being due to cross-linking between lysine-complexed and free RNA double helices, facilitated by the additional $O_{2'}$, hydroxyl and by the A conformation.

If DNA is complexed with mono-, oligo-, or polymeric arginine, it retains the B form under all humidity conditions ranging from 0% to saturation (1238,1240). With polylysine, however, a conformational change of the DNA double helix is observed upon complexation—the nature of the change is still under debate and has been described as A form, C form (1238,1245,1246), and distorted B form (1241). A similar behavior is also found with polyhistidine (1247).

Why do polyarginine and polylysine exert a different influence on the structural properties of DNA complexes? The reason might be that, as outlined in Section 18.2, complexes between phosphate and terminal ammonium or guanidinium groups have grossly different geometries, suggesting that arginine binds more tightly to DNA, in agreement with equilibrium dialysis studies (1248).

Interaction of DNA and single-stranded polynucleotides with aromatic amino acid side chains. The bookmark hypothesis. In these systems, only spectroscopic and hydrodynamic data are available. Using model di- and tripeptides in which one aromatic amino acid is joined to (dipeptide) or flanked (tripeptide) by either arginine or lysine, the association with double- and single-stranded polynucleotides has been studied in order to determine possibilities for the interaction of aromatic amino acid residues in basic peptides (histones, protamines) with double-helical DNA (1195, 1249–1251). In general, the reaction with single-stranded polynucleotides is primarily through charged end groups and the stacking of aromatic amino acid side chains with bases. The association constants in the case of poly(A) decrease in the order tryptophan > tyrosine > phenylalanine (1250).

For double-stranded DNA, the complex formation with aromatic amino acid side chains displays selectivity. A fast, strong electrostatic binding ($K_{ass} \sim 10^5$ moles/liter) is followed by a slow, weaker intercalation ($K_{ass} \approx 10^2$ moles/liter) of the aromatic side chain between base-pairs (1251a). Because the insertion is only partial, the aromatic residue acts as a wedge and causes bending of the double helix (1195,1251) (Figure 18-8). This bookmarking process is accompanied by a certain selectivity for A–T base-pairs which, as described in Chapter 6, are less strongly stacked than G–C base-pairs. For a series of dipeptides Lys–X-amide, with X representing different aromatic residues, additional selectivity decreasing in the order tryptophan > phenylalanine > tyrosine > leucine has been observed (1251).

By and large, complex formation of the type described considerably increases the melting temperatures T_m of single- and of double-stranded polynucleotides (1195,1249–1251). This is surprising, at least in double-helical DNA, and suggests that the energetically unfavorable kinking, which should lower the T_m, is overcompensated by favorable charge–charge interactions between phosphate and positively charged amino acid side chains.

α Helix–double helix interaction between protamine and tRNA. When

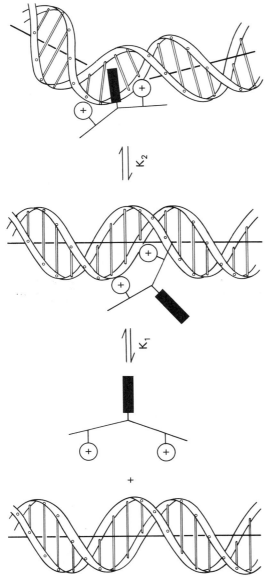

Figure 18-8. Interaction of tripeptides Lys–X–Lys (X = aromatic amino acid) with double-stranded DNA. A first, fast association at the outside of the DNA helix is stabilized by charge–charge attraction. This process is followed by slow, partial intercalation of the amino acid side chain between base-pairs (bookmarking), leading to bends in the DNA. From (1195).

salmon sperm protamine AI, a polypeptide mixture of average molecular weight 4200 and composed mainly of lysine and arginine (Box 18-2), was diffused into tRNA$^{Phe}_{yeast}$ orthorhombic crystals, it settled in niches formed by minor grooves of adjacently located tRNA molecules (1252). The overall structure of the complexed protamine did not resemble the stretched-out picture displayed in Figure 18-7 but rather that of a protein α helix. Assuming the sequence of protamine AI is representative of bulk salmon sperm protamine, it is striking to find that it consists of four stretches exhibiting four to six consecutive arginines. The electron density distribution in the tRNA–protamine complex suggested that protamine forms four short α helices composed of arginine which are connected by short peptides containing other amino acids.

The arginine guanidinium groups are at the periphery of the short helices and interact with tRNA phosphates. Based on this geometry, a model for DNA–protamine interaction was proposed with protamine α helices associated with the major grooves of DNA—occupation of the minor groove as proposed from X-ray fiber diffraction (1243,1244) could not be ruled out. Owing to the radially extended arginine side chains, cross-links to adjacent DNA molecules can occur, and this is consistent with electron micrographs of reconstituted nucleoprotamine showing a network of interconnected DNA molecules (1253). Convincing as this proposal might be, it has been disputed (1254) and, based on electron micrographs, a comb- or star-shaped protamine structure was proposed around which DNA would wind in supercoil form. As we are not in a position to judge which model describes nucleoprotamine more closely, we have to wait for more evidence. But we should keep in mind that α helix–double helix interaction as observed in this tRNA–protamine complex probably occurs as well in DNA–protein interactions (see section 18.6).

18.4 Nucleotides and Single-Stranded Nucleic Acids Adopt Extended Forms When Binding to Proteins

As we have found in previous chapters, nucleotides occurring as monomeric units or incorporated into polynucleotide chains with helical, ordered configuration adopt a rather confined "rigid" or "standard" geometry with the salient features:

- base is oriented *anti* with respect to sugar, i.e., torsion angle about glycosyl $C_{1'}$–N link is in the range $-90° \leq \chi \leq -160°$;
- sugar puckering mode is either $C_{2'}$-*endo* or $C_{3'}$-*endo;*
- torsion angle γ describing rotation about $C_{4'}$–$C_{5'}$ bond is preferentially in $+sc$ range ($\sim 60°$);
- C–O torsion angles β and ϵ are in *trans* range ($\sim 180°$);

- in helical polynucleotides with right-handed screw sense, P–O torsion angles α and ζ are in $-sc$ range ($\sim300°$).

These structural features are, in general, also maintained if DNA or RNA double helices interact with polypeptides or if DNA is condensed by the addition of certain reagents or histones (Chapter 19). With mononucleotides and single-stranded, nonhelical polynucleotides, however, the situation is different if binding occurs to proteins.

When binding to recognition sites in proteins, nucleotides adopt an extended, "open" configuration. The standard structure of a nucleotide is also more or less preserved as it is recognized and bound to the active site of a protein—with one important difference: the orientation about the $C_{4'}$–$C_{5'}$ bond, γ, changes from $+sc$ to $-sc$ or ap, corresponding to a rotation of the $O_{5'}$ and attached phosphate away from the furanose ring. This movement opens up the nucleotide unit and exposes its sugar, base, and phosphate groups to the surface of the protein to facilitate and optimize mutual recognition and binding contacts.

Among the collection of crystallographically determined nucleotide conformations in protein–nucleotide complexes given in Table 18-3, there are only two examples, the binding of GDP to elongation factor Tu (1261) and the nicotinamide–ribotide unit of NAD^+ bound to the active site of glyceraldehyde-3-phosphate dehydrogenase from lobster (1256), which exhibit γ in the usual $+sc$ range. In the latter, adenine in two of the four subunits (the "green" ones, Table 18-3) is also rotated into *syn* range whereas it is *anti* in all other NAD^+ molecules listed in Table 18-3.

A *syn* conformation is also probable for the central nucleotide of the triplet bound to the subunit of tobacco mosaic virus (1266,1267). It is remarkable that the 8-bromoadenosine derivatives listed in Table 18-3 are all in *anti* form although the 8-bromo substituent tends to force the purine ring into *syn* (Section 4.7). This means that the binding energy is sufficient to compensate for a sterically unfavorable geometry of the nucleotide, a factor also coming into play in the unusual orientation of torsion angle γ.

18.5 Nature of Protein–Nucleotide and Nucleic Acid Interaction and Recognition

In most of the examples cited in Table 18-3, the geometry of both the nucleic acid and the protein are so well established that details concerning mutual interactions can be given. The reader should recall, however, that in protein structures—and these are considered here—resolution tends to be in the range 2–3 Å and therefore fine details such as hydrogen atoms cannot be seen. Yet the information obtained is generally sufficient to develop a picture of protein–nucleotide complexes which allows us to rec-

Table 18-3. Conformations of Nucleotides when Associated with Binding Sites of Proteins

Protein	Nucleotide[a]		Torsion angles (°)					Sugar pucker	Reference
			α	β	γ	ζ^{b}	χ^{c}		
Lactate dehydrogenase	5'-AMP			177	178		−91	$C_{3'}$-endo	(1255)
Lactate dehydrogenase	ADP		165	168	−168		−92	$C_{3'}$-endo	(1255)
Lactate dehydrogenase	NAD⁺	A	103	−176	−72	156	−96	$C_{3'}$-endo	(1255)
		N	109	178	−87	37	−90	$C_{3'}$-endo	
Lactate dehydrogenase	NAD⁺-pyruvate	A	109	−156	−78	156	−91	$C_{3'}$-endo	(1255)
		N	87	−171	−53	49	−74	$C_{3'}$-endo	
Glyceraldehyde-3-phosphate dehydrogenase	NAD⁺ green subunit	A	+ac	+ac	ap	ap	syn	$C_{3'}$-endo	
		N	+sc	+ac	+sc	ac	syn	$C_{2'}$-endo	
	red subunit	A	+ac	ap	−sc	ap	anti	$C_{2'}$-endo	(1256)
		N	+sc	+sc	+sc	+ac	syn	$C_{2'}$-endo	
Malate dehydrogenase	NAD⁺ subunit 1	A	58	−167	−78	−166	−70	$C_{3'}$-endo	(1257)
		N	147	160	−40	88	−48	$C_{2'}$-endo	
Liver alcohol dehydrogenase	ADP-ribose	A	177	167	159	39	−102	$C_{3'}$-endo	(1258)
		R	51	152	154	112	—		

								Pucker	Ref
Liver alcohol dehydrogenase	8-Br-ADP-ribose	A	165	167	171	39	−80	$C_{2'}$-endo	(1258)
		R	51	152	154	112	—		
Dihydrofolate reductase	NADPH	A	77	122	161	−112	−93	$C_{3'}$-endo	(1259)
		N	−74	−156	−68	−74	−39	$C_{2'}$-endo	
Tyrosyl tRNA synthetase	Tyrosinyl-5'-adenylate		170	140	−sc		anti	$C_{3'}$-exo	(1260)
Elongation factor Tu	GDP		60	160	+sc	30	anti	$C_{3'}$-endo	(1261)
Hexokinase	8-Br-ATP Cobalt complex				ap		anti	$C_{2'}$-endo	(1262)
Pancreatic ribonuclease S	UpcA (uridylyl-3'5'-5'-methyleneadenosine)	U	—	—	—	+ac	anti	$C_{3'}$-endo	(1263)
		A	+sc	ap	ap	—	anti	$C_{3'}$-endo	
Pancreatic ribonuclease S	Cp(2',5')A (cytidylyl-2'5'-adenosine)	C			270	−107		$C_{3'}$-endo	(1264)
		A	−42	173	153	121	−82	$C_{3'}$-endo	
Staphylococcal nuclease	pdTp (deoxythymidine-3'5'-diphosphate)				ap		anti	$C_{2'}$-endo	(1265)
Tobacco mosaic virus[d]	RNA (three nucleotides per protein subunit)	1	145	220	170	70	−70	$C_{2'}$-endo	(1266)
		2	40	185	195	275	−10	$C_{2'}$-endo	
		3	150	135	270	290	−75	$C_{2'}$-endo	

[a] A, C, N, R indicate adenosine, cytidine, nicotinamide ribose and ribose moieties.

[b] In NAD^+, NADPH and ADP, ζ denotes tortion angle $O_{5'}$–P–O–P.

[c] 180° subtracted from original data defined as $O_{4'}$–$C_{1'}$–N–C(8 or 6) in order to conform with our nomenclature.

[d] $\delta = 135°$ for all three riboses, torsion angles ϵ are (1), 250°; (2), 210°; (3), 210°. In a more recent publication, $C_{3'}$-endo puckering mode was reported for all three nucleotides (1267).

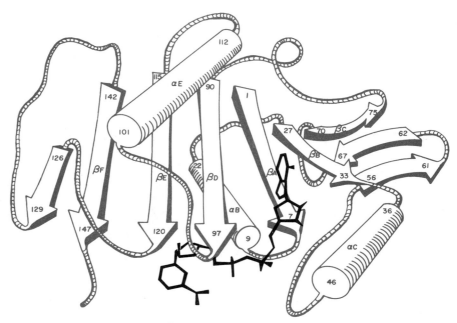

Figure 18-9. Structure of the NAD⁺ binding domain of glyceraldehyde phosphate
dehydrogenase. The folding of the very complex protein structure is represented
symbolically; the β structure is indicated by arrows pointing from the N to the C
terminus and α helices, by cylinders. Numbers give amino acid positions in sequence. (An even more abstract description is given in Figure 18-10.) Note that the
"Rossmann fold," marked by domains βxβxβ where x stands for β or α structure,
binds to NAD⁺, with adenosine 5'-phosphate interacting with β_A, α_B, and β_B, and
nicotinamide ribose 5'-phosphate with β_D, α_E, and β_E. From (1271).

ognize features of interest. As we shall see, protein–nucleotide interactions depend on two main characteristics—the overall topology of the
two partners and, in fine detail, interactions between nucleotide and protein main-chain or side-chain atoms.

   **A general topological feature in nucleotide-binding proteins: The Ross-
mann fold.** A number of proteins which interact with nucleotide coenzymes (NAD⁺, NADP) or nucleoside di- and triphosphates bind the nucleotide units by means of a characteristic topological sequence of α
helices and β strands. This nucleotide binding domain is built up of a six-
stranded parallel sheet and four connecting α helices, an arrangement displaying in a broad sense pseudo-twofold symmetry (Figures 18-9, 18-10).
Structural comparisons between NAD⁺-dependent dehydrogenase–
holoenzyme complexes inspired Rossmann and his colleagues to propose
an evolutionary conservation of at least the central four β sheets and asso-
ciated α helices, the $\beta\alpha\beta \cdot \beta\alpha\beta$ or $(\beta\alpha\beta)_2$ domain structure (1269,1270).
The dinucleotide NAD⁺ is bound near the twofold axis such that each
$(\beta\alpha\beta)$ nucleotide binding domain binds one nucleotide unit.

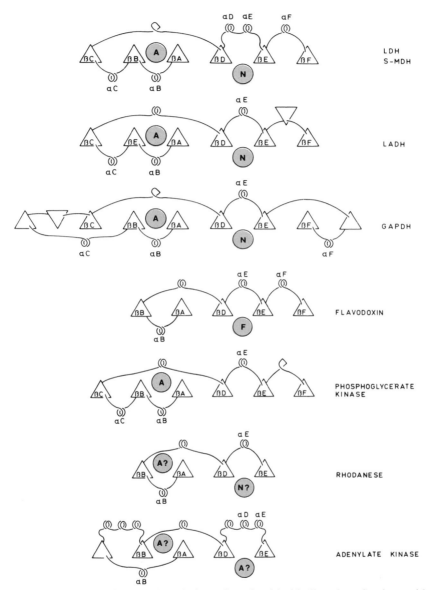

Figure 18-10. Very abstract description of nucleotide binding domains in a wide range of proteins. β strands and α helices are indicated by triangles and curls, respectively. N and A stand for nicotinamide–ribotide and adenosine 5'-phosphate. (LDH, S-MDH, LADH, GAPDH = lactate-, soluble malate-, liver alcohol-, glyceraldehyde-3-phosphate dehydrogenase) In each case the nucleotide units are bound by the $\beta\alpha\beta$ sequence. From (1270).

Table 18-4. Description of Nucleotide-protein Interaction in Two Complexes Studied at High Resolution. In this table, protein side chains are designated Thr64, Ile102, etc., and peptide carbonyl as CO-Gly17, peptide amide as NH-Ala6, etc. Salt bridge and hydrogen bond acceptors or donors on the nucleotide side are given in parentheses [e.g., $(O_{3'})$, (phosphate)].

Nucleotide group	Bound to protein via		
	Salt bridge	Hydrogen bond	Hydrophobic
Dihydrofolate Reductase and NADPH (1259)			
Adenine			Leu62, His64, Thr63, Ile102, His77, Asp78
Adenine 2'-mononucleotide ribose	Arg43 (phosphate)	Thr63 (phosphate) His64 (phosphate) Glu101 $(O_{3'})$	Ile102, Gly42
Pyrophosphate	Arg44	NH-Thr63, NH-Ala100, Thr45, Thr126	Gly99
Nicotinamide mononucleotide ribose		CO-Gly17 $(O_{2'})$, CO-His $(O_{3'})$, Ser48 $(O_{2'})$	Ile13, Gly14
Nicotinamide		NH-Ala6 (O_7), CO-Ala6 (N_7), CO-Ile13 (N_7)	Trp5, Ala6, Ile13, Leu19, Trp21, CO-Ala97
Ribonuclease S and Cp(2',5')A (1263)			
Cytidine		Thr45 (N_3)	Thr45, Phe120
Ribose		Lys41 $(O_{4'})$, His119 $(O_{5'})$	
Phosphate		Gln11 (O), His12 (O), NH-Phe120 (O)	Phe120, His119
Adenine		Glu111 (N_1), Gln69 (N_6), Asn71 (N_6)	Glu111, Ala109
Ribose		His119 $(O_{4'})$	His119

Subsequently, comparable $\beta\alpha\beta$ domains were observed in a number of other nucleotide binding proteins (Figure 18-10) so that the notion of a general topology designed specifically to recognize nucleotides emerged. Basically, this means that first, a specific overall three-dimensional geometry is required for constructing a nucleotide binding site and, second, certain amino acid side chains must be associated with it to confer specificity.

Nucleotide–protein interactions are versatile and multifunctional. A glance at Table 18-4 and Figures 18-11 and 18-12 (where a few of the known nucleotide–protein interactions are given in more detail) indicates that by no means are there predetermined, exclusive and specific base–side chain interactions. Rather, a full palette of possibilities is employed.

Nevertheless, we will find some general trends when discussing a few examples more closely.

Specific and unspecific "hydrophobic pockets." NADPH binds to dihydrofolate reductase, an enzyme catalyzing the NADPH-dependent reduction of dihydrofolate to tetrahydrofolate. It is remarkable that in this binding adenine is located in a hydrophobic pocket and is not involved in any hydrogen bonding (Table 18-4). Although His64 is stacked with adenine, this interaction cannot be specific because His64 is not invariant in this class of enzymes. In contrast, the $O_{2'}$-phosphate of NADPH forms hydrogen bonds with Thr63, His64, and Arg43, the latter being invariant in all dihydrofolate reductases, and thus provides for a general and specific interaction. Contacts with peptide groups of the main chain are evident for pyrophosphate and nicotinamide ribose, additionally supported by hydrogen bonding and charge attraction to side chains, especially to Arg44. Of importance are invariant residues Gly42 and Gly99 which appear to control the fitting of the coenzyme to the protein matrix and, if replaced by amino acids with larger side groups, would cause steric interference. In contrast to the adenine pocket, the nicotinamide binding site is more specific, with main-chain amide groups recognizing the 3-carboxamide group on the nicotinamide ring. This latter binding, it should be noted, has no resemblance with NAD^+-dependent dehydrogenases where the nicotinamide residue is recognized primarily by interactions with side chains (1271).

Tight binding to phosphates and unspecific hydrophobic trough in staphylococcal nuclease. Other well-studied examples for nucleotide–protein interactions are cytidylyl-2',5'-adenosine (Cp(2',5')A)·pancreatic ribonuclease S (1264) and the ternary complex Ca^{2+}-deoxythymidine-3',5'-diphosphate (pdTp)·staphylococcal nuclease (1265). In the latter, illustrated schematically in Figure 18-12, pdTp acts as an inhibitor and Ca^{2+} as an activator for the nuclease which hydrolyzes, both in the DNA and in the RNA series, the $O_{5'}$–P bond to release a nucleotide or nucleic acid fragment with a free $O_{5'}$–H group. The inhibitor is well anchored in the active site, with one hydrogen bond linking the phenolic hydroxyl of Tyr85 with the 3'-phosphate. The 5'-phosphate is involved in salt bridges with Arg87 and Arg35 which both form bidentate hydrogen bonds of the type described in Figure 18-6. Additionally, the latter phosphate is bound to Ca^{2+} which is octahedrally coordinated by Asp40, Asp21, the main-chain carbonyl of Val41, and two water molecules. This rather firm, well-defined binding of the two phosphates should not apply to the furanose and base moieties because the enzyme is not specific either for base or for sugar type (RNA or DNA). We find, indeed, that the thymine residue is located in a shallow hydrophobic trough at the surface of the nuclease and all contacts between protein and specific nucleotide sites—O_2, O_4, N_3H—are only via water bridges, a binding that obviously allows the enzyme to recognize all four kinds of bases.

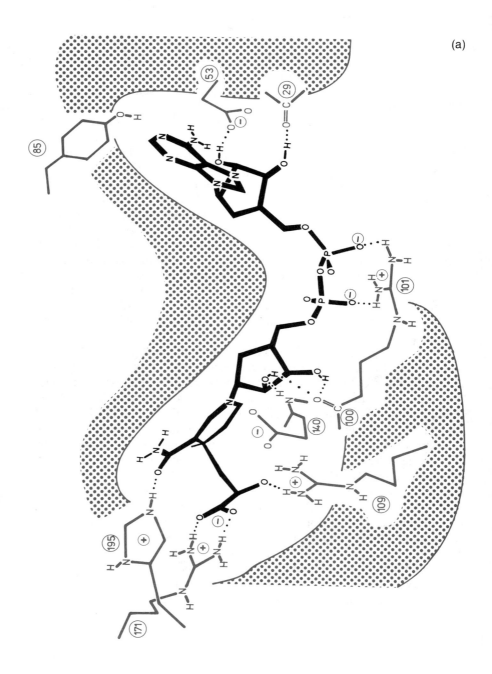

(a)

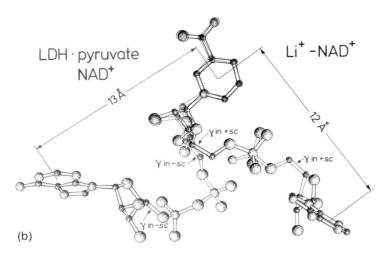

(b)

Figure 18-11. (a, opposite). Schematic representation of coenzyme binding as appears in the ternary lactate dehydrogenase·NAD⁺–pyruvate complex. Note extended configuration of NAD⁺–pyruvate and salt bridges between pyrophosphate and Arg101 and pyruvate and Arg 171. Adenine is located in a hydrophobic pocket with no obvious hydrogen bonding, whereas the carboxamide oxygen of nicotinamide is in contact with His195. All ribose $O_{2'}H$ and $O_{3'}H$ hydroxyls are hydrogen bonded by main-chain amide groups and by Asp53. From (1255). **(b, top).** Comparison of conformations of NAD⁺ at the active site of lactate dehydrogenase (color) and in the crystalline Li⁺-complex. The two molecules are oriented such that nicotinamides superimpose. This shows that nicotinamides and adenines are nearly perpendicular to each other and separated by $\sim$12 Å, i.e., the overall structural features of the NAD⁺ molecules are similar. In the Li⁺-complex all torsion angles are normal, with orientations γ about the $C_{4'}-C_{5'}$ bonds in the $+sc$ range. In the complex with protein, however, γ is changed to $-sc$, directing the pyrophosphate group away from the nicotinamide and adenine bases and exposing the phosphates to the active site of the protein. In contrast, in Li⁺-NAD⁺ the pyrophosphate is more hidden between the nucleoside moieties. Drawn with coordinates in (709,1255).

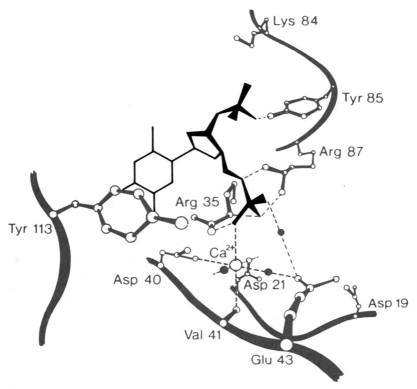

Figure 18-12. Schematic description of the nucleotide binding site in staphylococcal nuclease. The 3′-phosphate group of the inhibitor pdTp is bound rather weakly
to Tyr85 whereas the 5′-phosphate is tightly bound by two arginine side chains
Arg35 and Arg87, via a water molecule (black dot), to Glu43 and further coordinated with Ca^{2+}. The $P–O_{5'}$ bond is cleaved in the enzymatic hydrolysis of nucleic
acids (DNA as well as RNA). There is no well-defined binding to the base which is
located in a shallow trough, rendering staphylococcal nuclease not base-specific at
all. From (1265).

**Weak binding of phosphate and recognition of pyrimidine bases in pan-
creatic RNase S.** As with staphylococcal nuclease, pancreatic RNase S is
also a 5′-hydrolase; yet the mechanisms of cleavage are totally different.
RNase S acts without Ca^{2+} and requires the $O_{2'}H$ hydroxyl to form a 2′,3′-
O,O-cyclophosphate intermediate—therefore, it is specific for RNA. In
addition, RNase S cleaves exclusively at pyrimidine nucleotides to release RNA fragments with 3′-terminal pyrimidine 3′-phosphate.

Cytidylyl-2′,5′-adenosine (Cp(2′,5′)A) and uridylyl-3′,5′-methylene-
adenosine (UpcA) are modified at the phosphodiester link and are inhibitors for RNase S. In contrast to the tight phosphate binding in staphylococcal nuclease, the interactions in RNase S are rather soft, involving
His12, His119, and Gln11 (Table 18-4). Adenine, although not a specific

base for this enzyme, is hydrogen bonded through the N_6 amino group and N_1. For pyrimidine bases located in the pyrimidine binding site in slightly different orientations, the peptide NH of Thr45 hydrogen bonds with O_2 of UpcA but not with O_2 of Cp(2′,5′)A (distance N $\cdots$ $O_2 \sim 4$ Å). The hydroxyl group of the Thr45 side chain donates a hydrogen bond to N_3 of C but acts as acceptor from N_3H of U (Figure 18-13). Discrimination of purine versus pyrimidine bases by only the Thr45 hydroxyl and NH of the adhering peptide group appears marginal in view of the similar disposition of functional groups. However, a peptide chain (containing Thr45) running in front of N_3 and O_2 of pyrimidine bases would interfere sterically with purines if phosphate and sugar are held in positions as found for UpcA and Cp(2′,5′)A complexes with RNase S. In this respect it is of interest that 8-oxoadenosine 3′-phosphate binds in *syn* form to RNase S in

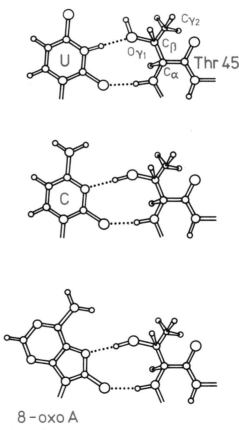

8 - oxo A

Figure 18-13. Diagrammatic description of hydrogen-bonding interactions between Thr45 in RNase S and pyrimidine bases uracil and cytosine (top) and also with 8-oxoadenine rotated into *syn* orientation (bottom). Circles with increasing radii indicate H, C, N, O.

the pyrimidine site, with the 8-oxo group mimicking pyrimidine O_2 whereas adenine N_7 resembles cytosine N_3 (1272).

Specific recognition of guanine via main-chain and side-chain interactions in ribonuclease T_1. In contrast to ribonuclease A from mammalian source, the prokaryotic ribonuclease T_1 (RNase T_1) extracted from the fungus *Aspergillus oryzae* exhibits very specific recognition for only one base, guanine. The crystalline complex between this enzyme and the inhibitor 2′-guanylic acid (2′-GMP) provides for a detailed picture of combined hydrogen-bonding and stacking interactions involving protein main- and side-chain atoms which envelope and bind the guanine base (1272a). As Figure 18-14 suggests, hydrogen bonding is to main-chain atoms of the enzyme (Asn43)NH---N_7, (Asn44)NH---O_6, (Tyr45)NH---O_6, (guanine)N_2H---O=C(Asn98), and to the side-chain carboxylate oxygen atoms

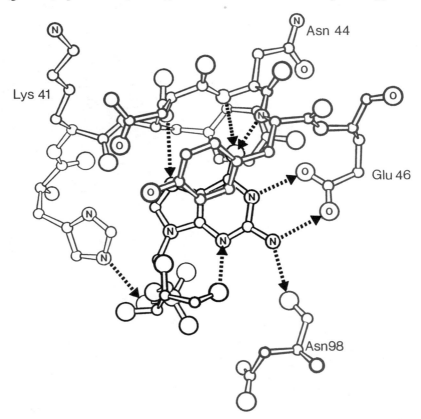

Figure 18-14. The enzyme ribonuclease T_1 cleaves RNA specifically after the 3′-phosphate of guanosine. Shown here is a section of the crystal structure of the complex between ribonuclease T_1 and the inhibitor 2′-guanylic acid which binds in the *syn* form. Recognition is by hydrogen bonds (dashed) between main-chain and side-chain functional groups of the enzyme and functional groups of guanine. In addition, the phenoxy groups of Tyr45 and of Tyr42 sandwich the guanine plane. From (1272a, with modifications after refinement of the crystal structure at 1.9 Å resolution).

of Glu46, with guanine N_1H and N_2H as donors. In addition, guanine is sandwiched between the phenolic side-chains of Tyr42, Tyr45. All these interactions discriminate G from the other three bases but permit recognition of inosine, which has all functional groups of G except the N_2 amino group. In addition, however, the side chain of Tyr45 swings over and stacks at a 3.5 Å distance with G, an interaction certainly favoring purine over pyrimidine bases (see Chapter 6 and previous sections), and guanine is in *syn* orientation not easily accessible for pyrimidine bases.

The α helix acts as a dipole and attracts phosphate groups at its N terminus. Because all consecutive peptide bonds and all hydrogen interactions point in the same direction in α helices, a rather strong dipole moment along the helix axis builds up, with a positive pole at the N terminus. The potential at about 5 Å distance from the N terminus of an α helix of 10 Å length (approximately two turns) has been estimated to be about 0.5 V. Based on this figure, then, binding of a single negative charge (a phosphate) contributes an attractive energy of 0.5 eV or about 12 kcal/mole. Assuming that water molecules that have to be replaced consume about 0.2 kcal/mole, the net attraction is still considerable (1273).

Such kinds of α helix–phosphate interactions have been observed in a great number of proteins. Of major interest in this connection is the "Rossmann fold" in dehydrogenases with $\beta\alpha\beta$ configuration and binding of phosphate groups in dihydrofolate reductase, triose phosphate isomerase, tyrosyl-tRNA synthetase, and p-hydroxybenzoate hydroxylase (1259,1260).

RNA is in extended form when bound to tobacco mosaic virus. In viruses, identical protein subunits are arranged in spherical or helical form to provide for a capsule which encloses DNA or RNA depending on virus type. Because in spherical viruses, the outer shell symmetry does not concur with that of the nucleic acid located in the interior, the latter appears as an averaged-out smear of electron density which could not yet be interpreted (1274). In the rod-like tobacco mosaic virus (TMV), however, folding of the single-stranded RNA follows the helical subunit arrangement and therefore the RNA chain can be seen, although bases are in "random" sequence and appear as averaged disks.

TMV consists of 2140 protein subunits, each of molecular weight 17,420 and arranged on a helix with $16^1/_3$ subunits per turn of pitch height 23 Å. Close to the "empty" interior of the helix, a single strand of 6400-nucleotide-long RNA is wound. The TMV subunit contains 158 amino acid residues, of which 60 are found in four prominent α helices called left, right slewed (LL, LR) and left, right radial (RL, RR), all oriented more or less radially with respect to the helix axis (Figure 18-15) (1266–1268, 1275). RNA runs as a single strand on a helix with radius 40 A and is held primarily only by the LR α helix to which a nucleotide triplet is attached. The most prominent components of the binding site are Asp115 and Asp116 which interact with ribose $O_{2'}H$ groups of residues 1 and 3, a motif also apparent in NAD^+-dehydrogenase complexes. The correspond-

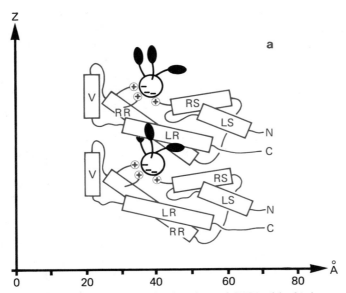

Figure 18-15. (a) Schematic drawing of RNA (black) interacting with two tobacco mosaic virus (TMV) subunits (color). The latter are in helical arrangement
(cylinders) and only a section of the virus is displayed here. The RNA binding site
is primarily at residues 114 to 123 of helix LR (see b, opposite), where it is grasped
by the bases drawn as black tabs. Another strong binding occurs between RNA
phosphates and Arg90, Arg92 of the RR helix belonging to the subunit immediately below (indicated by ⊕). (b) Another view of the RNA binding site in TMV,
looking along the helix axis. Displayed are two repeat units of RNA and LR helix.
Note extended configuration of the RNA chain and embracing of LR helix by
bases of nucleotides 1 and 3 drawn as elliptical figures. Purine nucleotide 2 is
forced out on a wider radius because nucleotides 1 and 3 are held so close together. From (1268).

ing bases are pocketed in hydrophobic areas and clasp against the α helix;
base 2 in *syn* form (whereas 1 and 3 are *anti*) is at a larger radius and could
hydrogen bond to Ser123. The tight binding of nucleotides 1 and 3 to the
LR helix contracts the RNA chain so that the phosphate residues approach each other suitably to form salt bridges with Arg90 and Arg92 from
the RR helix belonging to the subunit directly underneath.

**Some general conclusions about protein–nucleic acid interactions can be
drawn.** As stated at the beginning of this section, it is obvious that a wide
range of protein–nucleotide interactions does exist. Not only do we observe contacts to amino acid side chains but peptide groups contribute as
well, as both acceptors and donors of hydrogen bonds. It appears even
that interactions between nucleic acid bases and peptide functional
groups are more preferred compared to amino acid side chains. Considering the widely varying character of the latter from positively to nega-

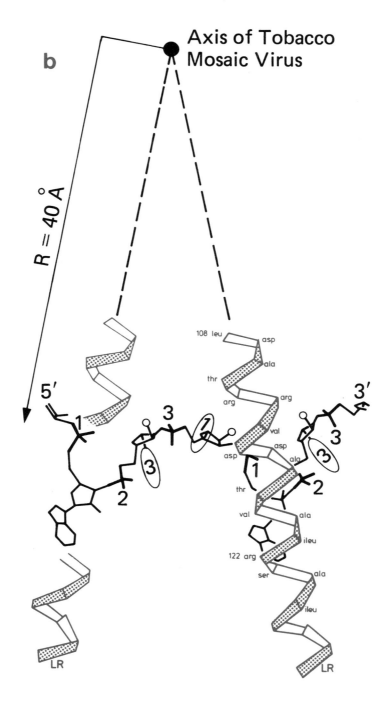

tively charged, this is surprising but can be explained with geometrical requirements that must be met during the recognition process. Thus the side chains are rather flexible and only rarely confined to fixed positions whereas the main chain, in general, is held in a certain well-determined configuration. It therefore provides for a matrix which a base with complementary functional groups can scan readily. In addition, twists and bends within the main chain and its actual position and orientation at the active site are very powerful instruments for modulation and fine tuning of the interactions which in the case of amino acid side chains with bifunctional guanidinium, carboxylate, and carboxamide groups in fixed, predetermined geometry would not be possible to such an extent.

In contrast, phosphate groups appear to prefer salt bridges with side chains such as Arg, Lys, and His although hydrogen-bonding contacts to the other groups, including main chain, are observed as well. Moreover, phosphates are attracted by the dipole moment of α helices. For ribose hydroxyls, again no general scheme can be derived, yet it appears that hydrogen bonds to carboxylate groups are especially favored. Stacking interactions between bases and aromatic side chains are found in only a few cases and the often quoted hydrophobic pocket has in most instances really pocket-like character, without any obvious specificity and direction.

18.6 Proteins Binding to DNA Double Helix and Single Strands

Thus far, five different DNA binding proteins have been studied by X-ray crystallographic techniques and their three-dimensional structures are known: the *cro* and λ repressors, the catabolite gene activator protein (CAP), the gene 5 product of fd phage, and the oligopeptide netropsin. Because all these proteins were investigated per se without attached DNA, the presumed protein–DNA binding interactions are still a matter of discussion, yet some preliminary conclusions can be drawn which are of interest here.

Netropsin, an oligopeptide binding preferentially to A–T-rich regions of B-DNA. The antibiotic netropsin binds to A–T stretches of B-DNA with binding constants around 10^8 liters/mole [measured for poly(dA)·poly (dT) (1276)] and covers about five base-pairs (1277). The molecular structure of netropsin, Figure 18-16, displays an extended, bow-shaped configuration with all potential hydrogen bond donor (N-H) groups oriented towards the concave side. Model building studies fitting netropsin to B-DNA suggested that binding occurs to the minor groove of DNA, with the antibiotic approximately parallel to the helix axis and donating hydrogen bonds from NH to pyrimidine O_2 and purine N_3 sites which, in B-DNA, are in near-isomerophons places so that for A-T and T-A base-pairs, binding is equally probable.

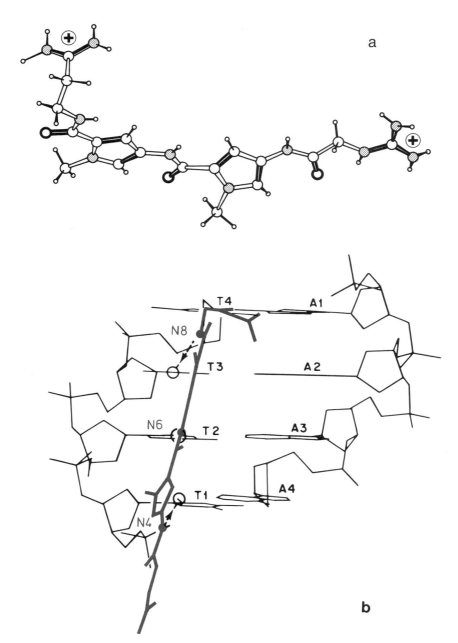

Figure 18-16. (a). Molecular structure of the DNA binding antibiotic netropsin. All potential hydrogen bond donor sites (NH groups stippled) point to the same (concave) side. (b). Model fitting of netropsin (black) to the minor groove of B-DNA drawn in color. The terminal guanidinium group is not involved in binding and the propionamidine bridges over to an adjacent DNA strand. From (1277).

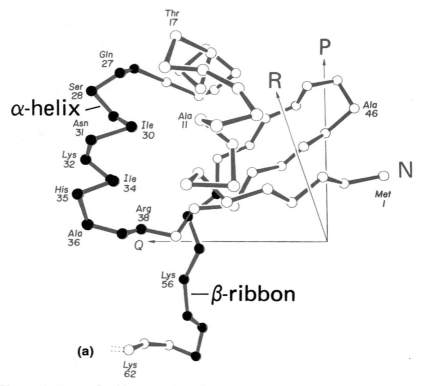

Figure 18-17. (a). Backbone tracing of *cro* repressor protein, connecting only α-carbon atoms. Presumed parts binding to DNA are helix residues 27 to 38 and C-terminal residues 53 to 62 (filled black spheres), the latter forming an antiparallel β-pleated ribbon with another *cro* molecule related by dyad Q. Two other dyads, P and R, augment this dimer to form a tetramer. Tail residues 63 to 66 "wag" and cannot be seen in electron density. They may act as "feelers" for recognizing DNA. (b,c). Model fitting of *cro* repressor dimer to B-DNA. Dyad axis Q coincides with dyad axis in B-DNA. (b). Helices and β-pleated ribbon interacting with B-DNA grooves are drawn in heavy lines and dyad axes P, Q, and R are indicated; view is along Q, with protein "behind" DNA helix. (c). Model is rotated by 90°, dyad Q is now in the plane of the paper, and N termini of the two *cro* molecules are marked. From (1278).

Interaction of B-DNA with α-helix and β-pleated sheet in *cro* repressor (1278). The *cro* protein from bacteriophage λ, a 66-amino acids-long polypeptide, recognizes certain near-palindromic operator regions of specific sequence within DNA of the phage, binds to them and thus prevents expression of the *cro* gene. In solution, *cro* repressor exists as dimer or tetramer, in the crystal a tetramer with near-222 point group symmetry is found and in the complex with DNA, *cro* probably also binds as dimer or tetramer. The subunit of *cro* consists of three strands of anti-parallel β-pleated sheet (residues 2–6, 39–45, 48–55) and three α-helices (residues

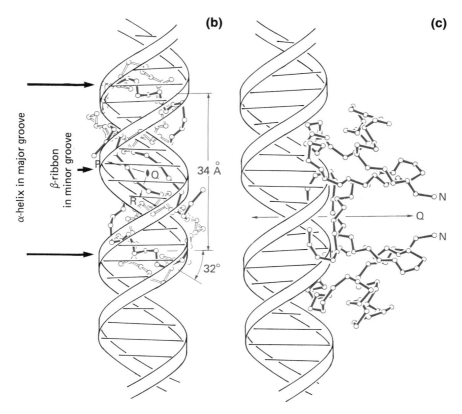

(b)   **(c)**

7–14, 15–23 and 27–36). The carboxyl-terminus forms a tail which interacts with that of another *cro* molecule (related by dyad Q, Figure 18-17) via β-pleated sheet hydrogen bonding. Another dyad, R, produces the tetramer which is seen in the crystal.

If looking along the Q dyad of the dimer, (Figure 18-17, b), it strikes one that helices formed by residues 27–36 are in antiparallel orientation, about 34 Å apart, and inclined to the vertical by 32°, with the β-pleated sheet formed by the carboxy termini in between, and ideally suited to bind to the B-DNA double helix. In the model of the complex, *cro* binds to DNA along one face and occupies the major groove sites with its α helices. The β-pleated sheet is accommodated in the minor groove, in agreement with chemical studies (1279). The *cro* dyad axis Q coincides with the intrinsic dyad of the B-DNA double helix. That is, the inherent dyad symmetry in the near-palindromic operator region to which *cro* repressor binds is matched by the symmetry of the repressor dimer. The interactions between nucleic acid double helix and structure elements of proteins, α helix and β sheet, can now be looked at more closely, but we recall that we can only model-build one—still unknown—partner, DNA.

Nevertheless, the model proposes (1279) that Ser28, Ala29, Asn31, Lys32, Ala33, and His35 of the α helix are in a position to interact with the

DNA major groove. Lys 32 and nearby Arg 38 and Lys39 could assist in binding to phosphates. In addition, the antiparallel β-pleated ribbon formed by the tail sequences Glu54, Val55, and Lys56 probably contacts the minor groove in a manner described in Section 18.1. In the electron density of the *cro* repressor the four terminal residues of the tail are disordered, expressing increased mobility which also could prevail in solution. One could infer that these flexible termini, equipped with two lysines (Lys62, Lys63), act as feelers to initiate and guide binding to DNA.

Having a well-fitted model for the *cro*-repressor interacting with DNA and knowing the respective amino acid and nucleotide sequences, it was tempting to look for specific recognition. It was found that *cro* side chains of Arg38, Lys32, Ser28, Gln27 and Tyr26 can form hydrogen bonds to the major groove sides (Figure 18-3) of guanines and adenines (1279a). The model explains chemical data, yet final proof will only be achieved if the *cro*-repressor·operator complex has been crystallized and analyzed by X-ray diffraction methods.

Operator-binding domain of λ-repressor displays similar structural features as *cro* (1279b, 1279c). The intact repressor from phage λ is a protein 236 amino acids long and consists of two domains. If mildly hydrolyzed with papain, a 92-residue N-terminal fragment is cleaved off which binds specifically to the operator DNA just as λ-repressor does. The crystal structure analysis of this fragment reveals that it is folded into five α-helices, of which two (helices 2 and 3) are in near-perpendicular arrangement comparable to helices 2 and 3 (residues 15–23 and 27–36) in *cro* and to helices E and F in CAP protein (*vide infra*). In *cro* and λ, helices 3 can be fitted into the major grooves of B-DNA (Figure 18-17) and in λ, N-terminal arms can additionally embrace DNA such that they are inserted in the major groove at the "backside" of the double helix.

Repressors display sequence homology at the putative DNA binding amino terminus. The amino-terminal sequence of *cro* repressor comprising the putative DNA binding site was compared with amino-terminal sequences of *cro* repressors from phage λ, phage 434, the cI and cII proteins from phage λ, the repressor protein from Salmonella phage P22, and *lac* repressor (1279d). With both amino acid and corresponding gene sequences, a good homology was found, suggesting that the amino termini of all these repressors are also similar in secondary structure and evolved from a common ancestor. If this hypothesis is correct, chemical protection and genetic modification experiments carried out with *lac* repressor can be explained. Moreover, it clarifies why the tetrameric *lac* repressor binds simultaneously to two operators.

Catabolite gene activator protein (CAP): binding of α helices to right- or left-handed B-DNA? (1280, 1280a). Whereas *cro* and λ repress transcription of DNA, CAP protein switches on gene operons for which it is specific, with cyclic AMP acting as allosteric effector. The CAP activator protein consists of a single polypeptide chain of 201 amino acids, occurs in its active form as a dimer, and was crystallized in the presence of the effector, cyclic AMP. The three-dimensional structure of the cyclic AMP–CAP complex (Figure 18-18) is considerably more complex than that of *cro* repressor and indicates folding into two structural domains. The larger one, bearing the amino terminus, consists of two short helices A and B and an antiparallel β structure folded into an eight-stranded β roll (Swiss roll). The long, dominating helix C leads into the carboxy-terminal domain composed of helices D, E, and F with the carboxy tail ill-defined in the electron density map.

The two CAP molecules in the dimer are in near-parallel arrangement and in direct contact via their long C helices. The two molecules are related by a pseudo-dyad axis running approximately between them (Figure 18-18). Since the binding site of CAP on the DNA contains a palindromic sequence (Box 18-1), again the protein–DNA interaction is best modeled if the two dyad axes coincide. Following these lines, one finds that structural elements of the CAP dimer which could interact suitably with DNA double helix are the F helices. These are ~ 22 Å long and separated by ~ 34 Å. However, they are inclined 65° to 70° to the line connecting their centers, an angle deviating greatly from the $\sim 32°$ characteristic of the slope of the chains in B-DNA. This suggests that the CAP dimer does not bind to a right-handed but rather to a left-handed B-DNA double helix, of a type derived on the basis of theoretical considerations (1281, 1281a). Such kind of complex formation not only neatly satisfies stereochemical criteria but also explains genetic and chemical findings (1280).

If the model is correct, it suggests that cyclic AMP as allosteric effector directs the relative orientations of the two CAP subunits within the dimer. Its removal would alter the overall CAP dimer structure, thereby destroying the DNA binding site. In addition, the change of B-DNA from right- to left-handed helical screw sense could activate the transcription process by destabilizing (or melting) the local right-handed B-DNA structure within the promoter region. This conclusion is consistent with the finding that CAP is most effective when it can remove negative supercoils from closed circular DNA, which is exactly what a right→left interchange of the DNA double helix would imply. The interchange would reduce the twist number T_w by -2 and, since the DNA is circularly closed, increase the (negative) writhing number W_r by $+2$, consistent with relaxation of the negatively supercoiled DNA (see Section 19-7 for definitions of T_w and W_r).

In a more recent proposal (1281b), it was suggested that the two symmetry related F-helices of dimeric CAP do not bind into consecutive grooves of the same (left-handed) DNA double helix. On the contrary, the notion was put forward that one α-helix is attached to the groove of one (right-handed) DNA and the other α-helix associates with the groove of another DNA being oriented parallel to the first one. The two parallel DNA helices can be part of a left-handed, narrow coil of the kind illustrated in Figure 19-13, a picture also consistent with electron microscopic

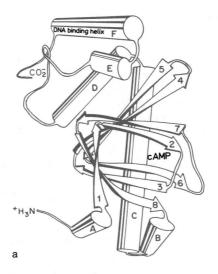

a

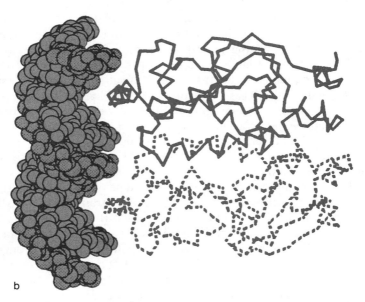

b

Figure 18-18. (a). Molecular structure of catabolite gene activator protein (CAP), showing α helices as cylinders and β-pleated "Swiss roll" as arrows, and binding site of cyclic AMP. CAP forms a dimer, the two molecules being related by a dyad running approximately vertical, and intermolecular contacts are mainly due to near-parallel, close alignment of long helices C. (b). Fitting of CAP dimer to left-handed B-DNA (not Z-DNA, Chapter 12), with helices F interacting with major grooves. Left-handed B-DNA is drawn with coordinates given in (1281); the DNA helix is tilted by about 65° from the vertical to allow a view along the F helices. One CAP molecule is drawn in solid lines, the other in dashed lines. More recent data suggest that this hypothetical model might be wrong and that CAP actually binds to right-handed DNA, see also (1281b). From (1280).

studies. As a consequence, the overall topology of the DNA supercoil is altered if in such an arrangement, in agreement with above mentioned experimental findings that CAP is most effective when circularly closed DNA is (partly) relaxed. The question now opens whether CAP activates RNA polymerase by direct protein–protein binding or by local changes in DNA conformation. In the latter case, CAP acts as an allosteric effector of topological changes of supercoiled DNA whereas repressor molecules such as *cro* can be considered as negative allosteric effectors (1281b).

A final statement as to the correctness of one or the other or none of the CAP/DNA binding models cannot be given at present. It looks, however, as if the latter proposal of CAP/right-handed-DNA corresponds more to reality. The CAP/left-handed-DNA model is certainly more speculative and received support mainly because left-handedness has been made so fashionable by Z-DNA. Carrying on with these thoughts would lead to philosophical considerations of scientific results and their interpretation which really is beyond the scope of this text and is left as a take-home lesson for the reader.

A two-helix motif in proteins binding to double helical DNA. If the tertiary structures of *cro* and λ repressors and CAP are compared, a common two-helix motif emerges which involves the presumed DNA-binding α-helices (1281c). They protude from the bulk of the respective proteins and comprise consecutive α-helices 2 and 3 in the repressors and E and F in the CAP activator. Looking from the bound DNA toward the proteins, the α-helices binding to the major groove of DNA (helices 3 in the repressors and helix F in CAP) are connected at their N-termini with other short helices arranged behind them at about right angles, namely helices 2 in the repressors and helix E in CAP. It was proposed that this two-helix motif is a general feature of proteins binding to DNA double helices, and sequence homology data for a series of repressors support this view (1279c, 1279d). The main structural differences between repressors on one side and the activator CAP on the other are found in the orientations of the binding helices as described above as well as in their location in protein sequence—toward the amino terminus in repressors and toward the carboxyl terminus for the activator.

Gene 5 product of fd phage binds cooperatively to single-stranded DNA: An unwinding protein. In living organisms and *in vitro*, expression of DNA is accompanied by unwinding in order to facilitate interaction of enzymes with DNA as a substrate. Unwinding and stabilization of a DNA single strand is best achieved if the polynucleotide chains are complexed with DNA binding proteins. Gene 5 protein of phage fd belongs to this class of widely distributed proteins. It is an 87-amino acid-long polypeptide chain produced in about 75,000 copies per infected *E. coli* cell (1282).

The X-ray analysis of crystalline gene 5 protein revealed a T-shaped molecule composed entirely of β-pleated sheet structure as displayed in Figure 18-19(a) (1190, 1190a). Starting with the N-terminus, the first ten

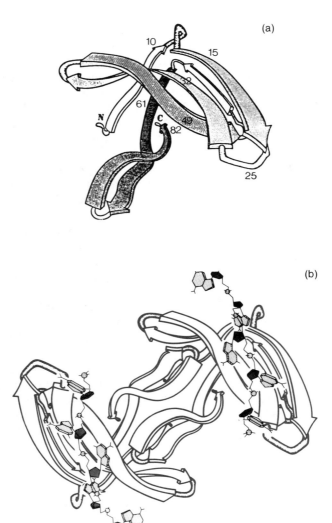

Figure 18-19 (a) to (c). Schematic drawing of the course of the polypeptide backbone of phage fd gene 5 product. (a) Representation of the gene 5 monomer. The T-shaped molecule consists of β-structure with the "DNA binding loop" formed by amino acid residues 15–32 and the "dyad loop" by residues 61–82. (b) In the dimer structure actually adopted by gene 5 protein, a dyad relates two molecules such that dyad loops interact, the T-bars and the DNA binding loops being about 35 Å apart and in antiparallel orientation. The fitting of a DNA single strand to the DNA binding loop is indicated. (c) The DNA strand composed of a pentanucleotide (dA)$_5$ drawn in the same orientation as in (b), with interacting amino acid side chains given. All aromatic side chains are stacked with bases, Tyr 26 being even sandwiched between two adenines. Drawings kindly supplied by authors before publication. To appear in (1190a).

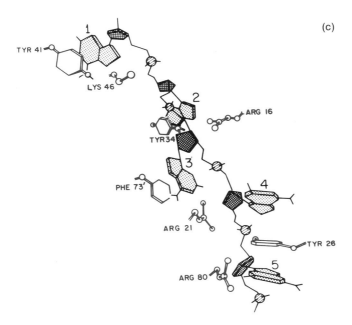

residues are arranged in a short β-strand, followed by the large "DNA binding loop," residues 15 through 32, which is continued by the "complex loop," residues 33 through 49. After a short β-strand, residues 52–59, the polypeptide chain leaves the bar of the T and forms the stem with the large "dyad loop" represented by residues 61 through 82.

In solution as well as in the crystalline state and when binding to DNA, gene 5 protein exists as a dimer, with the two molecules related by a dyad such that the "dyad loops" are aggregated [Figure 18-19(b)] and the T-bars are in antiparallel orientation. Along the T-bar there is a DNA binding channel about 10 Å wide and 35 Å long formed by residues 15–32. This channel is long enough to accommodate about five nucleotides of a DNA single strand in extended conformation, as displayed in Figure 18-19(c). Along the trough of the channel, several amino acid side chains are located such that they can interact with an inserted DNA single strand if some slight reorientations of the flexible side chains of the gene 5 structural model are allowed for. These studies suggest two kinds of interaction, via charge–charge attractions and via hydrophobic stacking between aromatic amino acid side chains and bases [Figure 18-19(c)]. The former involve DNA phosphate groups and positively charged Arg 16, 21, 80 and Lys 46, with Lys 24, 69 probably contributing as well. The latter are due to Tyr 26, 34, 41 and Phe 73′, the prime indicating a residue from the symmetry-related gene 5 protein in the dimer (1190a, 1283–1285).

The predominant distribution of Lys and Arg within the cleft suggests that the negatively charged sugar–phosphate backbone of DNA is recognized first and bound to the protein surface via charge interactions. This is followed by conformational changes which open up the base–base stacks in DNA and rotate aromatic amino acid side chains into a position where they can slip as bookmark between bases and anchor the protein–DNA complex (section 18.3). Within the crystal lattice, gene 5 protein occurs as a dimer which also prevails in solution. When binding to double-stranded DNA, gene 5 protein dimers probably initially separate the anti-parallel DNA chains and subsequently aggregate linearly in a side-by-side manner along the DNA chain. They build up a new helical arrangement in which six gene 5 protein dimers per turn of diameter ~ 100 Å and pitch 80–90 Å constitute a protein core. Around this core the separated DNA strands are spooled (1190,1286) at a radius between 21 and 23 Å as estimated from neutron scattering studies (1287), indicating that DNA is not at the outer periphery of the protein helix but rather is inserted in a notch. The whole process is promoted by the highly cooperative nature of the lateral protein–protein assembly which presents a powerful tool to pry the DNA double helix apart.

18.7 Prealbumin–DNA Interaction: a Hypothetical Model

The X-ray structure analysis of the first fully characterized hormone binding and transporting protein, the thyroid hormone binding prealbumin from human plasma shows that it consists of four identical subunits 127 amino acids long which are arranged with tetrahedral symmetry. The structural model of prealbumin displays a surface that is complementary in size and shape to a fragment of B-DNA 10–12 base-pairs long. It exhibits, at both sides of a semicylindrical groove, two helically disposed arms which could act as feelers (1288). The bottom of the groove is lined by a symmetry-related pair of β-pleated sheets and, if complexed with the DNA, the outer feelers fit snugly into the major grooves of the DNA double helix.

Again, the complex with DNA, if it really exists, would have the two-fold symmetry mentioned before in relation to other DNA–protein complexes, and the DNA binding site should be palindromic (Box 18-1). The DNA–prealbumin molecular model was constructed without detailed fitting of protein side chains; yet 10 lysines, 4 histidines, 8 glutamates, and 6 aspartates are distributed in positions where they could potentially interact with DNA.

Plausible as this model may be, there is no proof thus far that prealbumin actually interacts with DNA. In fact, experiments have disclosed that there is no binding (1289). If one looks at the surface potentials of DNA and of prealbumin, they are both negative and therefore complex formation is made impossible. This example demonstrates that even good model fitting suggestive of macromolecular interaction is not sufficient, and that besides structural complementarity the electronic complementarity is of prime importance (839a).

Summary

Protein–nucleic acid interactions can involve all parts of these macromolecules and include salt bridges, hydrogen bonds, stacking, and hydrophobic forces. Cocrystallizations between nucleotides and amino acids or their constituents give a first indication of interactions but larger model systems are necessary to further understand structural implications. Double-helical DNA can interact in several ways with polypeptides: (i) by hydrogen bond recognition between Watson–Crick base-pairs and amino acid side chains, (ii) by intercalation of aromatic side chains ("bookmarks") between base-pairs which displays some specificity, (iii) by direct binding of protein α helices or β-pleated sheets in the grooves of DNA. The latter two interactions were envisaged by model building studies of DNA double helices to *cro* repressor protein (α helix and β-pleated sheet in minor and major grooves, respectively) and to catabolite activator protein (CAP) where two models have been advanced with α helices binding to right- and left-handed B-DNA. On the basis of prealbumin crystal structure a complex with DNA was proposed hypothetically but not observed in reality. Polylysine and polyarginine interact cooperatively and irreversibly with DNA; yet there are differences in the structures of the respective complexes.

If nucleotides and single-stranded RNA and DNA bind to proteins, they usually "unfold"; i.e., the γ torsion angle adopts $-sc$ or ap range. This has been observed in NAD^+ binding to dehydrogenases (at the "Rossmann fold"), dinucleoside phosphate binding to ribonuclease A and S, and RNA binding to tobacco mosaic virus protein. Binding interactions can be of any type and involve any part of the two-partner molecules. In the case of the very specific guanylic acid·ribonuclease T_1 complex, for instance, guanine is recognized by hydrogen bond formation to peptide backbone and by stacking of tyrosine side chain. In general, interactions between nucleotides and protein main-chain (peptide) atoms appear to be more favored and specific than with amino acid side chains.

Strand separation of double-helical DNA is forced by binding to phage fd gene 5 protein which, as soon as (single stranded) DNA is complexed to its β-pleated sheet "active site," aggregates laterally to build up helical structures around which separated DNA single strands are wound.

Chapter 19

Higher Organization of DNA

Several forms of higher organization are known for DNA. RNA in single-
or double-stranded form does not assemble into any distinctly higher
ordered structure except if combined with coat proteins in viruses
(Figure 18-15). For DNA, the situation is quite different. In chromatin,
the DNA–protein complex constituting the chromosomes, DNA is
wound around globular histone octamers to form nucleosome cores.
Aligned along the DNA, these cores give chromatin the appearance of
beads on a string, the string being further compacted and organized into
a superhelical, solenoidal arrangement. Winding of DNA around histone
octamers is associated with superhelical twisting, the topological problems of which deserve special treatment (Section 19.7).

DNA does not necessarily need histone octamers to form ordered, compact structures. It has the intrinsic ability to do so on its own if, for instance, excess salt and inert polymers are added to an aqueous solution.
A spontaneous condensation due to lateral (side-by-side) aggregation of
DNA molecules takes place. This phenomenon is also observed if polyamines or ethanol are used. If, however, ethanol is added carefully, even
higher order is achieved and DNA arranges into lamellar microcrystals.
This kind of self-organization is of relevance to the packing of DNA in
phage heads and therefore has been of considerable interest (1161,1290–
1294). It is the purpose of this last chapter to describe the organization of
DNA in these various forms.

19.1 DNA Condensed into ψ-Form, Supercoils, Beads, Rods and Toroids

If increasing amounts of Mg(II) (1295), polyamines, basic polypeptides
(1295–1301), polyethylene oxide or similar inert polymers (1300), or ethanol (1302–1306) are added to an aqueous solution of DNA, the polyanion
shrivels into condensed forms displaying different morphologies. The
least well defined of these is the macroscopically dense phase of ψ-DNA,
formed in the presence of "inert" polymers and excess salt. Under other
conditions, notably with polyamines and ethanol, DNA adopts microscopic structures such as supercoiled or beaded fibers, or straight or circularly closed-up rods, depending mainly on the ionic strength and cations present. It should be stressed that in all these condensed
appearances, DNA by and large occurs in the standard B form.

Loose, near-parallel aggregation of molecules in ψ-DNA. If DNA dissolved in aqueous solution is treated with excess salt and a high concentration of ethanol or a macromolecule like polyethylene oxide, polyacrylate, or polyvinylpyrrolidone, it condenses cooperatively and reversibly into a macroscopic, compact state called ψ-DNA, the acronym ψ standing for "*p*olymer- and *s*alt-*i*nduced condensation" (1295,1296). X-ray scattering experiments suggested that double helices of standard B-DNA are oriented more or less parallel to each other at about 38 Å separation, relatively poor alignment being indicated by lack of higher-order reflections. If the concentration of the inert polymer is raised, the distance between adjacent molecules is reduced to 25 Å, a value closely related to the 23.9 Å observed in DNA microcrystals (Section 19.2).

Treatment of DNA in aqueous solution with varying amounts of NaCl and ethanol also produces ψ-DNA. As investigations using circular dichroism have demonstrated, however, the situation is rather complex with this and related systems described in Ref. (1296a). At low NaCl concentration and about 35% ethanol, calf thymus DNA condenses into a form displaying positive ellipticity in the 250–300 nm region (the ψ^+ state). If the NaCl concentration is increased, the ellipticity changes to negative (the ψ^- state) via several intermediate states. Presumably the B-DNA form is retained under all these conditions, yet the aggregation of the molecules differs from one state to another.

The three modes of DNA self-assembly obtained with ethanol precipitation depend on ionic strength. At ethanol concentrations between 50% and 95%, DNA shrivels into different morphologies dictated by ionic strength. This means that not only is the DNA double helix intrinsically able to self-assemble into ordered structures but also packing is determined by the electrostatic charge density around the polyelectrolyte (1305).

• *At low ionic strength, DNA condenses into a left-handed supercoil.* If precipitated from 1 mM Tris solution at pH 7.5 with 95% ethanol, DNA condenses about ninefold into a fiber structure with a nearly uniform diameter of 80 to 90 Å. The fiber displays a left-handed supercoil fine structure (Figure 19-1a), which probably arises from overwinding forced onto gradually dehydrated DNA (1305).

• *At intermediate ionic strength, DNA condenses into fibers and beads.* If 10 mM ammonium acetate or other salt is added to the DNA solution prior to ethanol precipitation, fibrous structures of about 150 Å diameter are obtained, corresponding to 30- to 35-fold compaction. The fibers sometimes display a beaded appearance (Figure 19-1b).

• *At high ionic strength, DNA folds into maggot- and doughnut-like particles.* If the ammonium acetate concentration lies between 0.15 and 0.5 M, precipitation with 95% ethanol leads to the maggot-like rod particles (Figure 19-1c). Because these are of constant length, about 2000 Å, while their cross section is proportional to the DNA molecular weight, it was concluded that in this case DNA is folded back and forth on itself in

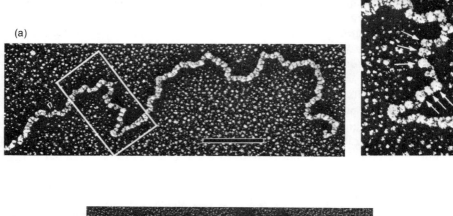

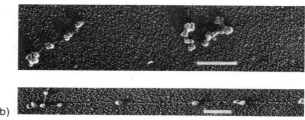

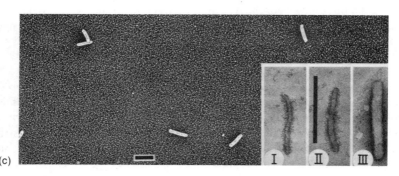

Figure 19-1. Condensation of DNA (phage λ) with 95% ethanol at different ionic strengths. Bar = 2000 Å. From (1305). **(a)** Coiled fibers obtained at low ionic strength; 1 m*M* Tris, pH 7.5. (Left) Low magnification; (right) examination at higher resolution shows left-handed supercoiling (arrows). **(b)** Beaded fibers obtained at intermediate ionic strength: 1 m*M* Tris, pH 7.5, 10 m*M* ammonium acetate. (Top) Unstretched fibers; (bottom) fibers that have been stretched during adsorption to carbon film to show beaded structure. **(c)** Rods obtained at high ionic strength: 1 m*M* Tris, pH 7.5, 0.5 *M* ammonium acetate. Insets I, II, and III display rods at larger magnifications. DNA is folded in 2000 Å-long stretches which are arranged *along* the rods.

2000 Å bends. Fibers of similar length but of increased diameter are also obtained if DNA is precipitated with polylysine or if spermidine replaces ethanol (1305).

The folding back of DNA on itself is reminiscent of synthetic polymers which form lamellar crystalline aggregates: folds extend from one lamellar face to the other (1307,1308). The same holds for DNA microcrystals (Section 19-2) and suggests that DNA condensed into rod-like particles represents crystalline fragments.

• *In addition to rod-like particles, toroidal or doughnut-shaped structures are observed* under the electron microscope. They result from intraparticular end-to-end fusion. The ratio of rod to toroid structures depends on the cation present in solution, divalent cations like Mg(II) strongly favoring toroids (1299–1301,1305). In essence, the toroids indicate that the rod-like particles are not stiff but rather flexible entities which coil up into annular structures if end-to-end aggregation is strong enough for stabilization.

19.2 Lamellar Microcrystals Formed by Fragmented DNA

If DNA in aqueous solution is exposed to ultrasonic radiation, it breaks down into pieces with average length centered on a broad distribution. Experiments with DNA fragments about 2000 Å long indicated that, in contrast to native DNA, they can be crystallized if about 50% ethanol is slowly added to form lamellar, hexagonal plates several micrometers in diameter and 1600 to 2400 Å thick (80,89) (Figure 19-2).

Crystals are one DNA molecule thick. Electron microscopic observations on freeze-fracture-etching replicas of DNA crystals as well as X-ray powder diffraction studies led to the conclusion that nearly straight DNA molecules of the B form are hexagonally closed packed at 23.9 Å lateral molecule-to-molecule separation and oriented essentially vertical to the hexagonal crystal face. These observations suggest that DNA molecules are generally not folded but span the crystal thickness from one hexagonal face to the opposite one (89). However, because the lengths of the DNA molecules vary between about 1000 and 5000 Å (88) the larger specimens must be folded at least once. The point was made that folding is probably not sharp but rather involves second- or third-neighbor locations in order to be in overall agreement with length persistence measurements (1309) or at least with the kink proposal discused in Sections 14.2 and 14.3. We should also recall here the back-and-forth folding of DNA in the rod-like particles of about 2000 Å length (Section 19.1), which indicates that 2000 Å or ~60 turns of B-DNA with 34 Å pitch appear to be a magic number of DNA self-assembly.

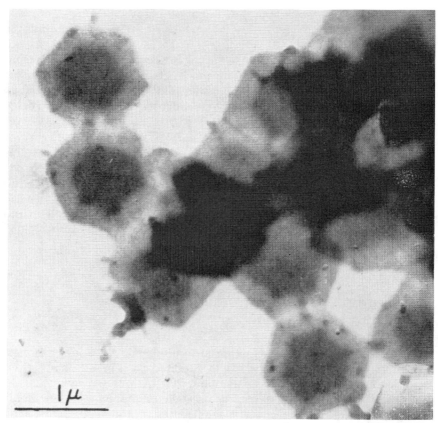

Figure 19-2. Electron micrograph of crystals obtained from about 2000-Å-long fragments of salmon sperm DNA. Conditions: 2–3 mg/ml sonicated DNA in 0.05 M sodium cacodylate buffer, pH 6.5. Ethanol was added dropwise until the onset of turbidity. Precipitate was dissolved by heating to 60–75°C; the solution was then cooled down slowly to yield crystals. From (88).

19.3 DNA in Cells Is Organized in the Form of Chromosomes

In organisms of progressively higher evolutionary level, the amount of genetic material, DNA, increases from about 10^4 up to nearly 10^{10} base-pairs per cell (Figure 1-4). In some cases, such as in certain bacteriophages of *E. coli*, DNA is a linear, double-helical molecule. Often, if not always in bacteria, it is either circularly closed or is at least organized so that its ends are held together, not free to rotate, and in a quasi-circular state (1310,1311). In bacteria, besides the main chromosomal DNA, small circular "plasmid" DNA exists which is also found in mitochondria and in chloroplasts of eukaryotic cells. Plasmid DNAs have typical molecular weights of only around 10×10^6 (1314). These small circular DNAs have recently found important applications as vectors in genetic engineering and they present some special topological problems to be discussed in Section 19.7.

Simple organization of DNA in bacterial chromosomes. In bacteria like *E. coli* the circular, chromosomal DNA is organized in a comparatively simple manner. The cell of diameter about 5 μm hosts the circular duplex DNA of contour length 1.36 mm, corresponding to 4×10^6 base-pairs with a total molecular weight 2.8×10^9. It is clear that the DNA must be highly folded. This is achieved by a core containing RNA and some protein which condenses the DNA into tight loops (Figure 19-3) (1313). It appears that in prokaryotic cells, chromosomal DNA occurs associated mainly with RNA and with basic proteins as in nucleated eukaryotic cells (1314a).

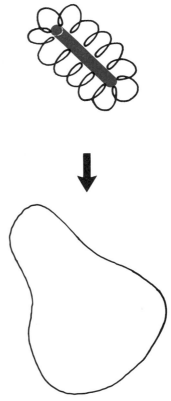

Figure 19-3. In prokaryotes like *E. coli*, the chromosome consists of a single large circular DNA, supercoiled and folded into compact form held together by a core containing RNA and some protein (color). Adapted from (1313).

Chromosomal DNA in eukaryotic cells combines with histones to form chromatin. The organization of the very long DNA in nucleated eukaryotic cells is much more complex because the volume of the nucleus, about 5 μm in diameter, matches that of the bacterial cell, yet the amount of DNA is considerably higher. Depending on the species, somatic cells contain several to many chromosomes. In human cells, there are 46 chromosomes with an average of 4 cm DNA. If all the DNA in a cell were lined

---

Box 19-1. Interphase and Metaphase Chromosomes

Depending on the life cycle of an eukaryotic cell, DNA occurs in two principal forms which can be macroscopically distinguished. One, unstructured with DNA filling virtually the whole nucleus, signifies *interphase,* the state when DNA is replicated and transcribed. The other form exists in a much shorter time, more transiently during cell division (mitosis, *metaphase*), and represents condensed DNA in its characteristic form, depicted for human chromosome 12 in the figure below.

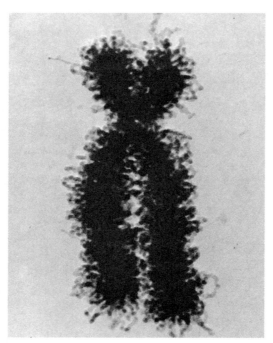

Electron micrograph of human chromosome 12. It consists of chromatin arranged into fibers 200 to 300 Å in diameter and 700 to 800 μm in length, containing one DNA duplex molecule about 4 cm long. From (1314).

---

up end to end, it would be 2 m long and would have a molecular weight of 4×10^{12}, equivalent to 5.5×10^9 base-pairs (Box 19-1; Figure 1-4).

The DNA in eukaryotic chromosomes is associated with some RNA and with proteins. Among the proteins, there are the structure-forming histones of basic character in weight ratio DNA/histone ~ 1. Moreover, there are mainly acidic, non-histone proteins with more varying amounts, DNA/protein ratio 0.5 to 1.5, depending on type and physiological state of the cell.

The DNA–histone complex is usually called chromatin. In it, histones do not uniformly surround DNA as originally assumed (1315). Rather, electron micrographs show that they form small, globular particles about 100 Å in diameter which are distributed along the DNA at distances of 30 to 70 Å. These particles, called nucleosomes (1316) or v bodies (1317), contain five different histones. Of these, H2A, H2B, H3, and H4 occur pairwise to form an octamer (1318,1319) *around* which DNA is wrapped (1320) in almost two turns of left-handed supercoil (1321). At the points of entry and exit, DNA is locked by histone H1 presumably located outside the particle and *on* the two DNA turns (1322–1330) (Box 19-2.)

Box 19-2. Characterization of Histones

Histones are proteins containing a large proportion of basic amino acid residues. Histones H2A, H2B, H3, and H4 are mainly hydrophobic at the C terminus and hydrophilic (basic) at the N terminus and are assumed to display secondary (helical) structure in the hydrophobic part. Histone H1 does not show these features, has different molecular weight, and associates directly with DNA whereas the other four histones form an octamer containing two of each histone. And *around* the octamer, DNA is wound in 1¾ left-handed superhelical turns.

Molecular Characteristics of Histones (1330)

| | | Amino acid residue | | | |
Histone	Molecular weight	Total	Basic[a]	Lys	Arg
H1	22,000	220	65	62	3
H2A	14,000	129	30	14	12
H2B	13,800	125	30	20	8
H3	15,300	135	33	13	18
H4	11,300	102	26	11	14

[a] Including Lys, Arg, and His.

19.4 Structure of the Nucleosome Core

If the metaphase chromosome is suspended in low-ionic-strength medium, it opens up to show the beads-on-a-string appearance. Depending on species and cell type, there is one nucleosome every 180 to 250 base-pairs, with a value of 200-base-pair DNA most frequently encountered (1318,1327). Extensive digestion of chromatin with micrococcal nuclease followed by purification yields particles devoid of histone H1 and containing 146 ± 1 base-pairs of DNA, a near-universal unit called nucleosome core (1332).

The location and assembly of the eight histones in these particles is approximately known. Cross-link experiments and association studies in solution have shown that a tetrameric unit (H3$_2$, H4$_2$) which constitutes the central core of the octamer is easily formed and that strong dimers (H2A, H2B) and (H2B, H4) and weakly bound dimers (H2A, H4) and (H2B, H3) exist (1323,1327,1330). The heart-shaped appearance of the histone octamer (Figures 19-4, 19-5) is due to the disk-like arrangement of the (H3$_2$, H4$_2$) tetramer which forms the center and lower tip of the heart and is associated with two (H2A, H2B) dimers producing the upper two tips (1330). The octamer is probably stabilized by interaction of the hydrophobic C-terminal ends of the histones (Box 19-2) while their N-terminal ends, rich in arginine and lysine, are located at the periphery and interact with DNA wound around the core in nearly two negative helical supercoil turns.

Nucleosome cores aggregate into arcs, helices, and single crystals. The nucleosome cores containing 146 base-pairs of DNA can be obtained in microcrystalline forms which, by electron microscopic studies, furnish overall dimensions of the nucleosome (1327). Under crystallization conditions, however, nucleosomes also aggregate and produce arc-like and cylindrical forms, the latter containing nucleosomes in a helical arrangement. Electron micrographs revealed that in all these forms, nucleosomes occur as wedge-shaped, stout cylinders about 57 Å in height and 110 Å in diameter which, as suggested by their fitting into arcs of variable curvature, are quite flexible either in shape or in intermolecular interaction and thus allow the arcs to adopt the proper radius. Moreover, these aggregates indicate that nucleosome cores join easily, with preferential contacts between top and bottom planes (1331).

If nucleosome cores are mildly cleaved by proteases, particles are obtained which still exhibit the original form but which can now be crystallized into specimens large enough for investigation by X-ray crystallographic methods. Combinations of X-ray and neutron diffraction as well as electron microscopic methods were employed in order to analyze the structure and to produce an electron density map at 20 Å resolution (1332, 1333) (Figure 19-4).

A heart-shaped histone octamer wrapped by 1¾ turns of left-handed

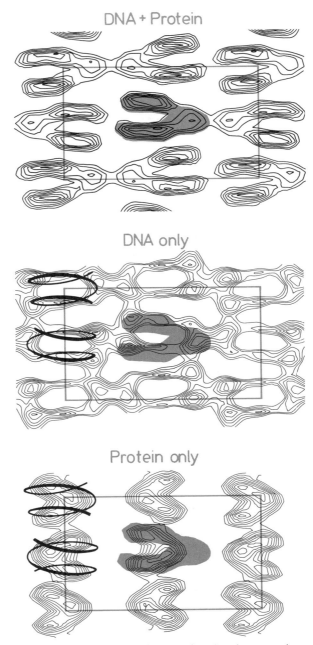

Figure 19-4. Combined X-ray and neutron diffraction results showing complete nucleosome cores (top), DNA only (center), and protein only (bottom). The top figure was obtained by X-ray diffraction; the others by using neutron radiation and applying the D_2O/H_2O contrast variation technique to compensate protein (39% D_2O, center) and DNA (65% D_2O, bottom). The outlines of the crystal unit cell are indicated by light-colored lines, and one nucleosome core, seen from the side, is shown in the same color. Note wedge shape and bipartite structure of particle. The DNA density (center) corresponds to 1¾ turns of the DNA superhelix. From (1327).

DNA supercoil yields a nucleosome core with dyad symmetry. In essence, it was found that the nucleosome core has a flat, wedge-shaped cylindrical form of the dimensions given above, which is strongly divided into two layers (Figure 19-4). Contrast variation experiments using different D_2O/H_2O ratios to equalize neutron scattering due to DNA (65% D_2O) and to protein (39% D_2O) showed clearly that the histone octamer itself is heart-shaped and surrounded by 1¾ turns of DNA with pitch height 27 Å, an organization also established for nucleosome cores in solution (1320, 1324) (Figures 19-5, 19-6).

Assuming the supercoiled DNA to be of the B form with a 20 Å diameter, we can estimate that the outer dimensions of the nucleosome core, 110 Å, correspond to 90 Å diameter of a DNA coil which would consist of about 80 base-pairs per turn. Combination of this information with enzymatic digestion studies using pancreatic deoxyribonuclease (DNase I) indicates that the nucleosome core has dyad symmetry (1333–1335) (Figure 19-5). The reason is that first, DNA is preferentially cleaved every 10 nucleotides as expected for the B form and, second, that frequent cutting sites are 80 nucleotides apart suggestive of 80 base-pairs per supercoil turn. In addition, the distribution of high- and low-frequency cutting sites discloses that the nucleosome core has an overall dyad symmetry.

DNA is locally overwound when wrapped around histone octamers. Anticipating the discussion in Section 19.7 on linking numbers, writhing, and twisting of circular DNA, one can show that DNA overwinds when coiling around the histone octamer. In an experiment, the small circular "naked" chromosomal DNA of simian virus 40 was incubated with histone octamers. It wrapped around the octamers in about two left-handed turns each and toroidal strain building up within the DNA was relaxed en-

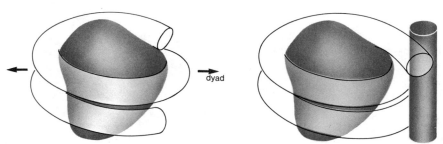

Figure 19-5. Model of the nucleosome core consisting of the histone octamer (H2A, H2B, H3, H4)$_2$ in the center, around which DNA 146 base-pairs long is wound in a left-handed supercoil. The dyad axis is indicated (left). In the complete nucleosome, DNA comprises two full turns 166 base-pairs long. Histone H1 is located at the periphery of the nucleosome and "seals off" DNA at entry and exit points (right).

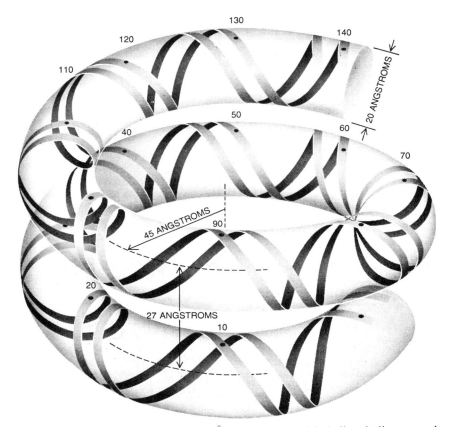

Figure 19-6. A close-up view of the 20-Å-wide DNA double helix winding around the histone octamer in 1¾ supercoil turns with external diameter 110 Å and pitch 27 Å. There are about 80 nucleotide pairs of DNA per turn. From (1327).

zymatically by the addition of topoisomerase I (1321). If, then, the histones were removed, a negatively supercoiled circular DNA remained with about − 1 to − 1¼ (average − 1⅛) supercoil turns per nucleosome. These data suggested that DNA wrapped around the histone octamers adopts a 10.0-fold helix. It is therefore overwound with respect to DNA free in solution which displays a 10.4-fold helix (Chapter 9).

Histone interaction facilitates DNA bending. The DNA double helix is generally assumed to behave as a rather stiff rod, stabilized by repulsion of negative charges along the sugar–phosphate backbone (1336,1337). However, theoretical calculations (836,1338) have suggested that smooth bending of DNA can account for its winding around the histone core and that the previously assumed kinking is not necessarily required (Chapter 14).

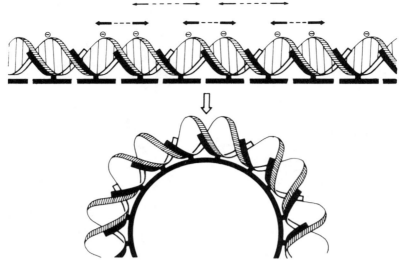

Figure 19-7. Schematic diagram illustrating how histones asymmetrically neutralize phosphate charges in DNA. Histones are indicated in black. Charge repulsion between phosphates in minor and major grooves symbolized by heavy and light arrows introduces bending, with histones located in the interior of the bend. From (1340).

This view holds even more because the basic amino acid side chains of the histone N-terminal arms tend to neutralize negative phosphate charges. As can be shown by cross-linking experiments, histone–DNA interactions involve preferentially the major groove side of the double helix (1339), and this is in agreement with the surface accessibility studies described earlier (Section 17.3). As illustrated (Figure 19-7), all histones (except histone H1) bind to and neutralize negative charges only from one side of the DNA double helix; on the opposite side, charge–charge repulsion is still present, thus facilitating bending of DNA around the histone core (1340).

Figure 19-8. The solenoidal structure of chromatin. If, as indicated at the bottom of the picture, histone H1 is absent, DNA enters and exits the nucleosome particle on opposite sides and no regular structure is formed. Histone H1 seals off the nucleosome, forcing DNA to enter and exit at the same side. With increasing salt concentration, H1-containing nucleosomes form a helical structure containing three to six nucleosomes per turn (*n*), with histones H1 aggregated in the central cavity of the helix, gluing the otherwise rather loose construction together. The screw sense of the helix can be right- or left-handed (the latter is shown here) with no regular order required. The external diameter is 250 to 300 Å, the pitch height 110 Å. For a more recent alternative model, see the text. From (1327, 1341).

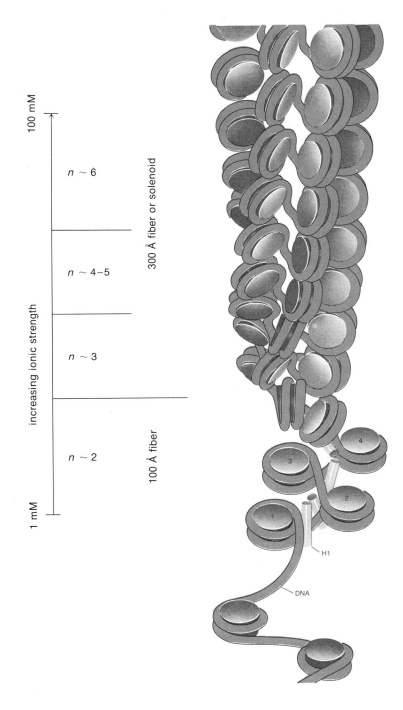

19.5 Organization of Nucleosomes into 100 Å and 300 Å Fibers: The Super-Superhelix or Solenoid

Because of the large size and complexity of structures involved, higher ordering of nucleosomes and of chromatin into chromosomes is not of strictly repetitive nature. Therefore, diffraction methods cannot be employed in their study and recourse must be made to electron microscopy in order to derive at least a macroscopic picture of the 100- and 300-Å fibers which are basic constituents of chromosome, schematically illustrated in Figure 19-8 (1341).

Chromatin devoid of histone H1 has no regular structure. If chromatin is depleted of histone H1 and kept at low ionic strength, nucleosomes are still distributed about every 200 base-pairs along the DNA matrix. Their appearance, however, is less regular and DNA does not enter and exit the core at the same point, but rather at opposite sides. If ionic strength is raised above 40 mM NaCl, this chromatin condenses into irregular clumps and not into the helical form depicted in Figure 19-8. These findings suggest that histone H1 has a structure-forming role and is essential for the higher organization of chromatin.

Nucleosome filament and 100 Å fiber: The low level of organization. If intact chromatin containing all histones is kept at low salt concentration, ~1 mM NaCl, it forms a rather open structure with well-defined nucleosomes showing DNA entering and exiting in opposite directions at virtually the same point. This form has approximate cross-dimensions of 100 Å and was termed the nucleosome filament (1341). If the ionic strength is increased to about 100 mM NaCl, the length of the spacer or linker DNA between nucleosomes is diminished, leading to more dense configurations with nucleosomes in zig-zag arrangement, a form usually called the 100 Å fiber or unit fiber (Figure 19-9).

"Solenoid" glued together by H1–H1 interactions. With a further increase in ionic strength and, very sensitively, with addition of small amounts of divalent cations (Mg^{2+}), the 100-Å fiber coils up to produce a 250-Å-thick fiber. Several models have been suggested for the molecular architecture of the 250-Å-thick fiber, of which only two are considered here. In one, it was concluded from electron micrographs that the 100 Å fiber is screwed into a helical organization containing three to six nucleosomes per turn, depending on salt concentration (Figure 19-8). The overall diameter of this solenoid (1342) or super-superhelix (1343) is the required 250 to 300 Å; the pitch height is 110 Å. The sense of the helix is not necessarily fixed and can be right- as well as left-handed. The latter is depicted in Figure 19-8.

An important feature of this model as well as of similar models devel-

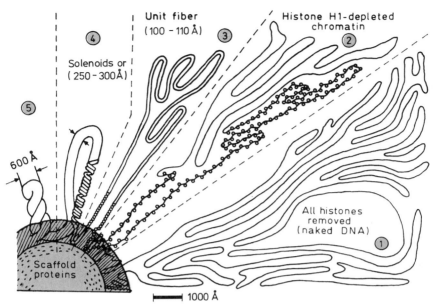

Figure 19-9. Schematic drawing to illustrate the gradual organization of DNA into highly condensed chromatin. DNA is fixed at a protein scaffold (**1**); see also Figure 1-5. Addition of histones H2A, H2B, H3, and H4 produces the loose, already condensed zig-zag structure (**2**) which, upon complexation with histone H1, aggregates into the 100-Å fiber (**3**), corresponding to the low-ionic-strength strand at the bottom of Figure 19-8. If ionic strength is increased, ''solenoids,'' ''super-superhelices,'' or ''superbeads'' with diameters of 250–300 Å are formed (**4**) which can further contract into a 600-Å knob-like organization (**5**). From (1322).

oped independently (1343) is that nucleosomes are arranged so that histones H1 are located in the center of the solenoidal helix and therefore in close contact with each other. This geometry is also evidenced by formation of H1 homopolymers using cross-linking reagents (1344,1345). Structurally, the H1–H1 interactions are obviously more important for the stabilization of the solenoid than individual nucleosome–nucleosome contacts. This is because the latter, if a crucial factor, would have to occur regularly in equivalent positions and require a well-defined helical (quasi-crystalline) arrangement, which is not observed. It is also quite clear that spacer DNA linking the nucleosomes cannot be a principal stabilization factor for the solenoidal structure because these can also be formed out of isolated nucleosome particles. (1342,1346).

In another, more recent model for the structure of the 250-Å-thick fiber, it was assumed that the repeat unit is not one nucleosome as in the solenoid but rather a dimer of two zig-zag folds of the 100-Å fiber. If this fiber is twisted about its long axis, a helical, cylindrical arrangement with diameter of 250–300 Å can be produced in which the center is occupied by condensed spacer DNA (1347). A groove running along the periphery of the helix would be suited to accommodate histone H1 which would stabilize the structure. Although both proposals have in common a helical appearance of the 250-Å-thick fiber, they differ strikingly in the position of histone H1: in the center of the helix in the solenoid and at the periphery in the twisted 100-Å fiber. A decision as to the correctness of one over the other cannot be made at present because direct experiments demonstrating the position of histone H1 are lacking.

19.6 Organization of Chromatin in Chromosomes: A Glimpse at Transcription

The organization forms of chromatin depicted schematically in Figure 19-9 are only part of the story. In chromosomes, chromatin is anchored to a protein scaffold which forms the central part of the whole structure. If chromosome is depleted of histones and viewed with an electron microscope, the scaffold with attached DNA forming a "halo" of some 4000 loops about 40,000 to 90,000 base-pairs long is clearly seen (Figure 1-5) (1322, 1348). In steps of increasing complexity, then, this DNA is associated with histones to form chromatin in the beads-on-a-string appearance. Further condensation produces 100 Å fibers, followed by 300 Å fibers (solenoids), which are even more highly organized into a final, probably again helical configuration (Figure 19-9). Starting from DNA, the overall condensation is 5000- to 8000-fold (1322), with estimates of up to 250,000-fold (1327).

How does transcription occur? For this process, only incomplete information is available. One purely hypothetical model is summarized schematically in Figure 19-10. The solenoid or super-superhelix unwinds partially and releases H1-containing, inactive nucleosomes. Depletion of H1 and binding of high-mobility-group proteins (HMG) activates nucleosomes. They unfold and, still associated with histones in the proper sequence, the DNA strands transiently separate. The noncoding strand, lined with HMG and histone proteins, is "inert" while the other strand is transcribed in an RNA–polymerase complex. The nascent RNA immediately folds into secondary, short helical structures and is protected from nuclease attack by combination with structural proteins to form a heteronuclear RNA–protein complex (hnRNP). This process takes place not only at one point but simultaneously in many places so that the message contained in DNA is rapidly transcribed, the RNAs being later processed (spliced) for proper translation at the ribosome (1349).

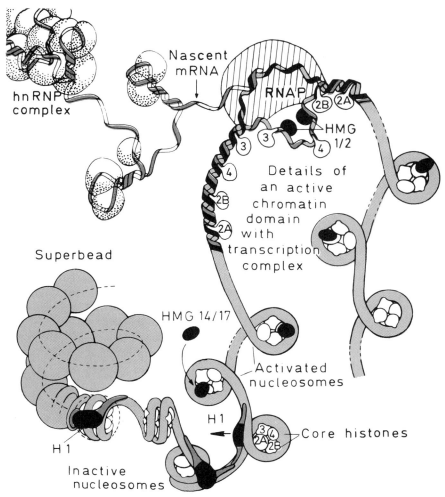

Figure 19-10. An artist's conception of DNA transcription. For details, see text. From (1322).

19.7 Topological Problems in Circularly Closed, Supercoiled DNA

Supercoiled DNA is the biologically active form (1349a). Linear, double-stranded DNA free in solution or drawn into fibers exists in a topologically relaxed state. As such, DNA is inactive or at least displays reduced activity for several biological processes like replication, transcription, and recombination which require DNA not in the relaxed, unstrained state but rather as a supercoiled entity with more or less well-defined three-dimensional tertiary structure (1350–1354). Supercoiling, sometimes called supertwisting or superhelix formation, presupposes DNA in circularly closed loops. Closure in that sense is not only achieved by physically linking the 3' and 5' ends of each DNA strand as found in bacterial and mitochondrial plasmids (1349a), but also by holding together DNA loops as observed, for instance, on the protein scaffold in eukaryotic chromosomes (Section 19.5; Figures 1-5, 19-9), or by simply restricting free rotation of the ends of very long, double-helical DNA molecules.

What is supercoiled DNA? If DNA in relaxed, normal B-form is closed into a circle, there is no supercoiling. Let us now assume that the DNA double helix is slightly unwound before joining its ends, and after joining it is allowed to adopt the normal B-form. It will do so by winding up by the amount of previous unwinding. Since our DNA, however, is circularly closed, strain is introduced and winding up is therefore compensated by twisting of the circle into a superhelical form, or supercoiling. If DNA was *un*wound before joining, it is said to be *negatively* supercoiled although the form of the super helix is *right*-handed, Figure 19-11. This is the normally observed DNA supercoil. If drugs are intercalated into DNA, a *positively* supercoiled DNA displaying *left*-handed superhelix is obtained, corresponding to *over*winding before ring closure (Figure 16-3).

A class of enzymes called topoisomerases control supercoiling of DNA. Supercoiling of DNA is regulated by enzymes called topoisomerases because they isomerize the topology or overall three-dimensional geometry of DNAs (1350–1358). Both in eukaryotes and in prokaryotes, topoisomerase I enzymes have been found which relieve supercoiling without requiring ATP or other input of energy. These enzymes are also named swivelases, DNA relaxing or unwinding enzymes, nicking-closing (N/C) enzymes, or ω protein [in *E. coli* (1356)]. They act by cutting one strand of the supercoiled double helix, rotating one strand about the other and then resealing the cut strand. The whole process is driven by the energy inherent in supercoiled DNA.

The reverse action, supercoiling of relaxed DNA, is performed in prokaryotes and in eukaryotes, by enzymes called gyrases or topoisomerase II (1357,1358). They accept double-stranded DNA as a substrate and "crank it up" until the required supercoiled state is reached—in general one negative supercoil (= one right-handed superhelical turn) per about 15 double-helical turns of the B-DNA (1354,1358). This isomerization, by a mechanism discussed in (1358), consumes energy and therefore gyrases require ATP and Mg(II) for proper functioning, with spermine stimulating the reaction.

Supercoiling as a means of energy storage. Supercoiling of DNA could not only be essential to prepare DNA in a form suitable for interaction with certain enzymes and to relax strain at the replication fork (1350). It could also represent a kind of energy storage for which a monetary analog has been suggested (1358): buy now (use supercoil energy), pay later (crank up circular DNA with ATP). Extending the analogy, we may ask about the currencies used in these transactions. On one side, we have ATP/Mg(II) and on the other, there is the linking number of supercoiled DNA. What is the meaning of that linking number?

Linking number, writhing, and twisting—The supercoiled state defined mathematically. In order to describe supertwisting of a closed ribbon, mathematicians have worked out an equation which has been applied to supercoiled DNA (1359–1362).

In this equation, the *linking number* L_k is an integer, invariant under all topological forms of the closed circle, and describes how often, in DNA, one polynucleotide strand winds about the other. Right-handed winding, as in double-helical DNA, produces positive L_k; negative L_k is associated with left-handed turns. In relaxed, unstrained, circular DNA, L_k equals the number of double-helical twists (or turns), T_w (Figure 19-11), and the shape of DNA corresponds to a flat circle,

$$L_k = T_w.$$

If the DNA double helix were able to over- or underwind easily by decreasing or increasing the rotation angle between one base-pair and its next neighbor, this relation could hold in all cases and no supercoiling would be observed. However, because DNA tends to retain the B form with about 10.5 base-pairs per turn, T_w is limited within narrow bounds and changes in linking number can either be compensated for by disrupting base-pairs (Figure 19-11) or, more favorably, by supercoiling of the DNA. Supercoiling is described by the writhing number W_r (1360) and the three parameters L_k, T_w, and W_r are related by the equation

$$L_k = T_w + W_r. \tag{19-1}$$

As already mentioned, L_k is an integer and invariant under all conditions, once the DNA is circularly closed. T_w and W_r can adopt any value and depend on the respective three-dimensional shape of the DNA. It should be stressed that L_k and T_w are defined positively for right-handed rotation or twist whereas W_r, if describing a superhelix, is negative for right-handed helix and positive for left-handed sense (see figure in Box 19-3).

Let us look at the two cases illustrated in Figure 19-11. If circular DNA is confined to planar arrangement, changes in linking number

$$\Delta L_k = \Delta T_w + \Delta W_r \tag{19-2}$$

are only compensated by equivalent changes in twist ΔT_w because W_r and ΔW_r are zero in that particular case. ΔT_w can either be accommodated by smoothly modifying all helical turns or, because DNA retains the B form, by disrupting some base-pairs. If, on the other hand, the double-helical

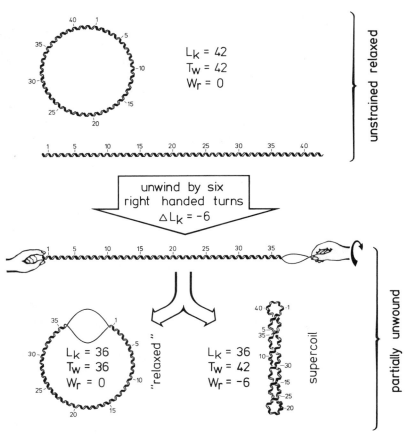

Figure 19-11. Description of interrelation between linking L_k, twisting T_w, and writhing W_r. Unstrained, relaxed B-DNA with 420 base-pairs has 42 turns of one strand wound about the other, $L_k = T_w = 42$ and $W_r = 0$ (top). If, now, the DNA is held tight at the left end, and the right end is rotated clockwise by six turns so that the double helix is partially *un*wound, L_k and consequently T_w reduce to 36. This DNA can again be circularly closed with $L_k = T_w = 36$ and $W_r = 0$; i.e., compared with the above picture, base-pairs of six turns are disrupted. Due to the tendency of DNA to retain the B-DNA form, however, T_w increases to the original $T_w = 42$ with 420 base-pairs. However, since L_k is a topological number, it is constant, $L_k = 36$ and in order to satisfy Eq. (19-1), the writhing must become $W_r = -6$, corresponding to a right-handed supercoil (superhelix) with six crossovers.

twist T_w is retained, ΔT_w is 0 and

$$\Delta L_k = \Delta W_r.$$

That is, the change in linking number is transformed almost entirely into superhelical twist ΔW_r [almost, because superhelix formation also influences the twist T_w (1362)]. In the example given in Figure 19-11, DNA is twisted in a *right*-handed sense corresponding to *un*winding, i.e., reduction in L_k and therefore *negative* $\Delta L_k (\Delta L_k = -6)$.

If the circle is now closed and allowed to writhe, it adopts a *right-handed* supercoiled (superhelical) structure with *negative* $\Delta W_r(-\Delta Wr)$, according to the definition given in Box 19-3. If, on the contrary, DNA is twisted in *left*-handed sense to produce an *over*wound double helix with *positive* ΔL_k, it compensates after cyclization with formation of a *left*-handed superhelix, the turns of which are now counted with *positive* ΔW_r.

In an alternative description, the relation

$$\alpha = \beta + \tau \qquad (19\text{-}3)$$

is used, with the *topological winding number* α equivalent to L_k, β to T_w, and τ to W_r. Frequently the term *superhelical density* $\sigma = \tau/\beta$ is used to define the degree of supercoiling. It gives a rough estimate of the number of superhelical turns per 10 base-pairs and is, typically, around $-(1/15)$ or -0.06 for supercoiled DNA obtained from cells and virions (1353,1363, 1364).

Z-DNA is of particular importance for supercoiling. If in B-DNA one single right-handed helical turn is changed into a left-handed Z-DNA turn, the twisting number changes by $\Delta T_w = -2$ in an ideal situation, i.e., without end effects. Since in a circularly closed DNA molecule L_k remains constant under all conditions (topological invariant), the writhing number has to change accordingly by $\Delta W_r = 2$. This means that B $\rightleftharpoons$ Z transformations of only small stretches of DNA can have a dramatic influence on the macroscopic topology of supercoiled DNA and might be involved in gene expression (see also Chapter 12).

How to determine supercoiling of DNA experimentally? The most direct route to demonstrate superhelical structure is, of course, visualization with the electron microscope (1350,1362).

More indirect, yet very sensitive evidence applicable to aqueous solutions of DNA is obtained by intercalation of drug or dye molecules into the double helix of circular DNA. As pointed out in Chapter 16 and illustrated in Figure 16-3, increasing drug concentration relaxes the (negatively supercoiled i.e. right-handed) DNA superhelix and, upon further addition of drug, a superhelix with opposite handedness (and opposite sign of writhing number) is obtained. If we apply again Eq. (19-1), L_k is constant over the whole experiment because the DNA remains circularly closed. Each intercalated drug molecule unwinds the DNA double helix by about 26° (1350,1362,1365); the reduction in twist, $-\Delta T_w$, must be compensated by an increase in ΔW_r (because L_k is constant). If W_r approaches 0, the circular DNA adopts an annular shape and upon addition of more intercalator $T_w < L_k$ and, therefore, W_r becomes positive, in agreement with opposite (left-handed) screw sense of the newly formed

Box 19-3. Definition of the Writing number W_r of a Superhelix

The verbal definition of the screw sense of a DNA superhelix, given by the writhing, is consistent with that used for double-helical DNA. The sign convention of the writhing number W_r, however, is reversed and requires some comments, aided by illustration in the figure below.

Let us assume that a rubber tube, as a model for DNA double helix, is wound around a cylinder in N *left*-handed turns. The free end of the tube is allowed to rotate freely during winding. Then both ends of the tube are fused together and the cylinder is removed. The tube jumps into another topological configuration and now describes an interwound double helix $\equiv$ superhelix. In this superhelix, the pitch angle α_2 is larger than α_1 in the original left-handed coil and there are approximately $N/2$ turns going up and $N/2$ turns going down. We do not observe exactly $N/2$ turns but only 3 turns because coiling (writhing) of the tube in the superhelix is partly compensated by internal twisting and by the two bends on both sides of the superhelix. Most surprisingly, the screw sense of the superhelix is *right*-handed. Nevertheless, since we started out with a left-handed coil, the definition of the number of superhelical turns is *negative*, $-N/2$ (1357,1365). The writhing number W_r counts all crossover points negatively, and in the example $W_r = -8$ (see legend to figure).

The tube (the DNA) in the superhelix is unwound by N turns ($\Delta T_w = -N$), as can be shown in another experiment. Holding the rubber tube straight with both hands, we rotate the end in the right hand in clockwise direction (in a right-handed sense). This corresponds to *un*winding of the DNA double helix, as can be demonstrated with a double-stranded, right-handed twisted cord or with a right-handed telephone cord. If, with our rubber tube, $N \times 360°$ right-handed rotations are performed, the linking number of our model DNA is reduced by $-\Delta L_k = N$. The two ends are now sealed and the tube is allowed to shrivel into superhelical form which is identical to the one derived from the left-handed coil wound around a cylinder (see figure, below).

These experiments bear on winding of DNA around histone octamers in nucleosome cores. As described in Section 19.4, the winding is left-handed and DNA is relaxed by topoisomerase I enzymes after the winding is complete. If, then, histones are removed, DNA supercoils into a right-handed superhelix with underwound DNA. The superhelix is formed because applying Eq. (19-1), we find that ΔT_w is close to zero since DNA tends to maintain its B form and underwinding ($-\Delta L_k$) has to be compensated by writhing ($-\Delta W_k$), equivalent to right-handed supercoiling.

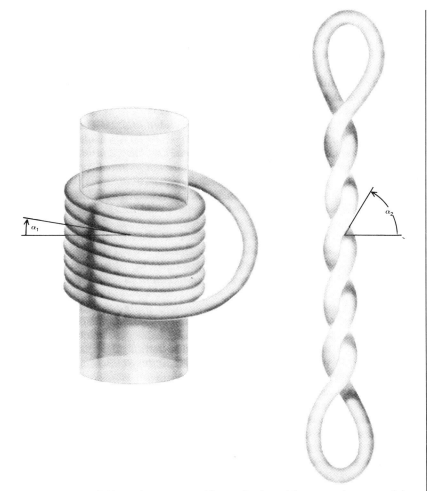

Definition of writing. Assume a rubber tube is, without strain, wound in eight *left*-handed turns around a cylinder. If, now, both ends of the tube are closed and the cylinder is removed, the tube jumps into a *right*-handed helical form, with eight crossovers (only 6 in this illustration because of end effects). Since we started out with a left-handed coil, the writhing number is defined as $W_r = -8$, both for the coil and for the closed helix although the latter is right-handed. From (1362).

DNA superhelix. It is surprising to see that the original DNA superhelix with underwound DNA relaxes if it is even further underwound by inter-calation—a process easily demonstrated with a rubber tube as DNA model.

The quantized behavior of the linking number L_k can be verified by agarose gel electrophoresis (1365). In this method, DNA molecules of the same length (and charge) are separated according to their shape and compactness. If circular DNA is treated with topoisomerase I and increasing amounts of the intercalator ethidium are added, a series of circular DNA molecules is obtained. Depending on the ethidium present in the reaction mixture, these display different linking numbers L_k and show up as well-separated, individual bands.

Another indication for superhelicity of circular DNA is obtained with several enzymes which interact preferentially with superhelical DNA (1350–1352).

Other topological isomerizations catalyzed by enzymes. Besides relaxing and "cranking up" of DNA double helices, there are other topological reactions performed by topoisomerase I and involving single-stranded DNA as well.

Topoisomerase I can act on single-stranded, circular or linear DNA

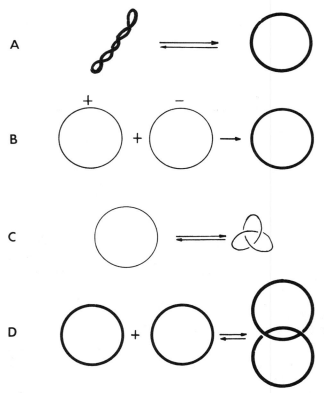

Figure 19-12. Schematic description of action of different topoisomerases on double-helical (thick lines) and single-stranded (thin lines) DNA. + and − in reaction (B) indicate the two complementary DNA strands. Reaction (C) also occurs with double-stranded DNA. Catenanes are formed in reaction (D). From (1375).

and fold it into "topological knots" of the form depicted in Figure 19-12. The reaction is reversible. Similarly, double-stranded circular DNA is deconvoluted into its components and can again be reformed. Another, rather exotic DNA configuration is represented by interlocked, circular DNA molecules called *catenanes*. These forms of DNA can be prepared *in vitro* but they have also been isolated from several organisms (1354, 1366). Especially large quantities of concatenated DNA amounting to well above 50% of total DNA can be obtained from mitochondria in mammalian L cells kept in the stationary growth phase, treated with cycloheximide, or deprived of amino acids (1366–1368).

Supercoiled circular DNA has defined tertiary structure. Light scattering experiments, ultracentrifugation, and electron microscopic studies on supercoiled, circularly closed DNA suggested that it forms not only a straight, interwound superhelix but rather star-like patterns (1311,1369-1371). The existence of these three-dimensional DNA configurations depends largely on ionic strength and is particularly stabilized under conditions prevailing in the cell nucleus (371).

The reason for the development of such structures might reside in inhomogeneities in nucleotide sequence along the DNA chain. These are effective even at relatively small differences in energy; the Boltzmann distribution predicts 98% preference of one conformation over another if the energy difference is only 2.0 kcal/mole (1370). Thus, A/T-rich regions are "soft" areas and, upon supercoiling, are prone to form separate arms of the "star." On the other hand, palindromic sequences (Box 18-1) with their tendency to arrange into Gierer trees (see Box 18-2) are likely to form extra arms (1368), especially in underwound, negatively superhelical DNA where part of the strain can thus be compensated (1372). It is still a matter of discussion, however, whether palindromes, and especially the very long palindromic sequences observed in eukaryotic genomes (1373, 1374), are, under conditions found in the cell nucleus, integrated into normal DNA or whether they are looped out to present a well-marked recognition signal for certain proteins. In addition, mention should be made of attempts to arrange cruciform-like Gierer trees in three-dimensional space, with the two palindromic arms folding together into a quadruple, superhelical form (1375). Here, again, easily recognizable, characteristic structural elements can be created but the biological existence and significance are still a matter of speculation.

Summary

For double-helical DNA (not for RNA), several forms of higher organization are known. These include simple aggregation of DNA by divalent cations like Mg(II), by polyamines, and by basic polypeptides into bead-like structures which can coil up into toroids. DNA condensed by high salt and macromolecules like polyethylene oxide adopts cooperatively and reversibly a state called ψ-DNA. If DNA is sonicated into 2000 Å-long fragments, it can even be crystallized by slow addition of alcohol. The crystals are only one DNA length thick and not suitable for diffraction studies, reminiscent of other polymers where the long molecules are folded back and forth spanning the thickness of the crystals.

In cell nuclei, DNA is condensed by four different histones. These are polypeptides of basic character and molecular weights 14,000–22,000. Eight histones (two of each kind) aggregate to form the globular nucleosome core around which DNA is wound in 1¾ left-handed, superhelical turns comprising 146 base-pairs. Another histone, H1, is attached to the DNA and "seals it off" to produce a beads-on-a-string appearance with nucleosomes spaced 40 to 100 base-pairs apart along DNA ("100 Å fiber"). This kind of complex exists at low ionic strength and is called chromatin. If the ionic strength is increased or if divalent cations like Mg(II) are added, the string condenses into a helical "solenoidal" form with three to six nucleosomes per turn, held together by intermolecular contacts between H1 histones. These "solenoids" or "300 Å fibers" are then further condensed by coiling and constitute the chromosome: a complex formed, in humans, by a DNA duplex 2 m long, arranged on a protein scaffold with histones as condensing material and adopting the well-known X shape.

In chromosomes and in circularly closed (plasmid) DNA, the ends of the double helix are fixed either by coiling of the molecule or by factual covalent linkage. There are a number of enzymes called topoisomerases which can induce superhelical twists in these DNAs by "cranking up" (equivalent to energy storage) and they can also relax these twists, thereby regulating overall structure of DNA which is important for DNA expression. The topology of a closed ribbon (DNA) is described by the formula $L_k = W_r + T_w$, where linkage number L_k is an integer, invariant under all topological forms, and indicates how often, in DNA, one strand is wound about the other. Writhing W_r and twisting T_w are any numbers, the supercoiling being given by W_r ($= 0$ for flat circle). Formation of left-handed Z-DNA (change of L_k by -2 for each turn) can be important for DNA topological features. Other forms of single- and double-stranded DNA include knots, catenanes, and Gierer trees, the biological importance of which is still a matter of discussion.

References

1. S. Furberg (1951), The crystal structure of cytidine. *Acta Crystallogr.* **3,** 325–331.
2. C. R. Dekker, A. M. Michelson, and A. R. Todd (1953), Nucleotides. Part XIX. Pyrimidine deoxyribonucleoside diphosphates. *J. Chem. Soc.,* 947–951.
3. S. Zamenhof, G. Brawermann, and E. Chargaff (1952), On the desoxypentose nucleic acids from several microorganisms. *Biochim. Biophys. Acta* **9,** 402–405.
4. W. T. Astbury (1947), X-ray studies of nucleic acids. *Symp. Soc. Exp. Biol. (Nucleic Acids)* **1,** 66–76.
5. J. M. Gulland (1947), The structures of nucleic acids. *Cold Spring Harbor Symp. Quant. Biol.* **12,** 95–103.
6. M. H. F. Wilkins (1963), Molecular configuration of nucleic acids. *Science* **140,** 941–950; *Angew. Chem.* **75,** 429–439.
7. J. D. Watson and F. H. C. Crick (1953), A structure for deoxyribose nucleic acid. *Nature* **171,** 737–738.
8. F. H. C. Crick and J. D. Watson (1954), The complementary structure of deoxyribonucleic acid. *Proc. Roy. Soc. (London) Ser. A* **223,** 80–96.
9. J. D. Watson (1968), *The Double Helix.* Weidenfeld & Nicholson, London.
10. R. Britten and E. Davidson (1969), Gene regulation for higher cells: A theory. *Science* **165,** 349–357.
11. J. R. Paulson and U. K. Laemmli (1977), The structure of histone-depleted metaphase chromosomes. *Cell* **12,** 817–828.
12. IUPAC–IUB Commission on Biochemical Nomenclature (CBN) (1970), Abbreviations and symbols for nucleic acids, polynucleotides and their constituents. *Eur. J. Biochem.* **15,** 203–208; for corrections, see (1972) *Eur. J. Biochem.* **25,** 1–4.
13. IUPAC Commission on Macromolecular Nomenclature (1979), Stereochemical definitions and notations relating to polymers. *Pure Appl. Chem.* **51,** 1101–1121.
14. IUPAC, Definitive Rules for Nomenclature of Organic Chemistry. Section A. Hydrocarbons. Section B. Fundamental heterocyclic systems. *J. Amer. Chem. Soc.* **82,** 5545–5574.
15. B. Pullman, W. Saenger, V. Sasisekharan, M. Sundaralingam, and H. R. Wilson (1973), Recommendations of standard conventions and nomenclature for the description of the conformation of polynucleotide chains. *Jerus. Symp. Quant. Chem. Biochem.* **5,** 815–820.
16. N. C. Seeman, J. M. Rosenberg, F. L. Suddath, J. J. Park Kim, A. Rich (1976), A simplified alphabetical nomenclature for dihedral angles in the polynucleotide backbone. *J. Mol. Biol.* **104,** 142–143.
17. S. Arnott and D. W. L. Hukins (1969), Conservation of conformation in mono- and polynucleotides. *Nature* **224,** 886–888.

18. A. V. Lakshminarayanan and V. Sasisekharan (1970), Stereochemistry of nucleic acids and polynucleotides. II. Allowed conformations of the monomer unit for different ribose puckerings. *Biochim. Biophys. Acta* **204**, 49–53.

19. IUPAC–IUB Joint Commission on Biochemical Nomenclature (1983), Abbreviations and symbols for the description of conformations of polynucleotide chains. *Eur. J. Biochem.* **131**, 9–15.

20. R. J. Suhadolnik (1970), *Nucleoside Antibiotics*. Wiley, New York.

21. A. Bloch (Ed.) (1975), *Chemistry, Biology and Clinical Uses of Nucleoside Analogs. Ann. N. Y. Acad. Sci.* **225**.

22. IUPAC–IUB Commission on Biochemical Nomenclature (1970), Abbreviations and symbols for the description of the conformation of polypeptide chains. *Eur. J. Biochem.* **17**, 193–201.

23. W. Klyne and V. Prelog (1960), Description of steric relationships across single bonds. *Experientia* **16**, 521–523.

24. IUPAC (1971), Tentative rules for the nomenclature of organic chemistry. Section E. Fundamental stereochemistry. *Eur. J. Biochem.* **18**, 151–170.

25. L. D. Hall (1963), Conformations of some ribofuranosides. *Chem. Ind. (London)*, 950–951.

26. C. D. Jardetzky (1960), Proton magnetic resonance studies on purines, pyrimidines, ribose nucleosides and nucleotides. III. Ribose conformation. *J. Amer. Chem. Soc.* **82**, 229–233.

27. J. E. Kilpatrick, K. S. Pitzer, and R. Spitzer (1947), The thermodynamics and molecular structure of cyclopentane. *J. Amer. Chem. Soc.* **69**, 2483–2488.

28. K. S. Pitzer and W. E. Donath (1959), Conformations and strain energy of cyclopentane and its derivatives. *J. Amer. Chem. Soc.* **81**, 3213–3218.

29. L. D. Hall, P. R. Steiner, and C. Pedersen (1970), Studies of specifically fluorinated carbohydrates. Part VI. Some pentafuranosyl fluorides. *Can. J. Chem.* **48**, 1155–1165.

30. C. Altona, H. J. Geise, and C. Romers (1968), Conformation of nonaromatic ring compounds. XXV. Geometry and conformation of ring D in some steroids from X-ray structure determinations. *Tetrahedron* **24**, 13–32.

31. C. Altona and M. Sundaralingam (1972), Conformational analysis of the sugar ring in nucleosides and nucleotides. A new description using the concept of pseudorotation. *J. Amer. Chem. Soc.* **94**, 8205–8212.

32. M. Levitt and A. Warshel (1978), Extreme conformational flexibility of the furanose ring in DNA and RNA. *J. Amer. Chem. Soc.* **100**, 2607–2613.

33. J. Donohue and K. N. Trueblood (1960), Base pairing in DNA. *J. Mol. Biol.* **2**, 363–371.

34. A. E. V. Haschemeyer and A. Rich (1967), Nucleoside conformations: An analysis of steric barriers to rotation about the glycosidic bond. *J. Mol. Biol.* **27**, 369–384.

35. M. Sundaralingam (1973), The concept of a conformationally 'rigid' nucleotide and its significance in polynucleotide conformational analysis. *Jerus. Symp. Quant. Chem. Biochem.* **5**, 417–456.

36. P. Prusiner and M. Sundaralingam (1972), Stereochemistry of nucleic acids and their constituents. XXV. Crystal and molecular structure of allopurinol, a potent inhibitor of xanthine oxidase. *Acta Crystallogr. B* **28**, 2148–2152.

37. E. Shefter and K. N. Trueblood (1965), The crystal and molecular structure of D(+)Ba-uridine-5'-phosphate. *Acta Crystallogr.* **18,** 1067–1077.
38. S. Arnott (1970), The geometry of nucleic acids. *Prog. Biophys. Mol. Biol.* **21,** 267–319.
39. T. F. Lai and R. E. Marsh (1972), The crystal structure of adenosine. *Acta Crystallogr. B* **28,** 1982–1989.
40. E. A. Green, R. D. Rosenstein, R. Shiono, D. J. Abraham, B. L. Trus, and R. E. Marsh (1975), The crystal structure of uridine. *Acta Crystallogr. B* **31,** 102–107.
41. S. S. Tavale and H. M. Sobell (1970), Crystal and molecular structure of 8-bromoguanosine and 8-bromoadenosine, two purine nucleosides in the *syn* conformation. *J. Mol. Biol.* **48,** 109–123.
42. D. Suck and W. Saenger (1972), Molecular and crystal structure of 6-methyluridine. A pyrimidine nucleoside in the *syn* conformation. *J. Amer. Chem. Soc.* **94,** 6520–6526.
43. J. P. Glusker and K. N. Trueblood (1972), *Crystal Structure Analysis: A Primer.* Oxford Univ. Press, London.
44. G. H. Stout and L. H. Jensen (1968), *X-Ray Structure Determination—A Practical Guide.* McMillan, London.
45. M. J. Buerger (1970), *Contemporary Crystallography.* McGraw–Hill, New York.
46. M. M. Woolfson (1970), *An Introduction to X-Ray Crystallography.* Cambridge Univ. Press, London.
47. D. Sherwood (1976), *Crystals, X-Rays and Proteins.* Longman, London.
48. U. W. Arndt and B. T. M. Willis (1966), *Single Crystal Diffractometry.* Cambridge Univ. Press, London.
49. N. F. M. Henry and K. Lonsdale (Eds.) (1968), *International Tables for X-Ray Crystallography,* Vols. II–IV. Kynoch Press, Birmingham.
50. S.-I. Mizushima and T. Shimanouchi (1961), Possible polypeptide configurations of proteins from the viewpoint of internal rotation potential. *Adv. Enzymol.* **23,** 1–27.
51. G. N. Ramachandran, C. Ramakrishnan, and V. Sasisekharan (1963), Stereochemistry of polypeptide chain configurations. *J. Mol. Biol.* **7,** 95–99.
52. J. A. Schellman and C. Schellman (1964), The conformation of polypeptide chains in proteins. In: *The Proteins,* (H. Neurath, Ed.), 2nd ed., Vol. 2, pp. 1–139. Academic Press, New York.
53. G. Nemethy and H. A. Scheraga (1965), Theoretical determination of a polypeptide chain by a computer method. *Biopolymers* **3,** 155–184.
54. V. Sasisekharan, A. V. Lakshminarayanan, and G. N. Ramachandran (1967), Stereochemistry of nucleic acids and polynucleotides. I. Theoretical determination of the allowed conformation of the monomer unit. In: *Conformation of Biopolymers* (G. N. Ramachandran, Ed.), Vol. 2, pp. 641–654. Academic Press, London.
55. H. A. Scheraga (1968), Calculations of conformations of polypeptides. *Adv. Phys. Org. Chem.* **6,** 103–183.
56. G. N. Ramachandran and V. Sasisekharan (1968), Conformation of polypeptides and proteins. *Adv. Protein Chem.* **23,** 283–437.
57. A. V. Lakshminarayanan and V. Sasisekharan (1969), Stereochemistry of nucleic acids and polynucleotides. IV. Conformational energy of base-sugar units. *Biopolymers* **8,** 475–488.

58. N. Yathindra and M. Sundaralingam (1973), Correlation between the backbone and side chain conformations in 5'-nucleotides. The concept of a 'rigid' nucleotide conformation. *Biopolymers* **12**, 297–314.

59. O. E. Millner, Jr., and J. A. Andersen (1975), The conformational analysis of adenosine triphosphate by classical potential energy calculations. *Biopolymers* **14**, 2159–2179.

60. S. B. Broyde, R. M. Wartell, S. D. Stellman, B. Hingerty, and R. Langridge (1975), Classical potential energy calculations for ApA, CpC, GpG and UpU. The influence of the bases on RNA subunit conformations. *Biopolymers* **14**, 1597–1613.

61. W. K. Olson and P. J. Flory (1972), Spatial configurations of polynucleotide chains. II. Conformational energies and average dimensions of polyribonucleotides. *Biopolymers* **11**, 25–56.

62. A. I. Kitaigorodsky and V. G. Dashevsky (1968), Conformational analysis of overcrowded aromatic molecules. *Tetrahedron* **24**, 5917–5928.

63. G. Govil and A. Saran (1971), Quantum chemical studies of nucleic acids. I. Extended Hückel calculations on D-ribose phosphate. *J. Theor. Biol.* **30**, 621–630.

64. S. Lifson (1981), Potential energy functions for structural molecular biology. *In: Methods in Structural Molecular Biology,* (D. B. Davies, W. Saenger, and S. S. Danyluk Eds.), pp. 359–385. Plenum Press, London.

65. F. Jordan (1973), Lennard–Jones potential calculations of the barrier to rotation about the glycosidic C–N linkage in selected purine nucleosides and nucleotides. A direct comparison of the results of 6-12 potential calculations with results of semiempirical molecular orbital studies. *J. Theor. Biol.* **41**, 375–395.

66. G. Klopman and R. C. Evans (1977), The neglect-of-differential-Overlap methods of molecular orbital theory. In: *Electronic Structure Calculation* (G. A. Segal, Ed.), pp. 29–67. Plenum Press, New York.

67. B. Pullman and A. Pullman (1963), *Quantum Biochemistry,* Interscience, New York.

68. G. Del Re (1958). A simple MO-LCAO method for the calculation of charge distributions in saturated organic molecules. *J. Chem. Soc.,* 4031–4040.

69. G. Del Re (1964), Theoretical procedures for the study of biochemical σ-systems. In: *Electronic Aspects of Biochemistry* (B. Pullman, Ed.), pp. 221–235. Academic Press, New York/London.

70. A. Pullman and B. Pullman (1968), Aspects of the electronic structure of the purine and pyrimidine bases of the nucleic acids and of their interactions. *Adv. Quant. Chem.* **4**, 267–325.

71. F. Jordan and B. Pullman (1968), Molecular orbital calculations on the preferred conformation of nucleosides. *Theor. Chim. Acta (Berlin)* **9**, 242–252.

72. V. Renugopalakrishnan, A. V. Lakshminarayanan, and V. Sasisekharan (1971), Stereochemistry of nucleic acids and polynucleotides. III. Electronic charge distribution. *Biopolymers* **10**, 1159–1167.

73. R. A. Scott and H. A. Scheraga (1965), Method for calculating internal rotation barriers. *J. Chem. Phys.* **42**, 2209–2215.

74. A. Warshel (1977), The consistent force field and its quantum chemical extension. In: *Modern Theoretical Chemistry* (G. Segal, Ed.) Plenum Press, New York. Vol. **7**, pp. 133–172.

75. N. Trinajstić (1977), Hückel theory and topology. In: *Electronic Structure Calculation* (G. A. Segal, Ed.), pp. 1–28. Plenum Press, New York.

76. F. Jordan (1973), An extended Hückel molecular orbital approach to the study of the electronic structures and barriers to *syn–anti* interconversion in *syn* purine nucleosides. *Biopolymers* **12**, 243–255.

77. J.-P Malrieu (1977), The PCILO method. In: *Electronic Structure Calculation* (G. A. Segal, Ed.), pp. 69–103. Plenum Press, New York.

78. B. Pullman and A. Pullman (1974), Molecular orbital calculations on the conformation of amino-acid residues of proteins. *Adv. Protein Chem.* **28**, 348–526.

79. C. C. J. Roothaan (1951), New Developments in molecular orbital theory, *Rev. Mod. Phys.* **23**, 69–89.

80. T. L. Blundell and L. N. Johnson (1976), *Protein Crystallography*. Academic Press, New York.

81. K. C. Holmes and D. M. Blow (1966), *The Use of X-Ray Diffraction in the Study of Protein and Nucleic Acid Structure*. Interscience, New York.

82. R. E. Dickerson, M. L. Kopka, J. C. Varnum, and J. E. Weinzierl (1967), Bias, feedback and reliability in isomorphous phase analysis. *Acta Crystallogr.* **23**, 511–522.

83. F. M. Richards (1968), The matching of physical models to three-dimensional electron-density maps; a simple optical device. *J. Mol. Biol.* **37**, 225–230.

84. J. L. Chambers and R. M. Stroud (1979), The accuracy of refined protein structures: Comparison of two independently refined models of bovine trypsin. *Acta Crystallogr.* B **35**, 1861–1874.

85. C. D. Stout, H. Mizuno, S. T. Rao, P. Swaminathan, J. Rubin, T. Brennan, and M. Sundaralingam (1978), Crystal and molecular structure of yeast phenylalanyl transfer RNA. Structure determination, difference Fourier refinement, molecular conformation, metal and solvent binding. *Acta Crystallogr.* B **34**, 1529–1544.

86. J. L. Sussman, S. R. Holbrook, R. W. Warrant, and S.-H. Kim (1978), Crystal structure of yeast phenylalanine transfer RNA. I. Crystallographic refinement. *J. Mol. Biol.* **123**, 607–630.

87. A. Jack, J. E. Ladner, and A. Klug (1976), Crystallographic refinement of yeast phenylalanine transfer RNA at 2.5 Å resolution. *J. Mol. Biol.* **108**, 619–649.

88. G. Giannoni, F. Padden, and H. D. Keith (1969), Crystallization of DNA from dilute solution. *Proc. Nat. Acad. Sci. USA* **62**, 964–971.

89. L. S. Lerman, L. S. Wilkerson, J. H. Venable, Jr., and B. H. Robinson (1976), DNA packing in single crystals inferred from freeze-fracture-etch replicas. *J. Mol. Biol.* **108**, 271–293.

90. M. O. Kliya and V. D. Osika (1976), Morphology and growth of DNA cetavlate salt crystals. *Sov. Phys. Crystallogr.* **21**, 681–687. [in English]

91. O. F. Polivtsev, V. D. Osika, B. Ya Sukharevskii, and D. Ya. Tsvankin (1977), X-Ray diffraction from individual crystals of cetavlon salt of DNA. *Sov. Phys. Crystallogr.* **22**, 45–48. [in English]

92. W. Fuller, F. Hutchinson, M. Spencer, and M. H. F. Wilkins (1967), Molecular and crystal structure of double-helical RNA. I. An X-ray diffraction study of fragmented yeast RNA and a preliminary double-helical RNA model. *J. Mol. Biol.* **27**, 507–524.

93. N. F. M. Henry, H. Lipson, and W. A. Wooster (1961), *The Interpretation of X-ray Diffraction Photographs*. Macmillan, New York.

94. D. A. Marvin, M. Spencer, M. H. F. Wilkins, and L. D. Hamilton (1961), The molecular configuration of DNA. III. X-ray diffraction study of the C form of the lithium salt. *J. Mol. Biol.* **3**, 547–565.

95. D. A. Marvin and C. Nave (1981), X-Ray fiber diffraction. In: *Methods in Structural Molecular Biology* (D. B. Davies, W. Saenger and S. S. Danyluk, Eds.), pp. 3–43. Plenum Press, London.

96. W. Cochran, F. H. C. Crick, and V. Vand (1952), The structure of synthetic polypeptides. I. The transform of atoms on a helix. *Acta Crystallogr.* **5**, 581–586.

97. A. R. Stokes (1955), The theory of X-ray fiber diagrams. *Prog. Biophys.* **5**, 140–167.

98. A. Klug, F. H. C. Crick, and H. S. Wyckoff (1958), Diffraction by helical molecules. *Acta Crystallogr.* **11**, 199–213.

99. G. N. Ramachandran (1960), Analysis of the X-ray diffraction pattern of helical structures. *Proc. Indian Acad. Sci.* **52A**, 240–254.

100. S. Arnott (1973), Fiber diffraction analysis of biopolymer molecules. *Trans. Amer. Cryst. Assoc.* **9**, 31–56.

101. R. Langridge, D. A. Marvin, W. E. Seeds, H. R. Wilson, C. W. Hooper, M. H. F. Wilkins, and L. D. Hamilton (1960), The molecular configuration of deoxyribonucleic acid. II. Molecular models and their Fourier transforms. *J. Mol. Biol.* **2**, 38–64.

102. S. Arnott, S. D. Dover, and A. J. Wonacott (1969), Least-squares refinement of the crystal and molecular structures of DNA and RNA from X-ray data and standard bond lengths and angles. *Acta Crystallogr. B* **25**, 2192–2206.

103. R. A. Langridge and A. Rich (1963), Molecular structure of helical polycytidylic acid. *Nature* **198**, 725–728.

104. S. Arnott, R. Chandrasekharan, and A. G. W. Leslie (1976), Structure of the single-stranded polyribonucleotide polycytidylic acid. *J. Mol. Biol.* **106**, 735–738.

105. A. Rich (1958), The molecular structure of polyinosinic acid. *Biochim. Biophys. Acta* **29**, 502–509.

106. S. Arnott, R. Chandrasekharan, and C. M. Marttila (1974), Structures of polyinosinic acid and polyguanylic acid. *Biochem. J.* **141**, 537–538.

107. J. Donohue (1956), Hydrogen-bonded helical configurations of polynucleotides. *Proc. Nat. Acad. Sci. USA* **42**, 60–65.

108. T. T. Wu (1969), Secondary structures of DNA. *Proc. Nat. Acad. Sci. USA* **63**, 400–405.

109. J. Donohue (1969), Fourier analysis and the structure of DNA. *Science* **165**, 1091–1096.

110. J. Donohue (1969), Crystal symmetry of deoxyribonucleic acid. On the validity of the use of intensity statistics on limited numbers of data. *J. Mol. Biol.* **41**, 291–294.

111. J. Donohue (1970), Fourier series and difference maps as lack of structure proof: DNA is an example. *Science* **167**, 1700–1702.

112. G. A. Rodley, R. S. Scobie, R. H. T. Bates, and R. M. Lewitt (1976), A possible conformation for double-stranded polynucleotides. *Proc. Nat. Acad. Sci. USA* **73**, 2959–2963.

113. V. Sasisekharan, N. Pattabiraman, and Goutam Gupta (1978), Some implications of an alternative structure for DNA. *Proc. Nat. Acad. Sci USA* **75,** 4092–4096.
114. B. Cyrax and R. Gäth (1978), The conformation of double-stranded DNA. *Naturwissenschaften,* **65,** 106–108.
115. R. C. Hopkins (1981), Deoxyribonucleic acid structure. A new model. *Science* **211,** 289–291.
116. S. Arnott, M. H. F. Wilkins, L. D. Hamilton, and R. Langridge (1965), Fourier synthesis studies of lithium DNA. Part III. Hoogsteen models. *J. Mol. Biol.* **11,** 391–402.
117. M. H. F. Wilkins, S. Arnott, D. A. Marvin, and L. D. Hamilton (1970), Fourier analysis and the structure of DNA. Some misconceptions on Fourier analysis and Watson–Crick base pairing. *Science* **167,** 1693–1694.
118. S. Arnott (1970). Crystallography of DNA: Difference synthesis supports Watson–Crick base pairing. *Science* **167,** 1694–1670.
119. M. H. F. Wilkins, H. R. Wilson, and L. D. Hamilton (1970), Secondary structures of DNA. *Proc. Nat. Acad. Sci USA* **65,** 761–762.
120. S. Arnott (1971), Investigation of the crystal and molecular symmetry of DNA using X-ray intensity statistics. *J. Mol. Biol.* **59,** 381–384.
121. S. Arnott (1979), Is DNA really a double helix? *Nature* **278,** 780–781.
122. R. J. Greenall, W. J. Pigram and W. Fuller (1979), X-Ray diffraction from the side-by-side model of DNA. *Nature* **282,** 880–882.
123. V. A. Bloomfield, D. Crothers, and I. Tinoco, Jr. (1974), *Physical Chemistry of Nucleic Acids.* Harper & Row, New York.
124. G. L. Cantoni and D. R. Davies (Eds.) (1971), *Procedures in Nucleic Acid Research,* Vol. 2, pp. 3–296. Harper & Row, New York.
125. P. O. P. Ts'o (Ed.) (1974), *Basic Principles in Nucleic Acid Chemistry,* Vols. 1 and 2. Academic Press, New York.
126. B. Pullman (Ed.) (1978), *Nucleic Magnetic Resonance Spectroscopy in Molecular Biology. Jerus. Sympos. Quant. Chem. Biochem.* **11.**
127. I. Tinoco, Jr., and C. Bustamente (1981), Chiroptical methods and their applications to biomolecular systems. In: *Methods in Structural Molecular Biology* (D. B. Davies, W. Saenger, and S. S. Danyluk, Eds.), pp. 269–305. Plenum Press, London.
128. W. L. Peticolas (1981), Classical and resonance Raman spectroscopy of biological macromolecules. In: *Methods in Structural Molecular Biology* (D. B. Davies, W. Saenger, and S. S. Danyluk, Eds.), pp. 237–266. Plenum Press, London.
129. C. Altona (1981), High resolution NMR studies of nucleic acids. In: *Methods in Structural Molecular Biology* (D. B. Davies, W. Saenger and S. S. Danyluk, Eds.), pp. 161–213. Plenum Press, London.
130. D. B. Davies (1978), Conformations of nucleosides and nucleotides. *Prog. NMR Spectrosc.* **12,** 135–225.
131. S. S. Danyluk (1979), Nuclear magnetic resonance studies of nucleoside conformational properties. In: *Nucleoside Analogues. Chemistry, Biology, and Medical Application* (R. T. Walker, E. DeClercq, and F. Eckstein, Eds.), pp. 15–34. Plenum Press, New York.
132. M. M. Dhingra and R. H. Sarma (1979), Nuclear magnetic resonance spectroscopy. Chemical shifts, coupling constants and molecular geometry. In:

Stereodynamics of Molecular Systems (R. H. Sarma, Ed.), pp. 3–38. Pergamon, New York.

133. G. Giessner-Prettre and B. Pullman (1977), On the conformational dependence of the proton chemical shifts in nucleosides and nucleotides. I. Proton shifts in the ribose ring of pyrimidine nucleosides as a function of the torsion angle about the glycosyl bond; and II. Proton shifts in the ribose ring of purine nucleosides as a function of the torsion angle about the glycosidic bond. *J. Theor. Biol.* **65**, 171–188, 188–201.

134. C. Giessner-Prettre, B. Pullman, and J. Caillet (1977), Theoretical study on the proton chemical shifts of hydrogen bonded nucleic acid bases. *Nucleic Acids Res.* **4**, 99–116.

135. C. Giessner-Prettre, B. Pullman, P. N. Borer, L.-S. Kan, and P. O. P. Ts'o (1976), Ring-current effects in the NMR of nucleic acids: A graphical approach. *Biopolymers* **15**, 2277–2286.

136. M. Karplus (1959), Contact electron-spin coupling of nuclear magnetic moments. *J. Chem. Phys.* **30**, 11–15.

137. C. Altona and M. Sundaralingam (1973), Conformational analysis of the sugar ring in nucleosides and nucleotides. Improved method for the interpretation of proton magnetic resonance coupling constants. *J. Amer. Chem. Soc.* **95**, 2333–2344.

138. D. B. Davies and S. S. Danyluk (1974), Nuclear magnetic resonance studies of 5'-ribo- and deoxyribonucleotide structures in solution. *Biochemistry* **13**, 4417–4434.

139. W. Guschlbauer and Tran-Dinh Son (1975), Nucleoside conformations. 20. Determination of sugar conformations of nucleosides and nucleotides by a simple graphical method. *Nucleic Acids Res. Spec. Publ. No. 1*, S85–88.

140. F. E. Evans and R. H. Sarma (1974), The intramolecular conformation of adenosine-5'-monophosphate in aqueous solution as studied by Fast Fourier Transform ^{1}H and ^{1}H-$\{^{31}P\}$ nuclear magnetic resonance spectroscopy. *J. Biol. Chem.* **249**, 4754–4759.

141. A. Jaworski, I. Ekiel, and D. Shugar (1978), Coupling constants between cisoidal protons in pentose nucleosides. Limitations of range of application of Karplus relation, and solution conformations of β-arabinofuranosyl and β-xylofuranosyl nucleosides. *J. Amer. Chem. Soc.* **100**, 4357–4361.

142. W. P. Niemczura and F. E. Hruska (1980), A ^{13}CMR study of 2'-deoxynucleotides in the *syn* and *anti* conformation. Can. J. Chem. **58**, 472–478.

143. J. L. Alderfer and P. O. P. Ts'o (1977), Conformational properties of the furanose phosphate backbone in nucleic acids. A carbon-13 nuclear magnetic resonance study. *Biochemistry* **16**, 2410–2416.

144. I. C. P. Smith, H. H. Mantsch, R. D. Lapper, R. Deslauriers, and T. Schleich (1973), A study of the conformations of nucleic acids by carbon-13 and hydrogen nuclear magnetic resonance spectroscopy. *Jerus. Symp. Quant. Chem. Biochem.* **5**, 381–401.

145. F. S. Ezra, C. H. Lee, N. S. Kondo, S. S. Danyluk, and R. H. Sarma (1977), Conformational properties of purine–pyrimidine and pyrimidine–purine dinucleoside monophosphate. *Biochemistry* **16**, 1977–1987.

146. D. M. Cheng, M. M. Dhingra, and R. H. Sarma (1978), Spatial configuration of deoxyribotrinucleoside diphosphate in aqueous solution. *Nucleic Acids Res.* **5**, 4399–4416.

147. D. M. Cheng and R. H. Sarma (1977), Intimate details of the conformational characteristics of deoxyribodinucleoside monophosphates in aqueous solution. *J. Amer. Chem. Soc.* **99**, 7333–7348.

148. C. Chachaty, T. Zemb. G. Langlet, Tran-Dinh Son, H. Buc, and M. Morange (1976), A proton-relaxation-time study of the conformation of some purine and pyrimidine 5′-nucleotides in aqueous solution. *Eur. J. Biochem.* **62**, 45–53.

149. R. E. Schirmer, J. P. Davies, J. H. Noggle, and P. A. Hart (1972), Conformational analysis of nucleosides in solution by quantitative application of the nuclear Overhauser effect. *J. Amer. Chem. Soc.* **94**, 2561–2572.

150. J. H. Noggle and R. E. Schirmer (1971), *The Nuclear Overhauser Effect: Chemical Applications.* Academic Press, New York.

151. S. S. Danyluk and F. E. Hruska (1968), The effect of pH upon the nuclear magnetic resonance spectra of nucleosides and nucleotides. *Biochemistry* **7**, 1038–1043.

152. M. P. Schweizer, A. D. Broom, P. O. P. Ts'o, and D. P. Hollis (1968), Studies of mononucleotides by proton magnetic resonance. *J. Amer. Chem. Soc.* **90**, 1042–1055.

153. M. P. Schweizer and R. K. Robins (1973), NMR studies on the conformation of nucleosides and 3′,5′-nucleotides. *Jerus. Sympos. Quant. Chem. Biochem.* **5**, 329–344.

154. R. U. Lemieux, T. L. Nagabushan, and B. Paul (1972), Relationship of ¹³C to vicinal ¹H coupling to the torsion angle in uridine and related structures. *Can. J. Chem.* **50**, 773–776.

155. B. R. Reid (1981), NMR studies of RNA structure and dynamics. *Annu. Rev. Biochem.* **50**, 969–996.

156. D. J. Patel, A. Pardi and K. Itakura (1982), DNA conformation, dynamics, and interactions in solution. *Science* **216**, 581–590.

157. A. Bondi (1964), Van der Waals volumes and radii. *J. Phys. Chem.* **68**, 441–451.

158. N. L. Allinger (1976), Calculation of molecular structural energy by force-field methods. *Adv. Phys. Org. Chem.* **13**, 1–82.

159. R. C. Weast (Ed.) (1976), *Handbook of Chemistry and Physics,* p. D-178. CRC Press Cleveland, Ohio.

160. D. Suck, P. C. Manor, G. Germain, C. H. Schwalbe, G. Weimann, and W. Saenger (1973), X-Ray study of helix, loop and base pair stacking in trinucleoside diphosphate ApApA. *Nature New Biol.* **246**, 161–165.

161. G. N. Reeke and R. E. Marsh (1965), The crystal structure of 5-ethyl-6-methyluracil. *Acta Crystallogr.* **20**, 703–708.

162. S.-H. Kim, G. Quigley, F. L. Suddath, and A. Rich (1971), High resolution X-ray diffraction patterns of crystalline transfer RNA that show helical regions. *Proc. Nat. Acad. Sci. USA* **68**, 841–845.

163. F. H. Allen, S. Bellard, M. D. Brice, B. A. Cartwright, A. Doubleday, H. Higgs, T. Hummelink, B. G. Hummelink-Peters, O. Kennard, W. D. S. Motherwell, J. R. Rodgers, and D. G. Watson (1979), The Cambridge Crystallographic Data Centre: Computer-based search, retrieval, analysis and display of information. *Acta Crystallogr. B* **35**, 2331–2339.

164. A. Rich (1977), Three-dimensional structure of transfer RNA. *Acc. Chem. Res.* **10**, 388–402.

165. R. Langridge, H. R. Wilson, C. W. Hooper, M. H. F. Wilkins, and L. D. Hamilton (1960), The molecular configuration of DNA. I. X-Ray diffraction study of a crystalline form of the lithium salt. *J. Mol. Biol.* **2**, 19–37.

166. M. Sundaralingam (1969), Stereochemistry of nucleic acids and their constituents. IV. Allowed and preferred conformations of nucleosides, nucleoside mono-, di-, tri-, tetraphosphates, nucleic acids and polynucleotides. *Biopolymers* **7**, 821–860.

167. M. Sundaralingam (1976), Structure and conformation of nucleosides and nucleotides and their analogs as determined by X-ray diffraction. *Ann. N. Y. Acad. Sci.* **255**, 3–42.

168. H. R. Wilson (1973), Some aspects of nucleotide conformation in crystals and in nucleic acids. *Jerus. Symp. Quant. Chem. Biochem.* **5**, 261–270.

169. W. Saenger (1973), Structure and function of nucleosides and nucleotides. *Angew. Chem. Int. Ed. Engl.* **12**, 591–601; *Angew. Chem.* **85**, 680–690.

170. H. P. M. de Leeuw, C. A. G. Haasnoot, and C. Altona (1980), Empirical correlations between conformational parameters in β-D-furanoside fragments derived from a statistical survey of crystal structures of nucleic acid constituents. Full description of nucleoside molecular geometries in terms of four parameters. *Isr. J. Chem.* **20**, 108–126.

171. P. Murray-Rust and S. Motherwell (1978), Computer retrieval and analysis of molecular geometry. III. Geometry of the β-1′-aminofuranoside fragment. *Acta Crystallogr. B* **34**, 2534–2546.

172. E. Westhof and M. Sundaralingam (1980), Interrelationships between the pseudorotation parameters P and m and the geometry of the furanose ring. *J. Amer. Chem. Soc.* **102**, 1493–1500.

173. F. E. Hruska (1973), Mapping nucleoside conformations in aqueous solution —A correlation of some furanose structural parameters. *Jerus. Symp. Quant. Chem. Biochem.* **5**, 345–360.

174. P. A. Hart and J. P. Davis (1973), Purine nucleoside conformational analysis —Applications of the nuclear Overhauser effect. *Jerus. Symp. Quant. Chem. Biochem.* **5**, 297–310; Pyrimidine nucleoside conformational analysis. Nuclear Overhauser effect and circular dichroism correlations. *J. Amer. Chem. Soc.* **93**, 753–760.

175. R. H. Sarma and R. Mynott (1973), Manifestation of the conformation of pyridine coenzymes in ^{31}P and ^{1}H Fourier transform nuclear magnetic resonance spectroscopy. *Jerus. Symp. Quant. Chem. Biochem.* **5**, 591–626.

176. C. Altona (1975), Backbone conformation of several dinucleoside monophosphates in solution deduced from Fourier transform NMR spectroscopy at 270 MHz. In: *Structure and Conformation of Nucleic Acids and Protein–Nucleic Acid Interactions* (M. Sundaralingam and S. T. Rao, Eds.), pp. 613–630. Univ. Park Press, London.

177. G. J. Thomas, Jr. (1975), Structural studies of nucleic acids and polynucleotides by laser–Raman spectroscopy. *Ibid.*, pp. 253–281.

178. T. Schleich, B. P. Cross, B. J. Blackburn, and I. C. P. Smith (1975), Elucidation of nucleic acid conformation by carbon-13-nuclear magnetic resonance spectroscopy. *Ibid.*, pp. 223–252.

179. J. S. Kwiatkowski and B. Pullman (1975), Tautomerism and electronic structure of biological pyrimidines. *Adv. Heterocyc. Chem.* **18**, 199–335.

180. B. Pullman and A. Pullman (1971), Electronic aspects of purine tautomerism, *Adv. Heterocyc. Chem.* **13**, 77–159.

181. B. Pullman (1976), Proteins, nucleic acids and their constituents. In: *Quantum-Mechanics of Molecular Conformations* (B. Pullman, Ed.), pp. 295–383. Wiley, New York.

182. B. Pullman and A. Saran (1976), Quantum-mechanical studies on the conformation of nucleic acids and their constituents. *Prog. Nucleic Acid Res. Mol. Biol.* **18**, 215–322.

183. E. Harbers, G. F. Domagk, and W. Müller (1968), *Introduction to Nucleic Acids.* Reinhold, New York/Amsterdam/London.

184. J. Duchesne (Ed.) (1973), *Physico-Chemical Properties of Nucleic Acids,* Vols. 1 and 2. Academic Press, New York.

185. W. Guschlbauer (1976), *Nucleic Acid Structure.* Springer-Verlag, Heidelberg (Science Library).

186. J. N. Davidson (1976), *The Biochemistry of the Nucleic Acids,* 8th ed. (revised by R. L. P. Adams, R. H. Burdon, A. M. Campbell, and R. M. S. Smellie). Chapman & Hall, London.

187. S. Y. Wang (1976), *Photochemistry and Photobiology of Nucleic Acids,* Vol. I, *Chemistry,* Vol. II. *Biology.* Academic Press, New York.

188. W. E. Cohn (Ed.) (1963–1979), *Progress in Nucleic Acid Research and Molecular Biology,* Vols. 1–23. Academic Press, New York.

189. B. Pullman (Ed.) (1968), *Molecular Associations in Biology.* Academic Press, New York.

190. S. Neidle (Ed.) (1981), *Topics in Nucleic Acid Structure.* MacMillan, London.

191. R. Taylor and O. Kennard (1982), The molecular structures of nucleosides and nucleotides. 1. The influence of protonation on the geometries of nucleic acid constituents. *J. Mol. Struct., 78,* 1–28.

192. D. Voet and A. Rich (1970), The crystal structures of purines, pyrimidines and their intermolecular complexes. *Prog. Nucleic Acid Res. Mol. Biol.* **10**, 183–265.

193. H. G. Ringertz (1973), Crystal structure studies of purines. *Jerus. Symp. Quant. Chem. Biochem.* **4**, 61–72.

194. J. H. Lister (1979), Current views on some physicochemical aspects of purines. *Adv. Heterocyc. Chem.* **24**, 215–216.

195. S. T. Rao and M. Sundaralingam (1970), Stereochemistry of nucleic acids and their constituents. XIII. The crystal and molecular structure of 3'-O-acetyladenosine. Conformational analysis of nucleosides and nucleotides with *syn* glycosidic torsion angle. *J. Amer. Chem. Soc.* **92**, 4963–4970.

196. R. C. Weast (Ed.) (1976), *Handbook of Chemistry and Physics,* p. F-215. CRC Press Cleveland, Ohio.

197. L. E. Sutton (Ed.) (1965), *Tables of Interatomic Distances and Configuration in Molecules and Ions.* Chem. Soc., Burlington House, London.

198. L. Pauling (1978), *The Nature of the Chemical Bond.* Cornell Univ. Press, Ithaca, N.Y.

199. A. Pullman and B. Pullman (1958), Recherches sur la structure en chimothérapie anticancéreuse. II. Les bases puriques fondamentales . *Bull. Soc. Chim. (Pairs)*, 766–772. Part VI: Les pyrimidines naturelles fondamentales (1959), *Bull. Soc. Chim. (Paris)*, 594–597.

200. C. N. R. Rao (1973), Conformational studies on amides and related systems. *Jerus. Symp. Quant. Chem. Biochem.* **5**, 107–120.

201. M. Geller and B. Lesyng (1975), Barrier to rotation and conformation of the

$-NR_2$ group in cytosine and its derivatives. I. Theoretical studies of cytosine. *Biochim. Biophys. Acta* **417**, 407–419.

202. I. Kulakowska, M. Geller, B. Lesyng, K. Bolewska, and K. L. Wierzchowski (1975), Barrier to rotation and conformation of the $-NR_2$ group in cytosine and its derivatives. II. Theoretical dipole moments of methylated cytosines. *Biochim. Biophys. Acta,* **417**, 420–429.

203. K. G. Rao and C. N. R. Rao (1973), Restricted rotation about the exocyclic C–N bond in nucleic acid bases. *J. Chem. Soc. Perkin Trans. 2,* 889–891.

204. M. Raszka and N. O. Kaplan (1972), Association by hydrogen bonding of mononucleotides in aqueous solution. *Proc. Nat. Acad. Sci. USA* **69,** 2025–2029.

205. R. R. Shoup, H. T. Miles, and E. D. Becker (1972), Restricted rotation about the exocyclic carbon–nitrogen bond in cytosine derivatives. *J. Phys. Chem.* **76,** 64–70.

206. R. R. Shoup, E. D. Becker, and H. T. Miles (1971), Restricted rotation of the amino group in 1-methylcytosine. *Biochem. Biophys. Res. Commun.* **43,** 1350–1353.

207. J. K. Dattagupta, W. Saenger, K. Bolewska, and I. Kulakowska (1977), X-Ray study of $1,5,N_4,N_4$-tetramethylcytosine—An overcrowded molecule with planar structure. *Acta Crystallogr. B* **33,** 85–89.

208. B. Lesyng and W. Saenger (1978), Theoretical investigations on the conformation of $1,5,N_4,N_4$-tetramethylcytosine. *Biochim. Biophys. Acta* **544,** 215–224.

209. C. Singh (1965), Location of hydrogen atoms in certain heterocyclic compounds. *Acta Crystallogr.* **19,** 861–864.

210. M. Sundaralingam and L. H. Jensen (1965), Stereochemistry of nucleic acid constituents. II. A comparative study. *J. Mol. Biol.* **13,** 930–943.

211. A. Saran, D. Perahia, and B. Pullman (1973), Molecular orbital calculations on the conformation of nucleic acids and their constituents. VII. Conformation of the sugar ring in β-nucleosides: The pseudorotational representation. *Theor. Chim. Acta (Berlin)* **30,** 31–44.

212. V. Sasisekharan (1973), Conformation of polynucleotides. *Jerus. Symp. Quant. Chem. Biochem.* **5,** 247–260.

213. G. Govil and A. Saran (1971), Quantum chemical studies on the conformational structure of nucleic acids. II. EHT and CNDO calculations on the puckering of D-ribose. *J. Theor. Biol.* **33,** 399–406.

214. A. A. Lugovskoi and V. G. Dashevskii (1972), Conformations of β-D-ribose. *Mol. Biol. (USSR)* **6,** 354–360.

215. A. A. Lugovskoi, V. G. Dashevskii, and A. I. Kitaigorodskii (1972), Conformations of sugar residues in pyrimidine nucleosides. *Mol. Biol. (USSR)* **6,** 361–367; Conformations of sugar residues in purine nucleosides. *Mol. Biol. (USSR)* **6,** 494–499.

216. W. K. Olson and J. L. Sussman (1982), How flexible is the furanose ring? 1. A comparison of experimental and theoretical studies. *J. Amer. Chem. Soc.* **104,** 270–278; W. K. Olson (1982), How flexible is the furanose ring? 2. An updated potential energy estimate. *J. Amer. Chem. Soc.* **104,** 278–286.

217. M. Spencer (1959), The stereochemistry of deoxyribonucleic acid: I. Covalent bond lengths and angles. *Acta Crystallogr.* **12,** 59–65.

218. R. H. Sarma and R. J. Mynott (1973), Conformation of pyridine nucleotides studied by phosphorus-31 and hydrogen-1 fast Fourier transform nuclear

magnetic resonance spectroscopy. I. Oxidized and reduced mononucleo-tides. *J. Amer. Chem. Soc.* **95**, 1641–1649.

219. H. Dugas, B. J. Blackburn, R. K. Robins, R. Deslauriers, and I. C. P. Smith (1971), A nuclear magnetic resonance study of the conformation of β-cyanuric acid riboside. Further evidence for the *anti* rotamer in pyrimidine nucleosides. *J. Amer. Chem. Soc.* **93**, 3468–3470.

220. T. Schleich, B. J. Blackburn, R. D. Lapper, and I. C. P. Smith (1972), A nuclear magnetic resonance study of the influence of aqueous sodium perchlorate and temperature on the solution conformation of uracil nucleosides and nucleotides. *Biochemistry* **11**, 137–145.

221. F. E. Hruska and S. S. Danyluk (1968), Conformational changes of the ribose group in dinucleoside mono- and diphosphates. Temperature dependence. *J. Amer. Chem. Soc.* **90**, 3266–3267.

222. F. E. Hruska, A. A. Grey, and I. C. P. Smith (1970), A nuclear magnetic resonance study of the molecular conformation of β-pseudouridine in solution. *J. Amer. Chem. Soc.* **92**, 4088–4094.

223. O. Röder, H.-D. Lüdemann, and E. von Goldammer (1975), Determination of the activation energy for pseudorotation of the furanose ring in nucleosides by ^{13}C nuclear-magnetic-resonance relaxation. *Eur. J. Biochim.* **53**, 517–524.

224. R. Cushley, J. F. Codington, and J. J. Fox (1968), Nucleosides. XLIX. Nuclear magnetic resonance studies of 2'- and 3'-halogeno nucleosides. The conformations of the 2'-deoxy-2'-fluorouridine and 3'-deoxy-3'-fluoro-β-D-arabinofuranosyluracil. *Can. J. Chem.* **46**, 1131–1140.

225. I. Ekiel, E. Darżynkiewicz, L. Dudycz, and D. Shugar (1978), Solution conformation and relative acidities of the sugar hydroxyls of the O'-methylated derivatives of the antimetabolite 9-β-D-xylofuranosyladenine. *Biochemistry* **17**, 1530–1536.

226. M. Imazawa, T. Ueda, and T. Ukita (1975), Nucleosides and nucleotides. XII. Synthesis and properties of 2'-deoxy-2'-mercaptouridine and its derivatives. *Chem. Pharm. Bull.* **23**, 604–610.

227. H. P. M. de Leeuw, J. R. de Jager, H. J. Koeners, J. H. van Boom, and C. Altona (1977), Puromycin and some of its analogues: Conformational properties in solution. *Eur. J. Biochem.* **76**, 209–217.

228. H. Plach, E. Westhof, H.-D. Lüdemann, and R. Mengel (1977), Solution conformational analysis of 2'-amino-2'-deoxyadenosine, 3'-amino-3'-deoxyadenosine and puromycin by pulsed nuclear magnetic resonance methods. *Eur. J. Biochem.* **80**, 295–304.

228a.C.A.G. Haasnoot, F.A.A.M. de Leeuw, H.P.M. de Leeuw and C. Altona (1982), The relationship between proton–proton NMR coupling constants and substituent electronegativities II. Conformational analysis of the sugar ring in nucleosides and nucleotides in solution using a generalized Karplus equation. *Org. Magn. Res.* **15**, 43–52.

229. S. Uesugi, H. Miki, M. Ikehara, H. Iwahashi, and Y. Kyogoku (1979), A linear relationship between electronegativity of 2'-substituents and conformation of adenine nucleosides. *Tetrahedron Lett.* **42**, 4073–4076.

230. W. Guschlbauer and K. Jankowski (1980), Nucleoside conformation is determined by the electronegativity of the sugar substituent. *Nucleic Acids Res.* **8**, 1421–1433.

231. G. Klimke, I. Cuno, H.-D. Lüdemann, R. Mengel, and M. J. Robins (1979),

Ribose conformations of adenosine analogs modified at the 2′, 3′ or 5′ positions. *Z. Naturforsch. C* **34**, 1075–1084.

232. D. C. Rohrer and M. Sundaralingam (1970), Stereochemistry of nucleic acids and their constituents. XII. The crystal and molecular structure of α-D-2′-amino-2′-deoxyadenosine monohydrate. *J. Amer. Chem. Soc.* **92**, 4956–4962.

233. P. Narayanan and H. Berman (1975), A crystallographic determination of a chemical structure: 6-Amino-10-(β-D-ribofuranosylamino)pyrimido-[5,4-d]pyrimidine, an example of an unusual D-ribose conformation. *Carbohyd. Res.* **44**, 169–180.

233a. J. A. Gerlt and A. V. Youngblood (1980), The solution conformational preference of the sugar and sugar phosphate constituents of RNA and DNA. *J. Amer. Chem. Soc.* **102**, 7433–7438.

234. A. Rabczenko and D. Shugar (1972), Hydrogen bonding scheme involving ribose 2′-hydroxyls in polyribouridylic acid. *Acta Biochem. Pol.* **19**, 89–94.

235. D. J. Abraham (1971), Proposed detailed structural model for tRNA and its geometric relationship to a messenger. *J. Theor. Biol.* **30**, 83–91.

236. K. Kölkenbeck and G. Zundel (1975), The significance of the 2′-OH group and the influence of cations on the secondary structure of the RNA backbone. *Biophys. Struct. Mechanism* **1**, 203–219.

237. P. R. Young and N. R. Kallenbach (1978), Secondary structure in polyuridylic acid. Non-classical hydrogen bonding and the function of the ribose 2′-hydroxyl group. *J. Mol. Biol.* **126**, 467–479.

238. B. Zmudzka and D. Shugar (1970), Role of the 2′-hydroxyl in polynucleotide conformation. Poly 2′-O-methyl-uridylic acid. *FEBS Lett.* **8**, 52–54.

239. B. Zmudzka, C. Janion, and D. Shugar (1969), Poly 2′-O-methylcytidylic acid and the role of the 2′-hydroxyl in polynucleotide structure. *Biochem. Biophys. Res. Commun.* **37**, 895–901.

240. A. M. Bobst, F. Rottman, and P. A. Cerutti (1969), Effect of the methylation of the 2′-hydroxyl group in polyadenylic acid on its structure in weakly acidic and neutral solutions and on its capability to form ordered complexes with polyuridylic acid. *J. Mol. Biol.* **46**, 221–234.

241. J. Hobbs, H. Sternbach, M. Sprinzl, and F. Eckstein (1972), Polynucleotides containing 2′-chloro-2′-deoxyribose. *Biochemistry* **11**, 4336–4344.

242. M. Blandin, S. Tran-Dinh, J. C. Catlin, and W. Guschlbauer (1974), Nucleoside conformations. 16. Nuclear magnetic resonance and circular dichroism studies on pyrimidine-2′-fluoro-2′-deoxyribonucleosides . *Biochim. Biophys. Acta* **361**, 249–256.

243. J. C. Catlin and W. Guschlbauer (1975), Oligonucleotide conformations. III. Comparative optical and thermodynamic studies of uridylyl-3′,5′-nucleosides containing ribose, deoxyribose or 2′-deoxy-2′-fluororibose in the uridine moiety. *Biopolymers* **14**, 51–72.

244. W. Guschlbauer, M. Blandin, J. L. Drocourt, and M. N. Thang (1977), Poly-2′-deoxy-2′-fluoro-cytidylic acid: Enzymatic synthesis, spectroscopic characterization and interaction with polyinosinic acid. *Nucleic Acids Res.* **4**, 1933–1943.

245. J. Pitha (1972), Unexpected conformational stability of poly(2′-azido-2′-deoxyuridylic acid). *J. Amer. Chem. Soc.* **94**, 3638–3639.

246. P. H. Bolton and D. R. Kearns (1978), Hydrogen bonding of the 2′-OH in RNA. *Biochim. Biophys. Acta* **517**, 329–337.

247. P. H. Bolton and D. R. Kearns (1979), Intramolecular water bridge between the 2'-OH and phosphate groups of RNA. Cyclic nucleotides as a model system. *J. Amer. Chem. Soc.* **101**, 479–484.

248. P. O. P. Ts'o, J. C. Barrett, L. S. Kan, and P. S. Miller (1972), Proton magnetic resonance studies of nucleic acid conformation. *Annu. Rev. N.Y. Acad. Sci.* **222**, 290–306.

249. H. J. Geise, W. J. Adams, and L. S. Bartell (1969), Electron diffraction study of gaseous tetrahydrofuran. *Tetrahedron* **25**, 3045–3052.

250. A. Almenningen, H. M. Seip, and T. Willadsen (1969), Studies on molecules with five-membered rings. II. An electron diffraction investigation of gaseous tetrahydrofuran. *Acta Chem. Scand.* **23**, 2748–2754.

251. G. G. Engerholm, A. C. Luntz, W. D. Gwinn, and D. O. Harris (1969), Ring puckering in five-membered rings. *J. Chem. Phys.* **50**, 2446–2457.

252. D. Cremer and J. A. Pople (1975), Molecular orbital theory of the electronic structure of organic compounds. XXIII. Pseudorotation in saturated five-membered ring compounds. *J. Amer. Chem. Soc.* **97**, 1358–1367.

253. J. B. Hendrickson (1961), Molecular geometry. I. Machine computation of the common rings. *J. Amer. Chem. Soc.* **83**, 4537–4547; (1963), **85**, 4059.

254. J. D. Dunitz (1972), Approximate relationships between conformational parameters in 5- and 6-membered rings. *Tetrahedron* **28**, 5459–5467.

255. F. H. Westheimer (1956), Calculation of the magnitude of steric effects. In: *Steric Effects in Organic Chemistry* (M. S. Newman, Ed.), Chap. 12, pp. 523–555. Wiley, New York.

256. B. Hingerty and W. Saenger (1976), Topography of cyclodextrin inclusion complexes. 8. Crystal and molecular structure of the α-cyclodextrin–methanol pentahydrate complex. Disorder in a hydrophobic cage. *J. Amer. Chem. Soc.* **98**, 3357–3365.

257. M. Sundaralingam (1965), Conformations of the furanose ring in nucleic acids and other carbohydrate derivatives in the solid state. *J. Amer. Chem. Soc.* **87**, 599–606.

258. W. Saenger and K. H. Scheit (1970), A pyrimidine nucleoside in the *syn* conformation: molecular and crystal structure of 4-thiouridine-hydrate. *J. Mol. Biol.* **50**, 153–169.

259. B. Lesyng and W. Saenger (1981), Influence of crystal packing forces on molecular structure in 4-thiouridine. Comparison of *anti* and *syn* forms. *Z. Naturforsch.* C **36**, 956–960.

260. K. H. Scheit and W. Saenger (1969), The conformation of 4-thiouridine-5'-phosphate in single and double stranded polynucleotides. *FEBS Lett.* **2**, 305–308.

261. M. P. Schweizer, J. T. Witkowski, and R. K. Robins (1971), Nuclear Magnetic Resonance determination of *syn* and *anti* conformation in pyrimidine nucleosides. *J. Amer. Chem. Soc.* **93**, 277–279.

262. F. E. Hruska, K. K. Ogilvie, A. A. Smith, and H. Wayborn (1971), Molecular conformation of 4-thiouridine in aqueous solution. *Can. J. Chem.* **49**, 2449–2452.

263. M. A. Viswamitra, B. S. Reddy, G. H.-Y. Lin, and M. Sundaralingam (1971), Stereochemistry of nucleic acids and their constituents. XVII. Crystal and molecular structure of deoxycytidine 5'-phosphate monohydrate. A possible puckering for the furanoside ring in B-deoxyribonucleic acid. *J. Amer. Chem. Soc.* **93**, 4565–4572.

264. R.E. Dickerson, and H.R. Drew (1982). A kinematic model for B-DNA. *Proc. Nat. Acad. Sci. USA* **78**, 7318–7322.

265. G. H.-Y. Lin, M. Sundaralingam, and S. K. Arora (1971), Stereochemistry of nucleic acids and their constituents. XV. Crystal and molecular structure of 2-thiocytidine dihydrate, a minor constituent of transfer ribonucleic acids. *J. Amer. Chem. Soc.* **93**, 1235–1241.

266. A. Lo, E. Shefter, and T. G. Cochran (1975), Analysis of N-glycosyl bond length in crystal structures of nucleosides and nucleotides. *J. Pharm. Sci.* **64**, 1707–1710.

267. H.-D. Lüdemann, E. Westhof, and O. Röder (1974), A dynamic correlation between ribose conformation and glycosyl torsion angle of dissolved xanthosine studied by continuous wave mode and pulsed nuclear magnetic resonance methods. *Eur. J. Biochem.* **49**, 143–150.

268. P. O. P. Ts'o, N. S. Kondo, M. P. Schweizer, and D. P. Hollis (1969), Studies of the conformation and interaction in dinucleoside mono- and diphosphates by proton magnetic resonance. *Biochemistry* **8**, 997–1029.

269. N. S. Kondo, J. N. Fang, P. S. Miller, and P. O. P. Ts'o (1972), Influence of the furanose on the conformation of adenine dinucleoside monophosphates in solution. *Biochemistry* **11**, 1991–2003.

270. R. Deslauriers and I. C. P. Smith (1972), A proton magnetic resonance study of the influence of base ionization on the conformation of pseudouridine. *Can. J. Biochem.* **50**, 766–774.

271. D. W. Miles, L. B. Townsend, M. J. Robins, R. K. Robins, W. H. Inskeep, and H. Eyring (1971), Circular dichrosim of nucleoside derivatives. X. Influence of solvents and substituents upon the Cotton effects of guanosine derivatives. *J. Amer. Chem. Soc.* **93**, 1600–1608.

272. L. Dudycz, R. Stolarski, R. Pless, and D. Shugar (1979), A ¹H NMR study of the *syn–anti* dynamic equilibrium in adenine nucleosides and nucleotides with the aid of some synthetic model analogues with fixed conformations. *Z. Naturforsch. C.* **34**, 359–373.

273. G. Klimke, H.-D. Lüdemann, and L. B. Townsend (1979), Further evidence for a S-*syn* correlation in the purine(β)ribosides: The solution conformation of two tricyclic analogs of adenosine and guanosine. *Z. Naturforsch. C* **34**, 653–657.

274. D. W. Miles, L. B. Townsend, D. L. Miles, and H. Eyring (1979), Conformation of nucleosides: Circular dichroism study on the *syn–anti* conformational equilibrium of 2-substituted benzimidazole nucleosides. *Proc. Nat. Acad. Sci. USA* **76**, 553–556.

275. M. P. Schweizer, E. B. Banta, J. T. Witkowski, and R. K. Robins (1973), Determination of pyrimidine *syn, anti* conformational preference in solution by proton and carbon-13 nuclear magnetic resonance. *J. Amer. Chem. Soc.* **95**, 3770–3778.

276. L. M. Rhodes and P. R. Schimmel (1971), Nanosecond relaxation processes in aqueous mononucleoside solution. *Biochemistry* **10**, 4426–4433.

277. P. R. Hemmes, L. Oppenheimer, and F. Jordan (1974), Ultrasonic relaxation evaluation of the thermodynamics of *syn–anti* glycosidic isomerization of adenosine. *J. Amer. Chem. Soc.* **96**, 6023–6026.

278. J. Cadet, R. Ducolomb, and C. Taieb (1975), Analyse de RMN à 250 MHZ de

la méthyl-6-desoxy-2'-uridine—Conformation preferentielle en solution aqueuse. *Tetrahedron Lett.* **40,** 3455–3458.

279. A. L. George, F. E. Hruska, K. K. Ogilvie, and A. Holý (1978), Proton magnetic resonance studies of 2'-deoxynucleosides in the *syn* conformation. *Can. J. Chem.* **56,** 1170–1176.

280. R. K. Nanda, R. Tewari, G. Govil, and I. C. P. Smith (1974), The conformation of β-pseudouridine about the glycosidic bond as studied by ¹H homonuclear Overhauser measurements and molecular orbital calculations. *Can. J. Chem.* **52,** 371–375.

281. C. F. G. C. Geraldes (1979), Nuclear magnetic resonance study of the solution conformation of adenine mononucleotides using the lanthanide probe method. *J. Magn. Resonance* **36,** 89–98.

282. I. D. Bobruskin, M. P. Kirpichnikov, and V. L. Florent'ev (1980), NMR investigation of conformation of nucleotides. Oligonucleotides, and their analogs in solution. *Mol. Biol. (USSR)* **13,** 660–674.

283. H. R. Wilson and A. Rahman (1971), Nucleoside conformation and nonbonded interactions. *J. Mol. Biol.* **56,** 129–142.

284. H. Berthod and B. Pullman (1971), Molecular orbital calculations on the conformation of nucleic acids and their constituents. I. Conformational energies of β-nucleosides with $C_{3'}$- and $C_{2'}$-*endo* sugars. *Biochim. Biophys. Acta* **232,** 595–606.

285. S. Kang (1971), Stereochemistry of pyrimidine nucleosides: Conformational analysis of cytidine, thymidine, 4-thiouridine and deoxycytidine by the neglect of differential overlap molecular orbital method. *J. Mol. Biol.* **58,** 297–315.

286. S. Kang (1973), Stereochemistry of purine nucleosides. Conformational analysis of adenosine and guanosine. *Jerus. Symp. Quant. Chem. Biochem.* **5,** 271–282.

287. C. H. Schwalbe and W. Saenger (1973), 6-Azauridine, a nucleoside with unusual ribose conformation. *J. Mol. Biol.* **75,** 129–143.

288. W. Saenger, D. Suck, M. Knappenberg and J. Dirkx (1979), Theoretical drug design: 6-Azauridine-5'-phosphate—Its X-ray crystal structure, potential energy maps, and mechanism of inhibition of orotidine-5'-phosphate decarboxylase. *Biopolymers* **18,** 2015–2036.

289. P. Singh and D. J. Hodgson (1974), *High-anti* conformation in *o*-azanucleosides. The crystal and molecular structure of 6-azacytidine. *Biopolymers* **13,** 5445–5452.

290. M. Yasuniwa, R. Tokuoka, K. Ogawa, Y. Yamagata, S. Fujii, K.-I. Tomita, W. Limn, and M. Ikehara (1979), The crystal and molecular structure of 8-methyladenosine-3'-monophosphate dihydrate. *Biochim. Biophys. Acta* **561,** 240–247.

291. S. Fujii, T. Fujiwara, and K.-I. Tomita (1976), Structural studies on two forms of 8-bromo-2',3'-O-isopropylideneadenosine. *Nucleic Acids Res.* **3,** 1985–1996.

292. M. Ikehara, S. Uesugi, and K. Yoshida (1972), Studies on the conformation of purine nucleosides and their 5'-phosphates. *Biochemistry* **11,** 830–836.

293. R. H. Sarma, C.-H. Lee, F. E. Evans, N. Yathindra, and M. Sundaralingam (1974), Probing the interrelation between the glycosyl torsion, sugar pucker,

and backbone conformation in C_8-substituted adenine nucleotides by ^{1}H and ^{1}H-{^{31}P} fast Fourier transform nuclear magnetic resonance methods and conformational energy calculations. *J. Amer. Chem. Soc.* **96**, 7337–7348.

294. F. Jordan and H. Niv (1977), C-8 amino purine nucleosides. A well-defined steric determinant of glycosyl conformational preferences. *Biochim. Biophys. Acta* **476**, 265–271.

295. M. A. Abdallah, J. F. Biellmann, B. Nordström, and C.-I. Brändén (1975), The conformation of adenosine diphosphoribose and 8-bromoadenosine diphosphoribose when bound to liver alcohol dehydrogenase. *Eur. J. Biochem.* **50**, 475–481.

296. S. Neidle, M. R. Sanderson, A. Subbiah, J. B. Chattopadhyaya, R. Kuroda, and C. B. Reese (1979), 9-β-D-Arabinofuranosyl-8-n-butylaminoadenine, a C_8 substituted nucleoside in the *anti* conformation. *Biochim. Biophys. Acta* **565**, 379–386.

297. R. Pless, L. Dudycz, R. Stolarski, and D. Shugar (1978), Purine nucleosides and nucleotides unequivocally in the *syn* conformation: Guanosine and 5'-GMP with 8-*tert*-butyl and 8-(α-hydroxyisopropyl) substituents. *Z. Naturforsch. C* **33**, 902–907.

298. G. I. Birnbaum and D. Shugar (1978), A purine nucleoside unequivocally constrained in the *syn* form. Crystal structure and conformation of 8-(α-hydroxyisopropyl)-adenosine. *Biochim. Biophys. Acta* **517**, 500–510.

299. W. Guschlbauer (1972), Why is guanosine different? *Jerus. Symp. Quant. Chem. Biochem.* **4**, 297–310.

300. D. W. Young, P. Tollin, and H. R. Wilson (1974), Molecular conformation of deoxyguanosine-5'-phosphate. *Nature* **248**, 513–514; (1974), The crystal structure of disodium deoxyguanosine-5'-phosphate tetrahydrate. *Acta Crystallogr. B.* **30**, 2012–2128.

301. M. A. Viswamitra and T. P. Seshadri (1974), Molecular structure of deoxyguanosine-5'-phosphate. *Nature* **252**, 176–177.

302. H. Drew, T. Takano, S. Tanaka, K. Itakura, and R. E. Dickerson (1980), High-salt d(CpGpCpG): A left-handed Z' DNA double helix. *Nature* **286**, 567–573.

303. A. H.-J. Wang, G. J. Quigley, F. J. Kolpak, J. L. Crawford, J. H. van Boom, G. van der Marel, and A. Rich (1979), Molecular structure of a left-handed double helical DNA fragment at atomic resolution. *Nature* **282**, 680–686.

304. S. Arnott, R. Chandrasekaran, D. L. Birdsall, A. G. W. Leslie, and R. L. Ratcliff (1980), Left handed DNA helices. *Nature* **283**, 743–745.

305. W. Guschlbauer, I. Frič, and A. Holý (1972), Oligonucleotide conformations. Optical studies on GpU analogues with modified uridine residues. *Eur. J. Biochem.* **31**, 1–13.

306. Tran-Dinh Son, W. Guschlbauer, and M. Guéron (1972), Flexibility and conformations of guanosine monophosphates by the Overhauser effect. *J. Amer. Chem. Soc.* **94**, 7903–7911.

307. N. Yathindra and M. Sundaralingam (1973), Conformational studies on guanosine nucleotides and polynucleotides. The effect of the base on glycosyl and backbone conformations. *Biopolymers* **12**, 2075–2082.

308. W. K. Olson (1973), *Syn–anti* effects on the spatial configuration of polynucleotide chains. *Biopolymers* **12**, 1787–1814.

309. J.-N. Lespinasse, H. Broch, R. Cornillon, and D. Vasilescu (1976), Flexibility of deoxyguanosine 5'-monophosphate (5'-dGMP): An E.H.T. conformational analysis. *J. Theor. Biol.* **57**, 225–230.

310. H. Berthod and B. Pullman (1973), Complementary studies on the rigidity–flexibility of nucleotides. *FEBS Lett.* **33**, 147–150.

311. S. T. Rao and M. Sundaralingam (1969), Stereochemistry of nucleic acids and their constituents. V. The crystal and molecular structure of a hydrated monosodium inosine 5'-phosphate. A commonly occurring unusual nucleotide in the anticodons of tRNA. *J. Amer. Chem. Soc.* **91**, 1210–1217.

312. N. Nagashima and Y. Iitaka (1968), The crystal structures of barium inosine-5'-phosphate and disodium inosine-5'-phosphate. *Acta Crystallogr.* B **24**, 1136–1138.

313. F. E. Hruska, A. A. Smith, and J. G. Dalton (1971), A correlation of some structural parameters of pyrimidine nucleosides. A nuclear magnetic resonance study. *J. Amer. Chem. Soc.* **93**, 4334–4336.

314. A. V. Lakshminarayanan and V. Sasisekharan (1969), Stereochemistry of nucleic acids and polynucleotides. V. Conformational energy of a ribose-phosphate unit. *Biopolymers* **8**, 489–503.

315. A. Saran, B. Pullman, and D. Perahia (1973), Molecular orbital calculations on the conformation of nucleic acids and their constituents. IV. Conformation about the exocyclic $C_{4'}$–$C_{5'}$ bond. *Biochim. Biophys. Acta* **287**, 211–231. See also (1973), VI. Conformation about the exocyclic $C_{4'}$–$C_{5'}$ bond in α-nucleosides. *Biochim. Biophys. Acta* **299**, 497–499.

316. A. Saran and G. Govil (1971), Quantum chemical studies on the conformational structure of nucleic acids. III. Calculation of backbone structure by extended Hückel theory. *J. Theor. Biol.* **33**, 407–418.

317. J. L. Sussman, N. C. Seeman, S.-H. Kim, and H. Berman (1972), Crystal structure of a naturally occurring dinucleoside phosphate: Uridylyl-3',5'-adenosine phosphate. Model for RNA chain folding. *J. Mol. Biol.* **66**, 403–421.

318. J. Rubin, T. Brennan, and M. Sundaralingam (1972), Crystal and molecular structure of a naturally occurring dinucleoside monophosphate. Uridylyl-3', 5'-adenosine hemihydrate. Conformational "rigidity" of the nucleotide unit and models for polynucleotide chain folding. *Biochemistry* **11**, 3112–3128.

319. E. Egert, H.-J. Lindner, W. Hillen, and H. G. Gassen (1978), Crystal structure and conformation of 5-aminouridine. *Acta Crystallogr.* B **34**, 2204–2208.

320. R. Taylor and O. Kennard (1982), Crystallographic evidence for the existence of C–H $\cdots$ O, C–H $\cdots$ N, and C–H $\cdots$ Cl hydrogen bonds. *J. Amer. Chem. Soc.* **104**, 5063–5070.

321. J. Donohue (1968), Selected topics in hydrogen bonding. In: *Structural Chemistry and Molecular Biology* (A. Rich and N. Davidson, Eds.), pp.443–465. Freeman, San Francisco.

322. A. Goel and C. N. R. Rao (1971), Hydrogen bonds formed by C–H groups. *Trans. Faraday Soc.*, 2828–2832.

323. M. P. Schweizer, S. I. Chan, G. H. Helmkamp, and P. O. P. Ts'o (1964), An experimental assignment of proton magnetic resonance spectrum of purine. *J. Amer. Chem. Soc.* **86**, 696–700.

324. R. N. Maslova, E. A. Lesnik, and J. M. Warshavsky (1969), Influence of the

conformation on the rate of the slow isotopic exchange of hydrogen in poly(adenylic acid) and its complexes. *Mol. Biol.* (*USSR*) **3**, 575–582.

325. V. I. Bruskov, V. N. Bushnev, and V. I. Poltev (1980), Nuclear magnetic resonance study of C–H · · · O hydrogen bonds in nucleic acid base analogs. *Mol. Biol.* (*USSR*) **14**, 245–250. [in English]

326. G. L. Amidon, S. Anik and J. Rubin (1975), An energy partitioning analysis of base–sugar intramolecular C–H · · · O hydrogen bonding in nucleosides and nucleotides. In: *Structure and Conformation of Nucleic Acids and Protein–Nucleic Acid Interactions* (M. Sundaralingam and S. T. Rao, Eds.), pp. 729–744. Univ. Park Press, Baltimore.

327. Y.-S. Wong and S. J. Lippard (1977), X-Ray crystal structure of a 2:2 chloro-terpyridineplatinum(II)–adenosine-5′-monophosphate intercalation complex. *J. Chem. Soc. Commun.*, 824–825.

328. S. Neidle, G. Taylor, M. Sanderson, H.-S. Shieh, and H. Berman (1978), A 1:2 crystalline complex of ApA:proflavine: A model for binding to single-stranded regions in RNA. *Nucleic Acids Res.* **5**, 4417–4422.

329. M. A. Viswamitra, T. P. Seshadri, and M. L. Post (1975), An uncommon nucleotide conformation shown by molecular structure of deoxyuridine-5′-phosphate and nucleic acid stereochemistry. *Nature* **258**, 542–544

330. C. D. Barry, A. C. T. North, J. A. Glasel, R. J. P. Williams, and A. V. Xavier (1971), Quantitative determination of mononucleotide conformations in solution using lanthanide ion shift and broadening NMR probes. *Nature* **232**, 236–245.

331. F. E. Evans and R. H. Sarma (1976), Nucleotide rigidity. *Nature* **263**, 567–572.

332. H. Berthod and B. Pullman (1973), Nucleotides: Rigid or flexible? *FEBS Lett.* **30**, 231–235.

333. A. Jack, A. Klug, and J. E. Ladner (1976), "Non-rigid" nucleotides in tRNA: A new correlation in the conformation of a ribose. *Nature* **261**, 250–251.

334. M. Sundaralingam and E. Westhof (1979), The "rigid" nucleotide concept in perspective. *Int. J. Quant. Chem.: Quant. Biol. Symp.* **6**, 115–130.

335. D. W. Cruickshank (1961), The role of 3d-orbitals in π-bonds between (a) silicon, phosphorus, sulphur, or chlorine and (b) oxygen or nitrogen. *J. Chem. Soc.* (*London*), 5486–5504.

336. D. E. C. Corbridge (1974), *The Structural Chemistry of Phosphorus*, pp. 1–8. Elsevier, New York.

337. J. Matheja and E. T. Degens (1971), *Structural Molecular Biology of Phosphates*. Gustav Fischer Verlag, Stuttgart.

338. V. Shomaker and D. P. Stevenson (1941), Some revisions of the covalent radii and the additivity rule for the lengths of partially ionic single covalent bonds. *J. Amer. Chem. Soc.* **63**, 37–40.

339. P. A. Akishin, N. G. Rambidi, and E. Z. Zasorin (1959), An electron diffraction investigation of the structure of the phosphorus pentoxide molecule. *Sov. Phys. Crystallogr.* **4**, 334–338; *Kristallografiya* **4**, 360–364.

340. L. Pauling (1952), Interatomic distances and bond character in the oxygen acids and related substances. J. Phys. Chem. **56**, 361–365.

341. W. H. Baur (1974), The geometry of polyhedral distortions. Predictive relationships for the phosphate group. *Acta Crystallogr. B* **30**, 1195–1215.

342. I. D. Brown and R. D. Shannon (1973), Empirical bond-strength–bond-length curves for oxides. *Acta Crystallogr. A* **29**, 266–282.
343. F. Lipmann (1941), Metabolic generation and utilization of phosphate bond energy. *Adv. Enzymol.* **1**, 99–162.
344. W. Saenger, B. S. Reddy, K. Mühlegger, and G. Weimann (1977), X-Ray study of the lithium complex of NAD⁺. *Nature* **267**, 225–229.
345. D. B. Boyd and W. Lipscomb (1969), Electronic structures for energy-rich phosphates. *J. Theor. Biol.* **25**, 403–420.
346. J. I. Fernandez-Alonso (1964), Electronic structures in quantum biochemistry. *Adv. Chem. Phys.* **7**, 3–83.
347. D. G. Watson and O. Kennard (1973), The structure of "high-energy" phosphate compounds. II. An X-ray analysis of cyclohexylammonium phosphoenolpyruvate. *Acta Crystallogr. B* **29**, 2358–2364.
348. L. Stryer (1975), *Biochemistry*. Freeman, San Francisco.
349. A. L. Lehninger (1975), *Biochemistry*, 2nd ed. Worth, New York.
350. R. A. Alberty (1969), Standard Gibbs free energy, enthalpy, and entropy changes as a function of pH and pMg for several reactions involving adenosine phosphate. *J. Biol. Chem.* **244**, 3290–3302.
351. J. Emerson and M. Sundaralingam (1980), Structure of the potassium salt of the modified nucleotide dihydrouridine 3'-monophosphate hemihydrate: Correlation between the base pucker and sugar pucker and models for metal interactions with ribonucleic acid loops. *Acta Crystallogr. B* **36**, 537–543.
352. D. J. Patel (1976), Proton and phosphorus NMR studies of d-CpG(pCpG)ₙ duplexes in solution. Helix-coil transition and complex formation with actinomycin-D. *Biopolymers* **15**, 533–558.
353. C.-H. Lee, F. S. Ezra, N. S. Kondo, R. H. Sarma, and S. S. Danyluk (1976), Conformational properties of dinucleoside monophosphates in solution: Dipurines and dipyrimidines. *Biochemistry* **15**, 3627–3639.
354. C. M. Dobson, C. F. G. C. Geraldes, G. Ratcliffe, and R. J. P. Williams (1978), Nuclear-magnetic-resonance studies of 5'-ribonucleotide and 5'-deoxyribonucleotide conformations in solution using the lanthanide probe method. *Eur. J. Biochem.* **88**, 259–266.
355. F. Inagaki, M. Tasumi, and T. Miyazawa (1978), Structures and populations of conformers of nucleoside monophosphates in solution. I. General methods of conformation search with lanthanide-ion probes and spin-coupling constants and application to uridine-5'-monophosphate. *Biopolymers* **17**, 267–289.
356. J. M. Thornton and P. M. Bayley (1975), Conformational energy calculations for dinucleotide molecules. A study of the component mononucleotides adenosine 5'-monophosphate, nicotinamide mononucleotide and adenosine 3'-monophosphate, *Biochem. J.* **149**, 585–596.
357. B. Pullman, D. Perahia, and A. Saran (1972), Molecular orbital calculations on the conformation of nucleic acids and their constituents. III. Backbone structure of di- and polynucleotides. *Biochim. Biophys. Acta* **269**, 1–14.
358. N. Camerman, J. K. Fawcett, and A. Camerman (1976), Molecular structure of a deoxyribose-dinucleotide, sodium thymidylyl-5',3'-thymidylate-5' hydrate (pTpT), and a possible structural model for polythymidylate. *J. Mol. Biol.* **107**, 601–621.
359. M. Viswamitra, O. Kennard, P. G. Jones, G. M. Sheldrick, S. Salisbury, L.

Falvello, and Z. Shakked (1978), DNA double helical fragment at atomic resolution. *Nature* **273**, 687–688.

360. S.-H. Kim, H. M. Berman, N. C. Seeman, and M. D. Newton (1973), Seven basic conformations of nucleic acid structural units. *Acta Crystallogr. B* **29**, 703–710.

361. M. D. Newton (1973), A model conformational study of nucleic acid phosphate ester bonds. The torsional potential of dimethyl phosphate monoanion. *J. Amer. Chem. Soc.* **95**, 256–258.

362. D. Perahia, B. Pullman, and A. Saran (1974), Molecular orbital calculations on the conformation of nucleic acids and their constituents. IX. The geometry of the phosphate group: Key to the conformation of polynucleotides? *Biochim. Biophys. Acta* **340**, 299–313.

363. V. Sasisekharan and A. V. Lakshminarayanan (1969), Stereochemistry of nucleic acids and polynucleotides. VI. Minimum energy conformations of dimethyl phosphate. *Biopolymers* **8**, 505–514.

364. C. Tosi and G. Lipari (1981), Molecular orbital computations on the conformational energy of ethyl methyl phosphate. *Theoret. Chim. Acta* **60**, 41–51.

365. R. U. Lemieux (1971), Effects of unshared pairs of electrons and their solvation on conformational equilibria. *Pure Appl. Chem.* **25**, 527–548.

366. S. Wolfe (1972), The *gauche* effect. Some stereochemical consequences of adjacent electron pairs and polar bonds. *Acc. Chem. Res.* **5**, 102–111.

367. L. Radom, W. J. Hehre, and J. A. Pople (1972), Molecular orbital theory and the electronic structure of organic compounds. XIII. Fourier component analysis of internal rotation potential functions in saturated molecules. *J. Amer. Chem. Soc.* **94**, 2371–2381.

368. T. K. Brunck and F. Weinhold (1979), Quantum-mechanical studies on the origin of barriers to internal rotation about single bonds. *J. Amer. Chem. Soc.* **101**, 1700–1709.

369. N. Yathindra and M. Sundaralingam (1974), Backbone conformations in secondary and tertiary structure units of nucleic acids. Constraint in the phosphodiester conformation. *Proc. Nat. Acad. Sci. USA* **71**, 3325–3328.

370. W. K. Olson and P. J. Flory (1972), Spatial configurations of polynucleotide chains. I. Steric interactions in polyribonucleotides: A virtual bond model. *Biopolymers* **11**, 1–23.

371. S. D. Stellman, B. Hingerty, S. D. Broyde, E. Subramanian, T. Sato, and R. Langridge (1973), Structure of guanosine-3',5'-cytidine monophosphate. I. Semi-empirical potential energy calculations and model-building. *Biopolymers* **12**, 2731–2750.

372. G. Govil (1973), Nucleic acid conformations and biological activity. *Jerus. Sympos. Quant. Chem. Biochem.* **5**,283–295.

373. R. Tewari, R. K. Nanda, and G. Govil (1974), Quantum chemical studies on the conformational structure of nucleic acids. IV. Calculation of backbone structure by CNDO method. *J. Theor. Biol.* **46**, 229–239.

374. D. Perahia, B. Pullman, and A. Saran (1974), Molecular orbital calculations on the conformation of nucleic acids and their constituents. XI. The backbone structure of 3',5' and 2',3'-linked diribose monophosphates with different sugar puckers. *Biochim. Biophys. Acta* **353**, 16–27.

375. G. Govil (1976), Conformational structure of polynucleotides around the O–

P bonds: Refined parameters for CPF calculations. *Biopolymers* **15**, 2303–2307.

376. A. R. Srinivasan, N. Yathindra, V. S. R. Rao, and S. Prakash (1980), Preferred phosphodiester conformations in nucleic acids. A virtual bond torsion potential to estimate lone-pair interactions in a phosphodiester. *Biopolymers* **19**, 165–171.

377. K. Kitamura, A. Wakahara, H. Mizuno, Y. Baba, and K.-I. Tomita (1981), Conformationally "concerted" changes in nucleotide structures. A new description using circular correlation and regression analyses. *J. Amer. Chem. Soc.* **103**, 3899–3904.

378. K. Kitamura, A. Wakahara, H. Mizuno, T. Amizaki, Y. Baba, and K.-I. Tomita (1982), A quantitative description of conformational change in nucleic acid helices. Submitted for publication.

379. S. Yayaraman and N. Yathindra (1982), Theoretical evidence for the occurrence of sequence-dependent helical variations in B-DNA dodecamer. A backbone near-neighbour bond correlation between sugar residue and phosphodiester. *Nucleic Acids Res.* Submitted for publication.

380. W. K. Olson (1976), The spatial configuration of ordered polynucleotide chains. I. Helix formation and base stacking. *Biopolymers* **15**, 859–878.

381. N. Yathindra and M. Sundaralingam (1976), Analysis of possible helical structures of nucleic acids and polynucleotides. Application of (n-h) plots. *Nucleic Acids Res.* **3**, 729–747.

382. M. Sundaralingam and N. Yathindra (1977), Probing possible left- and right-handed polynucleotide helical conformations from n-h plots. Glycosyl and backbone torsional variation on handedness of helix. *Int. J. Quant. Chem.: Quant. Biol. Symp.* **4**, 285–303.

383. V. Sasisekharan and N. Pattabiraman (1978), Structure of DNA predicted from stereochemistry of nucleoside derivatives. *Nature* **275**, 159–162.

383a. H. Einspahr, W. J. Cook and C. E. Bugg (1981), Conformational flexibility in single-stranded oligonucleotides: Crystal structure of a hydrated calcium salt of adenylyl-(3′,5′)-adenosine. *Biochemistry* **20**, 5788–5794.

384. G. Gupta, M. Bansal, and V. Sasisekharan (1980), Conformational flexibility of DNA: Polymorphism and handedness. *Proc. Nat. Acad. Sci USA* **77**, 6486–6490.

385. P. De Santis, S. Morosetti, A. Pallesche, and M. Savino (1981), Conformational and structural constraints in double-helical polynucleotides. *Biopolymers* **20**, 1707–1725.

386. G. Gupta, M. Bansal, and V. Sasisekharan (1980), Polymorphism and conformational flexibility of DNA: Right and left handed duplexes. *Int. J. Biol. Macromol.* **2**, 368–379.

387. V. I. Poltev, L. A. Molova, B. S. Zhorov, and V. A. Govyrin (1981), Simulation of conformational possibilities of DNA via calculation of nonbonded interactions of complementary dinucleoside phosphate complexes. *Biopolymers* **20**, 1–15.

388. Y. Mitsui, R. Langridge, B. E. Shortle, C. R. Cantor, R. C. Grant, M. Kodama, and R. D. Wells (1970), Physical and enzymatic studies on poly d(I-C) · poly d(I-C), an unusual double-helical DNA. *Nature* **228**, 1166–1169.

389. R. Malathi and N. Yathindra (1981), Virtual bond probe to study ordered and

random coil conformations of nucleic acids. *Int. J. Quant. Chem.* **20,** 241–257.

390. S. Yayaraman and N. Yathindra (1981), Probing possible left- and right-handed poly(dinucleotide) helical conformations from (n-h) plots. Models for polysequential nucleotides. *Int J. Quant. Chem.* **20,** 211–230.

391. V. Sasisekharan, G. Gupta, and M. Bansal (1981), Sequence-dependent molecular conformations of polynucleotides: Right and left-handed helices. *Int. J. Biol. Macromol.* **3,** 2–8.

392. S. Arnott and D. W. L. Hukins (1972), The dimensions and shapes of the furanose rings in nucleic acids. *Biochem. J.* **130,** 453–465.

393. F. E. Evans and N. O. Kaplan (1976), 8-Alkylaminoadenyl nucleotides as probes of dehydrogenase interactions with nucleotide analogs of different glycosyl conformation. *J. Biol. Chem.* **251,** 6791–6797.

394. A. Pohorille, D. Perahia, and B. Pullman (1978), Molecular orbital studies on the conformations of 8-amino- and 8-dimethylaminoadenosine 5′-monophosphate. *Biochim. Biophys. Acta* **517,** 511–516.

395. E. Fluck and K. Maas (1973). *Themen zur Chemie des Phosphors,* p. 45. A Hüthig Verlag, Heidelberg.

396. S. R. Holbrook, J. L. Sussman, R. W. Warrant, and S.-H. Kim (1978), Crystal structure of yeast phenylalanine transfer RNA. II. Structural features and functional implications. *J. Mol. Biol.* **123,** 631–660.

397. J. Konnert, I. L. Karle, and J. Karle (1970), The structure of dihydrothymidine. *Acta Crystallogr. B* **26,** 770–778.

398. A. Grand and J. Cadet (1978), Crystal and molecular structure of (−)-(5S)-5-hydroxy-5,6-dihydrothymidine. *Acta Crystallogr. B* **34,** 1524–1528.

399. P. Tougard (1973), Structure cristalline et moléculaire de la 1-β-D-arabino-fuanosyl-thymine, isomorphe de la 1-β-D-arabinofuranosyl-5-bromo-uracile. *Acta Crystallogr. B* **29,** 2227–2232.

400. L. E. Sutton (1961), *Chemische Bindung und Molekülstruktur.* Springer-Verlag, Heidelberg.

401. J. Clauwaert and J. Stockx (1968), Interactions of polynucleotides and their components. I. Dissociation constants of the bases and their derivatives. *Z. Naturforsch. B.* **23,** 25–30.

402. G. D. Fasman (Ed.) (1975), *Handbook of Biochemistry and Molecular Biology,* Vol. I, *Nucleic Acids,* pp. 76–206. Chem. Rubber Co., Cleveland, Ohio.

403. H. A. Sober, R. A. Harte, and E. K. Sober (1970), *Handbook of Biochemistry. Selected Data for Molecular Biology,* pp. G-3 to G-98. Chem. Rubber Co., Cleveland, Ohio.

404. J. G. Kirkwood and F. H. Westheimer (1938), The electrostatic influence of substituents on the dissociation constants of organic acids. I. *J. Chem. Phys.* **6,** 506–513.

405. C. L. Agnell (1961), An infrared spectroscopic investigation of nucleic acid constituents. *J. Chem. Soc.,* 504–515.

406. M. Tsuboi, Y. Kyogoku, and T. Shimanouchi (1962), Infrared absorption spectra of protonated and deprotonated nucleosides. *Biochim. Biophys. Acta* **55,** 1–12.

407. R. C. Lord and G. J. Thomas, Jr. (1967), Raman spectral studies of nucleic acids and related molecules. I. Ribonucleic acid derivatives. *Spectrochim. Acta* **A23,** 2551–2591.

408. J. J. Fox and D. Shugar (1952), Spectrophotometric studies of nucleic acid derivatives and related compounds as a function of pH. II. Natural and synthetic pyrimidine nucleosides. *Biochim. Biophys. Acta* **9**, 369–384.

409. C. D. Jardetzky and O. Jardetzky (1960), Investigation of the structure of purines, pyrimidines, ribose nucleosides and nucleotides by Proton Magnetic resonance, II. *J. Amer. Chem. Soc.* **82**, 222–229.

410. A. R. Katritzky and A. J. Waring (1962), Tautomeric azines. Part I. The tautomerism of 1-methyluracil and 5-bromo-1-methyluracil. *J. Chem. Soc.*, 1540–1544; (1963), Part III. The structure of cytosine and its mono-cation. *J. Chem. Soc.*, 3046–3051.

411. C. D. Poulter and R. B. Anderson (1972), Direct observation of uracil dication and related derivatives. *Tetrahedron Lett.* **36**, 3823–3826.

412. R. Wagner and W. von Philipsborn (1970), Protonierung von amino- und hydroxypyrimidinen. NMR-Spektren und Strukturen der mono- und dikationen. *Helv. Chim. Acta* **53**, 299–320.

413. H. M. Sobell and K. Tomita (1964), The crystal structures of salts of methylated purines and pyrimidines. IV. 9-Methylguanine hydrobromide. *Acta Crystallogr.* **17**, 126–131.

414. J. Iball and H. R. Wilson (1965), The crystal and molecular structure of guanine hydrochloride dihydrate. *Proc. Roy. Soc. A* **288**, 418–429.

415. R. F. Bryan and K. Tomita (1961), Crystal structure of salts of methylated purines and pyrimidines. *Nature* **192**, 812–814; (1962), The crystal structures of salts of methylated purines and pyrimidines. II. 9-Methyladenine dihydrobromide. *Acta Crystallogr.* **15**, 1179–1182.

416. H. M. Sobell and K. Tomita (1964), The crystal structures of salts of methylated purines and pyrimidines. III. 1-Methyluracil hydrobromide. *Acta Crystallogr.* **17**, 122–126.

417. R. M. Izatt, J. J. Christensen, and H. Rytting (1971), Sites and thermodynamic quantities associated with proton and metal ion interaction with ribonucleic acid, deoxyribonucleic acid, and their constituent bases, nucleosides, and nucleotides. *Chem. Rev.* **71**, 439–481.

418. B. Pullman (1959), Electronic structure, chemical reactivity and basicity of purines and pyrazolopyrimidines. *J. Chem. Soc.*, 1621–1623.

419. R. Bonaccorsi, A. Pullman, E. Scrocco, and J. Tomasi (1972), The molecular electrostatic potentials for the nucleic acid bases: adenine, thymine, and cytosine. *Theor. Chim. Acta* **24**, 51–60.

420. R. M. Izatt, L. D. Hansen, J. H. Rytting, and J. J. Christensen (1965), Proton ionization from adenine. *J. Amer. Chem. Soc.* **87**, 2760–2761.

421. G. I. Birnbaum, J. Giziewicz, C. P. Huber, and D. Shugar (1976), Intramolecular hydrogen bonding and acidities of nucleoside sugar hydroxyls. Crystal structure and conformation of $O^2,2'$-anhydro-1-α-D-xylofuranosyl-uracil. *J. Amer. Chem. Soc.* **98**, 4640–4644.

422. R. L. van Etten, G. A. Clowes, J. F. Sebastian, and M. L. Bender (1967), The mechanism of the cycloamylose accelerated cleavage of phenylesters. *J. Amer. Chem. Soc.* **89**, 3253–3262.

423. T.-F. Chin, P.-H. Chung, and J. L. Lach (1968), Influence of cyclodextrins on ester hydrolysis. *J. Pharm. Sci.* **57**, 44–48.

424. D. O. Jordan (1960), *The Chemistry of Nucleic Acids*. Butterworth, Washington, D.C.

425. R. Phillips, P. Eisenberg, P. George, and R. J. Rutman (1965), Thermodynamic data for the secondary phosphate ionizations of adenosine, guanosine, inosine, cytidine and uridine nucleotides and triphosphates. *J. Biol. Chem.* **240**, 4393–4397.

426. J. Jonàš and J. Gut (1962), Nucleic acid components and their analogues. XVI. Dissociation constants of uracil, 6-azauracil, 5-azauracil and related compounds. *Coll. Czech. Chem. Commun.* **27**, 716–723.

427. C. K. Ingold (1953), *Structure and Mechanism in Organic Chemistry*. Cornell Univ. Press, New York.

428. P. Beak (1977), Energies and alkylations of tautomeric heterocyclic compounds: Old problems—New answers. *Acc. Chem. Res.* **10**, 186–192.

429. J. Elguero, C. Marzin, A. R. Katritzky, and P. Linda (1976), *The Tautomerism of Heterocycles. Adv. Heterocycl. Chem. Suppl. I.*

430. M. Dreyfus, O. Bensaude, G. Dodin, and J. E. Dubois (1976), Tautomerism in cytosine and 3-methylcytosine. A thermodynamic and kinetic study. *J. Amer. Chem. Soc.* **98**, 6338–6349.

431. R. Stolarski, M. Remin, and D. Shugar (1977), Studies on prototropic tautomerism in neutral and monoanionic forms of pyrimidines by nuclear magnetic resonance spectroscopy. *Z. Naturforsch. C.* **32**, 894–900.

432. J. Lin, C. Yu, S. Peng, I. Akiyama, K. Li, Li Kao Lee, and P. R. LeBreton (1980), Ultraviolet photoelectron studies on the ground-state electronic structure and gas-phase tautomerism of hypoxanthine and guanine. *J. Phys. Chem.* **84**, 1006–1012.

433. R. S. Norton, R. P. Gregson, and R. J. Quinn (1980), ^{13}C–N.M.R. Spin-lattice relaxation time measurements determining the major tautomer of 1-methylisoguanosine in solution. *J. Chem. Soc. Chem. Commun.*, 339–341.

434. R. V. Wolfenden (1969), Tautomeric equilibria in inosine and adenosine. *J. Mol. Biol.* **40**, 307–310.

435. M. J. Nowak, K. Szczepaniak, A. Barski, and D. Shugar (1978), Spectroscopic studies on vapor phase tautomerism of natural bases found in nucleic acids. *Z. Naturforsch. C.* **33**, 876–883.

436. M. Pieber, P. A. Kroon, J. H. Prestegard, and S. I. Chan (1973), Erratum. Tautomerism of nucleic acid bases. *J. Amer. Chem. Soc.* **95**, 3408.

437. Y. P. Wong, K. L. Wong, and D. R. Kearns (1972), On the tautomeric states of guanine and cytosine. *Biochem. Biophys. Res. Commun.* **6**, 1580–1587.

438. R. Czermínski, B. Lesyng, and P. Pohorille (1979), Tautomerism of pyrimidine bases—Uracil, cytosine, isocytosine: Theoretical study with complete optimization of geometry. *Int. J. Quant. Chem.* **16**, 605–613; (1979), Tautomerism of oxopyridines and oxopyrimidines: Theoretical study with complete optimization of geometry. *Int. J. Quant. Chem.* **16**, 1141–1148.

439. W. Hillen, E. Egert, H. J. Lindner, H. G. Gassen, and H. Vorbrüggen (1978), 5-Methoxyuridine: The influence of 5-substituents on the keto–enol tautomerism of the 4-carbonyl group. *J. Carbohyd., Nucleosides, Nucleotides* **5**, 23–32.

440. M. A. Viswamitra, M. L. Post, and O. Kennard (1979), The crystal structure of the dipotassium salt of uridine-5'-diphosphate. *Acta Crystallogr. B* **35**, 1089–1094.

441. E. Cherbuliez and K. Bernhard (1932), Recherches sur la graine de croton. I.

Sur le crotonoside (2-oxy-6-amino-purine-d-riboside). *Adv. Chim. Acta* **15**, 464–471.

442. R. Purrmann (1940), Über die Flügelpigmente der Schmetterlinge. VII. Synthese des Leukopterins und Natur des Guanopterins. *Liebigs Ann. Chem.* **544**, 182–186.

443. J. Sepiol, Z. Kazimierczuk, and D. Shugar (1976), Tautomerism of isoguanosine and solvent-induced keto–enol equilibrium. *Z. Naturforsch. C* **31**, 361–370.

444. M.-T. Chenon, R. J. Pugmire, D. M. Grant, R. P. Panzica, and L. B. Townsend (1975), Carbon-13 magnetic resonance. XXVI. A quantitative determination of the tautomeric populations of certain purines. *J. Amer. Chem. Soc.* **97**, 4636–4642.

445. A. Psoda, Z. Kazimierczuk, and D. Shugar (1974), Structure and tautomerism of the neutral and monoanionic forms of 4-thiouracil derivatives. *J. Amer. Chem. Soc.* **96**, 6832–6839.

446. K. Berens and D. Shugar (1963), Ultraviolet absorption spectra and structure of halogenated uracils and their glycosides. *Acta Biochim. Pol.* **10**, 25–48.

447. K. Dimroth, C. Reichardt, T. Siepmann, and F. Bohlmann (1963), Über Pyridinium-N-Phenol-Betaine und ihre Verwendung zur Charakterisierung der Polarität von Lösungsmitteln. *Liebigs Ann. Chem.* **661**, 1–37.

448. P. O. P. Ts'o (1974), Bases, nucleosides, and nucleotides. In: *Basic Principles in Nucleic Acid Chemistry* (P. O. P. Ts'o, Ed.), Vol. I, pp. 453–584. Academic Press, New York.

449. D. Riesner and R. Römer (1973), Thermodynamics and kinetics of conformational transitions in oligonucleotides and tRNA. In: *Physico-Chemical Properties of Nucleic Acids* (J. Duchesne, Ed.), pp. 237–318. Academic Press, New York.

450. S. N. Vinogradov and R. H. Linnell (1971), *Hydrogen Bonding*. Van Nostrand Reinhold, New York.

451. G. A. Jeffrey and S. Takagi (1978), Hydrogen-bond structure in carbohydrate crystals. *Acc. Chem. Res.* **11**, 264–270.

452. P. A. Kollman and L. C. Allen (1972), The theory of the hydrogen bond. *Chem. Rev.* **72**, 283–303.

453. W. Saenger and D. Suck (1973), The relationship between hydrogen bonding and base stacking in crystalline 4-thiouridine derivatives. *Eur. J. Biochem.* **32**, 473–478.

454. J. Donohue (1969), On N–H · · · S hydrogen bonds. *J. Mol. Biol.* **45**, 231–235.

455. H. A. Staab (1962), *Einführung in die theoretische organische Chemie*, p. 270. Verlag Chemie, Weinheim.

456. J. Kroon, J. A. Kanters, J. G. C. M. van Duijneveldt-van de Rijdt, F. B. van Duijneveldt, and J. A. Vliegenthart (1975), O–H · · · O hydrogen bonds in molecular crystals. A statistical and quantum-chemical analysis. *J. Mol. Struct.* **24**, 109–129.

456a. E. Kaun, H. Rüterjans and W. E. Hull (1982), ¹H NMR study of ¹⁵N labeled tRNA. *FEBS Lett.* **141**, 217–221.

457. J. Donohue (1956), Hydrogen-bonded helical configurations of polynucleotides. *Proc. Nat. Acad. Sci. USA* **42**, 60–65.

458. A. Rich, D. R. Davies, F. H. C. Crick, and J. D. Watson (1961), The molecular structure of polyadenylic acid. *J. Mol. Biol.* **3**, 71–86.

459. D. Bode, M. Heinecke, and U. Schernau (1973), An IR-investigation of the helic-coil conversion of poly(U). *Biochem. Biophys. Res. Commun.* **52**, 1234–1240.

460. A. Rabczenko and D. Shugar (1971), Studies in the conformation of nucleosides, dinucleoside monophosphates and homopolynucleotides containing uracil or thymine base residues, and ribose, deoxyribose or 2'-O-methylribose. *Acta Biochim. Pol.* **18**, 387–402.

461. S. K. Mazumdar, W. Saenger, and K. H. Scheit (1974), Molecular structure of poly-2-thiouridylic acid, a double helix with non-equivalent polynucleotide chains. *J. Mol. Biol.* **85**, 213–229.

462. J. C. Wang (1979), Helical repeat of DNA in solution. *Proc. Natl. Acad. Sci. USA* **76**, 200–203.

463. D. Rhodes and A. Klug (1980), Helical periodicity of DNA determined by enzyme digestion. *Nature* **286**, 573–578.

464. R. Wing, H. Drew, T. Takano, C. Broka, S. Tanaka, K. Itakura, and R. E. Dickerson (1980), Crystal structure analysis of a complete turn of DNA. *Nature* **287**, 755–758.

465. N. C. Seeman, J. M. Rosenberg, F. L. Suddath, J. J. P. Kim, and A. Rich (1976), RNA double-helical fragment at atomic resolution. I. The crystal and molecular structure of sodium adenylyl-3',5'-uridine hexahydrate. *J. Mol. Biol.* **104**, 109–144.

466. J. M. Rosenberg, N. C. Seeman, R. O. Day, and A. Rich (1976), RNA double-helical fragment at atomic resolution. II. The crystal structure of sodium guanylyl-3',5'-cytidine nonahydrate. *J. Mol. Biol.* **104**, 145–167.

467. M. N. Frey, T. F. Koetzle, M. S. Lehmann, and W. C. Hamilton (1973), Precision neutron diffraction structure determination of protein and nucleic acid components. XII. A study of hydrogen bonding in the purine–pyrimidine base pair 9-methyladenine · 1-methylthymine. *J. Chem. Phys.* **59**, 915–924.

468. T. D. Sakore, S. S. Tavale, and H. M. Sobell (1969), Base-pairing configurations between purines and pyrimidines in the solid state. I. Crystal and molecular structure of a 1:2 purine-pyrimidine hydrogen-bonded complex: 9-Ethyladenine: 1-methyl-5-iodouracil. *J. Mol. Biol.* **43**, 361–374.

469. K. Hoogsteen (1963), The crystal and molecular structure of a hydrogen-bonded complex between 1-methylthymine and 9-methyladenine. *Acta Crystallogr.* **16**, 907–916.

470. L. Katz, K.-I. Tomita, and A. Rich (1965), The molecular structure of the crystalline complex ethyladenine: methyl-bromouracil. *J. Mol. Biol.* **13**, 340–350.

471. K.-I. Tomita, L. Katz, and A. Rich (1967), Crystal structure of the intermolecular complex 9-ethyladenine: 1-methyl-5-fluorouracil. *J. Mol. Biol.* **30**, 545–549.

472. R. R. Shoup, H. T. Miles, and E. D. Becker (1966), NMR evidence of specific base-pairing between purines and pyrimidines. *Biochem. Biophys. Res. Commun.* **23**, 194–201.

473. Y. Kyogoku, R. C. Lord, and A. Rich (1967), The effect of substituents on the hydrogen bonding of adenine and uracil derivatives. *Proc. Natl. Acad. Sci. USA* **57**, 250–257.

474. H. Iwahashi and Y. Kyogoku (1977), Detection of proton acceptor sites of hydrogen bonding between nucleic acid bases by the use of ^{13}C magnetic resonance. *J. Amer. Chem. Soc.* **99,** 7761–7765.

475. W. Haynes (1964), *The Genetics of Bacteria and Their Viruses.* Blackwell, Oxford.

476. G. M. Nagel and S. Hanlon (1972), Higher order associations of adenine and uracil by hydrogen bonding. I. Self-association of 9-ethyl-adenine and 1-cyclohexyluracil. *Biochemistry* **11,** 816–823; II. Formation of complexes in mixed solutions of 9-ethyladenine and 1-cyclohexyluracil. *Biochemistry* **11,** 823–830.

477. R. A. Newmark and C. R. Cantor (1968), Nuclear magnetic resonance study of the interactions of guanosine and cytidine in dimethyl sulfoxide. *J. Amer. Chem. Soc.* **90,** 5010–5017.

478. Y. Kyogoku, R. C. Lord, and A. Rich (1967), An infrared study of hydrogen bonding between adenine and uracil derivatives in chloroform solution. *J. Amer. Chem. Soc.* **89,** 496–504.

479. J. S. Binford, Jr., and D. M. Holloway (1968), Heats of base pair formation with adenine and uracil analogs. *J. Mol. Biol.* **31,** 91–99.

480. I. K. Yanson, A. P. Teplitsky, and L. F. Sukhodub (1979), Experimental studies of molecular interactions between nitrogen bases of nucleic acids. *Biopolymers* **18,** 1149–1170.

481. G. G. Hammes and A. C. Park (1968), Kinetic studies of hydrogen bonding. 1-Cyclohexyluracil and 9-ethyladenine. *J. Amer. Chem. Soc.* **90,** 4151–4157.

482. B. Pullman, P. Claverie, and J. Caillet (1966), van der Waals-London interactions and the configuration of hydrogen-bonded purine and pyrimidine pairs. *Proc. Nat. Acad. Sci. USA* **55,** 904–912.

483. B. Pullman (1968) Associations moléculaires en biologie: théorie et expérience. Propos d'introduction. In: *Molecular Associations in Biology* (B. Pullman, Ed.), pp. 1–19. Academic Press, New York.

484. Z. G. Kudritskaya and V. I. Danilov (1976), Quantum mechanical study of bases interactions in various associates in atomic dipole approximation. *J. Theor. Biol.* **59,** 303–318.

485. R. Rein (1973), On physical properties and interactions of polyatomic molecules: With application to molecular recognition in biology. *Adv. Quant. Chem.* **7,** 335–396.

486. D. Voet and A. Rich (1969), The structure of an intermolecular complex between cytosine and 5-fluorouracil. *J. Amer. Chem. Soc.* **91,** 3069–3075.

487. S.-H. Kim and A. Rich (1969), A non-complementary hydrogen-bonded complex containing 5-fluorouracil and 1-methylcytosine. *J. Mol. Biol.* **42,** 87–95.

488. S. Hanlon (1966), The importance of London dispersion forces in the maintenance of the deoxyribonucleic acid double helix. *Biochem. Biophys. Res. Commun.* **23,** 861–867.

489. H. DeVoe and I. Tinoco, Jr. (1962), The stability of helical polynucleotides: Base contributions. *J. Mol. Biol.* **4,** 500–517.

490. C. E. Bugg (1972), Solid-state stacking patterns of purine bases. *Jerus. Symp. Quant. Chem. Biochem.* **4,** 178–204.

491. C. E. Bugg, J. M. Thomas, M. Sundaralingam, and S. T. Rao (1971), Stereochemistry of nucleic acids and their constituents. X. Solid-state base-stack-

ing patterns in nucleic acid constituents and polynucleotides. *Biopolymers* **10**, 175–219.

492. D. G. Watson, D. J. Sutor, and P. Tollin (1965), The crystal structure of adenosine monohydrate. *Acta Crystallogr.* **19**, 111–124.

493. J. Kraut and L. H. Jensen (1963), Refinement of the crystal structure of adenosine-5'-phosphate. *Acta Crystallogr.* **16**, 79–88.

494. A. E. V. Haschemeyer and H. M. Sobell (1965), The crystal structure of a hydrogen bonded complex of adenosine and 5-bromouridine. *Acta Crystallogr.* **18**, 525–532.

495. C. E. Bugg, U. Thewalt, and R. E. Marsh (1968), Base stacking in nucleic acid components: The crystal structures of guanine, guanosine and inosine. *Biochem. Biophys. Res. Commun.* **33**. 436–440.

496. C. E. Bugg and U. Thewalt (1969), Effects of halogen substituents on base stacking in nucleic acid components: The crystal structure of 8-bromoguanosine. *Biochem. Biophys. Res. Commun.* **37**, 623–629.

497. E. J. O'Brien (1967), Crystal structures of two complexes containing guanosine and cytosine derivatives. *Acta Crystallogr.* **23**, 92–106.

498. G. A. Jeffrey and Y. Kinoshita (1963), The crystal structure of cytosine monohydrate. *Acta Crystallogr.* **16**, 20–28.

499. F. S. Mathews and A. Rich (1964), The molecular structure of hydrogen bonded complex of N-ethyl adenine and N-methyl uracil. *J. Mol. Biol.* **8**, 89–95.

500. R. F. Stewart and L. H. Jensen (1967), Redetermination of the crystal structure of uracil. *Acta Crystallogr.* **23**, 1102–1105.

501. J. Iball, C. H. Morgan, and H. R. Wilson (1968), The crystal and molecular stucture of a 1:1 complex of 5-bromouridine and dimethylsulphoxide. *Proc. Roy. Soc. (London), Ser. A* **302**, 225–236.

502. D. R. Harris and W. M. MacIntyre, (1964), The crystal and molecular structure of 5-fluoro-2'-deoxy-β-uridine. *Biophys. J.* **4**, 203–225.

503. S. Furberg, C. S. Peterson, and C. Rømming (1965), A refinement of the crystal structure of cytidine. *Acta Crystallogr.* **18**, 313–320.

504. P. O. P. Ts'o, I. S. Melvin, and A. C. Olson (1963), Interaction and association of bases and nucleosides in aqueous solutions. *J. Amer. Chem. Soc.* **85**, 1289–1296.

505. N. I. Nakano and S. J. Igarashi (1970), Molecular interactions of pyrimidines, purines and some other heteroaromatic compounds in aqueous media. *Biochemistry* **9**, 577–583.

506. A. D. Broom, M. P. Schweizer, and P. O. P. Ts'o (1967), Interaction and association of bases and nucleosides in aqueous solution. V. Studies of the association of purine nucleosides by vapor pressure osmometry and by proton magnetic resonance. *J. Amer. Chem. Soc.* **89**, 3612–3622.

507. P. O. P. Ts'o, N. S. Kondo, R. K. Robins, and A. D. Broom (1969), Interaction and association of bases and nucleosides in aqueous solutions. VI. Properties of 7-methylinosine as related to the nature of the stacking interaction. *J. Amer. Chem. Soc.* **91**, 5625–5631.

508. D. Pörschke and F. Eggers (1972), Thermodynamics and kinetics of base-stacking interactions. *Eur. J. Biochem.* **26**, 490–498.

509. W. Gratzer (1969), Association of nucleic-acid bases in aqueous solution: A solvent partition study. *Eur. J. Biochem.* **10**, 184–187.

510. T. N. Solie and J. A. Schellman (1968), The interaction of nucleosides in aqueous solution. *J. Mol. Biol.* **33**, 61–77.

511. P. R. Mitchell and H. Sigel (1978), A proton nuclear magnetic resonance study of self-stacking in purine and pyrimidine nucleosides and nucleotides. *Eur. J. Biochem.* **88**, 149–154.

512. M. D. Topal and M. M. Warshaw (1976), Dinucleoside monophosphates. II. Nearest neighbor interactions. *Biopolymers* **15**, 1775–1793.

513. D. B. Davies (1978), Co-operative conformational properties of nucleosides, nucleotides and nucleotidyl units in solution. *Jerus. Sympos. Quant. Chem. Biochem.* **11**, 71–85.

514. C. A. G. Haasnoot and C. Altona (1979), A conformational study of nucleic acid phosphate ester bonds using phosphorus-31 nuclear magnetic resonance. *Nucleic Acids Res.* **6**, 1135–1149.

515. B. M. Baker, J. Vanderkooi, and N. R. Kallenbach (1978), Base stacking in a fluorescent dinucleoside monophosphate: ApϵA. *Biopolymers* **17**, 1361–1372.

516. N. S. Kondo and S. S. Danyluk (1976), Conformational properties of adenylyl-3′,5′-adenosine in aqueous solution. *Biochemistry* **15**, 756–767.

517. D. Pörschke (1973), The dynamics of nucleic-acid single-strand conformation changes. Oligo- and polyriboadenylic acids. *Eur. J. Biochem.* **39**, 117–126.

518. C. S. M. Olsthoorn, C. A. G. Haasnoot, and C. Altona (1980), Circular dichroism studies of 6-N-methylated adenylyladenosine and adenylyluridine and their parent compounds. Thermodynamics of stacking. *Eur. J. Biochem.* **106**, 85–95.

519. C. S. M. Olsthoorn, L. J. Bostelaar, J. F. M. de Rooij, J. H. van Boom, and C. Altona (1981), Circular dichroism study of stacking properties of oligodeoxyadenylates and polydeoxyadenylate. A three-state conformational model. *Eur. J. Biochem.* **115**, 309–321.

520. C. Altona, A. J. Hartel, C. S. M. Olsthoorn, H. P. M. de Leeuw, and C. A. G. Haasnoot (1978), The quantitative separation of stacking and self-association phenomena in a dinucleoside monophosphate by means of NMR concentration–temperature profiles: 6-N-(Dimethyl)-adenylyl-(3′,5′)-uridine. *Jerus. Symp. Quant. Chem. Biochem.* **11**, 87–101.

521. C. Reich and I. Tinoco, Jr. (1980), Fluorescence - detected circular dichroism of dinucleoside phosphates. A study of solution conformations and the two-state model. *Biopolymers* **19**, 833–848.

522. J. T. Powell, E. G. Richards, and W. B. Gratzer (1972), The nature of stacking equilibria in polynucleotides. *Biopolymers* **11**, 235–250.

523. T. T. Herskovits (1963), Nonaqueous solutions of DNA: Denaturation by urea and its methyl derivatives. *Biochemistry* **2**, 335–340.

524. T. T. Herskovits (1962), Nonaqueous solutions of DNA: Factors determining the stability of the helical configuration in solution. *Arch. Biochem. Biophys.* **97**, 474–484.

525. H. Edelhoch and J. C. Osborne, Jr. (1976), The thermodynamic basis of the stability of proteins, nucleic acids, and membranes. *Adv. Protein Chem.* **30**, 183–250.

526. W. Kauzmann (1959), Some factors in the interpretation of protein denaturation. *Adv. Protein Chem.* **14**, 1–63.

527. C. Tanford (1978), The hydrophobic effect and the organization of living matter. *Science* **200**, 1012–1018.
528. H. A. Scheraga (1978), Interactions in aqueous solution. *Acc. Chem. Res.* **12**, 7–14.
529. J. H. Hildebrand (1979), Is there a "hydrophobic effect"? *Proc. Nat. Acad. Sci. USA* **76**, 194.
530. H. S. Frank and M. W. Evans (1945), Free volume and entropy in condensed systems. *J. Chem. Phys.* **13**, 507–532.
531. F. Franks (1975), The hydrophobic interaction. In: *Water, a Comprehensive Treatise* (F. Franks, Ed.), pp. 1–94. Plenum Press, New York.
532. G. Némethy and H. A. Scheraga (1962), The structure of water and hydrophobic bonding in proteins. *J. Phys. Chem.* **66**, 1773–1789.
533. G. Némethy (1967), Hydrophobe Wechselwirkungen. *Angew. Chem.* **79**, 260–271.
534. O. Sinanoglu (1968), Solvent effects on molecular associations. In: *Molecular Associations in Biology* (B. Pullman, Ed.), pp. 427–445. Academic Press, New York.
535. O. Sinanoglu and S. Abdulnur (1965), Effects of water and other solvents on the structure of biopolymers. *Fed. Proc. 24,* Suppl. 15, S12–S23.
536. D. M. Crothers and D. I. Ratner (1968), Thermodynamic studies of a model system for hydrophobic bonding. *Biochemistry* **7**, 1823–1827.
537. I. Tazawa, T. Koike, and Y. Inoue (1980), Stacking properties of a highly hydrophobic dinucleotide sequence, N^6,N^6-dimethyladenylyl(3′-5′)N^6,N^6-dimethyladenosine, occurring in 16–18S ribosomal RNA. *Eur. J. Biochem.* **109**, 33–38.
538. H. Sapper and W. Lohmann (1978), Stacking interactions of nucleobases: NMR investigations. III. Molecular aspects of the solvent dependence. *Biophys. Struct. Mechanism* **4**, 327–335.
539. B. I. Sukhorukov, I. Ya. Gukovsky, A. I. Petrov, A. S. Gukovskaya, A. A. Mayevsky, and N. M. Gusenkova (1980), Interactions and self-organization of nucleic bases, nucleotides, and polynucleotides into ordered structures. Effect of ionization and solvent salt composition. *Int. J. Quant. Chem.* **17**, 339–359.
540. F. London (1930), Über die Eigenschaften und Anwendungen der Molekularkräfte. *Z. Phys. Chem. Abt. B* **11**, 222–251.
541. D. F. Waugh (1954), Protein–protein interactions. *Adv. Protein Chem.* **9**, 326–437.
542. R. L. Ornstein, R. Rein, D. L. Breen, and R. D. MacElroy (1978), An optimized potential function for the calculation of nucleic acid interaction energies. I. Base stacking. *Biopolymers* **17**, 2341–2360.
543. J. Langlet, P. Claverie, F. Caron, and J. C. Boeuve (1981), Interaction between nucleic acid bases in hydrogen bonded and stacked configurations: The role of the molecular charge density. *Int. J. Quant. Chem.* **19**, 299–338.
544. D. Pörschke (1977), Elementary steps of base recognition and helix-coil transitions in nucleic acids. *Mol. Biol. Biochem. Biophys.* **24**, 191–218.
545. D. Pörschke (1971), Cooperative nonenzymic base recognition. II. Thermodynamics of the helix coil transition of oligoadenylic + oligouridylic acids. *Biopolymers* **10**, 1989–2013.
546. V. V. Filimonov and P. L. Privalov (1978), Thermodynamics of base interaction in $(A)_n$ and $(A \cdot U)_n$. *J. Mol. Biol.* **122**, 465–470.

547. U. Schernau, S. Marcinowski, and Th. Ackerman (1979), Infrared spectroscopic studies of the interaction between polyuridylic acid and adenosine mono- and oligonucleotides. *Z. Phys. Chem. N. F.* **117**, 11–18.
548. D. Pörschke and M. Eigen (1971), Cooperative nonenzymic base recognition. III. Kinetics of the helix-coil transition of the oligoribouridylic oligoriboadenylic acid system and of oligoriboadenylic acid alone at acidic pH. *J. Mol. Biol.* **62**, 361–381.
549. J. Marmur and P. Doty (1962), Determination of the base composition of deoxyribonucleic acid from its thermal denaturation temperature. *J. Mol. Biol.* **5**, 109–118.
550. H. C. Spatz and D. M. Crothers (1969), The rate of DNA unwinding. *J. Mol. Biol.* **42**, 191–219.
551. W. S. Yen and R. D. Blake (1980), Analysis of high-resolution melting (thermal dispersion) of DNA. Methods. *Biopolymers* **19**, 681–700.
552. M. D. Frank-Kamenitskii and A. V. Vologodskii (1977), The nature of the fine structure of DNA melting curves. *Nature* **269**, 729–730.
553. G. Steger, H. Müller, and D. Riesner (1980), Helix-coil transitions in double-stranded viral RNA. Fine resolution melting and ionic strength dependence. *Biochim. Biophys. Acta* **606**, 274–284.
554. P. L. Privalov and V. V. Filimonov (1978), Thermodynamic analysis of transfer RNA unfolding. *J. Mol. Biol.* **122**, 447–464.
555. O. Gotoh and Y. Takashira (1981), Stabilities of nearest-neighbor doublets in double-helical DNA determined by fitting calculated melting profiles to observed profiles. *Biopolymers* **20**, 1033–1042.
556. P. N. Borer, B. Dengler, I. Tinoco, Jr., and O. C. Uhlenbeck (1974), Stability of ribonucleic acid double-stranded helices. *J. Mol. Biol.* **86**, 843–853.
557. I. Tinoco, Jr., P. N. Borer, B. Dengler, M. D. Levine, O. C. Uhlenbeck, D. M. Crothers, and J. Gralla (1973), Improved estimation of secondary structure in ribonucleic acids. *Nature New Biol.* **246**, 40–41.
558. R. D. Blake and S. G. Lefoley (1978), Spectral analysis of high resolution direct-derivative melting curves of DNA for instantaneous and total base composition. *Biochim. Biophys. Acta* **518**, 233–246.
559. T. R. Fink and H. Krakauer (1975), The enthalpy of the "bulge" defect of imperfect nucleic acid helices. *Biopolymers* **14**, 433–436.
560. J. M. Pipas and J. E. McMahon (1975), Method for predicting RNA secondary structure. *Proc. Nat. Acad. Sci. USA* **72**, 2017–2021.
561. J. Gralla and D. M. Crothers (1973), Free energy of imperfect nucleic acid helices. III. Small internal loops resulting from mismatches. *J. Mol. Biol.* **78**, 301–319.
562. S. M. Coutts (1971), Thermodynamics and kinetics of G–C base pairing in the isolated extra arm of serine-specific transfer RNA from yeast. *Biochim. Biophys. Acta* **232**, 94–106.
563. P. H. von Hippel and K.-Y. Wong (1971), Dynamic aspects of native DNA structure: Kinetics of the formaldehyde reaction with calf thymus DNA. *J. Mol. Biol.* **61**, 587–613.
564. C. R. Woese (1967), *The Genetic Code.* Harper & Row, New York.
565. E. Freese (1959), The difference between spontaneous and base-analogue induced mutations of phage T4. *Proc. Nat. Acad. Sci. USA* **45**, 622–633.
566. J. W. Drake, E. F. Allen, S. A. Forsberg, R.-M. Preparata, and E. O. Greening (1969), Spontaneous mutation. *Nature* **221**, 1128–1132; J. W. Drake

(1969), Appendix: Comparative rates of spontaneous mutation. *Nature* **221**, 1132.

567. A. Kornberg (1974), *DNA Synthesis*. Freeman, San Francisco.

568. D. Brutlag and A. Kornberg (1972), Enzymatic synthesis of deoxyribonucleic acid. XXXVI. A proofreading function for the 3′→5′ exonuclease activity in deoxyribonucleic acid polymerase. *J. Biol. Chem.* **247**, 241–248.

569. A. R. Fersht (1980), Enzymic editing mechanisms in protein synthesis and DNA replication. *Trends Biochem. Sci.* **5**, 262–265.

570. M. D. Topal and J. R. Fresco (1976), Complementary base pairing and the origin of substitution mutations. *Nature* **263**, 285–289.

571. V. I. Poltev and V. I. Bruskov (1978), On molecular mechanisms of nucleic acid synthesis. Fidelity aspects. 1. Contribution of base interactions. *J. Theor. Biol.* **70**, 69–83.

572. F. H. C. Crick (1966), Codon–anticodon pairing: The wobble hypothesis. *J. Mol. Biol.* **19**, 548–555.

573. D. Söll, J. D. Cherayil, and R. M. Bock (1967), Studies on polynucleotides. LXXV. Specificity of tRNA for codon recognition as studied by the ribosomal binding technique. *J. Mol. Biol.* **29**, 97–112.

574. D. Söll and U. L. RajBhandary (1967), Studies on polynucleotides. LXXVI. Specificity of tRNA for codon recognition as studied by amino acid incorporation. *J. Mol. Biol.* **29**, 113–124.

575. D. A. Kellogg, B. P. Doctor, J. E. Loebel, and M. W. Nirenberg (1966), RNA codons and protein synthesis, IX. Synonym codon recognition by multiple species of valine-, alanine- and methionine-sRNA. Proc. Nat. Acad. Sci. USA **55**, 912–919.

576. A. J. Lomant and J. R. Fresco (1975), Structural and energetic consequences of noncomplementary base opposition in nucleic acid helices. *Prog. Nucl. Acid Res. Mol. Biol.* **15**, 185–218.

577. Cold Spring Harbor Symp. Quant. Biol. (1966), Vol. XXXI.

578. M. Sprinzl, F. Grueter, A. Spelzhaus and D. H. Gauss (1980), Compilation of tRNA sequences. *Nucl. Acids Res.* **8**, r1-r22.

579. W. Fuller and A. Hodgson (1967), Conformation of the anti-codon loop in tRNA. *Nature* **215**, 817–821.

580. M. D. Topal and J. R. Fresco (1976), Base pairing and fidelity in codon-anticodon interaction. *Nature* **263**, 289–293.

581. T. D. Sakore and H. M. Sobell (1969), Crystal and molecular structure of a hydrogen-bonded complex containing adenine and hypoxanthine derivatives: 9-ethyl-8-bromoadenine - 9-ethyl-8-bromohypoxanthine. *J. Mol. Biol.* **43**, 77–87.

582. A. G. Lezius and E. Domin (1973), A wobbly double helix. *Nature* **244**, 169–170.

583. P. J. Romaniuk, D. W. Hughes, R. J. Gregorie, T. Neilson and R. A. Bell (1979), Contribution of a G · U base pair to the stability of a short RNA helix. *J. Chem. Soc. Chem. Commun.* 559–560.

584. F. H. C. Crick (1970), Central dogma of molecular biology. *Nature* **227**, 561–563.

585. C. Bresch and R. Hausmann (1970), *Klassische und molekulare Genetik*. Springer-Verlag, Berlin, p. 243.

586. R. T. Walker, E. De Clercq, and F. Eckstein (Eds.) (1979), *Nucleoside Ana-*

logues, Chemistry, Biology, and Medical Applications. Plenum Press, New York.

587. R. J. Suhadolnik (1970), *Nucleoside Antibiotics;* (1979), *Nucleosides as Biological Probes*. Wiley, New York.

588. R. E. Harmon, R. K. Robins, and L. B. Townsend (Eds.) (1978), *Chemistry and Biology of Nucleosides and Nucleotides*. Academic Press, New York.

589. R. H. Hall (1971), *The Modified Nucleosides in Nucleic Acids,* Columbia Univ. Press, New York.

590. P. F. Agris (1980), *The Modified Nucleosides of Transfer RNA,* A. R. Liss, New York.

591. D. Suck and W. Saenger (1973), The crystal and molecular structure of $O_{2,2'}$-cyclouridine. Influence of $O(2)-C(2')$ cyclization on the sugar conformation of pyrimidine nucleosides. *Acta Crystallogr. B* **29,** 1323–1330.

592. L. T. J. Delbaere and M. N. James (1973), The crystal and molecular structure of 2,2'-anhydro-1-β-D-arabinofuranosyl uracil. *Acta Crystallogr. B* **29,** 2905–2912.

593. Y. Yamagata, J. Yoshimura, S. Fujii, T. Fujiwara, and K.-I. Tomita (1980), Structural studies of pyrimidine cyclonucleoside derivatives. V. Structure of 2,2'-anhydro-1-β-D-arabinofuranosyl-2-thiouracil. *Acta Crystallogr. B* **36,** 343–346.

594. Y. Yamagata, Y. Suzuki, S. Fujii, T. Fujiwara, and K.-I. Tomita (1979), Structural studies of O-cyclocytidine derivatives. III. The crystal and molecular structure of 2,2'-anhydro-1-β-D-arabinofuranosylcytosine-3',5'-diphosphate monohydrate. *Acta Crystallogr. B* **35,** 1136–1140.

595. Y. Yamagata, S. Fujii, T. Kanai, K. Ogawa, and K.-I. Tomita (1979), Structural studies of O-cyclonucleoside derivatives. I. The crystal and molecular structure of 6,2'-anhydro-1-β-D-arabinofuranosyl-6-hydroxycytosine. *Acta Crystallogr. B* **35,** 378–382.

596. S. Neidle, G. L. Taylor, and P. C. Cowling (1979), The crystal structure of 8,2'-cycloadenosine trihydrate. *Acta Crystallogr. B* **35,** 708–712.

597. K. Tanaka, S. Fujii, T. Fujiwara, and K.-I. Tomita (1979), Structural studies of S-cycloadenosine derivatives. I. The crystal and molecular structure of 8,-2'-anhydro-8-mercapto-9-β-D-arabinofuranosyladenine-5'-monophosphate trihydrate (8,2'-S-cyclo 5'-AMP). *Acta Crystallogr. B* **35,** 929–933.

598. M. Yoneda, K. Tanaka, T. Fujiwara, and K.-I. Tomita (1979), Structural studies of S-cycloadenosine derivatives. II. The crystal and molecular structure of 8,3'-anhydro-8-mercapto-9-β-D-xylofuranosyladenine (8,3'-S-Cyclo A). *Acta Crystallogr. B* **35,** 2355–2358.

599. P. C. Manor, W. Saenger, D. B. Davies, K. Jankowski, and A. Rabczenko (1974), Conformation of nucleosides: The comparison of an X-ray diffraction and proton NMR study of 5',2-O-cyclo-2',3'-O-isopropylidene uridine. *Biochim. Biophys. Acta* **340,** 472–483.

600. L. T. J. Delbaere and M. N. James (1974), The crystal and molecular structure of 2,5'-anhydro-2',3'-isopropylidene cyclouridine. *Acta Crystallogr. B* **30,** 1241–1248.

601. Y. Yamagata, S. Fujii, T. Fujiwara, and K.-I. Tomita (1980), Structural studies of pyrimidine cyclonucleoside derivatives. IV. Structure of 2,5'-anhydro-1-(2',3'-O-isopropylidene-β-D-ribofuranosyl)-2-thiouracil. *Acta Crystallogr. B* **36,** 339–343.

602. S. Sprang, D. C. Rohrer, and M. Sundaralingam (1978), The crystal structure and conformation of 2',3'-O-isopropylideneadenosine: The coexistence of a planar and a puckered ribofuranose ring. *Acta Crystallogr. B* **34**, 2803–2810.

603. A. J. de Kok, C. Romers, H. P. M. de Leeuẃ, C. Altona, and J. H. van Boom (1977), Crystal structure and molecular conformation of 2',3'-O-methoxymethylene-uridine. X-ray and nuclear magnetic resonance investigation. *J. Chem. Soc. Perkin Trans. 2*, 487–493.

604. T. Brennan and M. Sundaralingam (1973), Molecular structure of 2,2'-anhydro-1-β-D-arabinofuranosyl cytosine hydrochloride (cyclo ara-C): Highly rigid nucleoside. *Biochem. Biophys. Res. Commun.* **52**, 1348–1353.

605. A. Gaudemer, F. Nief, R. Pontikis, and J. Zylber (1977), Etude conformationelle par résonance magnétique nucléaire protonique des dérivés de la O-isopropylidène-2',3'-adénosine. *Org. Magn. Reson.* **10**, 135–145.

606. G. T. Rogers and T. L. V. Ulbricht (1970), Optical properties and conformation of pyrimidine nucleosides in solution. *Biochem. Biophys. Res. Commun.* **39**, 414–418; Optical properties and conformation of purine nucleosides in solution. **39**, 419–422.

607. T. Ueda and S. Shibuya (1974), Nucleosides and nucleotides. IX. Synthesis of sulfur-bridged cyclonucleosides: (S)-2,2'-, 2,3'-, and 2,5'-cyclo-2-thiouridines. *Chem. Pharm. Bull.* **22**, 930–937.

608. M. Ikehara, M. Kaneko, Y. Nakahara, S. Yamada, and S. Uesugi (1971), Studies of nucleosides and nucleotides. XLIII. Purine cyclonucleotides (10). Optical rotatory dispersion and circular dichroism of adenine 8-cyclonucleosides. *Chem. Pharm. Bull.* **19**, 1381–1388.

609. B. P. Cross and T. Schleich (1973), Determination of the molecular conformation of β-D-O$_{2,2'}$-cyclouridine in aqueous solution by proton magnetic resonance spectroscopy. *Biopolymers* **12**, 2381–2389.

610. R. Stolarski, L. Dudycz, and D. Shugar (1980), NMR studies on the $syn \rightleftharpoons anti$ dynamic equilibrium in purine nucleosides and nucleotides. *Eur. J. Biochem.* **108**, 111–121.

611. A. Hoshi, F. Kanzawa, K. Kuretani, M. Saneyoshi, and T. Arai (1971), 2,2'-O-Cyclocytidine, an antitumor cytidine analog resistant to cytidine deaminase. *Gann* **62**, 145–146.

612. F. M. Richards and H. W. Wyckoff (1971), in *The Enzymes* (P.D. Boyer Ed.) 3rd ed. Vol. 4, pp. 647–806. Academic Press, New York.

613. G. A. Robison, R. W. Butcher, and E. W. Sutherland (1971), *Cyclic AMP.* Academic Press, New York.

614. H. Cramer and J. Schultz (Eds.) (1977). *Cyclic 3',5'Nucleotides: Mechanism of Action.* Wiley, New York.

615. P. Greengard and G. A. Robison (Eds.) (1980), *Advances in Cyclic Nucleotide Research,* Vols. 1–13. Raven Press, New York.

616. A. Ullmann and A. Danchin (1980), Role of cyclic AMP in regulatory mechanisms in bacteria. *Trends Biochem. Sci.,* **5**, 95–96.

617. N. D. Goldberg, M. K. Haddox, E. Dunham, C. Lopez, and J. W. Hadden (1974), In: *Cold Spring Harbor Symposium, Control of Proliferation in Animal Cells,* Vol. I, pp. 609–625. Cold Spring Harbor Laboratory, Cold Spring Harbor, New York.

618. C. L. Coulter (1973), Structural chemistry of cyclic nucleotides. II. Crystal

and molecular structure of sodium β-cytidine-2',3'-cyclic phosphate. *J. Amer. Chem. Soc.* **95**, 570–575.

619. B. S. Reddy and W. Saenger (1978), Molecular and crystal structure of the free acid of cytidine-2',3'-cyclophosphate. *Acta Crystallogr. B* **34**, 1520–1524.

620. W. Saenger and F. Eckstein (1970), Stereochemistry of a substrate for pancreatic ribonuclease. Crystal and molecular structure of the triethylammonium salt of uridine 2',3'-O,O-cyclophosphorothioate. *J. Amer. Chem. Soc.* **92**, 4712–4718.

621. W.-J. Kung, R. E. Marsh, and M. Kainosho (1977), Crystal and solution structure of 2',5'-arabinosylcytidine monophosphate. Influence of P–O–C bond angles on the proton–phosphorus vicinal coupling constants in the P–O–C–H fragments. *J. Amer. Chem. Soc.* **99**, 5471–5477.

622. A. Ku Chwang and M. Sundaralingam (1974), The crystal and molecular structure of guanosine 3',5'-cyclic monophosphate (cyclic GMP) sodium tetrahydrate. *Acta Crystallogr. B* **30**, 1233–1240.

623. C. L. Coulter (1969), The crystal and molecular structure of the triethylammonium salt of cyclic uridine-3',5'-phosphate. *Acta Crystallogr. B* **25**, 2055–2065; (1970) **26**, 441.

624. M. Sundaralingam and J. Abola (1972), Stereochemistry of nucleic acids and their constitutents. XXVII. The crystal structure of 5'-methyleneadenosine-3',5'-cyclic monophosphonate monohydrate, a biologically active analog of the secondary hormonal messenger cyclic adenosine-3',5'-monophosphate. Conformational "rigidity" of the furanose ring in cyclic nucleotides. *J. Amer. Chem. Soc.* **94**, 5070–5076.

625. D. K. Lavallee and C. L. Coulter (1973), Structural chemistry of cyclic nucleotides. III. Proton magnetic resonance studies of β-pyrimidine nucleotides. *J. Amer. Chem. Soc.* **95**, 576–581.

626. D. K. Lavallee and R. B. Myers (1978), Role of the pyrimidine base in ribonuclease A hydrolysis of RNA. Determination of the conformation of cyclic β-cytidine 2',3'-phosphate and cyclic β-uridine 2',3'-phosphate in solution. *J. Amer. Chem. Soc.* **100**, 3907–3912.

627. C. F. G. C. Geraldes and R. J. P. Williams (1978), Conformational studies of some 2',3'-cyclic mononucleotides in solution by different nuclear-magnetic-resonance methods. *Eur. J. Biochem.* **85**, 471–478.

628. R. D. Lapper and I. C. P. Smith (1973), A ^{13}C and ^{1}H nuclear magnetic resonance study of the conformations of 2',3'-cyclic nucleotides. *J. Amer. Chem. Soc.* **95**, 2880–2884.

629. A. Saran, H. Berthod, and B. Pullman (1973), Molecular orbital calculations on the conformation of nucleic acids and their constitutents. VIII. Conformations of 2',3'-and 3',5'-cyclic nucleotides. *Biochim. Biophys. Acta* **331**, 154–164.

630. N. Yathindra and M. Sundaralingam (1974), Conformations of cyclic 2',3'-nucleotides and 2',3'-cyclic-5'-diphosphates. Interrelation between the phase angle of pseudorotation of the sugar and the torsions about the glycosyl $C_{1'}$-N and the backbone $C_{4'}$-$C_{5'}$ bonds. *Biopolymers* **13**, 2061–2076.

631. G. M. Lipkind (1976), Theoretical conformational analysis of cytidine 2',3'-cyclophosphate. *Mol. Biol.* (*Moscow*) **10**, 490–493.

632. K. Watenpaugh, J. Dow, L. H. Jensen and S. Furberg (1968), Crystal and molecular structure of adenosine 3′,5′-cyclic phosphate. *Science* **159**, 206–207.

633. M. E. Druyan, M. Sparagana, and S. W. Peterson (1976), The molecular structure of the free acid of guanosine 3′,5′-cyclic monophosphate (cyclic GMP). *J. Cyclic Nucleotide Res.* **2**, 373–377.

634. W. S. Sheldrick and E. Rieke (1978), 8-[(2-Aminoethyl)amino]-adenosine cyclic 3′,5′-monophosphate tetrahydrate. *Acta Crystallogr. B* **24**, 2324–2327.

635. M. G. Newton, N. S. Pantaleo, G. S. Bajwa, and W. G. Bentrude (1977), The crystal structure of thymidine 3′,5′-cyclic N,N-dimethylphosphoroamidate. *Tetrahedron Lett.* **51**, 4457–4460.

636. W. Depmeier, J. Engels, and K. H. Klaska (1977), The crystal and molecular structure of 2′-acetyluridine-3′,5′-cyclophosphate benzyl triester, a pyrimidine nucleotide with *syn* conformation. *Acta Crystallogr. B* **33**, 2436–2440.

637. P. Hemmes, L. Oppenheimer, and F. Jordan (1976), Ultrasonic relaxation evidence for a two-state glycosyl conformational equilibrium in aqueous solution of adenosine-3′-5′-cyclic monophosphate. *J. Chem. Soc. Chem. Commun.*, 929–930.

638. C.-H. Lee and R. H. Sarma (1976), Aqueous solution conformation of rigid nucleosides and nucleotides. *J. Amer. Chem. Soc.* **98**, 3541–3548.

639. M. Kainosho and K. Ajisaka (1975), Conformational study of cyclic nucleotides. Lanthanide ion assigned analysis of the hydrogen-1 nuclear magnetic resonance spectra. *J. Amer. Chem. Soc.* **97**, 6839–6843.

640. M. McCoss, F. S. Ezra, M. J. Robins, and S. S. Danyluk (1978), Proton magnetic resonance studies of 9-(β-D-xylofuranosyl)adenine 3′,5′-cyclic monophosphate and 9-(β-D-arabinofuranosyl)adenine 2′,5′-cyclic monophosphate. *Carbohyd. Res.* **62**, 203–212.

641. M. McCoss, C. F. Ainsworth, G. Leo, F. S. Ezra, and S. S. Danyluk (1980), Conformational characteristics of rigid cyclic nucleotides. 3. The solution conformation of β-lyxonucleoside cyclic 2′,3′-and 3′,5′-monophosphates and of α-arabinonucleoside cyclic 2′,3′-monophosphates. Implications for evaluation of the solution properties of nucleoside analogues. *J. Amer. Chem. Soc.* **102**, 7353–7361.

642. I. D. Bobruskin, N. N. Gulyaev, M. P. Kirpichnikov, E. S. Severin, V. A. Tunitskaya, and V. L. Florent'ev (1979), Nuclear magnetic resonance study of conformation of nucleotides, oligonucleotides, and their analogs in solution. I. Conformation of adenosine-3′,5′-cyclophosphate and its analogs in aqueous solutions. *Mol. Biol. (Moscow)* **13**, 87–95.

643. C. L. Coulter (1975), Structural chemistry of cyclic nucleotides. IV. Crystal and molecular structure of tetramethylene phosphoric acid. *J. Amer. Chem. Soc.* **97**, 4084–4088.

644. J. M. Sturtevant, J. A. Gerlt, and F. H. Westheimer (1973), Enthalpy of hydrolysis of simple phosphate diesters. *J. Amer. Chem. Soc.* **95**, 8168–8169.

645. J. A. Gerlt, F. H. Westheimer, and J. M. Sturtevant (1975), The enthalpies of hydrolysis of acyclic, monocyclic and glycoside cyclic phosphate diesters. *J. Biol. Chem.* **250**, 5059–5067.

646. D. G. Gorenstein, D. Kar, B. A. Luxon, and R. K. Momii (1976), Conforma-

tional study of cyclic and acyclic phosphate esters. CNDO/2 calculations of angle strain and torsional strain. *J. Amer. Chem. Soc.* **98**, 1668–1673.

647. W. E. G. Müller, H. J. Rohde, R. Beyer, A. Maidhof, M. Lachmann, H. Taschner, and R. K. Zahn (1975), Mode of action of 9-β-D-arabinofuranosyladenine on the synthesis of DNA, RNA and protein *in vivo* and *in vitro*. *Cancer Res.* **35**, 2160–2168.

648. E. C. Herrmann, Jr. (Ed.) (1977), *Third Conference on Antiviral Substances*. *Ann. N.Y. Acad. Sci.* **284**

649. D. L. Miles, D. W. Miles, P. Redington, and H. Eyring (1977), A conformational basis for the selective action of ara-adenine. *J. Theor. Biol.* **67**, 499–514.

650. J. G. Dalton, A. L. George, F. E. Hruska, T. N. McGaig, K. K. Ogilvie, J. Peeling, and D. J. Wood (1977), Comparison of arabinose and ribose nucleosides conformation in aqueous and dimethylsulfoxide solution. *Biochim. Biophys. Acta* **478**, 261–273.

651. M. Remin, E. Darżynkiewicz, I. Ekiel, and D. Shugar (1976), Conformation in aqueous medium of the neutral, protonated and anionic forms of 9-β-D-arabinofuranosyladenine. *Biochim. Biophys. Acta* **435**, 405–416.

652. A. Saran, B. Pullman, and D. Perahia (1974), Molecular orbital calculations on the conformation of nucleic acids and their constituents. X. Conformation of β-D-arabinosyl nucleosides. *Biochim. Biophys. Acta* **349**, 189–203.

653. N. Yathindra and M. Sundaralingam (1979), Conformational analysis of arabinonucleosides and nucleotides. A comparison with the ribunucleosides and nucleotides. *Biochim. Biophys. Acta* **564**, 301–310.

654. F. Dinglinger and P. Renz (1971), 9-α-D-Ribofuranosyladenin (α-Adenosin), das Nucleosid des Corrinoidfaktors C_x aus *Propionibacterium shermanii*. *Hoppe Seyler's Z. Physiol. Chem.* **352**, 1157–1161.

655. H. Seto, N. Otake, and H. Jonehara (1972), The structures of pentopyranine A and C, two cytosine nucleosides with α-L-configuration. *Tetrahedron Lett.*, 3991–3994.

656. K. Suzuki, H. Nakano, and S. Suzuki (1967), Natural occurrence and enzymatic synthesis of α-nicotinamide adenine dinucleotide phosphate. *J. Biol. Chem.* **242**, 3319–3325.

657. R. Bonnett (1963), The chemistry of the vitamin B_{12} group. *Chem. Rev.* **63**, 573–605.

658. S. W. Hawkinson, C. L. Coulter, and M. L. Greaves (1970), The structure of vitamin B_{12}. VIII. The crystal structure of vitamin B_{12}-5'-phosphate. *Proc. Roy. Soc. Ser. A* **318**, 143–167.

659. A. Holy (1973), Nucleic acid components and their analogues. CLIV. Nucleoside and nucleotide derivatives of α-uridine, 2'-deoxy-α-uridine and 2'-deoxy-α-cytidine, and their affinity towards nucleolytic enzymes. *Coll. Czech. Chem. Commun.* **38**, 100–114.

660. T. Kulikowski and D. Shugar (1974), 5-Alkylpyrimidine nucleosides. Preparation and properties of 5-ethyl-2'-deoxycytidine and related nucleosides. *J. Med. Chem.* **17**, 269–273.

661. M. C. Henry, R. K. Morrison, D. E. Brown, M. Marlow, R. D. Davis, and D. A. Clooney (1973), Preclinical toxicologic studies of β-thioguanine deoxyriboside (NSC-71261). *Cancer Chem. Ther. Rep. Part 3* **4**, 41–49.

662. G. H. Milne and L. B. Townsend (1974), Synthesis and antitumor activity of α-and β-2'-deoxy-6-selenoguanosine and certain related derivatives. *J. Med. Chem.* **17**, 263–268.

663. M. Sundaralingam (1971), Stereochemistry of nucleic acids and their constituents. XVIII. Conformational analysis of α-nucleosides by X-ray crystallography. *J. Amer. Chem. Soc.* **93**, 6644–6647.

664. S. J. Cline and D. J. Hodgson (1979), The crystal and molecular structure of 9-α-D-arbinofuranosyladenine. *Biochim. Biophys. Acta* **563**, 540–544.

665. M. Remin, I. Ekiel, and D. Shugar (1975), Proton-magnetic-resonance study of the solution conformation of the α and β anomers of 5-ethyl-2'-deoxyuridine. *Eur. J. Biochem.* **53**, 197–206.

666. J. Cadet, C. Taïeb, M. Remin, W. P. Niemczura, and F. E. Hruska (1980), Conformational studies of α-and β-pyrimidine 2'-deoxyribonucleosides in the *syn* and *anti* conformation. *Biochim. Biophys. Acta* **608**, 435–445.

667. J. A. McCloskey and S. Nishimura (1977), Modified nucleosides in transfer RNA. *Acc. Chem. Res.* **10**, 403–410.

668. M. Ya. Feldman (1977), Minor components in transfer RNA: The location–function relationships. *Prog. Biophys. Molec. Biol.* **32**, 83–102.

669. R. H. Hall and D. B. Dunn (1975), Natural occurrence of the modified nucleosides. In: *Handbook of Biochemistry and Molecular Biology*, (G. D. Fasman, Ed.), 3rd ed. pp. 216–250. CRC Press, Cleveland.

670. J. P. Helgeson (1968), The cytokinins. *Science* **161**, 974–981.

671. R. H. Hall (1973), Cytokinins as probe of developmental processes. *Annu. Rev. Plant Physiol.* **24**, 415–444.

672. H. Sternglanz and C. E. Bugg (1973), Conformation of N⁶-methyladenine, a base involved in DNA modification: Restriction processes. *Science* **182**, 833–834.

673. M. Soriano-Garcia and R. Parthasarathy (1975), Structure–activity relationship of cytokinins: Crystal structure and conformations of 6-furfurylaminopurine. *Biochem. Biophys. Res. Commun.* **64**, 1062–1068; (1977), Stereochemistry and hydrogen bonding of cytokinins: 6-Furfurylaminopurine (kinetin). *Acta Crystallogr. B* **33**, 2674–2677.

674. R. Parthasarathy, J. M. Ohrt, and G. B. Chheda (1977), Modified nucleosides and conformation of anticodon loops: Crystal structure of t⁶A and g⁶A. *Biochemistry* **16**, 4999–5008.

675. D. A. Adamiak, T. L. Blundell, I. J. Tickle, and Z. Kosturkiewicz (1975), The structure of the rubidium salt of N-(purin-6-ylcabamoyl)-L-threonine tetrahydrate, a hypermodified base in the anticodon loop of some tRNA's. *Acta Crystallogr. B* **31**, 1242–1246.

676. R. Parthasarathy, M. Soriano-Garcia, and G. B. Chheda (1976), Bifurcated hydrogen bonds and flip–flop conformation in a modified nucleic acid base, gc⁶Ade. *Nature* **260**, 807–808.

677. C. E. Bugg and U. Thewalt (1972), Crystal structure of N⁶-(Δ²-iso-pentenyl)-adenosine, a base in the anticodon loop of some tRNA's. *Biochem. Biophys. Res. Commun.* **46**, 779–784.

678. J. D. Engel and P. H. von Hippel (1974), Effects of methylation on the stability of nucleic acid conformations: Studies at the monomer level. *Biochemistry* **13**, 4143–4158.

679. R. Parthasarathy, S. L. Ginell, N. C. De, and G. B. Chheda (1974), Confor-

mation of N_4-acetylcytidine, a modified nucleoside of tRNA, and stereochemistry of codon–anticodon interaction. *Biochem. Biophys. Res. Commun.* **83**, 657–663.

680. S. Yokoyama, T. Miyazawa, Y. Iitaka, Z. Yamaizumi, H. Kasai, and S. Nishimura (1979), Three-dimensional structure of hyper-modified nucleoside Q located in the wobbling position of tRNA. *Nature* **282**, 107–109.

681. K. Morikawa, K. Torii, Y. Iitaka and M. Tsuboi (1974), Crystal and molecular structure of the methyl ester of uridine-5-oxyacetic acid: A minor constituent of *Escherichia coli* tRNAs. *FEBS Lett.* **48**, 279–282.

682. W. Hillen, E. Egert, H. J. Lindner, and H. G. Gassen (1978), Crystal and molecular structure of 2-thio-5-carboxy-methyluridine and its methyl ester: Helix terminator nucleosides in the first position of some anticodons. Biochemistry **17**, 5314–5320.

683. W. Hillen, E. Egert, H. J. Lindner, and H. G. Gassen (1978), Restriction or amplification of wobble recognition. The structure of 2-thio-5-methylaminomethyluridine and the interaction of odd uridines with the anticodon loop backbone. *FEBS Lett.* **94**, 361–364.

684. P. Calabresi and R. E. Parks, Jr. (1970), In *The Pharmacological Basis of Therapeutics.* (L. S. Goodman and A. Gilman, Eds.), p. 1371. Macmillan, New York.

685. W. Saenger and D. Suck (1971), Kristall- und Molekülstruktur von 2,4-Dithiouridin-monohydrat. *Acta Crystallogr. B* **27**, 1178–1186.

686. U. Thewalt and C. E. Bugg (1972), Effects of sulfur substituents on base stacking and hydrogen bonding. The crystal structure of 6-thioguanosine monohydrate. *J. Amer. Chem. Soc.* **94**, 8892–8898.

687. C. E. Bugg and U. Thewalt (1970), The crystal and molecular structure of 6-thioguanine. *J. Amer. Chem. Soc.* **92**, 7441–7445.

688. A. Müller and J. Hüttermann (1973), Structure and formation of free radicals in irradiated single crystals of nucleic acid constituents. *Ann. N.Y. Acad. Sci.* **222**, 411–431.

689. I. L. Karle (1976), Crystal and molecular structure of photo-products from nucleic acids. In: *Photochemistry and Photobiology of Nucleic Acids* (S. Y. Wang, Ed.), Vol. I, pp. 483–519. Academic Press, New York.

690. J. Cadet, R. Ducolomb, and F. E. Hruska (1979), Proton magnetic resonance studies of 5,6-saturated thymidine derivatives produced by ionizing radiation. Conformational analysis of 6-hydroxylated diastereoisomers. *Biochim. Biophys. Acta* **563**, 206–215.

691. D. J. Hodgson and P. Singh (1976), Intra- and intermolecular interactions in azanucleosides. *Jerus. Sympos. Quant. Chem. Biochem.* **8**, 343–354.

692. B. Rada and J. Doscočil (1980), Azapyrimidine nucleosides. *Pharmacol. Ther.* **9**, 171–217.

693. A. Saran, C. Mitra, and B. Pullman (1978), Molecular orbital studies on the structure of nucleoside analogs. I. Conformation of 8-azapurine nucleosides. *Biochim. Biophys. Acta* **517**, 255–264.

694. C. Mitra and B. Pullman (1978), Molecular orbital studies on nucleoside analogs. II. Conformation of 6-azapyrimidine nucleosides. *Biochim. Biophys. Acta* **518**, 193–204.

695. S. Sprang, R. Scheller, D. Rohrer, and M. Sundaralingam (1978), Conformational analysis of 8-azanucleosides. Crystal and molecular structure of 8-aza-

tubercidin monohydrate, a nucleoside analogue exhibiting the "high-anti" conformation. *J. Amer. Chem. Soc.* **100**, 2867–2872.

696. W. Saenger and D. Suck (1973), 6-Azauridine-5'-phosphoric acid: Unusual molecular structure and functional mechanism. *Nature* **242**, 610–612.

697. J. R. Knowles (1980), Enzyme-catalyzed phosphoryl transfer reactions. *Annu. Rev. Biochem.* **49**, 777–819.

698. F. Eckstein (1979), Phosphorothioate analogues of nucleotides. *Acc. Chem. Res.* **12**, 204–210.

699. W. Saenger, D. Suck, and F. Eckstein (1974). On the mechanism of ribonuclease A. Crystal and molecular structure of uridine-3'-O-thiophosphate methyl ester triethylammonium salt. *Eur. J. Biochem.* **46**, 559–567.

700. D. Perahia, B. Pullman, and A. Saran (1972), A molecular orbital probe into the conformation of ATP. *Biochem. Biophys. Res. Commun.* **47**, 1284–1289.

701. Y.-F. Lam, G. P. P. Kuntz, and G. Kotowycz (1974), ^{13}C relaxation studies on the manganese(II)–adenosine 5'-triphosphate complex in solution. *J. Amer. Chem. Soc.* **96**, 1834–1839.

702. Y. H. Mariam and R. B. Martin (1979), Proximity of nucleic base and phosphate groups in metal ion complexes of adenine nucleotides. *Inorg. Chim. Acta* **35**, 23–28.

703. R. Basosi, N. Niccolai, E. Tiezzi, and G. Valensin (1978), EPR and NMR combined analysis of the metal–ATP interaction. *J. Amer. Chem. Soc.* **100**, 8047–8050.

704. W. Saenger, B. S. Reddy, K. Mühlegger, and G. Weimann (1977), Conformation of NAD$^+$ in solution, in holoenzymes and in the crystalline Li$^+$-complex. In: *Pyridine Nucleotide-Dependent Dehydrogenases* (H. Sund, Ed.), pp. 222–236. de Gruyter, Berlin.

705. T. J. Williams, A. P. Zens, J. C. Wisowaty, R. R. Fisher, R. B. Dunlap, T. A. Bryson, and P. D. Ellis (1976). Nuclear magnetic resonance studies on pyridine dinucleotides. The pH dependence of the carbon-13 nuclear magnetic resonance and NAD$^+$ analogs. *Arch. Biochem. Biophys.* **172**, 490–501.

706. M. Blumenstein and M. A. Raftery (1973), Natural abundance ^{13}C nuclear magnetic resonance spectra of nicotinamide adenine dinucleotide and related nucleotides. *Biochemistry* **12**, 3585–3590.

707. J. Jacobus (1971), Conformation of pyridine dinucleotides in solution. *Biochemistry* **10**, 161–164.

708. J. M. Thornton and P. M. Bayley (1977), Conformational energy calculations for dinucleotide molecules: A study of the nucleotide coenzyme nicotinamide adenine dinucleotide (NAD$^+$). *Biopolymers* **16**, 1971–1986.

709. B. S. Reddy, W. Saenger, K. Mühlegger, and G. Weimann (1981), Crystal and molecular structure of the lithium salt of nicotinamide-adenine-dinucleotide dihydrate (NAD$^+$, DPN$^+$, Cozymase, Codehydrase I). *J. Amer. Chem. Soc.* **103**, 907–914.

710. M. Sundaralingam and S. K. Arora (1970), Stereochemistry of nucleic acids and their constitutents. IX. The conformation of the antibiotic puromycin dihydrochloride pentahydrate. *Proc. Natl. Acad. Sci USA* **64**, 1021–1026.

711. M. Sundaralingam and S. K. Arora (1972), Crystal structure of the aminoglycosyl antibiotic puromycin dihydrochloride pentahydrate. Models for the terminal 3'-aminoacyladenosine moieties of transfer RNA's and protein–nucleic acid interactions. *J. Mol. Biol.* **71**, 49–70.

712. W. Saenger and D. Suck (1969), Kristall- und Molekülstruktur von 3'-O-Ace-tyl-4-thiothymidin. *Acta Crystallogr. B* **27**, 2105–2112.
713. R. A. G. de Graaff, G. Admiraal, E. H. Koen, and C. Romers (1977), Nucleic acid constituents, VI. The crystal and molecular structure of 3',5'-di-O-ace-tyluridine at −170°C. *Acta Crystallogr. B* **33**, 2459–2464.
714. P. Tougard and O. Lefebvre-Soubeyran (1974), Stéréochimie de la 1-β-D-ara-binofuranosyl-cytosine. *Acta Crystallogr. B* **30**, 86–89.
715. G. Bunick and D. Voet (1974), Crystal and molecular structure of 9-β-D-ara-binofuranosyladenine. *Acta Crystallogr. B* **30**, 1651–1660.
716. M. L. Post, G. I. Birnbaum, C. P. Huber, and D. Shugar (1977), α-Nucleo-sides in biological systems. Crystal structure and conformation of α-cytidine. *Biochim. Biophys. Acta* **479**, 133–142.
717. S. J. Cline and D. J. Hodgson (1979), The crystal and molecular structure of 9-α-D-arabinofuranosyladenine. *Biochim. Biophys. Acta* **563**, 540–544.
717a. D.C. Rohrer and M. Sundaralingam (1970), Stereochemistry of nucleic acids and their constituents. XII. The crystal and molecular structure of α-D-2'-amino-2'-deoxyadenosine monohydrate. *J. Amer. Chem. Soc.* **92**, 4956-4962.
718. J. W. Gibson and I. L. Karle (1971), Structure of the *cis–syn* photodimer of 6-methyluracil crystallized with a molecule of water. *J. Cryst. Mol. Struct.* **1**, 115–121.
719. J. Konnert and I. L. Karle (1971), Crystal structures of the *cis–anti* photo-dimer of uracil. *J. Cryst. Mol. Struct.* **1**, 107–114.
720. D. Suck, W. Saenger, and K. Zechmeister (1971), Molecular structure of the tRNA minor constituent dihydrouridine. *Acta Crystallogr. B* **28**, 596–605.
721. O. Kennard, N. W. Isaacs, W. D. S. Motherwell, J. C. Coppola, D. L. Wampler, A. C. Larson, and D. G. Watson (1971), The crystal and molecular structure of adenosine triphosphate. *Proc. Roy. Soc. London Ser. A* **325**, 401–436.
722. F. H. Westheimer (1968), Pseudo-rotation in the hydrolysis of phosphate esters. *Acc. Chem. Res.* **1**, 70–78.
723. G. L. Eichhorn (Ed.) (1973), *Inorganic Biochemistry*. Elsevier, Amsterdam.
724. I. Sissoëff, J. Grisvard, and E. Guillé (1976), Studies on metal ions–DNA in-teractions: Specific behavior of reiterative DNA sequences. *Prog. Biophys. Molec. Biol.* **31**, 165–199.
725. H. Pezzano and F. Podo (1980), Structure of binary complexes of mono- and polynucleotides with metal ions of the first transition group. *Chem. Rev.* **80**, 365–401.
726. A. S. Mildvan and L. A. Loeb (1979), The role of metal ions in the mecha-nism of DNA and RNA polymerases. *CRC Crit. Rev. Biochem.* **6**, 219–244.
727. S. R. Holbrook, J. L. Sussman, R. W. Warrant, G. M. Church, and S. H. Kim (1977), RNA–ligand interactions: (1) Magnesium binding sites in yeast tRNA[Phe]. *Nucl. Acids Res.* **4**, 2811–2820.
728. T. Lindahl, A. Adams, and J. R. Fresco (1966), Renaturation of transfer ri-bonucleic acids through site binding of magnesium. *Proc. Nat. Acad. Sci. USA* **55**, 941–948.
729. G. L. Eichhorn and Y. A. Shin (1968), Interaction of metal ions with polynu-cleotides and related compounds. XII. The relative effect of various metal ions on DNA helicity. *J. Amer. Chem. Soc.* **90**, 7323–7328.
730. N. Davidson, J. Widholm, U. S. Nandi, R. Jensen, B. M. Olivera, and J. C. Wang (1965), Preparation and properties of native crab dAT. *Proc. Nat. Acad. Sci. USA* **53**, 111–118.

731. C. A. Thomas (1954), The interaction of $HgCl_2$ with sodium thymonucleate. *J. Amer. Chem. Soc.* **76**, 6032–6034.

732. R. M. K. Dale, E. Martin, D. C. Livingston, and D. C. Ward (1975), Direct covalent mercuration of nucleotides and polynucleotides. *Biochemistry* **14**, 2447–2457.

733. R. H. Jensen and N. Davidson (1966), Spectrophotometric, potentiometric, and density gradient ultracentrifugation studies of the binding of silver ion by DNA. *Biopolymers* **4**, 17–32.

734. S. K. Arya and J. T. Yang (1975), Optical rotation, dispersion and circular dichroism of silver(I): Polyribonucleotide complexes. *Biopolymers* **14**, 1847–1861.

735. G. L. Eichhorn, E. Tarien, and J. J. Butzow (1971), Interaction of metal ions with nucleic acids and related compounds. XVI. Specific cleavage effects in the depolymerization of ribonucleic acids by zinc(II) ions. *Biochemistry* **10**, 2014–2019.

736. J. J. Butzow and G. L. Eichhorn (1971), Interaction of metal ions with nucleic acids and related compounds. XVII. On the mechanism of degradation of polyribonucleotides and oligoribonucleotides by zinc(II) ions. *Biochemistry* **11**, 2019–2027.

736a. R. S. Brown, B. E. Hingerty, J. C. Dewan, and A. Klug (1983), Pb(II)-catalyzed cleavage of the sugar–phosphate backbone of yeast $tRNA^{Phe}$—implications for lead toxicity and self-splicing RNA. *Nature* **303**, 543–546.

737. B. Rosenberg, L. Van Camp, J. E. Trosko, and V. H. Mansour (1969), Platinum compounds: A new class of potent antitumour agents. *Nature* **222**, 385–386.

738. R. C. Harrison and C. A. McAuliffe (1978), Platinum(II) complexes of DNA constituents. *Inorg. Perspect. Biol. Med.* **1**, 261–288.

739. A. W. Pestayko, S. T. Crooke, and S. K. Carter (Eds.) (1980), *Cisplatin: Current Status and New Developments.* Academic Press, New York.

740. M. M. Millard, J. P. Macquet, and T. Theophanides (1975), X-ray photoelectron spectroscopy of DNA·Pt complexes. Evidence of O_6 (Gua)·N_7 (Gua) chelation of DNA with *cis*-dichlorodiammine platinum(II). *Biochim. Biophys. Acta* **402**, 166–170.

741. A. T. M. Marcelis, C. G. van Kralingen, and J. Reedijk (1980), The interaction of *cis*- and *trans*-diammineplatinum compounds with 5'-guanosine monophosphate and 5'-deoxyguanosine monophosphate. A proton NMR investigation. *J. Inorg. Biochem.* **13**, 213–222.

742. R. M. Wing, P. Pjura, H. R. Drew, and R. E. Dickerson (1982), X-ray single crystal study of the interaction of cisplatin with the DNA dodecamer CGCGAATTCGCG. Private communication.

743. J. J. Roberts and A. J. Thomson (1979), The mechanism of action of antitumor platinum compounds. *Prog. Nucl. Acids Mol. Biol.* **22**, 71–133.

744. T. A. Connors and J. J. Roberts (Eds.) (1977), *Platinum Coordination Complexes in Cancer Chemotherapy.* Springer-Verlag, New York.

745. *The Proceedings of the Third International Symposium of Platinum Coordination Complexes in Cancer Chemotherapy (Dallas, Texas, 1976)*; *J. Clin. Hematol. Oncol.* (Wadley Medical Bull.) **7**, (1977).

746. K. Aoki (1981), Interaction between transition metal ions and nucleic acid bases, nucleosides, and nucleotides. (Jap.) *J. Cryst. Soc. Japan* **23**, 309–327.

747. D. J. Hodgson (1977), The stereochemistry of metal complexes of nucleic acid constituents. *Prog. Inorg. Chem.* **23**, 211–254.

747a. T. G. Spiro (Ed.) (1980), *Metal Ions in Biology.* Vol. I. Nucleic Acid-Metal Ion Interactions. John Wiley & Sons, New York.

748. L. G. Marzilli and T. J. Kistenmacher (1977), Stereoselectivity in the binding of transition-metal chelate complexes to nucleic acid constituents: Bonding and nonbonding effects. *Acc. Chem. Res.* **10**, 146–152.

749. R. W. Gellert and R. Bau (1979), X-ray structural studies of metal–nucleoside and metal–nucleotide complexes. In: *Metal Ions in Biological Systems* (H. Sigel, Ed.), Vol. VIII, pp. 1–55. Dekker, New York.

750. V. Swaminathan and M. Sundaralingam (1979), The crystal structures of metal complexes of nucleic acids and their constituents. *CRC Crit. Rev. Biochem.* **6**, 245–336.

751. S. J. Lippard (1978), Platinum complexes: probes of polynucleotide structure and antitumor drugs. *Acc. Chem. Res.* **11**, 211–217.

752. V. Kleinwächter (1978), Interaction of platinum(II) coordination complexes with deoxyribonucleic acid. *Stud. Biophys.* **73**, 1–17.

753. R. S. J. Phillips (1966), Adenosine and the adenine nucleotides. Ionization, metal complex formation and conformation in solution. *Chem. Rev.* **66**, 501–527.

754. H. Sigel (1975), Stability, structure and reactivity of ternary Cu^{2+}-complexes. *Angew. Chem. Int. Ed. Engl.* **14**, 394–403.

755. A. T. Tu and M. J. Heller (1973), Structure and stability of metal–nucleotide complexes. In: *Metal Ions in Biological Systems* (H. Sigel, Ed.), Vol. I. Dekker, New York.

756. C. M. Frey and J. Stuehr (1973), Kinetics of metal ion interactions with nucleotides and base-free phosphates. In: *Metal Ions in Biological Systems* (H. Sigel, Ed.), Vol. 1. Dekker, New York.

757. M. Daune (1975), Interactions of metal ions with nucleic acids. In: *Metal Ions in Biological Systems* (H. Sigel, Ed.), Vol. III. Dekker, New York.

758. L. G. Marzilli (1977), Metal–ion interactions with nucleic acids and nucleic acid derivatives. *Prog. Inorg. Chem.* **23**, 255–378.

759. R. B. Martin and Y. H. Miriam (1979), The interactions between nucleotides and metal ions in solution. In: *Metal Ions in Biological Systems* (H. Sigel, Ed.), Vol. VIII. Dekker, New York. 57–124.

760. S. Ahrland, J. Chatt, and N. K. Davies (1958), The relative affinities of ligand atoms for acceptor molecules and ions. *Quart. Rev. Chem. Soc.* **12**, 265–276.

761. R. G. Pearson (1966), Acids and bases. *Science* **151**, 172–177.

762. D. M. L. Goodgame, I. Jeeves, C. D. Reynolds, and A. C. Skapski (1975), Multiplicity of cadmium binding sites in nucleotides: X-ray evidence for the involvement of $O_{2'}$ and $O_{3'}$ as well as phosphate N_7 in inosine 5'-monophosphate. *Nucl. Acids Res.* **2**, 1375–1379.

763. W. J. Cook and C. E. Bugg (1977), Structures of calcium–carbohydrate complexes. *Jerus. Symp. Quant. Chem. Biochem.* **9**, 231–256.

764. E. A. Brown and C. E. Bugg (1980), Calcium-binding to nucleotides: Structure of a hydrated calcium salt of inosine-5'-monophosphate. *Acta Crystallogr. B* **36**, 2597–2604; for a strontium salt, see *J. Cryst. Mol. Struct.* **10**, 19–30.

765. E. Sletten (1971), Crystal structure of a copper-nucleoside analogue. *Chem. Commun.*, 558.

766. G. P. P. Kuntz and G. Kotowycz (1975), A nuclear magnetic resonance relaxation time study of the manganese(II)–inosine-5'-triphosphate complex in solution. *Biochemistry* **14,** 4144–4150.

767. K. Aoki (1975), Crystallographic studies of interaction between nucleotides and metal ions. I. Crystal structures of the 1:1 complexes of cobalt and nickel with inosine 5'-phosphate. *Bull. Chem. Soc. Japan* **48,** 1260–1271.

768. A. Jack, J. E. Ladner, D. Rhodes, R. S. Brown, and A. Klug (1977), A crystallographic study of metal-binding to yeast phenylalanine transfer RNA. *J. Mol. Biol.* **11,** 315–328.

769. J. J. Rosa and P. B. Sigler (1974), The site of covalent attachment in the crystalline osmium-tRNA^{fMet} isomorphous derivative. *Biochemistry* **13,** 5102–5110.

770. J. F. Conn, J. J. Kim, F. L. Suddath, P. Blattman, and A. Rich (1974), Crystal and molecular structure of an osmium bispyridine ester of adenosine. *J. Amer. Chem. Soc.* **96,** 7152–7153.

771. T. J. Kistenmacher, L. G. Marzilli, and M. Rossi (1976), Conformational properties of the osmium tetraoxide bispyridine ester of 1-methylthymine and a comment on the linearity of the *trans* O=Os=O group. *Bioinorg. Chem.* **6,** 347–364.

772. S. Neidle and D. I. Stuart (1976), The crystal and molecular structure of an osmium bispyridine adduct of thymine. *Biochim. Biophys. Acta* **418,** 226–231.

773. B. E. Fischer and R. Bau (1978), Coordination sites in nucleotides. *Inorg. Chem.* **17,** 27–34.

774. G. Y. H. Chu, R. E. Duncan, and R. S. Tobias (1977), Heavy metals–nucleosides interactions. *Inorg. Chem.* **16,** 2625–2636.

775. G. V. Fazakerley, D. Hermann, and W. Guschlbauer (1980), Proton NMR study of the interaction of *cis*-dichloro-diammine-platinum(II) with poly(I) and poly(I)·poly(C). *Biopolymers* **19,** 1299–1309.

776. C. Miller, C. M. Frey, and J. E. Stuehr (1972), Interactions of divalent metal ions with inorganic and nucleoside phosphates. I. Thermodynamics. *J. Amer. Chem. Soc.* **94,** 8898–8904.

777. A. Szent-Györgyi (1957), *Bioenergetics,* Chap. 10. Academic Press, New York.

778. S. Estrada-Parra and E. Garcia-Ortigoza (1972), Immunochemical determination of the molecular conformation of nucleotides. *Immunochemistry* **9,** 799–807.

779. Y. H. Mariam and R. B. Martin (1979), Proximity of nucleic base and phosphate groups in metal ion complexes of adenine nucleotides. *Inorg. Chim. Acta* **35,** 23–28.

780. J. Granot, H. Kondo, R. N. Armstrong, A. S. Mildvan, and E. T. Kaiser (1979), Nuclear magnetic resonance studies of the conformation of tetraammine–cobalt(III)–ATP bound at the active site of bovine heart protein kinase. *J. Amer. Chem. Soc.* **18,** 2339–2345.

781. G. V. Fazakerley and D. G. Reid (1979), Determination of the interaction of ADP and dADP with copper(II), manganese(II) and lanthanide(III) ions by nuclear-magnetic-resonance spectroscopy. *Eur. J. Biochem:,* **93,** 535–543.

782. Tran-Dinh Son, M. Roux, and M. Ellenberger (1975), Interaction of Mg^{2+} with nucleoside triphosphates by phosphorus magnetic resonance spectroscopy. *Nucl. Acids Res.* **2,** 1101–1110.

783. W. W. Cleland and A. S. Mildvan (1979), Chromium(III) and cobalt(III) nucleotides as biological probes. In: *Advances in Inorganic Biochemistry* (G. L. Eichhorn and L. G. Marzilli, Eds.), Vol. I. pp. 163–191. Elsevier, New York.

784. R. D. Cornelius, P. A. Hart and W. W. Cleland (1977), Phosphorus-31 NMR studies of complexes of adenosine triphosphate, adenosine diphosphate, tripolyphosphate, and pyrophosphate with cobalt(III) ammines. *Inorg. Chem.* **16**, 2799–2805.

785. E. A. Merritt, M. Sundaralingam, R. D. Cornelius, and W. W. Cleland (1978), X-ray crystal and molecular structure and absolute configuration of (dihydrogen tripolyphosphato)tetramminecobalt(III) monohydrate, $Co(NH_3)_4H_2P_3O_{10} \cdot H_2O$. A model for a metal–nucleoside polyphosphate complex. *Biochemistry* **17**, 3274–3278.

786. R. D. Cornelius and W. W. Cleland (1978), Substrate activity of (adenosine triphosphato)tetramminecobalt(III) with yeast hexokinase and separation of diastereomers using the enzyme. *Biochemistry* **17**, 3279–3286.

786a. P. Orioli, R. Cini, D. Donati and S. Mangani (1981), Crystal and molecular structure of the ternary complex bis [(adenosine-5'-triphosphato)(2,2'-bipyridine)zinc(II)] tetrahydrate. *J. Amer. Chem. Soc.* **103**, 4446–4452.

786b. W. S. Sheldrick (1981), Charge-Transfer-Wechselwirkungen zwischen den Liganden eines ternären ATP-Cu^{2+}-Phenanthrolin-Komplexes. *Angew. Chem.* **93**, 473–474.

787. D. A. Adamiak and W. Saenger (1980), Structure of the monopotassium salt of adenosine-5'-diphosphate dihydrate, $KADP \cdot 2H_2O$. *Acta Crystallogr. B* **36**, 2585–2589.

788. F. Guay and A. L. Beauchamp (1979), Model compounds for the interaction of silver(I) with polyuridine. Crystal structure of a 1:1 silver complex with 1-methylthymine. *J. Amer. Chem. Soc.* **101**, 6260–6263.

789. G. R. Clark and J. D. Orbell (1975), Transition metal–nucleotide complexes. X-ray and molecular structures of the cobalt(II) and cadmium(II) complexes of cytidine-5'-monophosphate, $[Co(CMP)(H_2O)]$ and $[Cd(CMP)(H_2O)] \cdot H_2O$. *J. Chem. Soc. Chem. Commun.*, 697–698.

790. S. Arnott (1981), The secondary structures of polynucleotide chains as revealed by X-ray diffraction analysis of fibers. In S. Neidle, ed. Topics in Nucleic Acid Structure, Macmillan Publ. Ltd., London, 65–82.

791. R. E. Dickerson, H. R. Drew, B. N. Conner, R. M. Wing, A. V. Fratini, and M. L. Kopka (1982), The anatomy of A-, B-, and Z-DNA. *Science* **216**, 475–485.

792. R. D. Wells, T. C. Goodman, W. Hillen, G. T. Horn, R. D. Klein, J. E. Larson, U. R. Müller, S. K. Neuendorf, N. Panayotatos, and S. M. Stirdivant (1980), DNA structure and gene regulation. *Prog. Nucl. Acid Res. Mol. Biol.* **24**, 167–267.

793. N. E. Kallenbach and H. M. Berman (1977), RNA structure. *Quart. Rev. Biophys.* **10**, 138–236.

794. Yu. A. Neifakh and V. G. Tumanyan (1980), Possible conformations of DNA. *Biophysics* **24**, 577–585.

795. V. I. Ivanov, Yu. P. Lysov, G. G. Malenkov, L. E. Minchenkova, E. E. Minyat, A. K. Schyolkina, and V. B. Zhurkin (1976), Conformational possibilities of double-helical DNA. *Stud. Biophys. Berlin* **55**, 5–13.

796. V. I. Poltev, L. A. Milova, B. S. Zhorov, and V. A. Govyrin (1981), Simula-

tion of conformational possibilities of DNA via calculation of non-bonded interactions of complementary dinucleoside phosphate complexes. *Biopolymers* **20**, 1–15.

797. F. G. Calascibetta, M. Dentini, P. de Santis, and S. Morosetti (1975), Conformational analysis of polynucleotide chains. Double stranded structures. *Biopolymers* **14**, 1667–1684.

798. R. Cornillon, J.-N. Lespinasse, H. Broch, and D. Vasilescu (1975), Influence of the sugar configuration on the structure of DNA by conformational analysis of the deoxyribose-phosphate unit. *Biopolymers* **14**, 1515–1529.

799. W. K. Olson (1978), Spatial configuration of ordered polynucleotide chains. V. Conformational energy estimates of helical structure. *Biopolymers* **17**, 1015–1040.

800. P. A. Kollman, P. K. Weiner, and A. Dearing (1981), Studies of nucleotide conformations and interactions. The relative stabilities of double helical B-DNA sequence isomers. *Biopolymers* **20**, 2583–2621.

801. C. Tosi, E. Clementi, and O. Matsuoka (1978), Conformational studies on polynucleotide chains. II. Analysis of steric interactions and derivation of potential functions for internal rotations. *Biopolymers* **17**, 51–66; III. Intramolecular energy maps and comparison with experiments. *Biopolymers* **17**, 67–84.

802. O. Matouka, C. Tosi, and E. Clementi (1978), Conformational studies on polynucleotide chains. I. Hartree-Fock energies and description of non-bonded interactions with Lennard-Jones potentials. *Biopolymers* **17**, 33–49.

803. K. J. Miller (1979), Interactions of molecules with nucleic acids. I. An algorithm to generate nucleic acid structures with an application to the B-DNA structure and a counterclockwise helix. *Biopolymers* **18**, 959–980.

804. G. Govil, C. Fisk F. B. Howard, and H. T. Miles (1977), Structure of poly-8-bromo-adenylic acid: Conformational studies by CPF energy calculations. *Nucl. Acids Res.* **4**, 2573–2592.

805. W. K. Olson (1975), Configurational statistics of polynucleotide chains. A single virtual bond treatment. *Macromolecules* **8**, 272–275.

806. M. S. Waterman and T. F. Smith (1978), RNA secondary structure: a complete mathematical analysis. *Math. Biosci.* **42**, 257–266.

807. I. A. Il'Icheva, V. G. Tumanyan, A. E. Kister, and V. G. Dashevsky (1980), Conformational studies of double-helical polynucleotides by the method of pair-wise potential functions. *Int. J. Quant. Chem.* **17**, 321–326.

808. S. B. Zimmerman (1977), Polynucleotide models. II. Prediction of the radial position and tilt of the bases from the helical parameters. *Biopolymers* **16**, 749–763.

809. C. K. Mitra, M. H. Sarma and R. H. Sarma (1981), Plasticity of the DNA double helix. *J. Amer. Chem. Soc.* **103**, 6727–6737.

810. R. E. Depew and J. C. Wang (1975), Conformational fluctuations of DNA helix. *Proc. Nat. Acad. Sci. USA* **72**, 4275–4279.

811. M. E. Hogan and O. Jardetzky (1979), Internal motions in DNA. *Proc. Nat. Acad. Sci. USA* **76**, 6341–6345.

812. D. J. Patel, A. Pardi and K. Itakura (1982), DNA conformation, dynamics, and interactions in solution. *Science* **216**, 581–590.

813. D. P. Millar, R. J. Robbins, and A. H. Zewail (1981), Time-resolved spectroscopy of macromolecules: Effect of helical structure on the torsional dynamics of DNA and RNA. *J. Chem. Phys.* **74**, 4200–4201.

814. W. K. Olson (1979), The flexible DNA double helix. I. Average dimensions and distribution functions. *Biopolymers* **18**, 1213–1233; II. Superhelix formation. *Biopolymers* **18**, 1235–1260.

815. D. Jolly and H. Eisenberg (1976), Photon correlation spectroscopy, total intensity light scattering with laser radiation, and hydrodynamic studies of a well fractionated DNA sample. *Biopolymers* **15**, 61–95.

816. C. Frontali, E. Dore, A. Ferranto, E. Gratton, A. Bettini, M. R. Pozzan, and E. Valdevit (1979), An absolute method for the determination of the persistence length of native DNA from electron micrographs. *Biopolymers* **18**, 1353–1373.

817. K. Akasaka, A. Yamada, and H. Hatano (1977), ^{31}P magnetic relaxation in polynucleotides. *Bull. Chem. Soc. Japan* **50**, 2858–2862.

818. R. Tewari, R. K. Nanda, and G. Govil (1974), Spatial configurations of single-stranded polynucleotides. Calculations of average dimensions and nmr coupling constants. *Biopolymers* **13**, 2015–2035.

819. G. G. Malenkov, A. V. Gagua, and V. P. Timofeev (1979), Influence of intermolecular interactions on the DNA conformation. *Int. J. Quant. Chem.* **16**, 655–668.

820. D. R. C. Priore and F. S. Allen (1979), Comparisons between oriented film and solution tertiary structure of various nucleic acids. *Biopolymers* **18**, 1809–1820.

821. W. A. Baase and W. C. Johnson, Jr. (1979), Circular dichroism and DNA secondary structure. *Nucl. Acids Res.* **6**, 797–814.

822. C. A. G. Haasnoot, J. H. J. den Hartog, J. F. M. de Rooij, J. H. van Boom, and C. Altona (1980), Loop structures in synthetic oligodeoxynucleotides. *Nucl. Acids Res.* **8**, 169–181.

823. C. K. Mitra, R. H. Sarma, C. Giessner-Prettre, and B. Pullman (1980), Solution structure of DNA: The method of nuclear magnetic resonance spectroscopy. *Int. J. Quant. Chem: Quant. Biol. Symp.* **7**, 39–66.

824. A. G. W. Leslie, S. Arnott, R. Chandrasekaran, and R. L. Ratliff (1980), Polymorphism of DNA double helices, *J. Mol. Biol.* **143**, 49–72.

825. S. Bram (1971), Secondary structure of DNA depends on base composition. *Nature New Biol.* **232**, 174–176.

826. S. Bram and P. Tougard (1972), Polymorphism of natural DNA. *Nature New Biol.* **239**, 128–131.

827. J. Pilet, J. Blicharski, and J. Brahms (1975), Conformations and structural transitions in polydeoxynucleotides. *Biochemistry* **14**, 1869–1876.

828. D. M. Gray and J. G. Gall (1974), The circular dichroism of three *Drosophila virilis* satellite DNA's. *J. Mol. Biol.* **85**, 665–679.

829. S. Arnott (1975), The sequence dependence of circular dichroism spectra of DNA duplexes. *Nucl. Acids Res.* **2**, 1493–1502.

830. E. Selsing and S. Arnott (1976), Conformations of A-T rich DNA's. *Nucl. Acids Res.* **3**, 2443–2450.

831. E. Selsing, A. G. W. Leslie, S. Arnott, J. G. Gall, D. M. Skinner, E. M. Southern, J. H. Spencer, and K. Harbers (1976), Conformations of satellite DNA's. *Nucl. Acids Res.* **3**, 2451–2458.

832. S. Premilat and G. Albiser (1975), X-ray diffraction study of three DNA fibers with different base composition. *J. Mol. Biol.* **99**, 27–36.

833. J. Pilet and J. Brahms (1972), Dependence of B-A conformational change in DNA on base composition. *Nature New Biol.* **236**, 99–100.

834. P. Anderson and W. Bauer (1978), Supercoiling in closed circular DNA: Dependence upon ion type and concentration. *Biochemistry* **17**, 594–601.

835. A. Chan, R. Kilkuskie, and S. Hanlon (1979), Correlations between the duplex winding angle and the circular dichroism spectrum of calf thymus DNA. *Biochemistry* **18**, 84–91.

836. M. Levitt (1978), How many base-pairs per turn does DNA have in solution and in chromatin? Some theoretical calculations. *Proc. Nat. Acad. Sci. USA* **75**, 640–644.

837. C. W. Carter, Jr., and J. Kraut (1974), A proposed model for interaction of polypeptides with RNA. *Proc. Nat. Acad. Sci. USA* **71**, 283–287.

838. S. H. Kim, J. L. Sussman, and G. M. Church (1975), A model for recognition scheme between double stranded DNA and proteins. In: *Structure and Conformation of Nucleic Acids and Protein–Nucleic Acid Interactions* M. Sundaralingam and S. T. Rao, Eds.), Univ. Park Press, Baltimore. pp. 571–575.

839. V. I. Ivanov, L. E. Minchenkova, A. K. Schyolkina, and A. I. Poletayev (1973), Different conformations of double stranded nucleic acids in solution as revealed by circular dichroism. *Biopolymers* **12**, 89–100.

839a. P. K. Weiner, R. Langridge, J. M. Blaney, R. Schaefer, and P. Kollman (1982), Electrostatic potential molecular surfaces. *Proc. Nat. Acad. Sci.* **79**, 3754–3758.

839b. S. Corbin, R. Lavery, and B. Pullman (1982), The molecular electrostatic potential and steric accessibility of double-helical A-RNA. *Biochim. Biophys. Acta* **698**, 86–92.

839c. R. Lavery, A. Pullman, and B. Pullman (1982), The electrostatic field of B-DNA. *Theoret. Chim. Acta* (Berl.) **62**, 93–106.

840. S. D. Dover (1977), Symmetry and packing in B-DNA. *J. Mol. Biol.* **110**, 699–700.

841. S. B. Zimmerman and B. H. Pheiffer (1979), Helical parameters of DNA do not change when DNA fibers are wetted: X-ray diffraction study. *Proc. Nat. Acad. Sci. USA* **76**, 2703–2707.

842. W. Fuller, M. H. F. Wilkins, H. R. Wilson, and L. D. Hamilton (1965), The molecular configuration of deoxyribonucleic acid. IV. X-ray diffraction study of the A-form. *J. Mol. Biol.* **12**, 60–80.

843. S. Arnott and D. W. L. Hukins (1972), Optimised parameters for A-DNA and B-DNA. *Biochem. Biophys. Res. Commun.* **47**, 1504–1510.

844. S. Arnott and E. Selsing (1974), The structure of polydeoxyguanylic acid·polydeoxycytidylic acid. *J. Mol. Biol.* **88**, 551–552.

845. S. Arnott, R. Chandrasekaran, D. W. L. Hukins, P. J. C. Smith, and L. Watts (1974), Structural details of a double-helix observed for DNA's containing alternating purine–pyrimidine sequences. *J. Mol. Biol.* **88**, 523–533.

845a. A. Mahendrasingam, N. J. Rhodes, D. C. Goodwin, C. Nave, W. J. Pigram, W. Fuller, J. Brahms, and J. Vergne (1983), Conformational transitions in oriented fibers of the synthetic polynucleotide poly[d(AT)]·poly [d(AT)] double helix. *Nature* **301**, 535–537.

846. D. R. Davies and R. L. Baldwin (1963), X-ray studies of two synthetic DNA copolymers. *J. Mol. Biol.* **6**, 251–255.

847. E. Selsing, S. Arnott, and R. L. Ratliff (1975), Conformations of poly d(A–T–T)·poly d(A–A–T). *J. Mol. Biol.* **98**, 243–248.

848. S. Arnott, D. W. L. Hukins, S. D. Dover, W. Fuller, and A. R. Hodgson (1973), Structures of synthetic polynucleotides in the A-RNA and A'-RNA conformations: X-ray diffraction analyses of the molecular conformations of polyadenylic acid·polyuridylic acid and polyinosinic acid·polycytidylic acid. *J. Mol. Biol.* **81,** 107–122.

849. E. J. O'Brien and A. W. MacEwan (1970), Molecular and crystal structure of polynucleotide complex: Polyinosinic acid plus polydeoxycytidylic acid. *J. Mol. Biol.* **48,** 243–261.

850. G. Milman, R. Langridge, and M. J. Chamberlin (1967), The structure of a DNA–RNA hybrid. *Proc. Nat. Acad. Sci. USA* **57,** 1804–1810.

851. M. J. Chamberlin and D. L. Patterson (1965), Physical and chemical characterization of the ordered complexes formed between polyinosinic acid, polycytidylic acid and their deoxyribo-analogues. *J. Mol. Biol.* **12,** 410–428.

852. S. Arnott, D. W. L. Hukins, and S. D. Dover (1972), Optimised parameters for RNA double-helices. *Biochem. Biophys. Res. Commun.* **48,** 1392–1399.

853. S. Arnott, P. J. Bond, E. Selsing, and P. J. C. Smith (1976), Models of triple-stranded polynucleotides with optimised stereochemistry. *Nucl. Acids Res.* **3,** 2459–2470.

854. S. Arnott, R. Chandrasekaran, A. W. Day, L. C. Puigjaner, and L. Watts (1981), Double helical structures for polyxanthylic acid. *J. Mol. Biol.* **149,** 489–505.

855. C. H. Chou, G. J. Thomas, Jr., S. Arnott, and P. J. Campbell Smith (1977), Raman spectral studies of nucleic acids. XVII. Conformational structures of polyinosinic acid. *Nucl. Acids Res.* **4,** 2407–2419.

856. S. Arnott and D. W. L. Hukins (1973), Refinement of the structure of B-DNA and implications for the analysis of X-ray diffraction data from fibers of biopolymers. *J. Mol. Biol.* **81,** 93–105.

857. S. Arnott and E. Selsing (1975), The conformation of C-DNA. *J. Mol. Biol.* **98,** 265–269.

858. M. A. Mokul'skii, K. A. Kapitanova, and T. D. Mokul'skaya (1972), Secondary structure of phage T2 DNA. *Mol. Biol. (Moscow)* **6,** 714–731.

859. S. Arnott and E. Selsing (1974), Structures of poly d(A)·poly d(T) and poly d(T)·poly d(A)·poly d(T). *J. Mol. Biol.* **88,** 509–521.

860. M. Laskowski (1972), The poly d(A–T) of crab. *Prog. Nucl. Acid Res. Mol. Biol.* **12,** 161–188.

861. J. G. Gall and D. D. Atherton (1974), Satellite DNA sequences in *Drosophila virilis. J. Mol. Biol.* **85,** 633–664.

862. C. Bostock (1980), A function for satellite DNA? *Trends Biochem. Sci.* **5,** 117–119.

863. M. Sundaralingam (1974), Principles governing nucleic acid and polynucleotide conformations. In: *Structure and Conformation of Nucleic Acids and Protein–Nucleic Acid Interactions* (M. Sundaralingam and S. T. Rao, Eds.), pp. 487–524. Univ. Park Press, Baltimore.

864. R. Nussinov and A. B. Jackson (1980), Fast algorithm for predicting the secondary structure of single-stranded RNA. *Proc. Nat. Acad. Sci. USA* **77,** 6309–6313.

865. M. Spencer and F. Poole (1965), On the origin of crystallizable RNA from yeast. *J. Mol. Biol.* **11,** 314–326.

866. V. Erdmann (1976), Structure and function of 5S and 5.8S RNA. *Prog. Nucl. Acid Res. Mol. Biol.* **18**, 45–90.

867. H. F. Noller and C. R. Woese (1981), Secondary structure of 16S ribosomal RNA. *Science* **212**, 403–411.

868. Zs. Schwarz and H. Kössel (1980), The primary structure of 16S rRNA from *Zea mays* chloroplast is homologous to *E. coli* 16S rRNA. *Nature* **283**, 739–742.

869. W. Fiers, R. Contreras, F. Duerinck, G. Haegeman, D. Iserentant, F. Merregaert, W. Min Jou, F. Molemans, A. Raeymaekers, A. Van den Berghe, G. Volckaert, and M. Ysebaert (1976), Complete nucleotide sequence of bacteriophage MS2 RNA: Primary and secondary structure of replicase gene. *Nature* **260**, 500–507.

870. F. Sanger (1971), Nucleotide sequences in bacteriophage ribonucleic acid. *Biochem. J.* **124**, 833–843.

871. N. J. Proudfoot and G. G. Brownlee (1974), Sequence at the 3′-end of globin mRNA shows homology with immunoglobulin light chain mRNA. *Nature* **252**, 359–362.

872. H. B. White III, B. E. Laux, and D. Dennis (1972), Messenger RNA structure: Compatibility of hairpin loops with protein sequences. *Science* **175**, 1264–1266.

873. T. Sato, Y. Kyogoku, S. Higuchi, Y. Mitsui, Y. Iitaka, M. Tsuboi, and K.-I. Miura (1966), A preliminary investigation on the nuclear structure of rice dwarf virus ribonucleic acid. *J. Mol. Biol.* **16**, 180–190.

874. R. Langridge and P. Gomatos (1963), The structure of RNA. *Science* **141**, 694–698.

875. S. Arnott, F. Hutchinson, M. Spencer, M. H. F. Wilkins, W. Fuller, and R. Langridge (1966), X-ray diffraction studies of double helical ribonucleic acid. *Nature* **211**, 227–232.

876. S. Arnott, M. H. F. Wilkins, W. Fuller, and R. Langridge (1967), Molecular and crystal structures of double-helical RNA. II. Determination and comparision of diffracted intensities for the α and β crystalline forms of reovirus RNA and their interpretation in terms of groups of three RNA molecules. *J. Mol. Biol.* **27**, 525–533; III. An 11-fold molecular model and comparison of the agreement between the observed and calculated three-dimensional diffraction data for 10- and 11-fold models. *J. Mol. Biol.* **27**, 535–548.

877. S. Arnott, M. H. F. Wilkins, W. Fuller, J. Venable, and R. Langridge (1967), Molecular and crystal structures of double-helical RNA. IV. Molecular packing in crystalline fibers. *J. Mol. Biol.* **27**, 549–562.

878. K.-I. Tomita and A. Rich (1964), X-Ray diffraction investigations of complementary RNA. *Nature* **201**, 1160–1163.

879. S. Arnott, W. Fuller, A. Hodgson, and I. Prutton (1968), Molecular conformations and structure transitions of RNA complementary helices and their possible biological significance. *Nature* **220**, 561–564.

880. R. D. Blake, J. Massoulié, and J. R. Fresco (1967), Polynucleotides VIII. A spectral approach to the equilibria between polyriboadenylate and polyribouridylate and their complexes. *J. Mol. Biol.* **30**, 291–308.

881. J. Massoulié (1968), Associations de poly(A) et poly(U) en milieu acide. Phénomènes irréversibles. *Eur. J. Biochem.* **3**, 439–447.

882. L. D. Hamilton, V. I. Babcock, and C. M. Southern (1969), Inhibition of herpes simplex virus by synthetic double-stranded RNA (polyriboadenylic

and polyribouridylic acids and polyriboinosinic and polyribocytidylic acids). *Proc. Nat. Acad. Sci. USA* **64**, 878–883.

883. A. K. Field, A. A. Tytell, G. P. Lampson, and M. R. Hillemann (1967), Inducers of interferon and host resistance. II. Multistranded synthetic polynucleotide complexes. *Proc. Nat. Acad. Sci. USA* **58**, 1004–1010.

884. M. Ikehara, M. Hattori, and T. Fukui (1972), Synthesis and properties of poly(2-methyladenylic acid). Formation of a poly(A)·poly(U) complex with Hoogsteen-type hydrogen bonding. *Eur. J. Biochem.* **31**, 329–334.

885. F. Ishikawa, J. Frazier, F. B. Howard, and H. T. Miles (1972), Polyadenylate·polyuridylate helices with non-Watson-Crick hydrogen bonding. *J. Mol. Biol.* **70**, 475–490.

886. T. Fukui and M. Ikehara (1979), Polynucleotides. XLVII. Synthesis and properties of poly(2-methylthio- and 1-ethylthioadenylic acid). Formation of non-Watson-Crick type complexes. *Biochim. Biophys. Acta* **562**, 527–533.

887. T. Hakoshima, T. Fukui, M. Ikehara, and K.-I. Tomita (1981), Molecular structure of a double helix with a non-Watson-Crick type base pairing formed by 2-substituted poly(A) and poly(U). *Proc. Nat. Acad. Sci. USA* **78**, 7309–7313.

888. C. T. Lu, K. I. Varughese and G. Kartha (1980), The crystal structure of ammonium adenylyl-3′,5′-uridine tetrahydrate. *Abstracts, Summer Meeting of American Crystallographic Association Series 2* **8**, 33.

889. B. Hingerty, E. Subramanian, S. D. Stellman, T. Sato, S. B. Broyde, and R. Langridge (1976), The crystal and molecular structure of a calcium salt of guanylyl-3′,5′-cytidine (GpC). *Acta Crystallogr. B* **32**, 2998–3013.

890. J. M. Rosenberg, N. C. Seeman, R. O. Day, and A. Rich (1976), RNA double helices generated from crystal structures of double helical dinucleoside phosphates. *Biochem. Biophys. Res. Commun.* **69**, 979–987.

891. N. C. Seeman, J. L. Sussman, H. M. Berman, and S.-H. Kim (1971), Nucleic acid conformation: Crystal structure of a naturally occurring dinucleoside phosphate (UpA). *Nature New Biol.* **233**, 90–92.

892. W. Fuller, M. H. F. Wilkins, H. R. Hamilton, and S. Arnott (1965), The molecular configuration of deoxyribonucleic acid. IV. X-Ray diffraction study of the A-form. *J. Mol. Biol.* **12**, 60–80.

892a. C. R. Calladine (1982), Mechanism of sequence-dependent stacking of bases in B-DNA. *J. Mol. Biol.* **161**, 343–352.

892b. R. E. Dickerson (1983), Base sequence and helix structure variation in B- and A-DNA. *J. Mol. Biol.*, **166**, 419–441.

893. B. N. Conner, T. Takano, S. Tanaka, K. Itakura, and R. E. Dickerson (1982), The molecular structure of d(ICpCpGpG), a fragment of right-handed double helical A-DNA. *Nature* 295, 294–299.

893a. A. H.-J. Wang, S. Fujii, J. H. van Boom and A. Rich (1982), Molecular structure of the octamer d(GGCCGGCC): modified A-DNA. *Proc. Nat. Acad. Sci. USA* **79**, 3968–3972.

894. Z. Shakked, D. Rabinovich, W. B. T. Cruse, E. Egert, O. Kennard, G. Sala, S. A. Salisbury, and M. A. Viswamitra (1981), Crystalline A-DNA: The X-ray analysis of the fragment d(G–G–T–A–T–A–C–C). *Proc. Roy. Soc. London Ser. B* **231**, 479–487.

894a. Z. Shakked and O. Kennard (1983), The A-form of DNA, in A. McPherson and F. Jurnak, Eds. *Structural Biology*, John Wiley & Sons, New York.

895. S. B. Zimmerman and B. H. Pheiffer (1979), A direct demonstration that

the ethanol-induced transition of DNA is between the A and B forms: an X-ray diffraction study. *J. Mol. Biol.* **135**, 1023–1027.

896. S. R. Holbrook, J. L. Sussman, and S.-H. Kim (1981), Absence of correlation between base-pair sequence and RNA conformation. *Science* **212**, 1275–1277.

897. J. A. Schellman (1974), Flexibility of DNA. *Biopolymers* **13**, 217–226.

898. D. P. Millar, R. J. Robbins, and A. H. Zewail (1980), Direct observation of the torsional dynamics of DNA and RNA by picosecond spectroscopy. *Proc. Nat. Acad. Sci. USA* **77**, 5593–5597.

899. R. E. Dickerson and H. R. Drew (1981), Structure of a B-DNA dodecamer. II. Influence of base sequence on helix structure. *J. Mol. Biol.* **149**, 761–786.

899a.H. R. Drew, R. M. Wing, T. Takano, C. Broka, S. Tanaka, K. Itakura and R. E. Dickerson (1981), Structure of a B-DNA dodecamer: conformation and dynamics. *Proc. Nat. Acad. Sci. (USA)* **78**, 2179–2183.

899b.W. Kabsch, C. Sander and E. N. Trifonov (1982), The ten helical twist angles of B-DNA. *Nucl. Acids Res.* **10**, 1097–1104.

900. A. Klug, A. Jack, M. A. Viswamitra, O. Kennard, Z. Shakked, and T. A. Steitz (1979), A hypothesis on a specific sequence-dependent conformation of DNA and its relation to the binding of the *lac*-repressor protein. *J. Mol. Biol.* **131**, 669–680.

900a.M. A. Viswamitra, Z. Shakked, P. G. Jones, G. M. Sheldrick, S. A. Salisbury and O. Kennard (1982), Structure of the deoxytetranucleotide d-pApT-pApT and a sequence-dependent model for poly(dA–dT). *Biopolymers* **21**, 513–533.

901. R. V. Hosur, G. Govil, M. V. Hosur, and M. A. Viswamitra (1981), Sequence effects in structures of the dinucleotides d-pApT and d-pTpA. *J. Mol. Struct.* **72**, 261–267.

902. H. R. Wilson and J. Al-Mukhtar (1976), Structure of thymidylyl-3′,5′-deoxyadenosine. *Nature* **263**, 171–172.

902a.M. M. Radwan and H. R. Wilson (1982), Fiber and molecular structure of thymidylyl-3′,5′-deoxyadenosine. *Int. J. Biol. Macromol.* **4**, 145–149.

903. H. Shindo and S. B. Zimmerman (1980), Sequence-dependent variations in the backbone geometry of a synthetic DNA fibre. *Nature* **283**, 690–691.

904. H. Shindo, R. T. Simpson, and J. S. Cohen (1979), An alternating conformation characterizes the phosphodiester backbone of poly d(A–T) in solution. *J. Biol. Chem.* **254**, 8125–8128.

905. D. J. Patel, L. L. Canuel and F. M. Pohl (1979), "Alternating B-DNA" conformation for the oligo (dG-dC) duplex in high-salt solution. *Proc. Nat. Acad. Sci. USA* **76**, 2508–2511.

906. I. E. Scheffer, E. L. Elson, and R. L. Baldwin (1968), Helix formation by d(A–T) oligomers. I. Hairpin and straight-chain helices. *J. Mol. Biol.* **36**, 291–304.

907. L. C. Lutter (1977), Deoxyribonuclease I produces staggered cuts in the DNA of chromatin. *J. Mol. Biol.* **117**, 53–69.

908. A. D. Riggs, S.-Y. Lin, and R. D. Wells (1972), *lac* repressor binding to synthetic DNAs of defined nucleotide sequence. *Proc. Nat. Acad. Sci. USA* **69**, 761–764.

909. L. J. Peck and J. C. Wang (1981), Sequence dependence of the helical repeat of DNA in solution. *Nature* **292**, 375–378.

910. D. Rhodes and A. Klug (1981), Sequence-dependent helical periodicity of DNA. *Nature* **292**, 378–380.

910a.F. Strauss, C. Gaillard and A. Prunell (1981), Helical periodicity of DNA, poly(dA)·poly(dT) and poly(dA–dT)·poly(dA–dT) in solution. *Eur. J. Biochem.* **118**, 215–222.

911. J. Brahms, J. Pilet, T.-T. P. Lan, and L. R. Hill (1973), Direct evidence of the C-like form of sodium deoxyribonucleate. *Proc. Nat. Acad. Sci. USA* **70**, 3352–3355.

911a.N. J. Rhodes, A. Mahendrasingam, W. J. Pigram, W. Fuller, J. Brahms, J. Vergne and R. A. J. Warren (1982), The C conformation is a low salt form of sodium DNA. *Nature* **296**, 267–269.

912. S. B. Zimmerman and B. H. Pheiffer (1981), A RNA·DNA hybrid that can adopt two conformations: An X-ray diffraction study of poly(A)·poly(dT) in concentrated solution or in fibers. *Proc. Nat. Acad. Sci. USA* **78**, 78–82.

913. S. Arnott and P. J. Bond (1973), Structures for poly(U)·poly(A)·poly(U) triple stranded helices. *Nature New Biol.* **244**, 99–101.

913a.A. H.-J. Wang, S. Fujii, J. H. van Boom, G. A. van der Marel, S. A. A. van Boeckel, and A. Rich (1982), Molecular structure of r(GCG)d(TATACGC): A DNA-RNA hybrid helix joined to double helical DNA. *Nature* **299**, 601–604.

914. F. M. Pohl and T. M. Jovin (1972), Salt-induced co-operative conformational change of a synthetic DNA: Equilibrium and kinetic studies with poly d(G–C). *J. Mol. Biol.* **67**, 375–396.

915. F. M. Pohl (1976), Polymorphism of a synthetic DNA in solution. *Nature* **260**, 365–366.

916. D. R. Davies and S. B. Zimmerman (1980), A new twist for DNA? *Nature* **283**, 11–12.

917. A. H.-J. Wang, G. J. Quigley, F. J. Kolpak, G. van der Marel, J. H. van Boom, and A. Rich (1980), Left-handed double helical DNA: Variations in the backbone conformation. *Science* **211**, 171–176.

918. H. R. Drew, R. E. Dickerson, and K. Itakura (1978), A salt-induced conformational change in crystals of the synthetic DNA tetramer d(CpGpCpG). *J. Mol. Biol.* **125**, 535–543.

919. J. L. Crawford, F. J. Kolpak, A. H.-J. Wang, G. J. Quigley, J. H. van Boom, G. van der Marel, and A. Rich (1980), The tetramer d(CpGpCpG) crystallizes as a left-handed double helix. *Proc. Nat. Acad. Sci. USA* **77**, 4016–4020.

920. H. R. Drew and R. E. Dickerson (1981), Conformation and dynamics in a Z'-DNA tetramer. *J. Mol. Biol.* **152**, 723–736.

921. C. K. Mitra, M. H. Sarma, and R. H. Sarma (1981), Left-handed deoxyribonucleic acid double helix in solution. *Biochemistry* **20**, 2036–2041.

922. M. Behe, S. B. Zimmerman, and G. Felsenfeld (1981), Changes in the helical repeat of poly(dG–m⁵dC)·poly(dG–m⁵dC) and poly(dG–dC)·poly(dG–dC) associated with the B–Z transition. *Nature* **293**, 233–235.

923. J. M. Neumann, W. Guschlbauer, and S. Tran-Dinh (1979), Conformation and flexibility of GpC and CpG in neutral aqueous solutions using ¹H nuclear-magnetic-resonance and spin-lattice-relaxation time measurements. *Eur. J. Biochem.* **100**, 141–148.

924. W. Saenger, H. Landmann, and A. G. Lezius (1973), X-Ray diffraction studies on fibers of poly d(A–s⁴T)·poly d(A–s⁴T). *Jerus. Sympos. Quant. Chem. Biochem.* **5**, 457–467.

924a. J. H. van de Sande and T. M. Jovin (1982), Z*DNA, the left-handed helical form of poly[d(G–C)] in $MgCl_2$-ethanol, is biologically active. *EMBO J.* **1**, 115–120.

925. R. M. Santella, D. Grunberger, I. B. Weinstein, and A. Rich (1981), Induction of the Z conformation in poly(dG–dC)·poly(dG–dC) by binding of N-2-acetylaminofluorene to guanine residues. *Proc. Nat. Acad. Sci. USA* **78**, 1451–1455.

925a. J. Nickol, M. Behe and G. Felsenfeld (1982), Effect of the B-Z transition in poly(dG–m⁵dC)·poly(dG–m⁵dC) on nucleosome formation. *Proc. Nat. Acad. Sci. (USA)* **79**, 1771–1775.

926. A Nordheim, M. L. Pardue, E. M. Lafer, A. Möller, D. Stollar, and A. Rich (1981), Antibodies to left-handed Z-DNA bind to interband regions of *Drosophila* polytene chromosomes. *Nature* **294**, 417–422.

927. M. Behe and G. Felsenfeld (1981), Effects of methylation on a synthetic polynucleotide: The B–Z transition in poly(dG–m⁵C)·poly(dG–m⁵C). *Proc. Nat. Acad. Sci. USA* **78**, 1619–1623.

928. A. Razin and A. D. Riggs (1980), DNA methylation and gene function. *Science* **210**, 604–610.

929. J. Klysik, S. M. Stirdivant, J. E. Larson, P. A. Hart, and R. D. Wells (1981), Left-handed DNA in restriction fragments and a recombinant plasmid. *Nature* **290**, 672–677.

929a. C. K. Singleton, K. Klysik, S. M. Stirdivant, and R. D. Wells (1982), Left-handed Z-DNA is induced by supercoiling in physiological ionic conditions. *Nature* **299**, 312–316.

930. B. Prescott, R. Gamache, J. Livramento, and G. J. Thomas, Jr. (1974), Raman studies of nucleic acids. XII. Conformations of oligonucleotides and deuterated polynucleotides. *Biopolymers* **13**, 1821–1845.

931. M. Leng and G. Felsenfeld (1966), A study of polyadenylic acid at neutral pH. *J. Mol. Biol.* **15**, 455–466.

932. H. Eisenberg and G. Felsenfeld (1967), Studies on the temperature-dependent conformation and phase separation of polyriboadenylic acid solutions at neutral pH. *J. Mol. Biol.* **30**, 17–37.

933. E. K. Achter and G. Felsenfeld (1971), The conformation of single-strand polynucleotides in solution: Sedimentation studies of apurinic acid. *Biopolymers* **10**, 1625–1634.

934. M. Edmonds and M. A. Winters (1976), Polyadenylate polymerases. *Prog. Nucl. Acid Res. Mol. Biol.* **17**, 149–179.

935. P. Ahlquist and P. Kaesberg (1978), Determination of the length distribution of poly(A) at the 3′-terminus of the virion RNAs of EMC virus, poliovirus, rhinovirus, RAV-61 and CPMV and of mouse globin mRNA. *Nucl. Acids Res.* **7**, 1195–1204.

936. E. O. Akinrimisi, C. Sander, and P. O. P. Ts'o (1963), Properties of helical polycytidylic acid. *Biochemistry* **2**, 340–344.

937. G. D. Fasman, C. Lindblow, and L. Grossman (1964), The helical conformation of polycytidylic acid: Studies on the forces involved. *Biochemistry* **3**, 1015–1021.

938. A. Adler, L. Grossman, and G. D. Fasman (1967), Single-stranded oli-

gomers and polymers of cytidylic and 2'-deoxycytidylic acids: comparative optical rotatory studies. *Proc. Nat. Acad. Sci. USA* **57**, 423–430.

939. W. Guschlbauer (1967), Protonated polynucleotide structures. I. The thermal denaturation of polycytidylic acid in solution. *Proc. Nat. Acad. Sci. USA* **57**, 1441–1448.

940. H. R. Mahler and G. Green (1970), Comparative study of polyribonucleotides in aqueous and glycol solutions. *Biochemistry* **9**, 368–387.

941. A. Gulik, H. Inoue, and V. Luzzati (1970), Conformation of single-stranded polynucleotides: Small-angle X-ray scattering and spectroscopic study of polyribocytidylic acid in water and in water–alcohol solutions. *J. Mol. Biol.* **53**, 221–238.

942. J. T. Yang and T. Samejima (1969), Optical rotatory dispersion and circular dichroism of nucleic acids. *Prog. Nucl. Acid Res. Mol. Biol.* **9**, 223–300.

943. B. Borah and J. L. Wood (1976), The cytidinium–cytidine complex: Infrared and Raman spectroscopic studies. *J. Mol. Struct.* **30**, 13–30.

944. R. E. Marsh, R. Bierstedt, and E. L. Eichhorn (1962), The crystal structure of cytosine-5-acetic acid. *Acta. Crystallogr.* **15**, 310–316.

945. T. J. Kistenmacher, M. Rossi, and L. G. Marzilli (1978), A model for the interrelationship between asymmetric interbase hydrogen bonding and base–base stacking in hemiprotonated polyribocytidylic acid: Crystal structure of 1-methylcytosine hemihydroiodide hemihydrate. *Biopolymers* **17**, 2581–2585.

946. J. Alderfer, I. Tazawa, S. Tazawa, and P. O. P. Ts'o (1975), Comparative studies on polydeoxyribocytidylate (dC), polyribocytidylate (rC) and poly-2'-O-methylribocytidylate (mC) and polydeoxyriboinosinate (dI), polyriboinosinate (rI) and poly-2'-O-methylriboinosinate (mI). *Biophys. J.* **15**, 29a.

947. A. G. W. Leslie and S. Arnott (1978), Structure of single-stranded polyribonucleotide poly(2'-O-methylcytidylic acid). *J. Mol. Biol.* **119**, 399–414.

948. S. D. Stellman, S. B. Broyde, and M. Wartell (1976), Influence of ribose-2'-O-methylation on GpC conformation by classical potential energy calculations. *Biopolymers* **15**, 1951–1964.

949. S. D. Stellman, B. Hingerty, S. Broyde, and R. Langridge (1975), Conformation of a rare nucleoside in the anticodon loop of tRNAs: Potential energy calculations for 2'-O-methyl cytidine. *Biopolymers* **14**, 2049–2060.

950. B. Hingerty, P. J. Bond, R. Langridge, and F. Rottman (1974), Conformation of 2'-O-methyl cytidine, a modified furanose component of ribonucleic acids. *Biochem. Biophys. Res. Commun.* **61**, 875–881.

950a. M. S. Broido and D. R. Kearns (1982), ^{1}H NMR evidence for a left-handed helical structure of poly(ribocytidylic acid) in neutral solution. *J. Amer. Chem. Soc.* **104**, 5207–5216.

951. W. M. Scovell (1978), Structural and conformational studies of polyribonucleic acid in neutral and acid solution. *Biopolymers* **17**, 969–984.

952. J. Brahms, A. M. Michelson, and K. E. Van Holde (1966), Adenylate oligomers in single- and double-stranded conformation. *J. Mol. Biol.* **15**, 467–488.

953. J. Witz and V. Luzzati (1965), La structure des acides polyadénylique et polyuridylique en solution: étude par diffusion centrale des rayons X. *J. Mol. Biol.* **11**, 620–630.

954. D. N. Holcomb and I. Tinoco, Jr. (1965), Conformation of polyriboadenylic acid: pH and temperature dependence. *Biopolymers* **3**, 121–133.

955. P. Thiyagarajan and P. K. Ponnuswamy (1978), Conformational characteristics of the RNA subunit ApA from energy minimization studies. *Biopolymers* **17**, 2143–2158.

956. W. K. Olson (1975), The spatial configuration of ordered polynucleotide chains. II. The poly(rA) helix. *Nucl. Acids Res.* **2**, 2055–2068.

957. W. Saenger, J. Riecke, and D. Suck (1975), A structural model for the polyadenylic acid single helix. *J. Mol. Biol.* **93**, 529–534.

958. C. S. M. Olsthoorn, L. J. Bostelaar, J. H. van Boom, and C. Altona (1980), Conformational characteristics of the trinucleoside diphosphate dApdApdA and its constituents from nuclear magnetic resonance and circular dichroism studies. *Eur. J. Biochem.* **112**, 95–110.

959. Z. E. Kahana and B. F. Erlanger (1980), Immunochemical study of the structure of poly(adenylic acid). *Biochemistry* **19**, 320–324.

960. J. C. Pinder, D. Z. Staynov, and W. B. Gratzer (1974), Properties of RNA in formamide. *Biochemistry* **13**, 5367–5373.

961. J. H. Strauss, Jr., R. B. Kelly, and R. L. Sinsheimer (1968), Denaturation of RNA with dimethyl sulfoxide. *Biopolymers* **6**, 793–807.

962. R. S. Morgan and R. Byrne (1959), "Alkaline" polyadenylic acid. *J. Mol. Biol.* **1**, 188–189.

963. S. B. Zimmerman, D. R. Davies, and M. A. Navia (1977), An ordered single-stranded structure for polyadenylic acid in denaturing solvents. An X-ray fiber diffraction and model building study. *J. Mol. Biol.* **116**, 317–330.

964. S.-H. Kim and J. L. Sussman (1976), π-Turn is a conformational pattern in RNA loops and bends. *Nature* **260**, 645–646.

965. J. R. Fresco (1959), Polynucleotides. II. The X-ray diffraction patterns of solutions of the randomly coiled and helical forms of polyriboadenylic acid. *J. Mol. Biol.* **1**, 106–110.

966. S. B. Zimmerman and N. F. Coleman (1972), An ordered precipitate of polyadenylic acid formed by freezing at acidic pH: Comparison of X-ray diffraction and other properties of the precipitate with those of fibers or direct acid-precipitates. *Biopolymers* **11**, 1943–1960.

967. J. T. Finch and A. Klug (1969), Two double helical forms of polyriboadenylic acid and the pH-dependent transition between them. *J. Mol. Biol.* **46**, 597–598.

967a. D. B. Lerner and D. R. Kearns (1981), Proton and phosphorus NMR investigation of the conformational states of acid polyadenylic double helix. *Biopolymers* **20**, 803–816.

968. D. Suck, P. C. Manor, and W. Saenger (1976), The structure of a trinucleoside diphosphate: Adenylyl-(3′,5′)-adenylyl-(3′,5′)-adenosine hexahydrate. *Acta Crystallogr. B* **32**, 1727–1737.

969. M. J. Lowe and J. A. Schellman (1972), Solvent effects on dinucleotide conformation. *J. Mol. Biol.* **65**, 91–109.

970. D. J. Wood, F. E. Hruska, and K. K. Ogilvie (1974), Proton magnetic resonance studies of 2′-deoxythymidine, its 3′- and 5′-monophosphates and 2′-deoxythymidylyl-(3′,5′)-2′-deoxythymidine in aqueous solution. *Can. J. Chem.* **52**, 3353–3366.

971. M. N. Lipsett (1960), Evidence for helical structure in polyuridylic acid. *Proc. Nat. Acad. Sci. USA* **46**, 445–446.

972. J. C. Thrierr, M. Dourlent, and M. Leng (1971), A study of polyuridylic acid. *J. Mol. Biol.* **58**, 815–830.

973. D. B. Millar and M. Mackenzie (1970), The properties of the helix and coil forms of polyribouridylic acid and its halogenated analogs. *Biochim. Biophys. Acta* **204**, 82–90.

974. S. B. Zimmerman (1976), The polyuridylic acid complex with polyamines: An X-ray fiber diffraction observation. *J. Mol. Biol.* **101**, 563–565.

975. W. Saenger, S. K. Mazumdar, D. Suck, and P. C. Manor (1974), Parallel and antiparallel homopolymer nucleic acid double helices. In: *Structure and Conformation of Nucleic Acids and Protein–Nucleic Acid Interactions* (M. Sundaralingam and S. T. Rao, Eds.), pp. 537–555. Univ. Park Press, Baltimore.

976. S. Arnott, R. Chandrasekaran, A. G. W. Leslie, L. C. Puigjaner, and W. Saenger (1981), Structure of poly(2-thiouridylic acid) duplex. *J. Mol. Biol.* **149**, 507–520.

977. W. Szer (1966), Interaction of polyribothymidylic acid with metal ions and aliphatic amines. *Acta Biochim. Polon.* **3**, 251–266.

978. W. Guschlbauer and J. F. Chantot (1976), Guanosine revisited. A short review of gel formation by guanosine and its derivatives. In: *Proceedings of the Conference of Synthesis, Structure and Chemistry of the tRNAs and Their Components* (M. Wiewiorowski, Ed.), pp. 96–114. Polish Acad. of Science, Dymaczewo near Poznan.

979. T. J. Pinnavaia, H. T. Miles, and E. D. Becker (1975), Self-assembled 5′-guanosine monophosphate. Nuclear magnetic resonance evidence for a regular, ordered structure and slow chemical exchange. *J. Amer. Chem. Soc.* **97**, 7198–7200.

980. W. Muarayama, N. Nagashima, and Y. Shimizu (1969), The crystal and molecular structure of guanosine-5′-phosphate trihydrate. *Acta Crystallogr. B* **25**, 2236–2245.

981. U. Thewalt, C. E. Bugg, and R. E. Marsh (1970), The crystal structure of guanosine dihydrate and inosine dihydrate. *Acta Crystallogr. B* **26**, 1089–1101.

982. M. Gellert, M. N. Lipsett, and D. R. Davies (1962), Helix formation by guanylic acid. *Proc. Nat. Acad. Sci. USA* **48**, 2013–2018.

983. V. Sasisekharan, S. B. Zimmerman, and D. R. Davies (1975), The structure of helical 5′-guanosine monophosphate. *J. Mol. Biol.* **92**, 171–179.

984. P. Tougard, J.-F. Chantot and W. Guschlbauer (1973), Nucleoside conformations. X. An X-ray fiber diffraction study of the gels of guanosine nucleosides. *Biochim. Biophys. Acta.* **308**, 9–16.

985. S. B. Zimmerman (1976), X-ray study by fiber diffraction methods of a self-aggregate of guanosine-5′-phosphate with the same helical parameters as poly r(G). *J. Mol. Biol.* **106**, 663–672.

986. J. F. Chantot and W. Guschlbauer (1972), Mechanism of gel formation by guanine nucleosides. *Jerus. Sympos. Quant. Chem. Biochem.* **4**, 205–214.

987. M. Borzo, C. Delletier, P. Laszlo, and A. Paris (1980), ^{1}H, ^{23}Na, and ^{31}P NMR studies of self-assembly of the 5′-guanosine monophosphate dianion

in neutral aqueous solution in the presence of sodium cations. *J. Amer. Chem. Soc.* **102**, 1124–1134.

988. T. J. Pinnavaia, C. L. Marshall, C. M. Mettler, C. L. Fisk, H. T. Miles, and E. D. Becker (1978), Alkali metal ion specificity in the solution ordering of a nucleotide, 5′-guanosine monophosphate. *J. Amer. Chem. Soc.* **100**, 3625–3627.

989. R. Hilgenfeld and W. Saenger (1981), Structural studies of ionophores and their complexes with cations. *Top. Curr. Chem.* **101**, 1–82.

990. D. Thiele and W. Guschlbauer (1973), The structures of polyinosinic acid. *Biophysik* **9**, 261–277.

991. P. K. Sarkar and J. T. Yang (1965), Optical activity of 5′-guanylic acid gel. *Biochem. Biophys. Res. Commun.* **20**, 346–351.

992. S. B. Zimmerman, G. H. Cohen, and D. R. Davies (1975), X-Ray fiber diffraction and model-building study of polyguanylic acid and polyinosinic acid. *J. Mol. Biol.* **92**, 181–192.

993. S. Arnott, R. Chandrasekaran, and C. M. Marttila (1974), Structures for polyinosinic acid and polyguanylic acid. *Biochem. J.* **141**, 537–543.

994. A. Rich (1958), The molecular structure of polyinosinic acid. *Biochim. Biophys. Acta* **29**, 502–509.

995. C. Souleil and J. Panijel (1968), Immunochemistry of polyribonucleotides. Study of polyriboinosinic and polyriboguanylic acids. *Biochemistry* **7**, 7–13.

996. F. B. Howard, J. Frazier, and H. T. Miles (1977), Stable and metastable forms of poly(G). *Biopolymers* **16**, 791–809.

997. S. B. Zimmerman (1975), An "acid" structure for polyriboguanylic acid observed by X-ray diffraction. *Biopolymers* **14**, 889–890.

998. A. Dugaiczyk, D. L. Robberson, and A. Ullrich (1980), Single-stranded double- and triple-stranded structures. *Biochemistry* **19**, 5869–5873.

999. M. Meselson and F. W. Stahl (1958), The replication of DNA in *Escherichia coli*. *Proc. Nat. Acad. Sci. USA* **44**, 671–682.

1000. W. F. Pohl land G. W. Roberts (1978), Topological considerations in the theory of replication of DNA. *J. Math. Biol.* **6**, 383–402.

1001. L. M. Fisher (1981), DNA supercoiling by DNA gyrase. *Nature* **294**, 607–610.

1002. R. H. T. Bates, R. M. Lewitt, C. H. Rowe, J. P. Day, and G. A. Rodley (1977), On the structure of DNA. *J. Roy. Soc. N.Z.* **7**, 273–301.

1003. V. Sasisekharan and N. Pattabiraman (1976), Double stranded polynucleotides: two typical alternative conformations for nucleic acids. *Curr. Sci.* **45**, 779–783.

1004. V. Sasisekharan and Goutam Gupta (1980), On the alternative structure of DNA: Role of *syn* conformation of the bases. *Curr. Sci.* **49**, 43–48.

1004a. F. Górski (1980), A comparison of the Watson–Crick DNA structural model with the new "side by side" model. *Folia Biol.* (*Krakow*) **28**, 211–224.

1005. G. A. Rodley, R. H. T. Bates, and S. Arnott (1980), Is DNA really a double helix? *Trends Biochem. Sci.* **5**, 231–234.

1006. U. H. Stettler, H. Weber, Th. Koller, and Ch. Weissmann (1979), Preparation and characterization of form V DNA, the duplex DNA resulting from association of complementary, circular single-stranded DNA. *J. Mol. Biol.* **131**, 21–40.

1007. F. H. C. Crick, J. C. Wang, and W. R. Bauer (1979), Is DNA really a double helix? *J. Mol. Biol.* **129**, 446–461.

1008. M. Noll (1974), Internal structure of the chromatin subunit. *Nucl. Acids Res.* **1**, 1573–1578.
1009. F. H. C. Crick and A. Klug (1975), Kinky helix. *Nature* **255**, 530–533.
1010. G. R. Pack, M. A. Muskavitch, and G. Loew (1977), Kinked DNA. Energetics and conditions favoring its formation. *Biochim. Biophys. Acta* **478**, 9–22.
1011. D. Vasilescu, H. Broch, and D. Cabrol (1978), Kinked helices in nucleic acids—A molecular orbital investigation. *Int. J. Quant. Chem. Quant. Biol. Sympos.* **5**, 345–354.
1012. H. M. Sobell, C.-C. Tsai, S. G. Gilbert, S. C. Jain, and T. D. Sakore (1976), Organization of DNA in chromatin. *Proc. Nat. Acad. Sci. USA* **73**, 3068–3072.
1013. H. M. Sobell, C.-C. Tsai, S. C. Jain, and S. G. Gilbert (1977), Visualization of drug–nucleic acid interactions at atomic resolution. III. Unifying structural concepts in understanding drug–DNA interactions and their broader implications in understanding protein–DNA interactions. *J. Mol. Biol.* **114**, 333–365.
1014. E. D. Lozansky, H. M. Sobell and M. Lessen (1979), Does DNA have two structures in solution that coexist at equilibrium? In: *Stereodynamics of Molecular Systems* (R. H. Sarma, Ed.), pp. 265–270. Pergamon Press, New York.
1014a. S. W. Englander, N. R. Kallenbach, A. J. Heeger, J. A. Krumhanse, and S. Litwin (1980). Nature of the open state in long polynucleotide double helices: Possibility of solution excitations. *Proc. Nat. Acad. Sci. (USA)*, **77**, 7222–7226.
1015. E. Selsing, R. D. Wells, C. J. Alden, and S. Arnott (1979), Bent DNA: Visualization of a base-paired and stacked A–B conformational junction. *J. Biol. Chem.* **254**, 5417–5422.
1016. E. Selsing and R. D. Wells (1979), Polynucleotide block polymers consisting of a DNA·RNA hybrid joined to a DNA·DNA duplex. *J. Biol. Chem.* **254**, 5410–5416.
1016a. J. W. Roberts (1976), Transcription termination and its control in E. coli. In R. Losick and M. Chamberlin, eds. RNA Polymerase. Cold Spring Harbor Laboratory, New York, p. 247–272.
1016b. E. Selsing, R. D. Wells, T. A. Early and D. R. Kearns (1978). Two contiguous conformations in a nucleic acid complex. *Nature* **275**, 249–250.
1017. J. C. Wang, J. H. Jacobsen, and J.-M. Saucier (1977), Physicochemical studies on interactions between DNA and RNA polymerease. Unwinding of the DNA helix by *Escherichia coli* RNA polymerase. *Nucl. Acids Res.* **4**, 1225–1241.
1018. G. E. F. Scherer, M. D. Walkinshaw, and S. Arnott (1978), A computer aided oligonucleotide analysis provides a model sequence for RNA polymerase promoter region in *E. coli*. *Nucl. Acids Res.* **5**, 3759–3773.
1019. S. B. Broyde, S. D. Stellman, and R. M. Wartell (1975), The A and B conformations of DNA and RNA subunits. Potential energy calculations for dGpdC. *Biopolymers* **14**, 2625–2637.
1020. S. Fujii and K.-I. Tomita (1976), Conformational analysis of polynucleotides. I. The favorable left-handed helical model for the poly(8,2′-S-cycloadenylic acid) with *high-anti* conformation. *Nucl. Acids Res.* **3**, 1973–1984.

1021. W. K. Olson and R. D. Dasika (1976), Spatial configuration of ordered polynucleotide chains. 3. Polycyclonucleotides. *J. Amer. Chem. Soc.* **98,** 5371–5380.

1022. W. K. Olson (1977), Spatial configuration of ordered polynucleotide chains: A novel double helix. *Proc. Nat. Acad. Sci. USA* **74,** 1775–1779.

1023. S. Uesugi, T. Tezuka, and M. Ikehara (1976), Polynucleotides. XXX. Synthesis and properties of oligonucleotides of cyclouridine phosphate. Hybridization with the oligomer of S-cycloadenosine phosphate to form left-handed helical complexes. *J. Amer. Chem. Soc.* **98,** 969–973.

1024. M. Ikehara and T. Tezuka (1973), Synthesis of cyclouridine oligonucleotides forming a double stranded complex of left-handedness with cycloadenosine oligonucleotides. *J. Amer. Chem. Soc.* **95,** 4054–4056.

1025. R. H. Sarma, M. M. Dhingra, and R. J. Feldmann (1979), The diverse spatial configurations of DNA. Evidence for a vertically stabilized double helix. In: *Stereodynamics of Molecular Systems* (R. H. Sarma, Ed.), pp. 251–264. Pergamon Press, New York.

1026. C. K. Mitra, M. M. Dhingra, and R. H. Sarma (1980), Experimental support for a right-handed vertical double helix. *Biopolymers* **19,** 1435–1450.

1027. D. Söll, J. N. Abelson, and P. R. Schimmel (Eds.) (1980), *Transfer RNA: Structure, Properties, and Recognition.* Cold Spring Harbor Monograph Series 9A.

1028. S. Altmann (Ed.) (1978), *Transfer RNA.* MIT Press, Cambridge.

1029. J. P. Goddard (1977), The structures and functions of transfer RNA. *Prog. Biophys. Mol. Biol.* **32,** 233–308.

1030. S.-H. Kim (1976), Three-dimensional structure of transfer RNA. *Prog. Nucl. Acid Res. Mol. Biol.* **17,** 181–216.

1031. S.-H. Kim (1978), Three-dimensional structure of transfer RNA and its functional implications. *Adv. Enzymol.* **46,** 279–315.

1032. S.-H. Kim (1981). Transfer RNA: Crystal structures. In: *Topics in Molecular and Structural Biology* (S. Neidle and W. Fuller, Eds.), Vol. 1, pp. 83–112.

1033. A. Rich and S.-H. Kim (1978), The three-dimensional structure of transfer RNA. *Sci. Amer.* **238,** 52–62.

1034. A. Rich and U. L. RajBhandary (1976), Transfer RNA: Molecular structure, sequence, and properties. *Annu. Rev. Biochem.* **45,** 805–860.

1035. B. F. C. Clark (1977), Correlation of biological activities with structural features of transfer RNA. *Prog. Nucl. Acid Res. Mol. Biol.* **20,** 1–19.

1036. P. B. Sigler (1973), An analysis of the structure of tRNA. *Annu. Rev. Biophys. Bioeng.* **4,** 477–527.

1037. R. P. Singhal and P. M. Fallis (1979), Structure, function, and evolution of transfer RNAs (with appendix giving complete sequences of 178 tRNAs). *Prog. Nucl. Acid Res. Mol. Biol.* **23,** 227–290.

1038. A. A. Baev and T. V. Venkstern (1977), Transfer ribonucleic acids. Structuro-functional aspects. *Mol. Biol. (Moscow)* **11,** 933–944.

1039. P. R. Schimmel and A. G. Redfield (1980), Transfer RNA in solution: Selected topics. *Annu. Rev. Biophys. Bioeng.* **9,** 181–221.

1040. D. J. Patel (1978), High resolution NMR studies of the structure and dynamics of tRNA in solution. *Annu. Rev. Phys. Chem.* **29,** 337–362.

1041. M. Sprinzl and D. H. Gauss (1982), Compilation of tRNA sequences. *Nucl. Acids Res.* **10,** r1–r55.

1042. J. E. Celis (1979), Collection of mutant tRNA sequences. *Nucl. Acids Res.* **6,** r21–r27.

1043. S.-H. Kim, G. J. Quigley, F. L. Suddath, A. McPherson, D. Sneden, J. J. Kim, J. Weinzierl, and A. Rich (1973), Three-dimensional structure of yeast phenylalanine transfer RNA: Folding of the polynucleotide chain. *Science* **179,** 285–288.

1044. F. Cramer (1971), Three-dimensional structure of tRNA. *Prog. Nucl. Acid Res. Mol. Biol.* **11,** 391–421.

1045. M. Levitt (1969), Detailed molecular model for transfer ribonucleic acid. *Nature* **224,** 759–763.

1046. J. L. Sussman, S. R. Holbrook, R. Wade Warrant, G. M. Church, and S.-H. Kim (1978), Crystal structure of yeast phenylalanine tRNA. I. Crystallographic refinement. *J. Mol. Biol.* **123,** 607–630.

1047. B. Hingerty, R. S. Brown, and A. Jack (1978), Further refinement of the structure of yeast tRNA^Phe. *J. Mol. Biol.* **124,** 523–534.

1048. D. Moras, M. B. Comarmond, J. Fischer, R. Weiss, J. C. Thierry, J. P. Ebel, and R. Giegé (1980), Crystal structure of yeast tRNA^Asp. *Nature* **288,** 669–674.

1049. R. W. Schevitz, A. D. Podjarny, N. Krishnamachari, J. J. Hughes, P. B. Sigler, and J. L. Sussman (1979), Crystal structure of a eukaryotic initiator tRNA. *Nature* **278,** 188–190.

1050. N. H. Woo, B. A. Roe, and A. Rich (1980), Three-dimensional structure of *E. coli* initiator tRNA^fMet. *Nature* **286,** 346–351.

1051. H. T. Wright, P. C. Manor, K. Beurling, R. L. Karpel, and J. Fresco (1979), The structure of baker's yeast tRNA^Gly: A second tRNA conformation. In: *Transfer RNA: Structure, Properties, and Recognition* (D. Söll, J. N. Abelson, and P. R. Schimmel, Eds.), Cold Spring Harbor Monograph Series 9A, pp. 145–160.

1052. S.-H. Kim (1975), Symmetry recognition hypothesis model for tRNA binding to aminocayl-tRNA synthetase. *Nature* **256,** 679–681.

1053. S.-H. Kim, J. L. Sussman, F. L. Suddath, G. J. Quigley, A. McPherson, A. H. J. Wang, N. C. Seeman, and A. Rich (1974), The general structure of transfer RNA molecules. *Proc. Natl. Acad. Sci. USA* **71,** 4970–4974.

1054. A. Klug, J. Ladner, and J. D. Robertus (1974), The structural geometry of co-ordinated base change in transfer RNA. *J. Mol. Biol.* **89,** 511–516.

1055. M. Sundaralingam (1979), The structure, conformation, and interaction of tRNA. In: *Transfer RNA: Structure, Properties, and Recognition* (D. Söll, J. N. Abelson, and P. R. Schimmel, Eds.), Cold Spring Harbor Monograph Series 9A, pp. 115–132.

1056. V. A. Erdmann, M. Sprinzl, and O. Pongs (1973), The involvement of 5S RNA in the binding of tRNA to ribosomes. *Biochem. Biophys. Res. Commun.* **54,** 942–946.

1056a. B. Pace, E. A. Matthews, K. D. Johnson, C. R. Cantor, and N. R. Pace (1982), Conserved 5S rRNA complement to tRNA is not required for protein synthesis. *Proc. Nat. Acad. Sci. USA* **79,** 36–40.

1057. A. Rich, G. J. Quigley, and A. H. J. Wang (1979), Conformational flexibility of the polynucleotide chain. In: *Stereodynamics of Molecular Systems* (R. H. Sarma, Ed.), pp. 315–330. Pergamon Press, New York.

1058. C. D. Stout, H. Mizuno, J. Rubin, T. Brennan, S. T. Rao, and M. Sundaralingam (1976), Atomic coordinates and molecular conformation of yeast phenylalanyl tRNA. An independent investigation. *Nucl. Acids Res.* **4**, 1111–1123.

1059. K. Kitamura, A. Wakahara, T. Hakoshima, H. Mizuno, and K.-I. Tomita (1978), Classifications of nucleotide conformations observed in yeast phenylalanine tRNA. Application of cluster analysis. *Nucl. Acids Res.* **5**, s373–s376.

1060. M. Sundaralingam, H. Mizuno, C. D. Stout, S. T. Rao, M. Liebman, and N. Yathindra (1976), Mechanism of chain folding in nucleic acids. The (ω',ω) plot and its correlation to the nucleotide geometry in yeast tRNA^{Phe}. *Nucl. Acids Res.* **3**, 2471–2484.

1061. A. R. Srinivasan and N. Yathindra (1977), A novel representation of the conformational structure of transfer RNAs. Correlation of the folding patterns of the polynucleotide chain with the base sequence and the nucleotide backbone torsions. *Nucl. Acids Res.* **4**, 3969–3979.

1062. A. R. Srinivasan and W. K. Olson (1980), Yeast tRNA^{Phe} conformational wheels: A novel probe of the monoclinic and orthorhombic models. *Nucl. Acids Res.* **8**, 2307–2329.

1063. D. Perahia, B. Pullman, D. Vasilescu, R. Cornillon, and H. Broch (1977), A molecular orbital investigation of the conformation of transfer RNA. *Biochim. Biophys. Acta* **478**, 244–259.

1064. C. W. Tabor and H. Tabor (1976), 1,4-Diaminobutane (putrescine), spermidine, and spermine. *Annu. Rev. Biochem.* **45**, 285–306.

1065. T. T. Sakai and S. S. Cohen (1976), Effects of polyamines on the structure and reactivity of tRNA. *Prog. Nucl. Acid. Res. Mol. Biol.* **17**, 15–42.

1066. J. R. Fresco, A. Adams, R. Ascione, D. Henley, and T. Lindahl (1966), Tertiary structure in transfer ribonucleic acids. *Cold Spring Harbor Symp. Quant. Biol.* **31**, 527–537.

1067. A. Stein and D. M. Crothers (1976), Equilibrium binding of magnesium (II) by *Escherichia coli* tRNA^{fMet}. *Biochemistry* **15**, 157–160.

1068. M. Bina-Stein and A. Stein (1976), Allosteric interpretation of Mg^{2+} binding to the denaturable *Escherichia coli* tRNA$_2^{Glu}$. *Biochemistry* **15**, 3912–3917.

1069. A. A. Schreier and P. R. Schimmel (1974), Interaction of manganese with fragments, complementary fragment recombinations, and whole molecules of yeast phenylalanine specific transfer RNA. *J. Mol. Biol.* **86**, 601–620.

1070. A. Danchin and M. Gueron (1970), Cooperative binding of manganese (II) to transfer RNA. *Eur. J. Biochem.* **16**, 532–536.

1071. R. L. Karpel, N. S. Miller, A. M. Lesk, and J. R. Fresco (1975), Stabilization of the native tertiary structure of yeast tRNA$_3^{Leu}$ by cationic metal complexes. *J. Mol. Biol.* **97**, 519–532.

1072. C. Jones and D. Kearns (1974), Investigation of the structure of yeast tRNA^{Phe} by nuclear magnetic resonance: Paramagnetic rare earth ion probes of structure. *Proc. Nat. Acad. Sci. USA* **71**, 4137–4240.

1073. J. Wolfson and D. Kearns (1975), Europium as a fluorescent probe of transfer RNA structure. *Biochemistry* **14**, 1436–1444.

1074. D. C. Lynch and P. R. Schimmel (1974), Cooperative binding of magnesium to transfer ribonucleic acid studied by a fluorescent probe. *Biochemistry* **13**, 1841–1852.

1075. G. Rialdi, J. Levy, and R. Biltonen (1972), Thermodynamic studies of transfer ribonucleic acids. I. Magnesium binding to yeast phenylalanine transfer ribonucleic acid. *Biochemistry* **11**, 2472–2479.

1076. D. M. Crothers and P. E. Cole (1978), Conformational changes of tRNA. In: *Transfer RNA* (S. Altman, Ed.), pp. 196–247. MIT Press, Cambridge.

1077. A. Jack, J. E. Ladner, D. Rhodes, R. S. Brown, and A Klug (1977), A crystallographic study of metal-binding to yeast phenylalanine transfer RNA. *J. Mol. Biol.* **111**, 315–328.

1078. S. R. Holbrook, J. L. Sussman, R. Wade Warrant, G. M. Church, and S.-H. Kim (1977), RNA–ligand interactions: (I) Magnesium binding sites in yeast tRNA^{Phe}. *Nucl. Acids Res.* **8**, 2811–2820.

1079. G. J. Quigley, M. M. Teeter, and A. Rich (1978), Structural analysis of spermine and magnesium ion binding to yeast phenylalanine transfer RNA. *Proc. Nat. Acad. Sci. USA* **75**, 64–68.

1080. P. H. Bolton and D. R. Kearns (1977), Effect of magnesium and polyamines on the structure of yeast tRNA^{Phe}. *Biochim. Biophys. Acta* **477**, 10–19.

1081. O. C. Uhlenbeck (1972), Complementary oligonucleotide binding to transfer RNA. *J. Mol. Biol.* **65**, 25–41.

1082. J. Eisinger, B. Feuer, and T. Yamane (1971), Codon–anticodon binding in tRNA^{Phe}. *Nature New Biol.* **231**, 127–128.

1083. C. J. Alden and S. Arnott (1973), Nucleotide conformations in codon–anticondon interactions. *Biochem. Biophys. Res. Commun.* **53**, 806–811.

1084. H. Prinz, A. Maelicke, and F. Cramer (1974), Unfolding of yeast transfer ribonucleic acid species caused by addition of organic solvents and studied by circular dichroism. *Biochemistry* **13**, 1322–1326.

1085. H. T. Wright (1980), A new mechanism for the hydrolytic editing function of aminoacyl-tRNA synthetases. Kinetic specificity for the tRNA substrate. *FEBS Lett.* **118**, 165–171.

1086. E. F. Gale, F. Cundliffe, P. E. Reynolds, M. H. Richmond, and M. J. Waring (1972), *The Molecular Basis of Antibiotic Action*. Wiley, London.

1087. J. W. Corcoran and F. E. Hahn (Eds.) (1974), *Antibiotics, III, Mechanism of Action of Antimicrobial and Antitumour Agents*. Springer-Verlag, Berlin.

1088. G. Hartmann, W. Behr, K.-A. Beissner, K. Honikel, and A. Sippel (1968), Antibiotics as inhibitors of nucleic acid and protein biosynthesis. *Angew. Chem. Int. Ed. Engl.* **7**, 693–701.

1089. M. J. Waring (1981), DNA modification and cancer. *Annu. Rev. Biochem.* **50**, 159–192.

1090. H. M. Berman and P. R. Young (1981), The interaction of intercalating drugs with nucleic acids. *Annu. Rev. Biophys. Bioeng.* **10**, 87–114.

1091. C. Heidelberger (1975), Chemical mutagenesis. *Annu. Rev. Biochem.* **44**, 78–121.

1092. L. S. Lerman (1961), Structural considerations in the interaction of DNA and acridines. *J. Mol. Biol.* **3**, 18–30.

1093. W. Fuller and M. Waring (1964), A molecular model for the interaction of ethidium bromide with deoxyribonucleic acid. *Ber. Bunsenges. Phys. Chem.* **68**, 805–809.

1094. D. M. Neville and D. R. Davies (1966), The interaction of acridine dyes with DNA: An X-ray diffraction and optical investigation. *J. Mol. Biol.* **17**, 57–74.

1095. H. Porumb (1978), The solution spectroscopy of drugs and the drug–nucleic acid interactions. *Prog. Biophys. Mol. Biol.* **34**, 175–195.

1096. H. J. Li and D. M. Crothers (1969), Relaxation studies of the proflavine–DNA complex: the kinetics of an intercalation reaction. *J. Mol. Biol.* **39**, 461–477.

1097. L. P. G. Wakelin and M. J. Waring (1980), Kinetics of drug–DNA inter-action. Dependence of the binding mechanism on structure of the ligand. *J. Mol. Biol.* **144**, 183–214.

1098. J. Cairns (1962), The application of autoradiography to the study of DNA viruses. *Cold Spring Harbor Symp. Quant. Biol.* **27**, 311–318.

1099. P. J. Bond, R. Langridge, K. W. Jennette, and S. J. Lippard (1975), X-Ray fiber diffraction evidence for the neighbor exlusion binding of a platinum metallointercalation reagent to DNA. *Proc. Nat. Acad. Sci. USA* **72**, 4825–4829.

1100. S. Arnott, P. J. Bond, and R. Chandrasekaran (1980), Visualization of an unwound DNA duplex. *Nature* **287**, 561–563.

1101. M. J. Waring and L. P. G. Wakelin (1974), Echinomycin: A bifunctional in-tercalating antibiotic. *Nature* **252**, 653–657.

1102. M. A. Viswamitra, O. Kennard, W. B. T. Cruse, E. Egert, G. M. Sheldrick, P. G. Jones, M. J. Waring, L. P. G. Wakelin, and R. K. Olsen (1981), Struc-ture of TANDEM and its implication for bifunctional intercalation into DNA. *Nature* **289**, 817–819.

1103. M. J. Waring (1970), Variation of the supercoils in closed circular DNA by binding of antibiotics and drugs: evidence of molecular models involving in-tercalation. *J. Mol. Biol.* **54**, 247–279.

1104. S. Neidle (1979), The molecular basis for the action of some DNA-binding drugs. *Prog. Medic. Chem.* **16**, 151–221.

1105. H. M. Sobell (1980), Structural and dynamic aspects of drug intercalation into DNA and RNA. In R. H. Sarma, ed., *Nucleic Acid Geometry and Dy-namics,* pp. 289–323, Pergamon Press, New York.

1106. H. M. Sobell (1973), The stereochemistry of actinomycin binding to DNA and its implications in molecular biology. *Prog. Nucl. Acid Res. Mol. Biol.* **13**, 153–190.

1107. H. S. Shieh, H. M. Berman, M. Dabrow and S. Neidle (1980), The structure of drug-deoxydinucleoside phosphate complex. Generalized conformational behavior of intercalation complexes with RNA and DNA fragments. *Nucl. Acids Res.* **8**, 85–97.

1108. S. Neidle, G. Taylor, and M. Sanderson (1978), A 1:2 crystalline complex of ApA:proflavine: A model for binding to single-stranded regions in RNA. *Nucl. Acids Res.* **5**, 4417–4422.

1109. E. Westhof and M. Sundaralingam (1980), X-ray structure of a cytidylyl-3′, 5′-adenosine-proflavine complex: A self-paired parallel-chain double helical dimer with an intercalated acridine. *Proc. Nat. Acad. Sci. USA* **7**, 1852–1856.

1110. N. C. Seeman, R. O. Day, and A. Rich (1975), Nucleic acid–mutagen inter-actions: Crystal structure of adenylyl-3′,5′-uridine plus 9-aminoacridine. *Nature* **253**, 324–326.

1111. T. R. Krugh and C. G. Reinhardt (1975), Evidence for sequence preferences in the intercalative binding of ethidium bromide to dinucleoside monophos-phates. *J. Mol. Biol.* **97**, 133–162.

1112. D. J. Patel and L. L. Canuel (1976), Ethidium bromide·(dC–dG–dC–dG)$_2$ complex in solution: Intercalation and sequence specificity of drug binding at the tetranucleotide duplex level. *Proc. Nat. Acad. Sci. USA* **73**, 3343–3347.

1113. D. J. Patel (1977), Mutagen–nucleic acid complexes at the polynucleotide duplex level in solution: Intercalation of proflavine into poly d(A–T) and the melting transition of the complex. *Biopolymers* **16**, 2739–2754.

1114. R. L. Ornstein and R. Rein (1979), Energetics of intercalation specificity. I. Backbone unwinding. *Biopolymers* **18**, 1277–1291.

1115. M. E. Nuss, F. J. Marsh, and P. A. Kollman (1979), Theoretical studies of drug–dinucleotide interactions. Empirical energy function calculations on the interaction of ethidium, 9-aminoacridine, and proflavine cations with the base-paired dinucleotides GpC and CpG. *J. Amer. Chem. Soc.* **101**, 825–833.

1116. K. J. Miller and J. F. Pycior (1979), Interaction of molecules with nucleic acids. II. Two pairs of families of intercalation sites, unwinding angles, and the neighbor-exclusion principle. *Biopolymers* **18**, 2683–2719.

1117. S. Broyde and B. Hingerty (1979), Conformational origin of the pyrimidine–(3′,5′)-purine base sequence preference for intercalation into RNAs. *Biopolymers* **18**, 2905–2910.

1118. P. M. Dean and L. P. G. Wakelin (1980), Electrostatic components of drug-receptor recognition. I. Structural and sequence analogues of DNA polynucleotides. *Proc. Roy. Soc. London Ser. B* **209**, 453–471.

1118a. K. J. Miller, R. Brodzinsky, and S. Hall (1980), Interactions of molecules with nucleic acids. IV. Binding energies and conformations of acridine and phenanthridine compounds in the two principal and in several unconstrained dimer-duplex intercalation sites. *Biopolymers* **19**, 2091–2122.

1119. C.-C. Tsai, S. C. Jain, and H. M. Sobell (1977), Visualization of drug–nucleic acid interactions at atomic resolution. I Structure of an ethidium/dinucleoside monophosphate crystalline complex, ethidium:5-iodouridylyl(3′-5′)adenosine. *J. Mol. Biol.* **114**, 301–305.

1120. H. M. Berman, W. Stallings, H. L. Carrell, J. P. Glusker, S. Neidle, G. Taylor, and A. Achari (1979), Molecular and crystal structure of an intercalation complex: Proflavine–cytidylyl-(3′,5′)-guanosine. *Biopolymers* **18**, 2405–2429.

1121. H. M. Berman and S. Neidle (1979), Modelling of drug–nucleic acid interactions. Intercalation geometry of oligonucleotides. In: *Stereodynamics of Molecular Systems* (R. H. Sarma, Ed.), pp. 367–382. Pergamon Press, New York.

1122. H. M. Berman, S. Neidle, and R. K. Stodola (1978), Drug–nucleic acid interactions: Conformational flexibility at the intercalation site. *Proc. Nat. Acad. Sci. USA* **75**, 828–832.

1123. G. J. Quigley, A. H.-J. Wang, G. Ughetto, G. van der Marel, J. H. van Boom, and A. Rich (1980), Molecular structure of an anti-cancer drug–DNA complex: Daunomycin plus d(CpGpTpApCpG). *Proc. Nat. Acad. Sci. USA* **77**, 7204–7208.

1124. E. J. Gabbay, D. Grier, R. E. Fingerle, R. Reimer, R. Levy, S. W. Pearce, and W. D. Wilson (1976), Interaction specificity of the anthracyclines with deoxyribonucleic acid. *Biochemistry* **15**, 2062–2070.

1125. M. E. Nuss, T. L. James, M. A. Apple, and P. A. Kollman (1980), An NMR study of the interaction of daunomycin with dinucleotides and dinucleoside phosphates. *Biochem. Biophys. Acta* **609**, 136–147.

1126. D. R. Phillips and G. C. K. Roberts (1980), Proton nuclear magnetic resonance study of the self-complementary hexanucleotide d(pTpA)$_3$ and its interaction with daunomycin. *Biochemistry* **19**, 4795–4801.

1127. D. W. Henry (1979), Structure-activity relationships among daunorubicin and adriamycin analogs. *Cancer Treat. Rep.* **63**, 845–854.

1128. F. Zunino, A. Di Marco and A. Zaccara (1979), Molecular structural effects involved in the interaction of anthracyclines with DNA. *Chem.–Biol. Interact.* **24**, 217–225.

1129. J. C. Wang (1974), The degree of unwinding of the DNA helix by ethidium. I. Titration of twisted PM2 DNA molecules in alkaline cesium chloride density gradients. *J. Mol. Biol.* **89**, 783–801.

1130. C. J. Alden and S. Arnott (1977), Stereochemical model for proflavine intercalation in A-DNA. *Nucl. Acids Res.* **4**, 3855–3861.

1131. C. J. Alden and S. Arnott (1975), Visualization of planar drug intercalation in B-DNA. *Nucl. Acids Res.* **2**, 1701–1717.

1132. B. S. Reddy, T. P. Seshadri, T. D. Sakore, and H. M. Sobell (1979), visualization of drug–nucleic acid interactions at atomic resolution. V. Structures of two aminoacridine-dinucleoside monophosphate crystalline complexes, proflavine–5-iodocytidylyl(3′-5′)guanosine and acridine orange–5-iodocytidylyl(3′-5′)guanosine. *J. Mol. Biol.* **135**, 787–812.

1133. D. J. Patel (1974), Peptide antibiotic–oligonucleotide interactions. Nuclear magnetic resonance investigations of complex formation between actinomycin D and d-ApTpGpCpApT in aqueous solution. *Biochemistry* **13**, 2396–2402.

1134. D. M. Crothers, N. Dattagupta, and M. Hogan (1979), DNA structure and its distortion by drugs. In: *Stereodynamics of Molecular Systems* (R. H. Sarma, Ed.), pp. 383–395. Pergamon Press, New York.

1135. M. Hogan, N. Dattagupta, and D. M. Crothers (1979), Transmission of allosteric effects in DNA. *Nature* **278**, 521–524.

1136. H. M. Sobell (1974), How actinomycin binds to DNA. *Sci. Amer.* **231**(2), 82–91.

1137. W. Müller and D. M. Crothers (1968), Studies of the binding of actinomycin and related compounds to DNA. *J. Mol. Biol.* **35**, 251–290.

1138. S. C. Jain and H. M. Sobell (1972), Stereochemistry of actinomycin binding to DNA. I. Refinement and further structural details of the actinomycin-deoxyguanosine crystalline complex. *J. Mol. Biol.* **68**, 1–20.

1139. H. M. Sobell and S. C. Jain (1972), Stereochemistry of actinomycin binding to DNA. II. Detailed molecular model of actinomycin–DNA complex and its implications. *J. Mol. Biol.* **68**, 21–34.

1140. H. M. Sobell (1973), Symmetry in protein–nucleic acid interaction and its genetic implications. *Adv. Genet.* **17**, 411–490.

1141. A. H.-J. Wang, J. Nathans, G. van der Marel, J. H. van Boom, and A. Rich (1978), Molecular structure of a double helical DNA fragment intercalator complex between deoxy CpG and a terpyridine platinum compound. *Nature* **276**, 471–474.

1142. A. H.-J. Wang, G. J. Quigley, and A. Rich (1979), Atomic resolution analysis of a 2:1 complex of CpG and acridine orange. *Nucl. Acids Res.* **6**, 3879–3890.

1143. S. C. Jain, C.-C. Tsai, and H. M. Sobell (1977), Visualization of drug–nucleic acid interactions at atomic resolution. II. Structure of an ethidium/dinucleoside monophosphate crystalline complex, ethidium:5-iodocyti-dylyl(3'-5')guanosine. *J. Mol. Biol.* **114**, 317–331.

1144. T. D. Sakore, B. S. Reddy, and H. M. Sobell (1979), Visualization of drug–nucleic acid interactions at atomic resolution. IV. Structure of an amino-acridine–dinucleoside monophosphate crystalline complex, 9-aminoacridine:5-iodocytidylyl(3'-5')guanosine. *J. Mol. Biol.* **135**, 763–785.

1145. S. C. Jain, K. K. Bhandary, and H. M. Sobell (1979), Visualization of drug–nucleic acid interactions at atomic resolution. VI. Structure of two drug–dinucleoside monophosphate crystalline complexes, ellipticine–5-iodo-cytidylyl(3'-5')guanosine and 3,5,6,8-tetramethyl-N-methyl phenan-throlinium–5-iodocytidylyl(3'-5')guanosine. *J. Mol. Biol.* **135**, 813–840.

1146. I. D. Kuntz, Jr., and W. Kauzmann (1974), Hydration of proteins and poly-peptides. *Adv. Protein Chem.* **28**, 239–345.

1147. C. Tanford (1968), Protein denaturation. *Adv. Protein Chem.* **23**, 121–282.

1148. A. J. Hopfinger (1977), *Intermolecular Interactions and Biomolecular Orga-nization,* pp. 159–169. Wiley, New York.

1149. J. Texter (1978), Nucleic acid–water interactions. *Prog. Biophys. Mol. Biol.* **33**, 83–97.

1150. S. Hanlon, S. Brudno, and B. Wolf (1975), Structural transitions of deoxyri-bonucleic acid in aqueous electrolyte solutions. I. Reference spectra of con-formational limits. *Biochemistry* **14**, 1648–1660.

1151. B. Wolf and S. Hanlon (1975), Structural transitions of deoxyribonucleic acid in aqueous electrolyte solutions. II. The role of hydration. *Biochemis-try* **14**, 1661–1670.

1152. G. Cohen and H. Eisenberg (1968), Deoxyribonucleate solutions: Sedimen-tation in a density gradient, partial specific volumes, density and refractive index measurements, and preferential interactions. *Biopolymers* **6**, 1077–1100.

1153. J. E. Hearst and J. Vinograd (1961), The net hydration of deoxyribonucleic acid. *Proc. Nat. Acad. Sci. USA* **47**, 825–830.

1154. M.-J. B. Tunis and J. E. Hearst (1968), On the hydration of DNA. I. Prefer-ential hydration and stability of DNA in concentrated trifluoracetate solu-tion. *Biopolymers* **6**, 1325–1344.

1155. M.-J. B. Tunis and J. E. Hearst (1968), On the hydration of DNA. II. Base composition dependence of the net hydration of DNA. *Biopolymers* **6**, 1345–1353.

1156. J. E. Hearst (1965), Determination of the dominant factors which influence the net hydration of native sodium deoxyribonucleate. *Biopolymers* **3**, 57–68.

1157. M. Falk, K. A. Hartman, Jr., and R. C. Lord (1962), Hydration of deoxyri-bonucleic acid. I. A gravimetric study. *J. Amer. Chem. Soc.* **84**, 3843–3846.

1158. M. Falk, K. A. Hartman, Jr., and R. C. Lord (1963), Hydration of deoxyri-bonucleic acid. II. An infrared study. *J. Amer. Chem. Soc.* **85**, 387–391.

1159. M. Falk, K. A. Hartman, Jr., and R. C. Lord (1963), Hydration of deoxyri-bonucleic acid. III. A spectroscopic study of the effect of hydration on the structure of deoxyribonucleic acid. *J. Amer. Chem. Soc.* **85**, 391–394.

1160. M. Falk, A. G. Poole, and C. G. Goymour (1970), Infrared study of the state of water in the hydration shell of DNA. *Can. J. Chem.* **48**, 1536–1542.

1161. A. C. T. North and A. Rich (1961), X-Ray diffraction studies of bacterial viruses. *Nature* **191**, 1242–1245.

1162. M.-J. B. Tunis-Schneider and M. F. Maestre (1970), Circular dichroism spectra of oriented and unoriented deoxyribonucleic acid films—A preliminary study. *J. Mol. Biol.* **52**, 521–541.

1163. S. Bram (1971), The secondary structure of DNA in solution and in nucleohistone. *J. Mol. Biol.* **58**, 277–288.

1164. M. Suwalsky and W. Traub (1972), A comparative X-ray study of a nucleoprotamine and DNA complexes with polylysine and polyarginine. *Biopolymers* **11**, 2223–2231.

1165. E. E. Minyat, V. I. Ivanov, A. M. Kritzyn, L. E. Minchenkova, and A. K. Schylokina (1978), Spermine and spermidine-induced B to A transition of DNA in solution. *J. Mol. Biol.* **128**, 397–409.

1166. Y. Huse, Y. Mitsui, Y. Iitaka, and K. Miyaki (1978), Preliminary X-ray studies of the interaction of salmon sperm DNA with spermine. *J. Mol. Biol.* **122**, 43–53.

1167. A. M. Liquori, L. Constantino, V. Crescenzi, V. Elia, E. Giglio, R. Puliti, M. De Santis Savino, and V. Vitagliamo (1967), Complexes between DNA and polymines: A molecular model. *J. Mol. Biol.* **24**, 113–122.

1168. H. R. Drew and R. E. Dickerson (1981), Structure of a DNA dodecamer. III. Geometry of hydration. *J. Mol. Biol.* **151**, 535–556.

1169. B. Lee and F. M. Richards (1971), The interpretation of protein structures: Estimation of static accessibility. *J. Mol. Biol.* **55**, 379–400.

1170. C. Chothia (1975), Structural invariants in protein folding. *Nature* **254**, 304–308.

1171. P. Thiyagarajan and P. K. Ponnuswamy (1979), Solvent accessibility study on tRNA[Phe]. *Biopolymers* **18**, 2233–2247.

1172. C. J. Alden and S.-H. Kim (1979), Solvent-accessible surfaces of nucleic acids *J. Mol. Biol.* **132**, 411–434.

1173. A. Goldblum, D. Perahia, and A. Pullman (1978), Hydration scheme of the complementary base-pairs of DNA. *FEBS Lett.* **91**, 213–215.

1174. B. Pullman (1975), Molecular orbital approach to hydration studies of nucleic acids and proteins. In: *Structure and Conformation of Nucleic Acids and Protein–Nucleic Acid Interactions* (M. Sundaralingam and S. T. Rao, Eds.), pp. 457–484. Univ. Park Press, Baltimore.

1175. D. Perahia, M. S. Jhon, and B. Pullman (1977), Theoretical study of the hydration of B-DNA. *Biochim. Biophys. Acta* **474**, 349–362.

1176. K. Kim and M. S. Jhon (1979), Theoretical study of hydration of RNA. *Biochim. Biophys. Acta* **565**, 131–147.

1177. E. Clementi and G. Corongiu (1979), Interaction of water with DNA single helix in the A conformation. *Biopolymers* **18**, 2431–2450.

1178. E. Clementi and G. Corongiu (1980), A theoretical study on the water structure for nucleic acids bases and base pairs in solution at T = 300K. *J. Chem. Phys.* **72**, 3979–3992.

1179. E. Clementi and G. Corongiu (1979), Interaction of water with DNA single and double helix in the A and B conformation. *Gazz. Chim. Ital.* **109**, 201–205.

1180. G. Corongiu and E. Clementi (1981), Simulations of the solvent structure for macromolecules. I. Solvation of B-DNA double helix at T = 300K. *Biopolymers* **20**, 551–571.

1181. B. Jacobson (1955), On the interpretation of dielectric constants of aqueous macromolecular solutions. Hydration of macromolecules. *J. Amer. Chem. Soc.* **77**, 2919–2926.

1182. I. M. Klotz (1965), Role of water structure in macromolecules. *Fed. Proc.* **24**, S-24–S-33.

1183. G. A. Jeffrey (1969), Water structure in organic hydrates. *Acc. Chem. Res.* **2**, 344–352.

1183a. M. L. Kopka, A. V. Fratini, H. R. Drew, and R. E. Dickerson (1982), Ordered water structure around a B-DNA dodecamer. A quantitative study *J. Mol. Biol.* **163**, 129–146.

1184. S. Neidle, H. M. Berman, and H. S. Shieh (1980), Highly structured water network in crystals of a deoxydinucleoside–drug complex. *Nature* **288**, 129– 133.

1185. W. Saenger (1979), Circular hydrogen bonds. *Nature* **279**, 343–344.

1186. W. Saenger and K. Lindner (1980). OH clusters with homodromic circular arrangement of hydrogen bonds. *Angew, Chem. Int. Ed. Engl.* **19**, 398–399.

1187. B. Lesyng and W. Saenger (1981), Theoretical investigations on circular and chain-like hydrogen bonded structures found in two crystal forms of α-cyclodextrin hexahydrate. *Biochim. Biophys. Acta* **678**, 408–413.

1188. W. Saenger, Ch Betzel, B. Hingerty, and G. M. Brown (1982), Flip-flop O–H $\cdots$ H–O hydrogen bonds in partially disordered systems. *Nature* **296**, 581–583.

1189. T.-S. Young, S.-H. Kim, P. Modrich, A. Beth, and E. Jay (1981), Preliminary X-ray diffraction studies of EcoRI restriction endonuclease–DNA complex. *J. Mol. Biol.* **145**, 607–610.

1190. A. McPherson, F. Jurnak, A. Wang, F. Kolpak, and A. Rich (1980), The structure of a DNA unwinding protein and its complexes with oligodeoxynucleotides by X-ray diffraction. *Biophys. J.* **32**, 155–173.

1190a. G. D. Brayer and A. McPherson (1983), Refined structure of the gene 5 DNA binding protein from bacteriophage fd. *J. Mol. Biol.,* submitted for publication.

1191. R. Giegé, B. Lorber, J.-P. Ebel, D. Moras, and J.-C. Thierry (1980), Cristallisation du complexe formé entre l'aspartate–tRNA de levure et son aminoacyl-tRNA synthétase spécifique. *Compt. Rend. Acad. Sci. Paris Ser. D* **291**, 393–396.

1192. P. H. von Hippel and J. D. McGhee (1972), DNA–protein interactions. *Annu. Rev. Biochem.* **41**, 231–300.

1193. M. Sundaralingam and S. T. Rao (Eds.) (1975), *Structure and Conformation of Nucleic Acids and Protein–Nucleic Acid Interactions.* Univ. Park Press, Baltimore.

1194. H. J. Vogel (Ed.) (1977), *Nucleic Acid–Protein Recognition.* Academic Press, New York.

1195. C. Hélène and J.-C. Maurizot (1981), Interactions of oligopeptides with nucleic acids. *CRC Crit. Rev. Biochem.* **10**, 213–258.

1196. G. Lancelot, R. Mayer, and C. Hélène (1979), Conformational study of the dipeptide arginylglutamic acid and of its complex with nucleic bases. *J. Amer. Chem. Soc.* **101**, 1569–1576.

1197. V. I. Bruskov and V. N. Bushuyev (1977), Investigation by the method of proton magnetic resonance of complexing between the nucleosides and compounds modelling the amino acid residues of proteins in dimethyl-sulphoxide. *Biophysics (Moscow)* **22**, 23–29.

1197a. G. Lancelot and R. Mayer (1981), The specific interaction of guanine with carboxylate ions in water. *FEBS Lett.* **130**, 7–11.

1198. V. Bruskov (1978), Specificity of interaction of nucleic acid bases with hydrogen bond forming amino acids. *Stud. Biophys. (Berlin)* **67**, S. 43–44 and microfiche 2/27–34.

1199. G. Lancelot, C. Hélène (1977), Selective recognition of nucleic acids by proteins: The specificity of guanine interaction with carboxylic acids. *Proc. Nat. Acad. Sci. USA* **74**, 4872–4875.

1200. H. Sellini, J.-C. Maurizot, J. L. Dimicoli, and C. Hélène (1973), Hydrogen bonding of amino acid side chains to nucleic acid bases. *FEBS Lett.* **30**, 219–224.

1201. G. Lancelot (1977), Hydrogen bonding of adenine derivatives to tyrosine side chains. *Biophys. J.* **17**, 243–254.

1202. N. Gresh and B. Pullman (1980), A theoretical study of the interaction of guanine and cytosine with specific amino acid side chains. *Biochim. Biophys. Acta* **608**, 47–53.

1203. V. I. Bruskov (1975), The recognition of nucleic acid bases by amino acids and peptides with the aid of hydrogen bonds. *Mol. Biol. (Moscow)* **9**, 245–249.

1204. N. C. Seeman, J. M. Rosenberg, and A. Rich (1976), Sequence-specific recognition of double helical nucleic acids by proteins. *Proc. Nat. Acad. Sci. USA* **73**, 804–808.

1205. A. Rich, N. C. Seeman, and J. M. Rosenberg (1977), Protein recognition of base-pairs in a double helix. In: *Nucleic Acid–Protein Recognition* (H. J. Vogel, Ed.), pp. 361–374. Academic Press, New York.

1206. R. Rein, R. Garduno, J. T. Egan, S. Columbano, Y. Coeckelenbergh, and R. D. Macelroy (1977), Elements of a DNA–polypeptide recognition code: Electrostatic potential around the double helix, and a stereospecific model for purine recognition. *Biosystems* **9**, 131–137.

1207. C. Chothia (1973), Conformation of twisted β-pleated sheets in proteins. *J. Mol. Biol.* **75**, 295–302.

1208. G. M. Church, J. L. Sussman, and S.-H. Kim (1977), Secondary structural complementarity between DNA and proteins. *Proc. Natl. Acad. Sci. USA* **74**, 1458–1462.

1209. P. J. Chou, A. J. Adler, and G. D. Fasman (1975), Conformational prediction and circular dichroism studies on the *lac* repressor. *J. Mol. Biol.* **96**, 29–45.

1209a. A. A. Ribeiro, D. Wemmer, R. P. Bray, and O. Jardetsky (1981), A folded structure for the lac-repressor headpiece. *Biochem. Biophys. Res. Commun.* **99**, 668–674.

1210. G. V. Gursky, V. G. Tumanyan, A. S. Zasedatelev, A. L. Zhuse, S. L. Grokhovsky, and B. P. Gottikh (1977), A code controlling specific binding of proteins to double-helical DNA and RNA. In: *Nucleic Acid–Protein Recognition* (H. J. Vogel, Ed.), pp. 189–217. Academic Press, New York.

1211. H. M. Sobell (1976), Symmetry in nucleic acid structure and its role in protein–nucleic acid interactions. *Annu. Rev. Biophys. Bioeng.* **5**, 307–335.

1212. C. Tamura, M. Yoshikawa, S. Sato, and T. Hata (1973), Crystal structure of cytidine–salicylic acid complex. *Chem. Lett. (Japan)*, 1221–1224.

1213. T. Hata, M. Yoshikawa, S. Sato, and C. Tamura (1975), 1:1 Cytidine–N-carbobenzoxyglutamic acid dihydrate. *Acta Crystallogr. B* **31,** 312–314.

1214. M. Ohki, A. Takenaka, H. Shimanouchi, and Y. Sasada (1980), Complexes between nucleotide base and amino acid. IV. Crystal and molecular structure of cytosine: N,N-Phthaloyl-DL-glutamic acid complex dihydrate. *Bull. Chem. Soc. Japan* **53,** 2724–2730.

1215. M. Ohki, A. Takenaka, H. Shimanouchi, and Y. Sasada (1976), Complexes between nucleotide base and amino acid. II. Crystal structure of 5-bromocytosine: N-tosyl-L-glutamic acid. *Bull. Chem. Soc. Japan* **49,** 3493–3497.

1216. M. Ohki, A. Takenaka, H. Shimanouchi, and Y. Sasada (1977), Complexes between nucleotide base and amino acid. III. Crystal structure of 5-bromocytosine: phthaloyl-DL-glutamic acid complex hemihydrate. *Bull. Chem. Soc. Japan* **50,** 90–96.

1217. D. Suck, W. Saenger, and W. Rohde (1974), X-Ray structure of thymidine—5′-carboxylic acid, an inhibitor of thymidine and thymidylate kinase: Preferred nucleobase–carboxylic acid hydrogen bonding scheme. *Biochim. Biophys. Acta* **361,** 1–10.

1218. P. Narayanan, H. M. Berman, and R. Rousseau (1976), Model compounds for protein nucleic acid interactions. 3. Synthesis and structure of a nucleoside dipeptide, N-(9-β-D-ribofuranosylpurin-6-yl)glycyl-L-alanine sesquihydrate. *J. Amer. Chem. Soc.* **98,** 8472–8475.

1219. T. Kaneda and J. Tanaka (1976), The crystal structure of the intermolecular complex between 9-ethyladenine and indole. *Bull. Chem. Soc. Japan* **49,** 1799–1804.

1220. M. Ohki, A. Takenaka, H. Shimanouchi, and Y. Sasada (1977), 3-(9-Adenyl)propionyltryptamine monohydrate. *Acta Crystallogr. B* **33,** 2954—2956.

1221. M. Ohki, A. Takenaka, H. Shimanouchi, and Y. Sasada (1977), Crystal and molecular structure of 3-(adenin-9-yl)propiontryptamide. *Bull. Chem. Soc. Japan* **50,** 2573–2578.

1222. F. Morita (1974), Molecular complex of tryptophan with ATP or its analogs. *Biochim. Biophys. Acta* **343,** 674–681.

1223. J. R. Herriott, A. Camerman, and D. D. Deranleau (1974), Crystal structure of 1-(2-indol-3-ylethyl)-3-carbamidepyridinium chloride, an intramolecular model of the nicotinamide adenine dinucleotide–tryptophan charge-transfer complex. *J. Amer. Chem. Soc.* **96,** 1585–1589.

1224. T. Ishida and M. Inoue (1981), An X-ray evidence for the charge-transfer interaction between adenine and indole rings: Crystal structure of 1,9-dimethyladenine–indole-3-acetic acid trihydrate complex. *Biochem. Biophys. Res. Commun.* **99,** 149–154.

1225. R. P. Ash, J. R. Herriott, and D. D. Deranleau (1977), Crystal structure of the ion pair 1-methyl-3-carbamidopyridinium N-acetyl-L-tryptophanate, a model for 1-substituted nicotinamide–protein charge-transfer complexes. *J. Amer. Chem. Soc.* **99,** 4471–4475.

1226. M. Inoue, M. Shibata, and T. Ishida (1980), X-Ray crystal structure of 7,8-dimethylisoalloxazine-10-acetic acid:tyramine (1:1) tetrahydrate complex. A model for flavin coenzyme–tyrosine residue charge transfer complexes in flavoproteins. *Biochem. Biophys. Res. Commun.* **93,** 415–419.

1227. T. Ishida, M. Inoue, T. Fujiwara, and K-I. Tomita (1979), X-Ray crystal structure of the 7,8-dimethylisoalloxazine-10-acetic acid–tryptamine complex. A model for flavin–indole charge-transfer complexes. *J. Chem. Soc. Chem. Commun.,* 358–360.

1228. T. Ishida, K.-I. Tomita, and M. Inoue (1980), An X-ray study on the interaction between indole ring and pyridine coenzymes: Crystal structure of 1-methyl-3-carbamoyl-pyridinium:indole-3-acetic acid (1:1) monohydrate charge-transfer complex. *Arch. Biochem. Biophys.* **200,** 492–502.

1229. M. A. Slifkin (1973), Charge transfer interactions of purines and pyrimidines. In: *Physico-Chemical Properties of Nucleic acids* (J. Duchesne, Ed.), pp. 67–98. Academic Press, New York.

1230. H. M. Berman, W. C. Hamilton, and R. J. Rousseau (1973), Crystal structure of an antiviral agent, 5-[N-(L-phenylalanyl)amino]uridine. *Biochemistry* **12,** 1809–1814.

1230a. M. Takimoto, A. Takenaka, and Y. Sasada (1981), Elementary patterns in protein–nucleic acid interaction. II. Crystal structure of 3-(adenin-9-yl)propionamide. *Bull. Chem. Soc. Japan* **54,** 1635–1639.

1231. M. Ohki, A. Takenaka, H. Shimanouchi, and Y. Sasada (1977), 3-(9-Adenyl)propionyltyramine dihydrate. *Acta Crystallogr. B* **33,** 2956–2958.

1232. P. Narayanan and H. M. Berman (1977), Model compounds for protein nucleic acid interactions. IV. Crystal structure of a nucleoside peptide with anti-viral properties: 5-[N-(L-leucyl)amino]uridine. *Acta Crystallogr. B* **33,** 2047–2051.

1233. S. Furberg and J. Solbakk (1974), On the stereochemistry of the interaction between nucleic acids and basic protein side chains. *Acta Chem. Scand. B* **28,** 481–483.

1234. N. H. Woo, N. C. Seeman, and A. Rich (1979), Crystal structure of putrescine diphosphate: A model system for amine–nucleic acid interactions. *Biopolymers* **18,** 539–552.

1235. W. Saenger and K. G. Wagner (1972), An X-ray study of the hydrogen bonding in the crystalline L-arginine phosphate monohydrate complex. *Acta Crystallogr. B* **28,** 2237–2244.

1236. M. Leng and G. Felsenfeld (1966), The preferential interactions of polylysine and polyarginine with specific base sequences in DNA. *Proc. Nat. Acad. Sci. USA* **56,** 1325–1332.

1237. J. A. Subirana, M. Chiva, and R. Mayer (1980), X-Ray diffraction studies of complexes of DNA with lysine and with lysine-containing peptides. In: *Biomolecular Structure, Conformation. Function and Evolution, I* (R. Srinivasan, N. Yathindra, and E. Subramanian, Eds.), pp. 431–440. Pergamon Press, New York.

1238. P. Suau and J. A. Subirana (1977), X-Ray diffraction studies of nucleoprotamine structure. *J. Mol. Biol.* **117,** 909–926.

1239. M. Tsuboi (1967), Helical complexes of poly-L-lysine and nucleic acids. In: *Conformation of Biopolymers* (G. N. Ramachandran, Ed.), Vol. II, pp. 689–702. Academic Press, New York.

1240. J. L. Campos, J. A. Subirana, J. Ayami, R. Mayer, E. Giralt, and E. Pedroso (1980), The conformational versatility of DNA in the presence of basic peptides. *Stud. Biophys.* **81,** 3–14.

1241. B. Prescott, C. H. Chou, and G. J. Thomas, Jr. (1976), A Raman spectroscopic study of complexes of polylysine with deoxyribonucleic acid and polyriboadenylic acid. *J. Phys. Chem.* **80,** 1164–1171.

1242. M. Haynes, R. A. Garrett, and W. B. Gratzer (1970), Structure of nucleic acid–poly base complexes. *Biochemistry* **22**, 4410–4416.
1243. M. Feughelman, R. Langridge, W. E. Seeds, A. R. Stokes, H. R. Wilson, C. W. Hooper, M. H. F. Wilkins, R. K. Barclay, and L. D. Hamilton (1955), Molecular structure of deoxyribose nucleic acid and protein. *Nature* **175**, 834–838.
1244. M. H. F. Wilkins (1956), Physical studies of the molecular structure of deoxyribose nucleic acid and nucleoprotein. *Cold Spring Harbor Symp. Quant. Biol.* **21**, 75–90.
1245. J. Liquier, M. Pinot-Lafaix, E. Taillandier, and J. Brahms (1975), Infrared linear dichroism investigations of deoxyribonucleic acid complexes with poly(L-arginine) and poly(L-lysine). *Biochemistry* **14**, 4191–4197.
1246. E. C. Ong, C. Snell, and G. D. Fasman (1976), Chromatin models. The ionic strength dependence of model histone–DNA interactions: Circular dichroism studies of lysine–leucine polypeptide–DNA complexes. *Biochemistry* **15**, 468–486.
1247. Ch. Zimmer, G. Burckhardt, and G. Luck (1973), Similarity of the conformational changes of DNA in the complex with poly-L-histidine and in the presence of polyethyleneglycol. *Stud. Biophys.* **40**, 57–62.
1248. K.-H. C. Standke and H. Brunnert (1975), The estimation of affinity constants for the binding of model peptides to DNA by equilibrium dialysis. *Nucl. Acids Res.* **2**, 1839–1849.
1249. P. E. Brown (1970), The interaction of basic dipeptide methyl esters with DNA. *Biochim. Biophys. Acta.* **213**, 282–287.
1250. M. Durand, J.-C. Maurizot, H. N. Borazan, and C. Hélène (1975), Interaction of aromatic residues of proteins with nucleic acids. Circular dichroism studies of the binding of oligopeptides to poly(adenylic acid). *Biochemistry* **14**, 563–570.
1251. E. J. Gabbay, K. Sanford, C. S. Baxter, and L. Kapicak (1973), Specific interaction of peptides with nucleic acids. Evidence for a "selective bookmark" recognition hypothesis. *Biochemistry* **12**, 4021–4029.
1251a. D. Pörschke and J. Ronnenberg (1981), The reaction of aromatic peptides with double helical DNA. Quantitative characterization of a two-step reaction scheme. *Biophys. Chem.* **13**, 283–290.
1252. R. W. Warrant and S.-H. Kim (1978), α-Helix–double helix interaction shown in the structure of a protamine-transfer RNA complex and a nucleoprotamine model. *Nature* **271**, 130–135.
1253. S. Inoue and M. Fuke (1970), An electron microscope study of deoxyribonucleoprotamines. *Biochim. Biophys. Acta* **204**, 296–303.
1254. D. P. Bazett-Jones and F. P. Ottensmeyer (1979), A model for the structure of nucleoprotamine. *J. Ultrastruct Res.* **67**, 255–266.
1255. K. Chandrasekhar, A. McPherson, Jr., M. J. Adams, and M. G. Rossmann (1973), Conformation of coenzyme fragments when bound to lactate dehydrogenase. *J. Mol. Biol.* **76**, 503–518.
1256. D. Moras, K. W. Olsen, M. N. Sabesan, M. Buehner, G. C. Ford, and M. G. Rossmann (1975), Studies of asymmetry in the three-dimensional structure of lobster D-glyceraldehyde–3-phosphate dehydrogenase. *J. Biol. Chem.* **250**, 9137–9162.
1257. L. E. Webb, E. J. Hill, and L. J. Banaszak (1973), Conformation of nicotin-

amide adenine dinucleotide bound to cytoplasmatic malate dehydrogenase. *Biochemistry* **12**, 5101–5109.

1258. B. Nordström and C.-I. Brändén (1975), The binding of nucleotides to horse liver alcohol dehydrogenase. In: *Structure and Conformation of Nucleic Acids and Protein–Nucleic Acid Interactions* (M. Sundaralingam and S. T. Rao, Eds.), pp. 387–395. Univ. Park Press, Baltimore.

1259. D. A. Matthews, R. A. Alden, S. T. Freer, N.-H. Xuong, and J. Kraut (1979), Dihydrofolate reductase from *Lactobacillus casei*. Stereochemistry of NADPH binding. *J. Biol. Chem.* **254**, 4144–4151.

1260. C. Montheilet and D. M. Blow (1978), Binding of tyrosine, adenosine triphosphate and analogues to crystalline tyrosyl transfer RNA synthetase. *J. Mol. Biol.* **122**, 407–417.

1261. K. Morikawa, T. F. M. La Cour, J. Nyborg, K. M. Rasmussen, D. L. Miller, and B. F. C. Clark (1978), High resolution X-ray crystallographic analysis of a modified form of the elongation factor Tu: Guanosinediphosphate complex. *J. Mol. Biol.* **125**, 325–338.

1262. M. Shohan and T. A. Steitz (1980), Crystallographic studies and model building of ATP at the active site of hexokinase. *J. Mol. Biol.* **140**, 1–14.

1263. F. M. Richards and H. W. Wyckoff (1973) Ribonuclease-S, In: *Atlas of Molecular Structures in Biology* (D. C. Phillips and F. M. Richards, Eds.), Vol. 1. Oxford Press, London.

1264. S. Y. Wodak, M. Y. Liu, and H. W. Wyckoff (1977), The structure of cytidylyl(2′,5′)adenosine when bound to pancreatic ribonuclease S. *J. Mol. Biol.* **116**, 855–875.

1265. F. A. Cotton, E. A. Hazen, and M. J. Legg (1979), Staphylococcal nuclease: Proposed mechanism of action based on structure of enzyme–thymidine 3′,5′-bisphosphate–calcium ion complex at 1.5 Å resolution. *Proc. Nat. Acad. Sci. USA* **76**, 2551–2555.

1266. G. Stubbs, S. Warren, and K. C. Holmes (1977), Structure of RNA and RNA binding site in tobacco mosaic virus from 4 Å map calculated from X-ray fibre diagrams. *Nature* **267**, 216–221.

1267. G. Stubbs and C. Stauffacher (1980), Protein–RNA interactions in tobacco mosaic virus. *Biophys. J.* **32**, 244–246.

1268. K. C. Holmes (1980), Protein–RNA interactions during the assembly of tobacco mosaic virus. *Trends Biochem. Sci.* **5**, 4–7.

1269. S. T. Rao and M. G. Rossmann (1973), Comparison of super-secondary structures in proteins. *J. Mol. Biol.* **76**, 241–256.

1270. M. G. Rossmann, A. Liljas, C.-I. Brändén, and L. J. Banaszak (1975), Evolutionary and structural relationships among dehydrogenases. In: *The Enzymes* Vol. XI, pp. 61–102. Academic Press, New York.

1271. M. Buehner (1975), The architecture of the coenzyme binding domain in dehydrogenases as revealed by X-ray structure analysis. In: *Protein–Ligand Interactions* (H. Sund and G. Blauer, Eds.), pp. 78–96. de Gruyter, Berlin.

1272. A. G. Pavlovskii, N. Sh. Padynkova, and M. Ya. Karpeiskii (1979), Structure of crystalline complexes of ribonuclease S with 8-substituted purine nucleotides. *Biophysics* (*Moscow*) **24**, 327–330; *Dokl. Akad. Nauk SSSR* **242**, 961–964 (1978).

1272a U. Heinemann and W. Saenger (1982). Specific protein-nucleic acid recognition in ribonuclease T₁-2′-guanylic acid complex: an X-ray study. *Nature* **299**, 27–31, and R. Arni, U. Heinemann, R. Tokuoka and W. Saenger (1988), Three-dimensional structure of the ribonuclease T₁*2′-GMP complex at 1.9 Å resolution. *J. Biol. Chem.* in press.

1273. W. G. J. Hol, P. T. van Duijnen, and H. C. Berendsen (1978), The α-helix dipole and the properties of proteins. *Nature* **273**, 443–446.

1274. S. C. Harrison (1980), Virus crystallography comes of age. *Nature* **286,** 558–559.

1275. A. C. Bloomer, J. N. Champness, G. Bricogne, R. Staden, and A. Klug (1978), Protein disk of tobacco mosaic virus at 2.8 Å resolution showing the interactions within and between subunits. *Nature* **276**, 362–368.

1276. Ch. Zimmer (1975), Effects of the antibiotics netropsin and distamycin A on the structure and function of nucleic acids. *Prog. Nucl. Acid Res. Mol. Biol.* **15,** 285–318.

1277. H. M. Berman, S. Neidle, Ch. Zimmer, and H. Thrum (1979), Netropsin, a DNA-binding oligopeptide. Structural and binding studies. *Biochim. Biophys. Acta* **561,** 124–131.

1278. W. F. Anderson, D. H. Ohlendorf, Y. Takeda, and B. W. Matthews (1981), Structure of the *cro* repressor from bacteriophage λ and its interaction with DNA. *Nature 290*, 754–758.

1279. M. Ptashne, A. Jeffrey, A. D. Johnson, R. Maurer, B. J. Meyer, C. O. Pabo, T. M. Roberts, and R. T. Sauer (1980), How the λ repressor and *cro* work. *Cell* **19**, 1–11.

1279a. B. W. Matthews, D. H. Ohlendorf, W. F. Anderson, R. G. Fisher, and Y. Takeda (1983), How does cro repressor recognize its DNA target sites? *Trends Biochem. Sci.* **8**, 25–29.

1279b. C. O. Pabo and M. Lewis (1982), The operator-binding domain of repressor: structure and DNA recognition. *Nature* **298,** 443–447.

1279c. M. Ptashne, A. D. Johnson, and C. O. Pabo (1982), A genetic switch in a bacterial virus. *Sci. Amer.* **247**(5), 106–120.

1279d. B. W. Matthews, D. H. Ohlendorf, W. F. Anderson, and Y. Takeda (1982), Structure of the DNA-binding region of *lac* repressor inferred from its homology with *cro* repressor. *Proc. Nat. Acad. Sci. USA* **79,** 1428–1432.

1280. D. B. McKay and T. A. Steitz (1981), Structure of catabolite gene activator protein at 2.9 Å resolution suggests binding to left-handed B-DNA. *Nature* **290,** 744–749.

1281. G. Gupta, M. Bansal, and V. Sasisekharan (1980), Conformational flexibility of DNA: Polymorphism and handedness. *Proc. Nat. Acad. Sci. USA* **77,** 6486–6490.

1281a. G. Albiser and S. Premilat (1982), A critical analysis of a left-handed double helix model for B-DNA fibers. *Nucl. Acids Res.* **10,** 4027–4034.

1281b. F. R. Salemme (1982), A model for catabolite activator protein binding to supercoiled DNA. *Proc. Nat. Acad. Sci. USA* **79,** 5263–5267.

1281c. T. A. Steitz, D. H. Ohlendorf, D. B. McKay, W. F. Anderson, and B. W. Matthews (1982), Structural similarity in the DNA-binding domains of catabolite gene activator and *cro* repressor proteins. *Proc. Nat. Acad. Sci. USA,* **79,** 3097–3100.

1282. A. Kornberg (1980), *DNA Replication* p. 485. Freeman, San Francisco, p. 485.

1283. L. A. Day (1973), Circular dichroism and ultraviolet absorption of a deoxyribonucleic acid binding protein of filamentous bacteriophage. *Biochemistry* **12,** 2529–5339.

1284. J. E. Coleman, R. A. Anderson, R. G. Ratcliffe, and I. M. Armitage (1976), Structure of gene 5 protein–oligodeoxynucleotide complexes as determined by ^{1}H, ^{19}F and ^{31}P nuclear magnetic resonance. *Biochemistry* **15**, 5419–5430.

1285. J. E. Coleman and H. L. Oakley (1980), Physical chemical studies of the structure and function of DNA binding (helix-destabilizing) proteins. *CRC Crit. Rev. Biochem.* **7**, 247–289.

1286. B. Alberts, L. Frey, and H. Delius (1972), Isolation and characterization of gene 5 protein of filamentous bacterial viruses. *J. Mol. Biol.* **68**, 139–152.

1287. J. Torbet, K. M. Gray, C. W. Gray, D. A. Marvin, and H. Siegrist (1981), Structure of fd DNA–gene 5 protein complex in solution. A neutron small-angle scattering study. *J. Mol. Biol.* **146**, 305–320.

1288. C. C. F. Blake and S. J. Oatley (1977), Protein–DNA and protein–hormone interactions in prealbumin: A model of the thyroid hormone nuclear receptor? *Nature* **268**, 115–120.

1289. G. E. Schulz and H. Schirmer (1979), *Principles of Protein Structure.* Springer Verlag, New York.

1289a. A. Gierer (1966), Model for DNA and protein interactions and the function of the operator. *Nature* **212**, 1480–1481.

1289b. T. M. Jovin (1976), Recognition mechanisms of DNA-specific enzymes. *Annu. Rev. Biochem.* **45**, 889–920.

1289c. D. M. J. Lilley (1981), Hairpin-loop formation by inverted repeats in super-coiled DNA is a local and transmissible property. *Nucl. Acids. Res.* **9**, 1271–1289.

1289d. N. Panayotatos and R. D. Wells (1981), Cruciform structures in super-coiled DNA. *Nature* **289**, 466–470.

1289e. V. I. Lim and A. L. Mazanov (1978), Tertiary structure for palindromic regions of DNA. *FEBS Lett.* **88**, 118–123.

1289f. A Stasiak and T. Klopotowski (1979), Four-stranded DNA structure and DNA methylation in the mechanism of action of restriction endonucleases. *J. Theor. Biol.* **80**, 65–82.

1289g. F. A. Momany, R. F. McGuire, A. W. Burgess, and H. A. Scheraga (1975), Energy parameters in polypeptides. VII. Geometric parameters, partial atomic charges, nonbonded interactions, hydrogen bond interactions, and intrinsic torsional potentials for the naturally occurring amino acids. *J. Phys. Chem.* **79**, 2361–2381.

1290. A. Rich (1962), A hypothesis concerning the folding of DNA. *Pontificiae Academiae Scientiarium Scripta Varia (Roma)* **22**, 271–284.

1291. K. E. Richards, R. C. Williams, and R. Calendar (1973), Mode of DNA packing within bacteriophage heads. *J. Mol. Biol.* **78**, 255–259.

1292. V. Virrankoski-Castrodeza, M. J. Fraser, and J. H. Parish (1981), Condensed DNA structures derived from bacteriophage heads. *J. Gen. Virol.* **58**, 181–190.

1293. W. C. Earnshaw, J. King, S. C. Harrison, and F. A. Eiserling (1978), The structural organization of DNA packaged within the heads of T4 wild-type, isometric and giant bacteriophages. *Cell* **14**, 559–568.

1294. W. C. Earnshaw and S. C. Harrison (1977), DNA arrangement in isometric phage heads. *Nature* **268**, 598–602.

1295. L. S. Lerman (1973), The polymer and salt-induced condensation of DNA. In: *Physico-Chemical Properties of Nucleic Acids* (J. Duchesne, Ed.), Vol. III, pp. 59–76. Academic Press, New York.

1296. T. Maniatis, J. H. Venable, and L. S. Lerman (1974), The structure of ψ-DNA. *J. Mol. Biol.* **84**, 37–64.

1296a. R. Huey and S. C. Mohr (1981), Condensed states of nuclei acids. III. ψ(+) and ψ(−) conformational transitions of DNA induced by ethanol and salt. *Biopolymers* **20**, 2533–2552.

1297. R. L. Scruggs and P. D. Ross (1964), Viscosity study of DNA. *Biopolymers* **2**, 593–609.

1298. J. T. Shapiro, M. Leng, and G. Felsenfeld (1969), Deoxyribonucleic acid–polylysine complexes. Structure and nucleotide specificity. *Biochemistry* **8**, 3219–3232.

1299. L. C. Gosule and J. A. Schellman (1976), Compact form of DNA induced by spermine. *Nature* **259**, 333–335.

1300. U. K. Laemmli (1976), Characterization of DNA condensates induced by poly(ethylene oxide) and polylysine. *Proc. Nat. Acad. Sci. USA* **72**, 4288–4292.

1301. D. E. Olins and A. L. Olins (1971), Model nucleohistones: The interaction of F1 and F2a1 histones with native T7 DNA. *J. Mol. Biol.* **57**, 437–455.

1302. J. C. Girod, W. C. Johnson, Jr., S. K. Huntington, and M. F. Maestre (1973), Conformation of deoxyribonucleic acid in alcohol solution. *Biochemistry* **12**, 5092–5101.

1303. D. Lang (1973), Regular superstructures of purified DNA in ethanolic solution. *J. Mol. Biol.* **78**, 247–254.

1304. D. Lang, T. N. Taylor, D. C. Dobyan, and D. M. Gray (1976), Dehydrated circular DNA: Electron microscopy of ethanol-condensed molecules. *J. Mol. Biol.* **106**, 97–107.

1305. T. H. Eickbush and E. N. Mondrianakis (1978), The compaction of DNA helices into either continuous supercoils or folded-fiber rods and toroids. *Cell* **13**, 295–306.

1306. S. A. Allison, J. C. Herr, and J. M. Schurr (1981), Structure of viral φ29 DNA condensed by simple triamines: A light-scattering and electron-microscopy study. *Biopolymers* **20**, 469–488.

1307. A. Keller (1968), Polymer crystals. *Rep. Prog. Phys.* **31**, 623–704.

1308. R. Wunderlich (1973), *Macromolecular Physics, Vol. I, Crystal Structure, Morphology, Defects.* Academic Press, New York.

1309. H. Eisenberg (1974), Hydrodynamic and thermodynamic studies. In: *Basic Principles in Nucleic Acid Chemistry* (P. O. P. Ts'o, Ed.), Vol. II. Academic Press, New York.

1310. A. I. Gragerov and S. K. Mirkin (1980), Influence of DNA superhelicity on major genetic processes in prokaryotes. *Mol. Biol. (Moscow)* **14**, 1–23.

1311. A. M. Campbell (1978), Straightening out the supercoil. *Trends Biochem. Sci.* **3**, 104–108.

1312. P. Broda (1979), *Plasmids.* Freeman, San Francisco.

1313. A. Worcel and E. Burgi (1972), On the structure of the folded chromosome of *Escherichia coli. J. Mol. Biol.* **71**, 127–147.

1314. A. L. Lehninger (1970), *Biochemistry,* p. 644. Worth, New York.

1314a. U. Hübscher, H. Lutz, and A. Kornberg (1980), Novel histone H2A-like proteins of *Escherichia coli. Proc. Nat. Acad. Sci. USA* **77**, 5097–5101.

1315. J. F. Pardon and M. H. F. Wilkins (1972), A super-coil model for nucleohistone. *J. Mol. Biol.* **68**, 115–124.

1316. P. Oudet, M. Gross-Bellard, and P. Chambon (1975), Electron microscopic and biochemical evidence that chromatin structure is a repeating unit. *Cell* **4**, 281–300.

1317. A. L. Olins and D. E. Olins (1974), Spheroid chromatin units (*v*-bodies). *Science* **183**, 330–332.

1318. R. D. Kornberg (1974), Chromatin structure: A repeating unit of histones and DNA. *Science* **184**, 868–871.

1319. R. D. Kornberg (1977), Structure of chromatin. *Annu. Rev. Biochem.* **46**, 931–954.

1320. J. P. Baldwin, P. G. Boseley, and E. M. Bradbury (1975), The subunit structure of the eukaryotic chromosome. *Nature* **253**, 245–249.

1321. J. E. Germond, B. Hirt, P. Oudet, M. Gross-Bellard, and P. Chambon (1975), Folding of the DNA double helix in chromatin-like structures from simian virus 40. *Proc. Nat. Acad. Sci. USA* **72**, 1843–1847.

1322. K.-P. Rindt and L. Nover (1980), Chromatin structure and function. *Biol. Zentralbl.* **99**, 641–673.

1323. I. Isenberg (1979), Histones. *Annu. Rev. Biochem.* **48**, 159–191.

1324. D. M. Lilley and J. F. Pardon (1979), Structure and function of chromatin. *Annu. Rev. Genet.* **13**, 197–233.

1325. G. Felsenfeld (1978), Chromatin. *Nature* **271**, 115–122.

1326. J. D. McGhee and G. Felsenfeld (1980), Nucleosome structure. *Annu. Rev. Biochem.* **49**, 1115–1156.

1327. R. D. Kornberg and A. Klug (1981), The nucleosome. *Sci. Amer.* **244**, 48–60.

1328. A. Klug, D. Rhodes, J. Smith, J. T. Finch, and J. O. Thomas (1980), A low resolution structure for the histone core of the nucleosome. *Nature* **287**, 509–516.

1329. F. Azorin, A. B. Martinez, and J. A. Subirana (1980), Organization of nucleosomes and spacer DNA in chromatin fibers. *Int. J. Biol. Macromol.* **2**, 81–92.

1330. A. D. Mirzabekov (1980), Nucleosome structure and its dynamic transitions. *Quart. Rev. Biophys.* **13**(2), 255–295.

1331. J. Dubochet and M. Noll (1978), Nucleosome arcs and helices. *Science* **202**, 280–286.

1332. J. T. Finch, L. C. Lutter, D. Rhodes, R. S. Brown, B. Bushton, M. Levitt, and A. Klug (1977), Structure of nucleosome core particles of chromatin. *Nature* **269**, 29–36.

1333. J. T. Finch, A. Lewit-Bentley, G. A. Bentley, M. Roth, and P. A. Timmins (1980), Neutron diffraction from crystals of nucleosome core particles. *Phil. Trans. Roy. Soc. London Ser.* **B290**, 635–638.

1334. H. Weintraub, A. Worcel, and B. Alberts (1976), A model for chromatin based on two symmetrically paired half-nucleosomes. *Cell* **9**, 409–417.

1335. L. C. Lutter (1979), Precise location of DNAse I cutting sites in the nucleosome core determined by high resolution gel electrophoresis. *Nucl. Acids. Res.* **6**, 41–56.

1336. G. S. Manning (1978), The molecular theory of polyelectrolyte solutions with applications to the electrostatic properties of polynucleotides. *Quant. Rev. Biophys.* **11**, 179–246.

1337. R. E. Harrington (1978), Opticohydrodynamic properties of high-molecular-weight DNA. III. The effects of NaCl concentration. *Biopolymers* **17**, 919–939.

1338. J. L. Sussman and E. N. Trifonov (1978), Possibility of nonkinked packing of DNA in chromatin. *Proc. Nat. Acad. Sci. USA* **75**, 103–107.

1339. A. D. Mirzabekov, V. V. Shick, A. V. Belyavsky, and S. G. Bavykin (1978), Primary organization of nucleosome core particle of chromatin: Sequence of histone arrangement along DNA. *Proc. Nat. Acad. Sci. USA* **75**, 4185–4188.

1340. A. D. Mirzabekov and A. Rich (1979), Asymmetric lateral distribution of unshielded phosphate groups in nucleosomal DNA and its role in DNA bending. *Proc. Nat. Acad. Sci. USA* **76**, 1118–1121.

1341. F. Thoma, Th. Koller, and A. Klug (1979), Involvement of histone H1 in the organization of the nucleosome and of the salt-dependent superstructures of chromatin. *J. Cell Biol.* **83**, 403–427.

1342. J. T. Finch and A. Klug (1976), Solenoidal model for superstructure in chromatin. *Proc. Nat. Acad. Sci. USA* **73**, 1897–1901.

1343. A. Worcel and C. Benyajati (1977), Higher order coiling of DNA in chromatin. *Cell* **12**, 83–100.

1344. T. Boulikas, J. M. Wiseman, and W. Garrard (1980), Points of contact between histone H1 and the histone octamer. *Proc. Nat. Acad. Sci. USA* **77**, 127–131.

1345. D. Ring and R. D. Cole (1979), Chemical cross-linking of H1 histone to the nucleosomal histones. *J. Biol. Chem.* **254**, 11688–11695.

1346. L. Sperling and A. Klug (1977), X-Ray studies on "native" chromatin. *J. Mol. Biol.* **112**, 253–263.

1347. A. Worcel, S. Strogatz, and D. Riley (1981), Structure of chromatin and the linking number of DNA. *Proc. Nat. Acad. Sci. USA* **78**, 1461–1465.

1348. U. K. Laemmli, S. M. Cheng, K. W. Adolph, J. R. Paulson, J. A. Brown, and W. R. Baumbach (1977), Metaphase chromosome structure: The role of nonhistone proteins. *Cold Spring Harbor Symp. Quant. Biol.* **42**, 351–360.

1349. F. H. C. Crick (1979), Split genes and RNA splicing. *Science* **204**, 264–271.

1349a. G. R. Smith (1981), DNA supercoiling: Another level for regulating gene expression. *Cell* **24**, 599–600.

1350. A. Kornberg (1980), *DNA Replication.* Freeman, San Francisco.

1351. A. I. Gagerov and S. M. Mirkin (1980), Influence of DNA superhelicity on major genetic processes in prokaryotes. *Mol. Biol. (Moscow)* **14**, 1–23.

1352. J. J. Champoux (1978), Proteins that affect DNA conformation. *Annu. Rev. Biochem.* **47**, 449–479.

1353. W. R. Bauer (1978), Structure and reactions of closed duplex DNA. *Annu. Rev. Biophys. Bioeng.* **7**, 287–313.

1354. W. Bauer and J. Vinograd (1974), Circular DNA. In: *Basic Principles of Nucleic Acid Chemistry* (P. O. P. Ts'o, Ed.), Vol. II, pp. 265–303. Academic Press, New York.

1355. J. C. Wang (1979), DNA: Bihelical structure, supercoiling and relaxation. *Cold Spring Harbor Symp. Quant. Biol.* **43**, 29–33.

1356. L. F. Lin, R. E. Depew, and J. W. Wang (1976), Knotted single-stranded DNA rings: A novel topological isomer of circular single-stranded DNA formed by treatment with *Escherichia coli* ω protein. *J. Mol. Biol.* **106**, 439–452.

1357. M. Gellert, K. Mizuuchi, M. H. O'Dea, and H. A. Nash (1976), DNA gyrase: An enzyme that introduces superhelical turns into DNA. *Proc. Nat. Acad. Sci. USA* **73**, 3872–3876.

1358. N. R. Cozzarelli (1980), DNA gyrase and the supercoiling of DNA. *Science* **207**, 953–960.

1359. J. H. White (1969), Self-linking and the Gauss intergral in higher dimensions. *Amer. J. Math.* **91**, 693–728.

1360. F. Brock Fuller (1971), The writhing number of a space curve. *Proc. Nat. Acad. Sci. USA* **68**, 815–819.

1361. F. H. C. Crick (1976), Linking numbers and nucleosomes. *Proc. Nat. Acad. Sci. USA* **73**, 2639–2643.

1362. W. R. Bauer, F. H. C. Crick, and J. H. White (1980), Supercoiled DNA. *Sci. Amer.* **243**, (1), 100–113.

1363. J. Vinograd, J. Lebowitz, and R. Watson (1968), Early and late helix–coil transitions in closed circular DNA. The number of superhelical turns in polyoma DNA. *J. Mol. Biol.* **33**, 173–197.

1364. W. Bauer and J. Vinograd (1968), The interaction of closed circular DNA with intercalative dyes. I. The superhelix density of SV40 DNA in the presence and absence of dye. *J. Mol. Biol.* **33**, 141–171.

1365. W. Keller (1975), Determination of the number of superhelical turns in simian virus 40 DNA by gel electrophoresis. *Proc. Nat. Acad. Sci. USA* **72**, 4876–4880.

1366. B. Hudson and J. Vinograd (1969), Sedimentation velocity properties of complex mitochondrial DNA. *Nature* **221**, 332–337.

1367. M. M. K. Nass (1969), Reversible generation of circular dimer and higher multiple forms of mitochondrial DNA. *Nature* **223**, 1124–1129.

1368. M. M. K. Nass (1969), Mitochondrial DNA: Advances, problems and goals. *Science* **165**, 25–35.

1369. A. M. Campbell (1976), The effect of DNA secondary structure on tertiary structure. *Biochem. J.* **159**, 615–620.

1370. A. M. Campbell (1976), Conformational analysis of DNA from PM2 bacteriophage. *Biochem. J.* **155**, 101–105.

1371. H. J. Vollenweiler, Th. Koller, J. Parello, and J. M. Sogo (1976), Superstructure of linear duplex DNA. *Proc. Nat. Acad. Sci. USA* **73**, 4125–4129.

1372. T.-S. Hsieh and J. C. Wang (1975), Thermodynamic properties of superhelical DNAs. *Biochemistry* **14**, 527–535.

1373. J. Engberg and H. Klenow (1977), Palindromic arrangement of specific genes in lower eukaryotes. *Trends Biochem. Sci.* **2**, 183–185.

1374. J. Engberg, P. Anderson, V. Leick, and J. Collins (1976), Free ribosomal DNA molecules from *Tetrahymena pyriformis* GL are giant palindromes. *J. Mol. Biol.* **104**, 455–470.

1375. J. C. Wang (1980), Superhelical DNA, *Trends Biochem. Sci.* **5**, 219–221.

1376. E. Keller (1980), SCHAKAL, a Plotting Program for Computers. Institut für Anorganische Chemie der Universität Freiburg, 7800 Freiburg, FRG.

1377. C. K. Johnson (1965) ORTEP, Oak Ridge Thermal Ellipsoids Plot Program. Oak Ridge National Laboratory Report, ORNL–3794, Oak Ridge, Tennessee (USA).

Index